Lecture Notes in Civil Engineering

Volume 182

Lecture Notes in Civil Engineering (LNCE) publishes the latest developments in Civil Engineering - quickly, informally and in top quality. Though original research reported in proceedings and post-proceedings represents the core of LNCE, edited volumes of exceptionally high quality and interest may also be considered for publication. Volumes published in LNCE embrace all aspects and subfields of, as well as new challenges in, Civil Engineering. Topics in the series include:

- Construction and Structural Mechanics
- Building Materials
- Concrete, Steel and Timber Structures
- Geotechnical Engineering
- Earthquake Engineering
- Coastal Engineering
- Ocean and Offshore Engineering; Ships and Floating Structures
- Hydraulics, Hydrology and Water Resources Engineering
- Environmental Engineering and Sustainability
- Structural Health and Monitoring
- Surveying and Geographical Information Systems
- Indoor Environments
- Transportation and Traffic
- Risk Analysis
- Safety and Security

To submit a proposal or request further information, please contact the appropriate Springer Editor:

– Pierpaolo Riva at pierpaolo.riva@springer.com (Europe and Americas);
– Swati Meherishi at swati.meherishi@springer.com (Asia - except China, and Australia, New Zealand);
– Wayne Hu at wayne.hu@springer.com (China).

All books in the series now indexed by Scopus and EI Compendex database!

More information about this series at https://link.springer.com/bookseries/15087

Nikolai Vatin · Svetlana Roshchina ·
Dmitrijs Serdjuks
Editors

Proceedings of MPCPE 2021

Selected Papers

Springer

Editors
Nikolai Vatin
Peter the Great St. Petersburg
Polytechnic University
Saint Petersburg, Russia

Svetlana Roshchina
Department of Building Constructions
Vladimir State University
Vladimir, Russia

Dmitrijs Serdjuks
Faculty of Civil Engineering
Riga Technical University
Riga, Latvia

ISSN 2366-2557 ISSN 2366-2565 (electronic)
Lecture Notes in Civil Engineering
ISBN 978-3-030-85235-1 ISBN 978-3-030-85236-8 (eBook)
https://doi.org/10.1007/978-3-030-85236-8

This Springer imprint is published by the registered company Springer Nature Switzerland AG
The registered company address is: Gewerbestrasse 11, 6330 Cham, Switzerland

Contents

Mathematical Model of Deformation of the River Channel in the Area of the Damless Water Intake

Dilshod Bazarov, Nikolai Vatin, Bekhzod Norkulov, Oybek Vokhidov, and Ikboloy Raimova

Abstract A two-dimensional mathematical model of unsteady uneven movement of a river flow, flowing in deformable channels, has been developed and numerically implemented. A numerical study of the unsteady two-dimensional movement of the Amu Darya river flow in the area of the head structure of the damless water intake to the head structure of the Karshi Main Canal was carried out. In the presented work, a two-dimensional mathematical model is proposed for calculating this kind of flows with a deformable channel, and its verification is carried out on the results of experimental and field studies.

Keywords Amu Darya · Karshi main canal · Downstream · Mathematical model · Two-dimensional

1 Introduction

The investigated section of the Amu Darya River is located in the area of a drain less water intake to the head structure of the Karshi Main Canal-KMC and is located 22 km above the water seal of the city Kerky. The total length of the investigated section is 10–12 km. This area has two characteristic sections: the upper one is 6 km upstream from the head water intake, and the lower one is located between the head water intake and the village of Kzylayak. The upper section has an unstable bed. The river bed here is composed of sandy-silty soils, and the banks are alluvial-loess-like loams. The main channel runs along the middle of the river or along the right bank, and sometimes deviates to the left bank.

D. Bazarov · B. Norkulov · O. Vokhidov (✉) · I. Raimova
Tashkent Institute of Irrigation and Agricultural Mechanization Engineers, Tashkent, Uzbekistan
e-mail: vohidov.oybek@bk.ru

N. Vatin
Peter the Great St. Petersburg Polytechnic University, St. Petersburg, Russian Federation, St. Petersburg, Russia

N. Vatin et al. (eds.), *Proceedings of MPCPE 2021*, Lecture Notes in Civil Engineering 182, https://doi.org/10.1007/978-3-030-85236-8_1

This channel position is due to the formation of an island along the right bank. This island is flooded in high water, but reappears in low water. The island to some extent deflects the flow in the direction of Cape Pulizindan [1–6]. The stream formed along the right bank at Cape Pulizindan has the greatest depths, reaching 8–12 m and sometimes up to 14 m. throughout its middle reaches. Coastal area of the village Kzylayak is located in the second section, i.e. 4–6 km below Cape Pulizindan on the left bank of the river. Amu Darya in the Kyzylayak section the floodplain of the river is two-sided. On the way to the left bank, the main stream flows along the left bank channel. In winter, the main channel of the river decreases to 0.25 km. During a flood, not formed underwater shoals are flooded, the width significantly increases towards the right bank and reaches 1.5 km. The floodplain of the river floods during the spring–summer flood—at a water flow rate of 5500–6000 m^3/s and more [7–9]. The main stream of the Amu Darya river in this section runs along the left, then along the right bank, and sometimes in the middle of the channel. When deviating to the right, the stream flows around Cape Pulizindan, to the left—if it is reflected from it. Here the river bed is composed of easily moving soils, the river banks are made of loess-like loam. The slope of the water surface in this area is 0.0002–0.00024. The water depth m of the right bank above Cape Pulizindan is 5–7 m and decreases downstream, and the average elevation of the river bottom in this place rises by 0.2–0.3 m due to the discharge of pulp into the river floodplain.

Forecasting channel processes in order to obtain more reliable information about the pattern of the flow of water in rivers and about their dynamics under the influence of the constructed hydraulic structures built in its channel is the most difficult task of channel hydraulics. To solve these problems, numerical or physical research methods are usually used [1] since the numerical study method requires less expenditure to study the hydraulic phenomenon than the experimental one and creates the possibility of multivariate calculations, it has recently become the main research method in channel hydraulics. It is for this justified reason that the use of numerical modeling processes for predictive calculations of channels is an urgent issue of our time [3].

Obviously, it is necessary that the mathematical model adequately reflect the real phenomenon and describe all the main effects inherent in this phenomenon. For the problems under consideration, an important condition is the possibility of taking into account in the model the process of reshaping the underwater slope, the generatrix of which is directed along the current velocity. In hydrotechnical research practice for modeling channel flows, approaches based on the numerical solution of the two-dimensional equations of Saint–Venant have shown high efficiency and sufficient accuracy. The derivation of these equations, algorithms for numerical integration and examples of calculations are given, for example, in [10–17]. However, for deformable channels, the Saint–Venant model turns out to be non-closed and requires supplementation, for example, by an equation or a system of equations to find the variables in time and space of the bottom marks.

2 Methods

The so-called two-dimensional sediment balance equation is often used for this purpose. In [1, 2], the need to introduce diffusion terms into the sediment balance equation was theoretically substantiated, and an expression was obtained for the diffusion coefficient of the bottom mark (which is taken by the spherical tensor), which is proportional to the value of the longitudinal (in the direction of the velocity vector) specific sediment discharge:

$$(1-\mathrm{p})\frac{\partial \mathrm{Z_b}}{\partial \mathrm{t}}+\frac{\partial US_*h}{\partial x}+\frac{\partial VS_*h}{\partial y}=\frac{\partial}{\partial x}D\frac{\partial Z_b}{\partial x}+\frac{\partial}{\partial y}D\frac{\partial Z_b}{\partial y} \tag{1}$$

where $D = ahS_*|\vec{U}|$, $|\vec{U}| = \sqrt{U^2+V^2}$, a—coefficient proportionality.

U, V—components of the depth-averaged water velocity vector along the x, y axes.

Model (3) was verified on experimental data.

In papers [9, 14] the proposed model. Its main differences from the previous one are that the diffusion term is written only for the direction orthogonal to the flow velocity vector, the additional term takes into account the influence of the streamline curvature on the change in the bottom elevations. In this case, the diffusion coefficient is taken to be proportional not to the flow rate, but to the value of the non-shearing flow rate. Using this model, interesting results were obtained on the calculation of channel reshaping.

In all the deformation models mentioned above, it is assumed that the concentration of sediment particles in the flow is close to equilibrium. At the same time, there are a number of problems for the solution of which this assumption is not applicable. For example, a clarified (without sediment) water flow usually enters the downstream of hydroelectric complexes; on the contrary, a flow oversaturated with sediments enters the reservoirs; when dredgers are operating, excessive turbidity occurs; in sharply unsteady and wave currents, the concentration of sediments in the flow also differs from the equilibrium. In [14], a model of sediment transport in an arbitrary flow (in a one-dimensional formulation), in which the concentration of particles generally does not coincide with the equilibrium one, was proposed and experimentally substantiated. It is shown that the specific discharge of sediment from the bottom into the flow is proportional to the value [8, 10]. Despite the fact that many scientists have created and are creating such mathematical models based on hydrodynamic and balance equations, adapted models have not yet been created that describe the movement of flow in channels passing on easily washed out soils [18–20].

In this work, an attempt is made to take into account the merits of a number of the above-mentioned models and to propose a mathematical model of sediment transport in uneven and unsteady river flows. As a result of the analysis and carrying out some mathematical transformations, the choice of the most suitable hydrodynamic equalization systems and the sediment balance equation describing the movement of the water flow in the channels passing on easily washed out soils, testing it for correct performance and carrying out calculations of natural objects is determined as the main goal and objectives of the study of this work.

3 Results and Discussion

As a result of the analysis of existing mathematical models and their verification to various research objects and taking into account the specific features of easily washed-out soils for numerical calculations of river flows with a deformable bottom, it is proposed to apply a mathematical model described by the system of two-dimensional equations of Saint–Venant [5–8] in partial derivatives and closing relations:

$$\frac{\partial hS}{\partial t} + \frac{\partial USh}{\partial x} + \frac{\partial VSh}{\partial y} = -K(S - S_*) \tag{2}$$

$$(1 - \mathrm{p})\frac{\partial Z_b}{\partial t} = K(S - S_*) + \frac{\partial}{\partial x}D\frac{\partial Z_b}{\partial x} + \frac{\partial}{\partial y}D\frac{\partial Z_b}{\partial y} \tag{3}$$

$$K = \begin{cases} \alpha U_* + (1 - \alpha)W, & U_* \geq W \\ W, & U_* \leq W \end{cases} \quad 0 \leq \alpha < 1; \tag{4}$$

$$D = \beta\tilde{S}hW; \tag{5}$$

$$S_* = \alpha_1 \frac{\lambda\rho}{2\rho_s}\frac{\left(|\vec{U}| - U_N\right)^2}{gh}\left(\frac{0.13}{tg\phi} + 0.01\frac{|\vec{U}|}{W}\right) \quad \lambda = 2gn^2h^{-1/3} \tag{6}$$

where t—time; h—flow depth; U, V—components of the flow velocity along the X and Y axes, respectively; $|\vec{U}| = \sqrt{U^2 + V^2}$; S—volumetric concentration of sediment particles in a stream; S_*—equilibrium volumetric concentration of particles (saturation concentration), taken according to the modified Bagnold formula; K—coefficient of intensity of vertical exchange of sediments between the bottom and the stream; p—soil porosity (the ratio of pore volume to the volume of the entire soil with pores); ρ_s—density of soil and water, respectively; ϕ—angle of internal friction of soil; W—hydraulic size of soil; U_*—dynamic speed; $|\vec{U}|$, U_N modulus of the vertical mean flow velocity and non-shearing velocity, respectively; λ—coefficient of hydraulic friction calculated by the Manning formula; n—roughness factor.

In the calculations, the non-shearing speed was taken according to the formula of T. E. Mirtskhulaeva, which, taking into account the standard values of the coefficients, has two formally equivalent recording types:

$$U_N = \lg \frac{8.8h}{d_{90}} \sqrt{\frac{2}{3.5}\left[(\rho_s - \rho)gd_{50} + 1.25C_y^H\right]} \tag{7}$$

$$U_N = 0.18\sqrt{\frac{2}{3.5\lambda}\left[(\rho_s - \rho)gd_{50} + 1.25C_y^H\right]} \tag{8}$$

where C_y^H—adhesion ground in t/m^2; d_{50}—average diameter of soil particles, d_{90}—diameter 90% of soil particles.

When calculating the diffusion coefficient (7), we used three versions of the formulas for $\tilde{S}$:

(a) by the total equilibrium concentration of traction and suspended sediment

$$\tilde{S} = S_* \tag{9}$$

(b) by bottom equilibrium concentration

$$\tilde{S} = \alpha_1 \frac{\lambda\rho}{2\rho_s} \frac{\left(|\vec{U}| - U_N\right)^2 \cdot 0.13}{ghtg\phi} \tag{10}$$

(c) by bottom concentration "without a square"

$$\tilde{S} = \alpha_1 \frac{\lambda\rho}{2\rho_s} \frac{\left(|\vec{U}| - U_N\right)|\vec{U}| \cdot 0.13}{ghtg\phi} \tag{11}$$

As the initial conditions, the initial bottom surface Z(x, y, 0), the corresponding instantaneous fields of velocity (x, y, 0), depth h (x, y, 0) and concentration S(x, y, 0).

As the boundary conditions at the "solid" boundaries, the no-flow condition is set, at the "liquid" boundaries, the flow rates or water levels are usually set. At the boundaries through which the flow enters the computational domain, the sediment concentration was specified. Sometimes, more complex types of boundary conditions were also used (relationship between costs and levels, "non-reflecting" boundary conditions, etc.).

The solution of the system of Eqs. (2) and (3) with respect to sediment concentration and bottom elevations was carried out by the finite volume method on mixed triangular-quadrangular grids together with the Saint–Venant equations.

The developed numerical scheme is consistent with the scheme for the equation of continuity of the liquid phase, which excludes the occurrence of the so-called "dipoles"—sources and sinks of mass. The scheme of the "directional differences" type excludes non-physical oscillations of the bottom topography, provides the property of "transportability" and the implementation of a difference analogue of mass conservation for the solid phase.

The system of Eqs. (2) and (3) is, in a sense, "minimal" for modeling the processes of bottom deformations in uneven and unsteady river flows. Indeed, the rejection of Eq. (2) leads, as noted above, to the postulation of the closeness of the concentration in the flow to the equilibrium one and does not allow setting the boundary conditions for the concentrations. The refusal in Eq. (3) from the terms describing the diffusion of the bottom marks leads to the impossibility of taking into account the processes of transformation (subsidence) of the underwater slopes. It is also clear that to describe three different physical processes (erosion-sedimentation, longitudinal sediment transport, transverse diffusion of bottom marks), at least three empirical coefficients are required at the present time (before the understanding of more general sediment transport laws) (in our model, this is). At the same time, the question of choosing the closing relations of model (6)—(11) remains open and requires further research.

In the test problem of the subsidence slope being considered below, 0.5 was assumed, the form of the closing relations (5)—(11) and the parameters and were varied in the course of calculations.

To test the model, we used laboratory experiments on the erosion of the slope by the current, the velocity vector of which is directed along the generatrix of this slope, carried out in the laboratory of Hydraulics NIIS [10].

Numerical calculations for identical conditions were carried out according to the model (2), (3). For a tray 18.0 m × 2.0 m, a rectangular grid of 1800 cells with a size of 0.1 m × 0.2 m was built. At the entrance boundary for the first row of cells, the bottom was taken not eroded, which corresponded to the entrance section of the tray reinforced with a cement crust in the experiment.

At the inlet boundary, a water flow rate of Q = 112 l/s and a clarified flow (S = 0) were set. The parameters of the numerical model and were selected in the calculation process from the condition of the best coincidence in the average diameter of the flume of the calculated profile of the eroded slope with the experimental data. The calculation results are shown in Figs. 1, 2 and 3.

The best results, as can be seen from the presented figures, are obtained by the first option (according to the bottom concentration (10) and the non-shearing velocity according to the formula (7): $\beta = 40$, $\alpha_1 = 0.25$).

Thus, it can be argued that the proposed mathematical model of unsteady uneven movement of a river flow is suitable for describing deformation processes, including the drift of dredging cuts. The model contains the minimum number of equations and empirical constants.

The channel process of the Amu Darya at this section of the head structure in the KMC, as practically throughout the Amu Darya, can be attributed to the type of channel multi-channel. This type is characterized by the fact that the geological structure of the floodplain and the channel is slightly different, therefore, further we will consider only channel-forming sediments, which can be both suspended and bottom. As is known, the channel-forming fraction is determined by the composition of bottom sediments. Since the main part of bottom sediments has a size from 0.25 to 0.05 mm, in the calculations their average size was set as 0.12 mm.

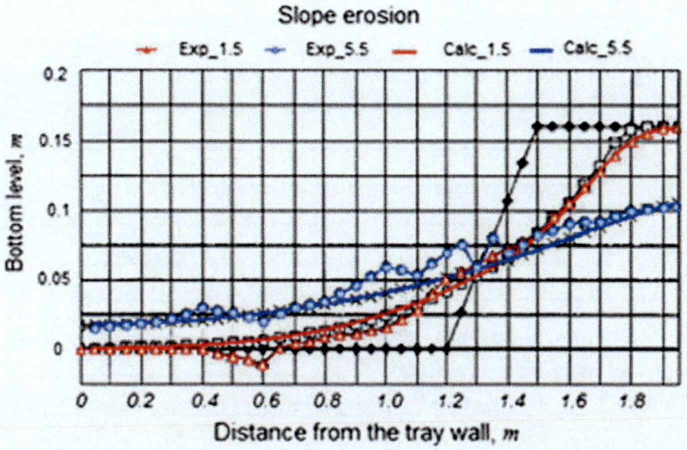

Fig. 1 Slope subsidence when calculating the diffusion coefficient (6) from the bottom concentration (12) and non-shearing velocity by the formula (9): $\beta = 40$, $\alpha_1 = 0.25$

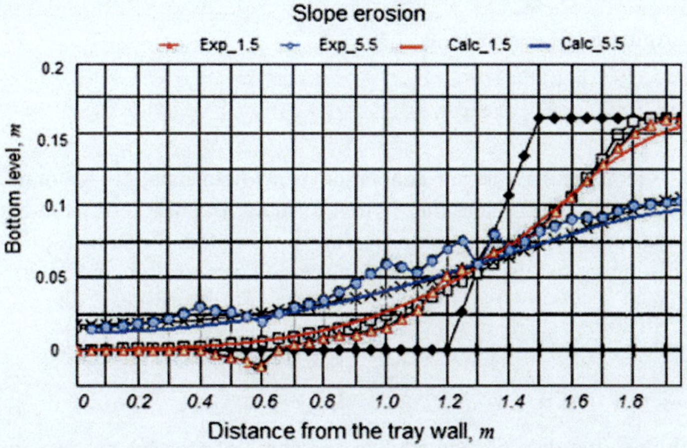

Fig. 2 Slope subsidence when calculating the diffusion coefficient from the bottom concentration (12) and the non-shearing velocity according to the formula (10): $\beta = 15$, $\alpha_1 = 0.25$

For calculations of a full-scale object, the above two-dimensional model was used. The use of Bagnold's formula as a formula for the transporting capacity is justified by the fact that, unlike other formulas, individual terms in it are due to bottom and suspended sediments. Therefore, having full-scale material on the flow rate and fractional composition of bottom and suspended sediments, it is easier to adapt.

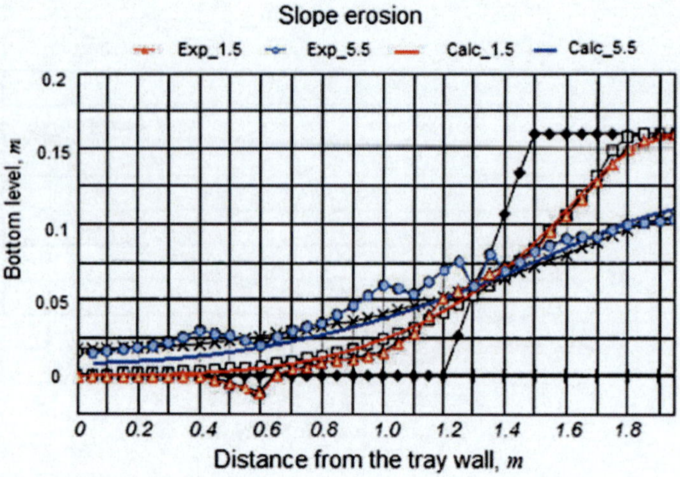

Fig. 3 Slope subsidence when calculating diffusion by the total concentration (8), (11) and non-shearing velocity by the formula (10): $\beta = 30$, $\alpha_1 = 0.25$

This formula can be written as:

$$S = 0.4\frac{U^2\lambda}{gh}\left(0.24 + \alpha_W\frac{U}{W}\right) \tag{12}$$

where α_W—coefficient requiring adaptation from field data, W—hydraulic size of the bed-forming fraction. Note that in this formula, the first term is due to traction (bottom) sediments, and the second term is weighted. Let the ratio of bottom sediment discharge to suspended sediment be α. Then for the coefficient we obtain the formula:

$$\alpha_W = \frac{W}{U}\frac{0.24}{\alpha} \tag{13}$$

According to the data of field studies, the share of the channel-forming fraction in the composition of suspended sediments in the area of the study object (α_p) is 0.31, the ratio of the discharge of bottom sediments to the discharge of all (and not only channel-forming) sediments (α_Σ) during the flood is about $0.1 \div 0.15$. Then we have that during the flood is $\alpha = \frac{\alpha_\Sigma}{\alpha_p} \approx 0.032 \div 0.049$. During summer floods, the characteristic flow rate is 1.5–2 m/s. According to [6] for the accepted average size, the hydraulic size can be taken equal to 0.01 m/s.

Moreover, it should be assumed that higher speeds correspond to a larger number α. Then: $\alpha_W = 0.036 \div 0.037$. This meaning α_W and was accepted in the calculations.

According to the developed two-dimensional mathematical model describing the flow movement on easily washed out soils, a series of calculations were performed, the results of which are given below.

In Figs. 4 and 5 show the computational grids and the initial marks of the bottom (in low-water periods and floods, respectively. Here and below the numbers of the computational points are plotted along the axes. The grid spacing along both axes is 50 m. flooded with water, a constant bottom elevation is set equal to 50 m., above which there can be no water level in the area. The rest of the area, depending on the flow rate and level, can either be flooded with water or dried. The boundary between an area with a given bottom elevation and an area with real elevations is characterized by the fact that it is either fortified so that the stream cannot erode the coast, or at this boundary—bedrock not eroded.

It is difficult to make comparisons based on bottom markings. Therefore, in this work, the relief with each other was compared in the following way. The topography calculated at the end of the year was recorded and the stationary current was calculated for two hydrological situations. The first one is the conditions close to the average low water level, with a mark in the downstream pool of 42.6 and a water discharge of 850 m^3/s. The second is the conditions of the average annual maximum flood with a mark in the downstream pool of 44 m and a discharge of 3700.

Figures 6 and 7 show the depths for these conditions on the initial bottom topography (in periods of high water and low water, respectively).

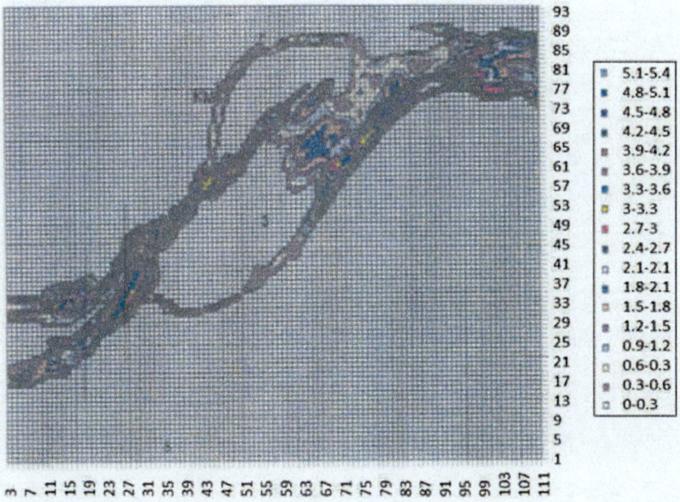

Fig. 4 Computational grid and initial marks of the bottom of the river bed in the area damless water intake at KMC

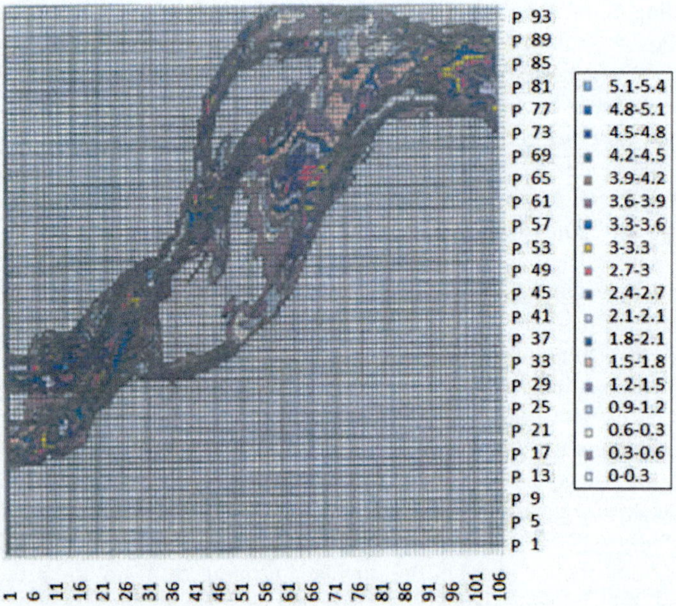

Fig. 5 Distribution of depths during flood on the initial topography of the river bed in the area of the damless water intake at the KMC

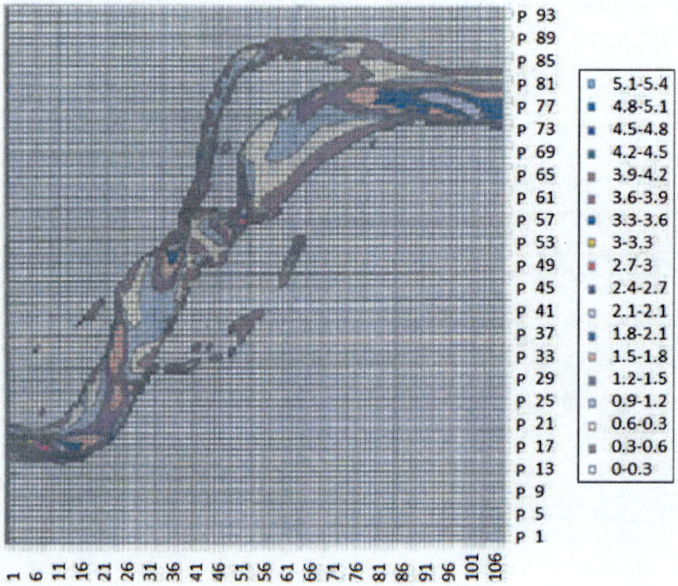

Fig. 6 Distribution of depths in low-water period after 1 year of observation (calculated)

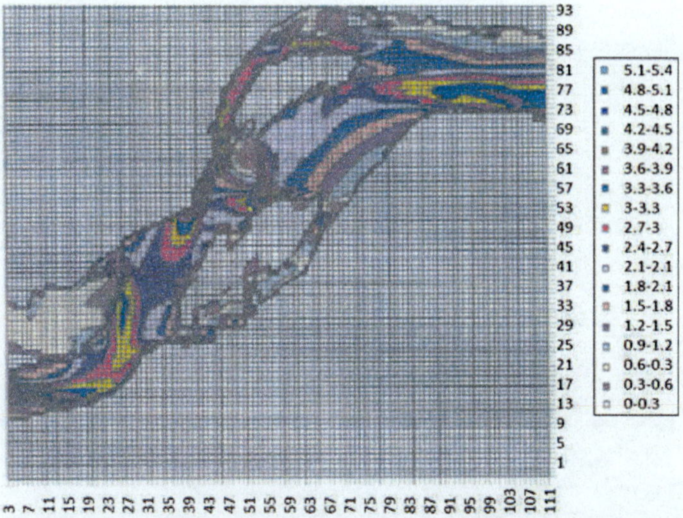

Fig. 7 Distribution of depths during flood after 1 year of observation (calculated)

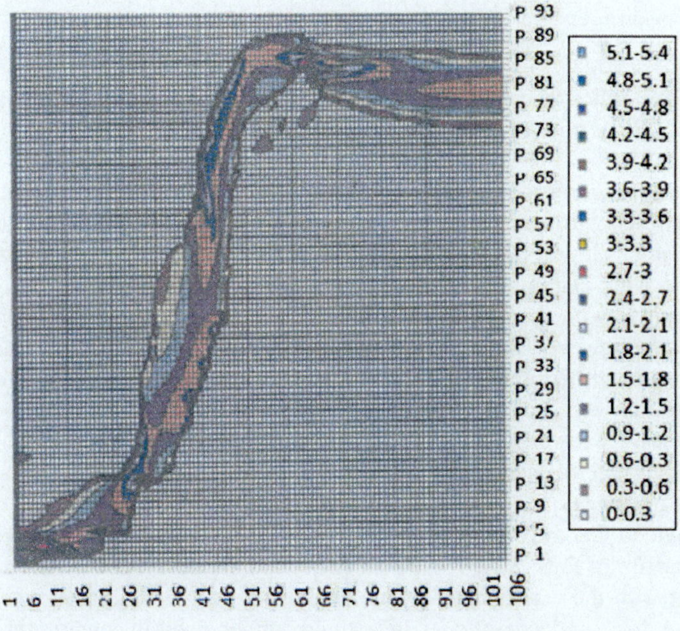

Fig. 8 Distribution of depths in low water after 10 years of observation (measured)

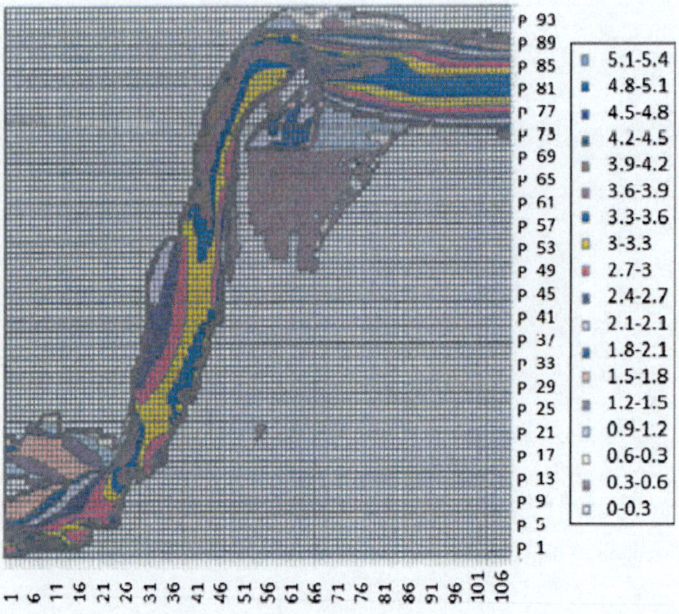

Fig. 9 Distribution of depths during flood after 10 years of observation (measured)

Figures 8 and 9 shows the depths for these conditions at the initial bottom topography (in periods of high water and low water, respectively), after 10 years of observation.

From these figures it can be seen that the channel is multi-armed, five branches are clearly distinguished, both in low-water periods and in floods. Due to the relatively small slopes, the Karshi main canal cannot pass all the sediments entering it from the Amu Darya.

Comparison with the data of experimental studies showed a good convergence of the results, however, the optimal values of the empirical coefficients for different experiments slightly differed from each other. Therefore, further work is required to obtain more universal closing relations for the model.

It is necessary to verify the model on reliable field data, as well as to extend it to hydraulically heterogeneous soils, which, apparently, can be done, since the concentration transfer Eq. (2) allows an independent calculation for fractions of various sizes (provided that the total concentration of particles in the flow is not very high).

Comparison of Figs. 8, 9, 10, and 11 show that the developed model gives good agreement with the results of field studies. For low-water conditions, the match is better than for flood conditions. In a flood, under natural conditions, there are slightly greater depths on the floodplain than in the calculations. It should be noted that both in calculations and in kind:

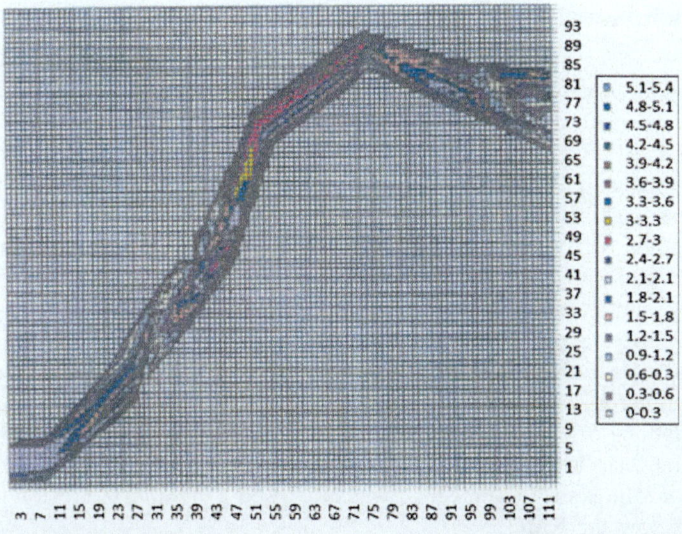

Fig. 10 Distribution of depths in low water after 10 years of observation (calculated)

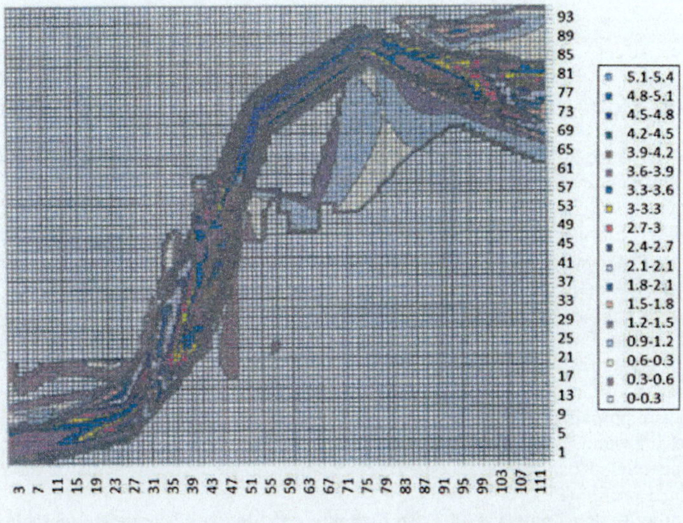

Fig. 11 Distribution of depths during flood after 10 years of observation (calculated)

- there was a significant erosion of the left bank below the head structure of the damless water intake at the KMC;
- the channel of the Amu Darya river in the section under consideration has turned from a multi-arm into a single-arm one.

4 Conclusions

1. According to the data of field studies, the calculated characteristics of soils and the connection of the level regime at the cross section Kerki with level mode near Cape Pulizindan;
2. There was a significant erosion of the left bank below the head structure of the damless water intake at the KMC;
3. The bed of the Amu Darya River in the area under consideration has turned from a multi-arm into a single-arm one;
4. Taking into account the fact that the hydrological regime in this section is quasi-steady, a transformation of variables has been found, the use of which makes it possible to drastically reduce the computer time when calculating channel deformations;
5. Calculations of channel deformations for a ten-year period were carried out in the Amu Darya section in the area of the head structure of the damless water intake of the Karshi main canal, which showed a significant erosion of the left bank below the KMC.
6. Comparison of the calculation results for the developed two-dimensional model with the measured data of field studies, showed satisfactory agreement.

References

1. Krutov A, Choriev R, Norkulov B, Mavlyanova D, Shomurodov A (2021) Mathematical modelling of bottom deformations in the kinematic wave approximation. IOP Conf Ser: Mater Sci Eng. https://doi.org/10.1088/1757-899X/1030/1/012147
2. Krutov A, Norkulov B, Mavlyanova D (2020) Simulation of spreading of non-conservative passive substances in water bodies. IOP Conf Ser: Mater Sci Eng. https://doi.org/10.1088/1757-899X/883/1/012028
3. Baranov EV, Gur'yev AP, Khanov NV (2020) Recommendations for hydraulic calculations of anti-erosion lining with the use of spatial geogrid with coarse fragmental soil. https://doi.org/10.1007/s10749-020-01115-z
4. Khanov NV, Martynov DY, Novichenko AI, Lagutina NV, Rodionova SM (2018) Outlook and special properties of earth anchors and screw piles in burial of modular protection dikes in nonrocky ground. https://doi.org/10.1007/s10749-018-0966-5
5. Kurbanov SO, Khanov NV (2004) To calculation of the critical depths of the canals with polygonal profile (PP)
6. Kurbanov SO, Khanov NV (2008) To hydraulic calculation of the most favorable sections of the power diversion canals (PDC) of a polygonal profile
7. Shomayramov M, Norkulov B, Rakhmanov J, Tadjiyeva D, Suyunov J (2019) Experimental researches of hydraulic vacuum breakdown devices of siphon outlets of pumping stations. E3S Web Conf 97:05099. https://doi.org/10.1051/e3sconf/20199705009
8. Shokirov B, Norkulov B, Nishanbaev K, Khurazbaev M, Nazarov B (2019) Computer simulation of channel processes. E3S Web Conf 97:05012. https://doi.org/10.1051/e3sconf/20199705012

9. Kattakulov F, Artikbekova F, Gafurov Z, Jumabaeva G, Musulmanov F (2021) Consideration of climatic factors in the operating mode of hydraulic facilities in the Amudarya river basin. E3S Web Conf 264:03068
10. Dilshod B, Markova I, Sultanov S, Kattakulov F, Baymanov R (2021) Dynamics of the hydraulic and alluvial regime of the lower reaches of the Amudarya after the commissioning of the Takhiatash and Tuyamuyun hydrosystems. IOP Conf Ser: Mater Sci Eng. https://doi.org/10.1088/1757-899X/1030/1/012110
11. Eshev S, Rakhimov A, Gayimnazarov I, Isakov A, Shodiev B, Bobomurodov F (2021) Dynamically stable sections of large soil canals taking into account wind waves. IOP Conf Ser: Mater Sci Eng. https://doi.org/10.1088/1757-899X/1030/1/012134
12. Eshev S, Rahmatov M, Khazratov A, Mamatov N, Sagdiyev J, Berdiev M (2021) Critical flow velocities in cohesive saline soils. E3S Web of Conf 264:03071
13. Matyakubov B, Begmatov I, Raimova I, Teplova G (2020) Factors for the efficient use of water distribution facilities. IOP Conf Ser: Mater Sci Eng. https://doi.org/10.1088/1757-899X/883/1/012025
14. Uralov B, Choriev R, Maksudova L, Sapaeva M, Shernaev A, Nurmatov P (2021) Substantiation of the influence of the channel shape and the roughness of machine canals on the pressure loss of irrigation pumping stations. IOP Conf Ser: Mater Sci Eng. https://doi.org/10.1088/1757-899X/1030/1/012148
15. Uralov B, Rakhmatov N, Khidirov S, Safarov G, Uljaev F, Raimova I (2021) Hydraulic modes of damless water intake. IOP Conf Ser: Mater Sci Eng. https://doi.org/10.1088/1757-899X/1030/1/012123
16. Kozlov D, Yurchenko A (2020) The role of inspection of hydraulic structures in the assessment of their technical condition. IOP Conf Ser: Mater Sci Eng. https://doi.org/10.1088/1757-899X/883/1/012049
17. Kozlov D, Ghebrehiwot A (2020) Integrated design and construction approach to hydrotechnical structures in Eritrea. IOP Conf Ser: Mater Sci Eng. https://doi.org/10.1088/1757-899X/869/7/072012
18. Kozlov DV, Kuleshov SL (2019) Multidimensional data analysis in the assessment of ice-jam formation in river basins. Water Resour 46:152–159. https://doi.org/10.1134/S0097807819020088
19. Lipin AA (2020) Surface telescopic water intake: A review. AlfaBuild. 15(3):1502–1502
20. Kovacs A, Parker G (1994) A new vectorial bedload formulation and its application to the time evolution of straight river channels. J Fluid Mech 267:153–183. https://doi.org/10.1017/S002211209400114X

Hydrodynamic Loads on the Water Chamber with Cavitating Dampers

Dilshod Bazarov, Bakhtiyor Obidov, Bekhzod Norkulov, Oybek Vokhidov, and Ikboloy Raimova

Abstract The data on the hydrodynamic loads on the water tank in the presence of cavitating non-erosion energy absorbers are presented.

It is shown that these loads increase in comparison with the cavitation-free regime, but despite this, the use of erosion-free dampers in the appropriate conditions is appropriate, different, providing favorable downstream conditions and reducing the volume of construction work and the cost of the structure.

As a result of cavitation and cavitation-erosion studies of erosion-free energy absorbers in vacuum-cavitation stands (in which simulations are carried out in compliance with gravitational similarity), moreover, for different stages of cavitation $\beta = K/K_{cr}$ ($K-$ cavitation parameter; $K_{cr}-$ its critical value characterizing the onset of cavitation), the drag coefficients C_x several types of erosion-free dampers (C_x—decreases with the development of cavitation β), the pulsation components of the horizontal hydrodynamic load on the damper, as well as the standards for the pulsations of the vertical hydrodynamic effects of the flow on the reservoir plate in the installation zone of the erosion-free dampers measured by "point" sensors $P' : \gamma v_1^2/2g$ (their maximum values are at $\beta \approx 0.5$, during supercavitation they decrease in comparison with those for the zero-cavitation regime), spectra and space–time transverse and longitudinal correlations of these pulsations.

Cavitation parameters were calculated according to the dependence

$$K = (H_{har} - H_{cr}) : \left(\frac{v_{har}^2}{2g}\right)$$

where $H_{char} = H_a + h$, (H_a—pressure above the free surface of the flow in the vacuum—cavitation stand, and for nature, atmospheric pressure; h—height of the water column above the damper; v_{har}—the characteristic speed of the flow on the damper, which was taken from the diagram of the velocity distribution in front of the damper at the level of the damper top; g—acceleration of free fall; v_1—velocity

D. Bazarov · B. Obidov · B. Norkulov · O. Vokhidov (✉) · I. Raimova
Tashkent Institute of Irrigation and Agricultural Mechanization Engineers, Tashkent, Uzbekistan
e-mail: vohidov.oybek@bk.ru

N. Vatin et al. (eds.), *Proceedings of MPCPE 2021*, Lecture Notes in Civil Engineering 182, https://doi.org/10.1007/978-3-030-85236-8_2

in a compressed section with a depth h_1 in front of the damper (the experiments were carried out with the outflow of water from under the shutter).

Keywords Cavitation parameters · Hydraulic structures · Cavitation erosion · Energy absorbers · Pulsation loads

1 Introduction

This article is devoted to one, relatively small issue of the dynamics of hydraulic structures—the determination of the hydrodynamic loads on the slabs of a high-pressure spillway in a cavitating flow in the presence of erosion-free dampers.

While working on the implementation of this research, the author simultaneously studied in the laboratory pulsating loads on real structures—elements of the downstream spillway devices of the projected lower Kafirnigansky hydrosystem.

In connection with the intensive construction of high-pressure and medium-pressure hydroelectric systems in mountainous areas, the spillways of which operate at high flow rates, a very urgent task is to develop reliable and economical downstream devices that provide intensive energy extinguishing with favorable uninterrupted flow regimes and the absence of cavitation erosion of streamlined elements. Traditional methods—stilling wells and walls do not always provide a solution to the problem. In a number of cases, they are additionally satisfied with elements as energy absorbers, which are an effective means of dealing with malfunctioning currents. However, most of the used types of absorbers have a serious drawback—they are destroyed in cavitation conditions.

Cavitation research N. P. Rozanov and his students (R. M. Razakov, A. T. Kaveshnikov, N. N. Rozanova) made it possible, on the basis of experiments, to develop several types of erosion-free or close to erosion-free dampers and obtain dependencies for determining the hydrodynamic loads on them at various stages of cavitation [1–3]. This made it possible to use energy absorbers at high flow rates, which was carried out at the spillways of the Shamkhor and Artyomov hydroelectric complexes.

At the same time, it should be noted that in cavitation studies carried out until recently, the effects of cavitation on energy absorbers were considered only from the point of view of the possibility of an erosion hazard and the effects of the flow on the absorbers themselves. There is no doubt, however, that the degree of development of cavitation affects the characteristics of the pressure pulsation in the cavitating flow not only on the surface of the dampers but also on the water tank. In cavitation-free modes, the pulsation loads on the water chamber have been studied in some detail for some types of absorbers, as for the loads on the water chamber under cavitation modes and erosion-free absorbers, they have not been studied. If we take into account that the cost of downstream attachment devices for high-pressure structures can be 20–30% of the cost of the entire structure, then it becomes obvious how important it is to correctly design the downstream device in order to ensure their long-term reliable operation. This is also required in the presence of erosion-free energy absorbers, which are promising, since they expand the scope of energy absorbers—devices that prevent unfavorable malfunctioning currents in the downstream. The main direction of this article is the study of hydrodynamic loads on slabs of a water face in the presence of erosion-free energy absorbers on it in conditions of various stages of cavitation and in its absence.

1.1 Hydrodynamic Effects of the Flow on the Fastening Elements in the Downstream in the Absence of Cavitation

Investigations of pressure pulsations in a hydraulic jump in the absence of cavitation are of interest mainly to the flow, which, in our opinion, allows us to draw a number of significant general conclusions.

Currently, two main approaches to the study of pressure pulsations have been outlined.

1. With the help of "point" sensors that register the pressure pulsation at individual points of the element surface and the propagation of its area by specifying multidimensional probability distribution functions and multidimensional correlation functions.
2. With the help of total sensors that directly measure the total load or moment on the investigated element.

The pressure pulsation measured by point sensors cannot be directly used to determine the load on a large area due to the lack of synchronicity of pulsations at individual points of the streamlined surface [4–8].

It should be noted that in all studies, conclusions were drawn about the possibility of modeling pressure pulsations if the Froude similarity criterion is met. In addition, the authors determined the lowest self-similarity zone, which corresponds to the Reynolds number $Re_m > (5 - 10) \cdot 10^3$. The conclusions made in these works substantiate the conduct of pressure pulsation studies in laboratory conditions on small-scale models. This undoubtedly expands the possibilities of the experiment for a better solution of the problems of hydrodynamic loads for specific objects.

1.2 Consequences Caused by the Action of Cavitation

The experience of operating high-pressure hydroelectric complexes shows that the force of the dynamic interaction of the flow and the elements of the downstream can lead to severe damage to the latter. These damages can be of two types: firstly, erosive from the action of cavitation, or, secondly, due to an increase in pulsating loads in the cavitation mode. The use of erosion-free absorbers, in principle, removes the issue of cavitation erosion of both absorbers and slabs. However, it should be borne in mind that the experience of the practical application of such structures is not yet great, therefore, it is necessary to exercise some caution in their design. When using erosion-free dampers, it is sometimes suggested to use solid walls in places where vertical vortices of cavitation torches can occur. In the event that individual cavitating vortices nevertheless break through to the surface of the pond, they should be made of materials with high cavitation resistance [5]. As for damages associated with an increase in hydrodynamic loads due to cavitation, it is not possible to avoid them by changing the design of cavitating absorbers. Failure to consider these loads can lead to serious damage downstream.

1.3 Feed Effects on the Elements of Hydraulic Structures in the Presence of Cavitation

As far as we know, the question of the wobble of cavitation on the pulsation characteristics of the flow acting on the slabs of the reservoir during the cavitation flow has not been practically studied. However, a qualitative understanding of the pressure pulsations behind the dampers can be obtained by the example of works studying various kinds of obstacles (protrusions, gates, etc.), since both are essentially sources of pressure pulsations.

In work [2, 6], laboratory tests of a flat valve operating under high-speed flow conditions with a head up to 200 m were carried out. One of the aspects of the work was to study the dynamic effect of a cavitating flow on the valve. According to the authors, the values of the pressure pulsation standards at the gate in the presence of developed cavitation are twice the values of the standards in its absence. In the supercavitation mode, data are also not given because the author of the work failed to obtain supercavitation.

In the works of V. M. Lyatkhera and L. V. Smirnov [2, 3, 5, 9–14] obtained data on the characteristics of pressure pulsation, according to which there is an increase in dispersion and a significant deformation of the pulsation spectrum towards high frequencies as cavitation develops. In the separation zone, the spectrum changes during cavitation, naturally, due to the fact that the most intense pulsations occur in the presence of cavitation.

In the first in hydraulic engineering, the force effects of a cavitating flow on erosion-free dampers were studied by N. N. Rozanov [15–19] based on cavitation studies, the author of the work managed to obtain quantitative regularities of horizontal averaged and pulsating loads as cavitation develops. In the course of the experiments, a decrease in drag coefficients with the development of cavitation was recorded. The noted decrease in the drag coefficient during cavitation, especially under the conditions of developing stages and supercavitation, the author of the work explains by the fact that with the development of cavitation the character of the pressure distribution curves on the streamlined body changes [20].

At the second stage, the author studies the pulsation loads on the damper at the maximum range. The analysis of the research results showed that when the absorbers operate in conditions of cavitation (initial and developed), an increase in the instantaneous pulsation component of the load occurs in comparison with the non-cavitation mode. For example, in non-cavitation mode, the ripple coefficient δ_p constant and equal to 0.14, and at the developed stage ($\beta = 0.5$) $\delta_p = 0.65$, that is, it increased by 4.6 times, and at $\beta < 0.5$ there is a tendency to decrease it.

2 Methods

To obtain data on the conditions of occurrence, development and impact of cavitating flow on the elements of spillway hydraulic structures, vacuum stands are used. Their main advantage lies in the fact that they allow for the creation of cavitation conditions on models that meet the Froude similarity criterion.

Studies of the force effects of the flow in the presence of cavitation were carried out in the vacuum test bench of the laboratory of hydraulic structures of the Moscow State Medical Institute.

The vacuum unit allows conducting cavitation studies of hydraulic structures elements in the presence of a free flow surface. The rather large dimensions of the installation provide for cavitation studies of flat and half-space models of sufficiently large dimensions, with direct observation of cavitation on absorbers located in a hydraulic jump.

It is known that when simulating the operation of energy absorbers in the downstream in the presence of cavitation and in its absence, it is necessary to observe the Froude similarity criterion (Fr = idem) and to carry out research in the self-similar region at Reynolds number $Re_m > Re_{gr}$.

To observe the approximate similarity of cavitation phenomena, the following conditions must be met:

$$K_n = \eta K_m$$

where $K_n\, and\, K_m$—parameters of cavitation for nature and model; η—correction factor of the model (we take $\eta = 1.0$ considering the large scale of the model $Re_m = 10^5 - 10^6$).

A hydraulic jump in the installation was created when water flowed out from under the shutter with a sharp loud sound.

The fragmentary model was a pond with two rows of damping devices: in the first row—erosion-free dampers, in the second—a pond wall.

The studies used the following instruments:

- Piezometers—to determine the average pressure and the bottom of the reservoir;
- Sensor–plate—for measuring the total vertical and moment loads in the longitudinal and transverse directions.

3 Results and Discussion

Averaged pressures were obtained (Fig. 1) $\frac{\overline{P_1}}{\gamma v_1^2/2g}$ at the bottom of the reservoir (in the cavitation-free mode) and the values of the standards of specific pulsation loads $P_v^{'} : \frac{\gamma v_1^2}{2g}$ and overturning moments $P_m^{'} : \frac{\gamma v_1^2}{2g}$ on a stilling slab at its different relative lengths $L : h_1$ (Fig. 2). The values of these ripple loads were obtained using a special sensor plate. They, like the values of $P^{'} : \frac{\gamma v_1^2}{2g}$ varied in depending on the value of β and were maximum at the developed stage of cavitation with $\beta \approx 0.5$ and during supercavitation they became less than in the non-cavitation mode. The specified sensor had a natural frequency in water of about 100 Hz.

As expected, a decrease in the relative length of the slab $L : h_1$ lead to an increase in the standard of pulsation of specific loads (Fig. 2); but for “point” sensors, they were obtained even more significant (for example, when $\beta = 0.5$ $P^{'} : \frac{\gamma v_1^2}{2g} \approx 0.3$—with a damper of the same type). The specific pulsation load on the slab with cavitating dampers, measured by areal sensors, turned out to be 22–28% more than the load according to the data of “point” sensors. With regard to one of the high-pressure hydroelectric complexes [6] at $v_1 = 24\text{m/sec}$ the volumes of concrete required for the downstream construction were calculated for the variants with a stilling well and with erosion-free dampers. In the latter version, the volume of concrete turned out to be 1.5–1.6 times less. In addition, in this version, the issues of preventing unfavorable malfunctioning currents in the downstream [1, 2, 4] were well resolved.

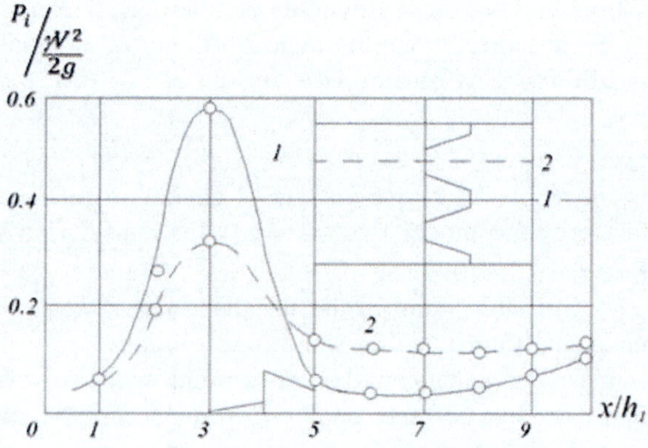

Fig. 1 Averaged pressures at the bottom of the water hole of erosion-free dampers 2 in sections 1 and 2 (jump flooding coefficient n = 1.0)

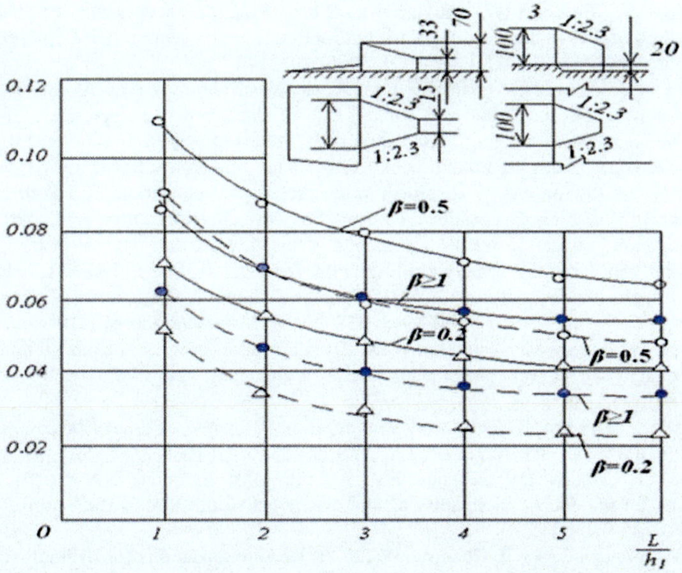

Fig. 2 Vertical detached loads *(1)* and loads equivalent *(2)* to the action of the tipping moment depending on the relative length of the plates L: h_1 and the cavitation stage $\beta = K$: K_{cr} for absorbers *2* and *3*

4 Conclusions

1. For several types of erosion-free energy absorbers, data were obtained on the averaged and pulsating vertical and horizontal loads on the absorbers and the slab, which makes it possible to carry out the required strength and stability calculations of the elements of the downstream devices.
2. Studies have shown that despite a slight decrease in the energy absorbing capacity of the absorbers during the development of cavitation (decrease in C_x) and an increase during cavitation of absorbers of the pulsation effects of the flow on the absorbers and the water column, in appropriate conditions they are rational, providing more favorable operating conditions for the downstream and reducing the volume of construction work and construction cost.

References

1. Ferrari A (2017) Fluid dynamics of acoustic and hydrodynamic cavitation in hydraulic power systems. Proc R Soc A Math Phys Eng Sci 473. https://doi.org/10.1098/rspa.2016.0345
2. Braun MJ, Hannon WM (2010) Cavitation formation and modelling for fluid film bearings: a review. Proc Inst Mech Eng Part J J Eng Tribol 224:839–863. https://doi.org/10.1243/13506501JET772

3. Shaazizov F, Shukurov D, Shukurov E (2021) System for ensuring the detection and elimination of fires in the building of the hydroelectric power station. IOP Conf Ser: Mater Sci Eng. https://doi.org/10.1088/1757-899X/1030/1/012142
4. Bazarov D, Vatin N, Bakhtiyor O, Oybek V, Rakhimov A, Akhmadi M (2021) Hydrodynamic effects of the flow on the slab of the stand in the presence of cavitation. IOP Conf Ser: Mater Sci Eng. https://doi.org/10.1088/1757-899X/1030/1/012116
5. Umurzakov U, Obidov B, Vokhidov O, Musulmanov F, Ashirov B, Suyunov J (2021) Force effects of the flow on energy absorbers in the presence of cavitation. E3S Web Conf
6. Obidov B, Vokhidov O, Tadjieva D, Saidkhodjaeva D, Kurbanova U, Isakov A (2021) Hydrodynamic effects on the flow elements of the downstream devices in the presence of cavitation. IOP Conf Seri: Mater Sci Eng. https://doi.org/10.1088/1757-899X/1030/1/012114
7. Budich B, Schmidt SJ, Adams NA (2018) Numerical simulation and analysis of condensation shocks in cavitating flow. J Fluid Mech 838:759–813. https://doi.org/10.1017/jfm.2017.882
8. Yangiev A, Gapparov F, Adjimuratov D (2019) Filtration process in earth fill dam body and its chemical effect on piezometers. E3S Web Conf. https://doi.org/10.1051/e3sconf/20199705032
9. Khidirov S, Norkulov B, Ishankulov Z, Nurmatov P, Gayur A (2020) Linked pools culverts facilities. IOP Conf Ser: Mater Sci Eng. https://doi.org/10.1088/1757-899X/883/1/012004
10. Gur'ev AP, Kozlov DV, Khanov NV, Abidov MM, Safonova NA (2020) Alternative solutions for the energy dissipation of idle discharges at the Rogun HPP. https://doi.org/10.1007/s10749-020-01157-3
11. Baranov EV, Gur'yev AP, Khanov NV (2020) Recommendations for hydraulic calculations of anti-erosion lining with the use of spatial geogrid with coarse fragmental soil. https://doi.org/10.1007/s10749-020-01115-z
12. Khanov NV, Martynov DY, Novichenko AI, Lagutina NV, Rodionova SM (2018) Outlook and special properties of earth anchors and screw piles in burial of modular protection dikes in nonrocky ground. https://doi.org/10.1007/s10749-018-0966-5
13. Kurbanov SO, Khanov NV (2004) To calculation of the critical depths of the canals with polygonal profile (PP)
14. Kurbanov SO, Khanov NV (2003) To hydraulic calculation of the most favorable sections of the power diversion canals (PDC) of a polygonal profile
15. Khanov NV (1999) Hydraulic characteristics of chamber-free tangential vortex flow generators. Hydrotechnical Constr 33:99–103. https://doi.org/10.1007/BF02769414
16. Khanov NV (1998) Hydraulic performance characteristics of a vortical spillway with a tangential vortex generator in the flow. Hydrotechnical Constr 32:253–258. https://doi.org/10.1007/BF02918697
17. Khanov NV (1997) Hydraulic operating conditions of an eddy tunnel outlet with an inclined shaft. Hydrotechnical Constr 31:694–698. https://doi.org/10.1007/bf02767223
18. Kurbanov SO, Khanov NV (2004) To calculation of the critical depths of the canals with polygonal profile (PP). Gidrotekhnicheskoe Stroitel'stvo 3:42–44
19. Volkov VI, Snezhko VL, Kozlov DV (2019) Prediction of safety level of low-head and ownerless hydraulic structures. https://doi.org/10.1007/s10749-019-01028-6
20. Gabaydulin DY, Grechneva MV (2012) Vozmozhnosti vosstanovleniya lopastey gidroturbin, povrezhdennykh kavitatsiyey. Vestnik Irkutskogo gosudarstvennogo tekhnicheskogo universiteta. 12(71)

Plane Bending Deformation of Structural Lumber for Construction with a Ring Structure of Annual Layers of Wood

Viktor Byzov, Vladimir Glukhikh, Vladimir Melekhov, Aleksandr Sergeevichev, and Anna Mihailova

Abstract The wood in the trunk of a growing tree has a ring structure of annual layers. The ring structure is the required strength characteristics. For the manufacture of load-bearing elements of building structures, strength is a critical property. In this regard, it is advisable to preserve the maximum amount of uncut annual wood in structural lumber. This is possible when using lumber of large sections with the presence of a core. The macrostructure of such lumber maximally repeats the restoration of the trunk of a growing tree. The purpose of this work is to solve the problem of calculating the strength characteristics of structural sawn timber, the macrostructure of the wood structure of which is a system of thin-walled shells coaxially strung on top of each other, simulating annual layers of wood. To solve this problem, a method is proposed for calculating the strength and deformities properties of a bar with an annular structure of the location of annual layers. The obtained solution makes it possible to study the stress–strain state of elements of load-bearing building structures with an annular structure of annual wood layers during transverse bending.

Keywords Load-bearing building structures · Structural lumber · Strength characteristics of lumber · Ring structure of annual layers of wood · The core beams

---

V. Byzov (✉) · V. Glukhikh
Saint Petersburg State University of Architecture and Civil Engineering, Saint Petersburg, Russia
e-mail: mapana@inbox.ru

V. Melekhov
Northern (Arctic) Federal University Named After M.V. Lomonosov, Arkhangelsk, Russia

A. Sergeevichev · A. Mihailova
Saint Petersburg State Forest Technical University, Saint Petersburg, Russia

N. Vatin et al. (eds.), *Proceedings of MPCPE 2021*, Lecture Notes in Civil Engineering 182, https://doi.org/10.1007/978-3-030-85236-8_3

1 Introduction

Softwood lumber is widely used in the manufacture of load-bearing building structures for buildings and structures for civil, agricultural and industrial purposes. Recently, the size and quality composition of round timber from which structural timber is obtained is constantly deteriorating. The amount of small diameter sawn timber is increasing. The number of saw logs with a diameter of 14–18 cm is more than half of the total amount of all round timber supplied to sawmills in the North European part of the Russian Federation [1].

Saw logs of small diameter are not used for the manufacture of sawn timber from which elements of load-bearing building structures are obtained. This is because a large number of knots are present in small diameter saw logs. As a result, it is believed that the lumber obtained from them has insufficient strength and is not allowed for the manufacture of elements of load-bearing building structures by the existing regulatory and technical documentation. Nevertheless, numerous studies of domestic and foreign scientists have established that the standards for knots poorly characterize the strength properties of saw logs [2–5]. Small diameter saw logs are obtained from the tops of tree trunks. The trunk of a coniferous tree is considered as a complex reinforced and equally strong complex structure capable of significant elastic deformations [6]. It is advisable to obtain from round timber of small diameters structural lumber in the form of beams in the center of the cross-section of which the core is located, the so-called core beams. In this case, the circular structure of the tree trunk is preserved as much as possible, and the amount of cut wood fibers is minimal. Investigations of the strength characteristics of such beams were carried out by numerous domestic and foreign scientists [7–17]. It was found that the strength characteristics of beams with a predominantly circular macrostructure of annual layers are not inferior to those inherent in the wood of a tree trunk.

Structural elements made of core beams generally operate under stress–strain state of buckling and compression along bending fibers. According to studies [18], the normal stresses arising in such beams with an annular structure of annual layers are 33–38% less than the normal stresses in sawn timber produced by traditional technology. Thus, the beams have the strength required for the manufacture of structural elements.

In this paper, a method is proposed for calculating the stresses arising in the cross-section of bending elements of load-bearing building structures that have an annular structure of annual layers of wood.

2 Methods

Let's consider an inhomogeneous orthotropic cylinder with an anisotropy axis coinciding with its longitudinal axis. Inhomogeneity consists in the variability of properties in the radial direction, which is a natural feature of wood. Periodicity of changes in properties is observed, taking into account the early and late wood of the annual layers. In some approximation, to simplify the solution of the problem, we can take the same width of both zones.

Let's suppose that the load is applied only to the ends and the law of its distribution along the ends is unknown. It is only known that this load is brought to the main moment, which is given. In our case, the case of pure bending is considered, when the stresses depend on only one coordinate. There are no forces on the outer surface, and the elastic coordinates β_{ij} depends on only one coordinate r.

When solving the problem, we used the differential equations obtained by S. G. Lekhnitsky [19]:

$$\begin{aligned}&\left(\frac{\partial^2}{\partial\theta}-r\frac{\partial}{\partial r}\right)\cdot\left(\frac{\beta_{11}}{r}\cdot\frac{\partial F}{\partial r}+\frac{\beta_{11}}{r^2}\cdot\frac{\partial^2 F}{\partial\theta^2}+\beta_{12}\frac{\partial^2 F}{\partial r^2}\right)\\&+r\frac{\partial^2}{\partial r^2}\left(\beta_{12}\frac{\partial F}{\partial r}+\frac{\beta_{12}}{r}\cdot\frac{\partial^2 F}{\partial\theta^2}+\beta_{22}\frac{\partial^2 F}{\partial r^2}\right)+\frac{\partial^2}{\partial r\partial\theta}\left(r\beta_{66}\frac{\partial^2}{\partial r\partial\theta}\left(\frac{F}{r}\right)\right)\\&=\left(\frac{\partial^2}{\partial\theta^2}-r\frac{\partial}{\partial r}\right)\left[\mu_{zt}(Ar\cos\theta+Br\sin\theta+C)\right]\\&+r^2\frac{\partial^2}{\partial r^2}\left[\mu_{zr}(Ar\cos\theta+Br\sin\theta+C)\right],\end{aligned}\tag{1}$$

где Θ—angle; β_{ij}—elastic coefficients, which are expressed in terms of elastic modules and Poisson's ratios as follows:

$$\begin{aligned}\beta_{11}&=\frac{1-\mu_{rz}\mu_{zr}}{E_r}\\\beta_{12}&=\frac{1-\mu_{rz}\mu_{zr}}{E_r}\\\beta_{22}&=\frac{1-\mu_{tz}\mu_{zt}}{E_t}\\\beta_{66}&=\frac{1}{G_{rt}}\end{aligned}\tag{2}$$

or in general

$$\beta_{ij}=a_{ij}-\frac{a_{i3}a_{j3}}{a_{33}}$$

$(i, j) = 1, 2, 3, 4, 5, 6\ldots$

$a_{i,j}$—multipliers in the formulas of the generalized Hooke's law;

A,B,C—integration constants determined from the conditions at the ends, where it will be necessary to equate the main moment of internal forces in the cylinder to a given moment *M*;
F—stress function satisfying differential Eq. (1).

The general solution to the differential Eq. (1) can be sought in the following form:

$$F(r,\theta) = f(r)\sin\theta \tag{3}$$

Substituting function (3) into Eq. (1), we obtain an inhomogeneous differential equation for finding the unknown function *f(r)*:

$$\begin{aligned}
&\frac{\partial}{\partial r^2}\left(r\beta_{22}\frac{\partial^2 f}{\partial r^2} + \beta_{12}\frac{\partial f}{\partial r} - \frac{\beta_{12}}{r}f\right) - \frac{\partial}{\partial r}\left(\beta_{12}\frac{\partial^2 f}{\partial r^2} + \frac{\beta_{11}}{r}\cdot\frac{\partial f}{\partial r} - \frac{\beta_{11}}{r^2}f\right) \\
&\quad - \frac{1}{r}\left(\beta_{12}\frac{\partial^2 f}{\partial r^2} + \frac{\beta_{11}}{r}\cdot\frac{\partial f}{\partial r} - \frac{\beta_{11}}{r^2}f\right) - \frac{1}{r}\frac{\partial f}{\partial r}\left(\beta_{66}\frac{\partial f}{\partial r} - \frac{\beta_{66}}{r}f\right) \\
&\quad = B\left[\frac{a_{13}}{a_{33}} - \frac{\partial^2}{\partial r^2}\left(\frac{a_{23}}{a_{33}}r^2\right) + \frac{\partial}{\partial r}\left(\frac{a_{13}}{a_{33}}r\right)\right]
\end{aligned} \tag{4}$$

For the case when all coefficients of elastic deformation are proportional to the same degree *r*, it is possible to obtain a solution to the differential Eq. (4), which takes the form of the Euler-Laplace equation:

$$a_{ij} = \alpha_{ij}\cdot r^{-n} \tag{5}$$

Taking into account the latter, S. G. Lekhnitsky from Eq. (4) obtained a differential equation for finding the function *f(r)*.

In the problem under consideration, the change in the elasticity coefficients along the radius of the section, taking into account the various properties of the early and late zones of annual layers, can be taken in the form of trigonometric functions, and then it becomes possible to represent the deformation coefficients in the form of power series using the expansion in a series of sine and cosine functions. In this case, it becomes possible to solve the problem of bending of an inhomogeneous cylindrically anisotropic solid by a constant moment.

The function of the deformation coefficient E_z can be represented:

$$E_z(r) = \frac{E_{zl} + E_{ze}}{2} + \left[\frac{E_{zl} - E_{ze}}{2}\right]^* \sin\left(\frac{\pi^* r}{\delta}\right), \tag{6}$$

where E_{zn}, E_{zr}—deformation coefficients of the late and early zones of the annual layer (Fig. 1); δ—the width of the annual layer, which is the sum of the widths of the layers of early and late wood (assumed to be equal in size in the paper).

In accordance with Fig. 1, function (6) can be rewritten as:

$$E_z(r) = E_{zm} + E_{za}\sin\left(\frac{\pi}{\delta}r\right), \tag{7}$$

where E_{zm}, E_{za}—mean and peak values of the deformation coefficient E_z.

Average values E_{tm}, E_{zm} are taken from tables [20].

Similarly, we can write for E_r and Poisson's ratios:

Differential equation (for exponent $n = 0$):

$$f_0^{IV} + \frac{2}{r}f_0^{III} - \alpha\frac{f_0^{II}}{r} + \alpha\left[\frac{f_0^{I}}{r^3} - \frac{f_0}{r^4}\right] = 2B_0\frac{\alpha_{13} - \alpha_{23}}{\gamma_{22}\alpha_{33}}r^{-1} \tag{8}$$

The general solution to this equation is:

$$f_0(r) = C_{10}r^{\lambda_0} + C_{20}r^{k_0} + C_{30}r + C_{40}r + 2Bl_{no}r^3, \tag{9}$$

where

$$\begin{Bmatrix}\lambda_0\\ k_0\end{Bmatrix} = 1 \pm \sqrt{1 + \alpha_{(0)}} \tag{10}$$

$$l_{no} = \frac{\mu_{zt}^m - \mu_{zr}^m}{\left[\frac{1}{E_{tm}} - \frac{(\mu_{zt}^m)^2}{E_{zm}}\right](3 - \alpha)} \tag{11}$$

The numerical values of the quantities used are obtained with the values of the elastic coefficients for Scotch fir specified in [20] are numerically equal to the values given in the Table 1.

Then we solve in a similar way for $m = 1$ and $m = 2$ and $m = 3$.

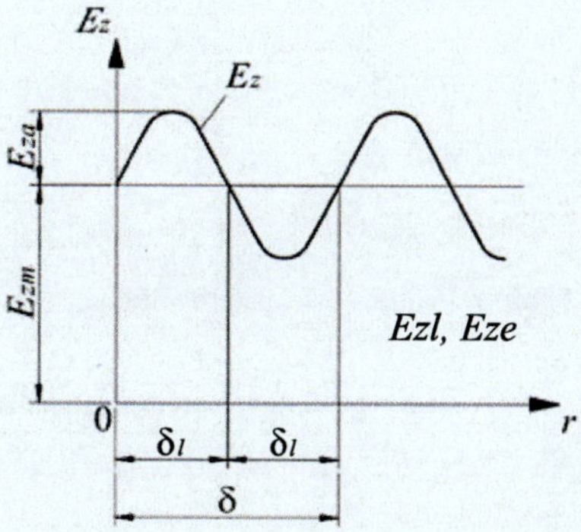

Fig. 1 Change in the deformation coefficient E_r along the radius of the cross section

3 Results of Researches

As a result of solving the equation for stresses in 4 problems: a problem with an exponent $n = 0$; problem for $n = 1$; the problem for $n = 3$ and the problem for $n = 5$ obtained the following.

In each problem, we found integration constants C_{1n}, C_{2n}, B_n (through C_{1n}, C_{2n}), solved the total problem from the third condition, found the constant B and solved the final stress problem.

Numerical solution to determine the coefficients:

$$\alpha_{33} = \frac{1}{E_{zm}} = \frac{1}{16,600} = 0.00006,$$

$$E_{zm} = E_a$$

If we accept

$$\frac{E_{zr}}{E_{zn}} = \upsilon_z, E_{zn} = E_{zr} + 2E_{za}, E_{za} = \frac{E_{zn} - E_{zr}}{2},$$

$$E_{za} = \frac{1}{2}[E_{zn} - E_{zn}\upsilon_z] = E_{zn}\frac{1 - \upsilon_z}{2}.$$

where E_{zn} and E_{zr}—modules of elasticity of late and early wood of the annual layer.

The same can be taken for the rest of the elasticity coefficients.

Common decision:

$$F = f(r)\sin\theta;$$

$$f = C_{1n}r^{\lambda n} + C_{2n}r^{k_n} + C_{3n}r^{1+n} + C_{4n}r + B_n l_n(n+2)r^{n+3};$$

$$\frac{\partial F}{\partial r} = \left[\lambda C_1 r^{\lambda-1} + kC_2 r^{k-1} + (1+n)C_3 r^n + C_4 + Bl_n(n+2)(n+3)r^{n+2}\right] \cdot \sin\theta;$$

$$\frac{\partial F}{\partial r^2} = \left[\lambda(\lambda-1)C_1 r^{\lambda-2} + k(k-1)C_2 r^{k-2} + (1+n)\cdot nC_3 r^{n-1} + Bl_n(n+2)(n+3)(n+2)r^{n+1}\right] \cdot \sin\theta;$$

Table 1 Values of elastic coefficients

Scotch fir	E_r	E_t	E_a	G_{rt}	G_{at}	G_{ar}	μ_{tr}/μ_{rt}	μ_{at}/μ_{ta}	μ_{ra}/μ_{ar}
	1124	582	16,600	187.6	690	1180	0.31/0.68	0.51/0.015	0.038/0.42

$$\frac{\partial F}{\partial \theta} = \left[C_1 r^{\lambda} + C_2 r^{k} + C_3 r^{1+n} + C_4 r + Bl_n(n+2)r^{n+3}\right] \cdot \cos\theta;$$

$$\frac{\partial^2 F}{\partial \theta^2} = \left[-C_1 r^{\lambda} - C_2 r^{k} - C_3 r^{1+n} - C_4 r - Bl_n(n+2)r^{n+3}\right] \cdot \sin\theta;$$

$$\begin{aligned}\sigma_r =& \frac{1}{r}\cdot\frac{\partial F}{\partial r} + \frac{1}{r^2}\cdot\frac{\partial^2 F}{\partial \theta^2}\\ =& \left[\lambda C_1 r^{\lambda-2} + kC_2 r^{k-2} + (1+n)\cdot nC_3 r^{n-1} + Bl_n(n+2)(n+3)r^{n+1}\right]\cdot\sin\theta\\ &+ \left[-C_1 r^{\lambda-2} - C_2 r^{k-2} - C_3 r^{n-1} - C_4\frac{1}{r} - Bl_n(n+2)r^{n+1}\right]\cdot\sin\theta;\end{aligned}$$

We get the prototype of this expression

$$\begin{aligned}\sigma_r =& \left[C_1 r^{\lambda-2}(\lambda-1) + C_2 r^{k-2}(k-1) + C_3 r^{n-1}(1+n-1) + C_4\frac{1}{r}\right.\\ &\left. + Bl_n(n+2)(n+3-1)r^{n+1}\right]\cdot\sin\theta;\end{aligned}$$

$$\begin{aligned}\sigma_\theta =& \frac{\partial^2 F}{\partial r^2} = \left[\lambda(\lambda-1)C_1 r^{\lambda-2} + k(k-1)C_2 r^{k-2} + (n+1)\cdot nC_3 r^{n-1}\right.\\ &\left. + Bl_n(n+2)(n+3)(n+2)r^{n+1}\right]\cdot\sin\theta;\end{aligned}$$

$$\begin{aligned}\sigma_z =& \mu_{zr}\sigma_r + \mu_{zt}\sigma_\theta = \mu_{zr}\left[C_1 r^{\lambda-2}(\lambda-1) + C_2 r^{k-2}(k-1) + C_3 r^{n-1}(1+n-1)\right.\\ &\left. + C_4\frac{1}{r} + Bl_n(n+2)(n+3-1)r^{n+1}\right]\cdot\sin\theta\\ &+ \mu_{zt}\left[\lambda(\lambda-1)C_1 r^{\lambda-2} + k(k-1)C_2 r^{k-2} + (n+1)\cdot nC_3 r^{n-1}\right.\\ &\left. + Bl_n(n+2)(n+3)(n+2)r^{n+1}\right]\cdot\sin\theta;\end{aligned}$$

$$\begin{aligned}\sigma_z =& \left[C_1 r^{\lambda-2}(\lambda-1)(\mu_{zr} + \lambda\mu_{zt}) + C_2 r^{k-2}(k-1)(\mu_{zr} + k\mu_{zt}) + C_3 r^{n-1} n(\mu_{zr} + \mu_{zt}(n+1))\right.\\ &\left. + Bl_n(n+2)(n+2)r^{n+1} + (\mu_{zr} + \mu_{zt}(n+3))\right]\cdot\sin\theta;\end{aligned}$$

$$\begin{aligned}\sigma_{rt} =& \frac{1}{r}\cdot\frac{\partial F}{\partial r\partial\theta} + \frac{1}{r^2}\cdot\frac{\partial^2 F}{\partial\theta^2} = \left[-C_1 r^{\lambda-2} - C_2 r^{k-2} - C_3 r^{1+n-2}\right.\\ &\left. - C_4\frac{1}{r} - Bl_n(n+2)r^{n+1}\right]\cdot\sin\theta\\ &- \left[\lambda C_1 r^{\lambda-2} + kC_2 r^{k-2} - (1+n)C_3 r^{n-1}\right.\\ &\left. + C_4\frac{1}{r} + Bl_n(n+2)(n+3)r^{n+1}\right]\cdot\cos\theta;\end{aligned}$$

First Condition:

(a) $\tau_{rt} = 0$, $\sigma_r = 0$ при $r = a$, $r = b$. a—core tube size, b—the diameter of the round log (Fig. 2). The size of the core tube was introduced to eliminate the singular point, that is, infinity in the center of the log (Fig. 3).

From the condition 1a we get:

$$C_1 a^{\lambda-2} + C_2 a^{k-2} + C_3 a^{n-1} + C_4 \frac{1}{a} + Bl_n(n+2)a^{n+1} = 0;$$

$$\lambda C_1 a^{\lambda-2} + kC_2 a^{k-2} + (n+1)C_3 a^{n-1} + C_4 \frac{1}{a} + Bl_n(n+2)(n+3)a^{n+1} = 0.$$

When $r = a$, the condition is automatically fulfilled according to the two upper equations obtained based on condition 1a. Then we obtain the following two equations with $r = b$, that is, $\tau_{rt} = 0$ if $r = b$.

$$C_1 b^{\lambda-2} + C_2 b^{k-2} + C_3 b^{n-1} + C_4 \frac{1}{b} + Bl_n(n+2)b^{n+1} = 0;$$

$$\lambda C_1 b^{\lambda-2} + kC_2 b^{k-2} + (1+n)C_3 b^{n-1} + C_4 \frac{1}{b} + Bl_n(n+2)(n+3)b^{n+1} = 0.$$

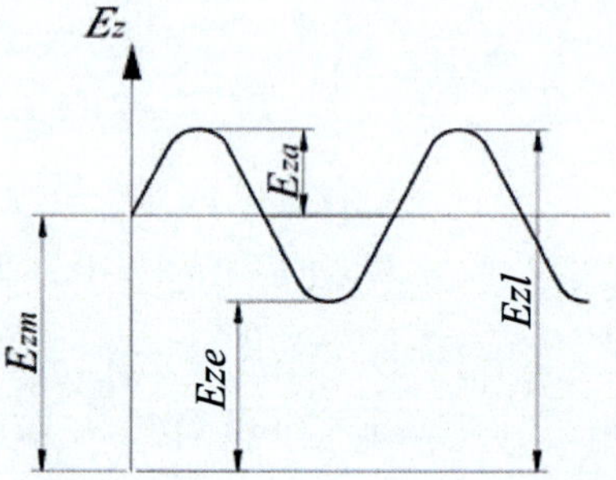

Fig. 2 Distribution of elastic modulus values

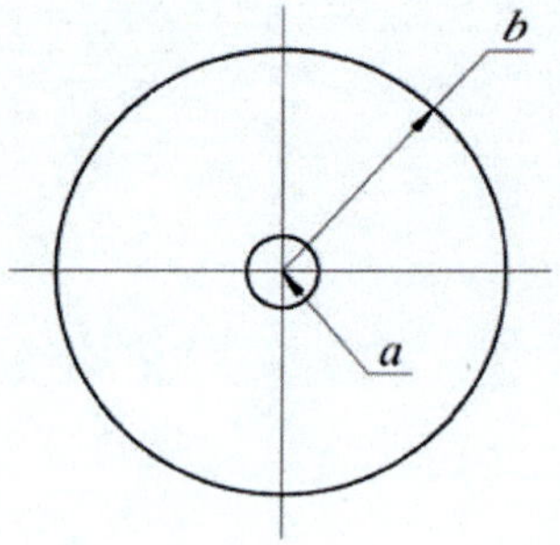

Fig. 3 Dimensions of the core tube and the diameter of the round log

Integration constants C_1, C_2, C_3, C_4 can be found from these four equations. It should be borne in mind that the constant C_4 will not enter the stress equation, and C_3 corresponds to multivalued variables, therefore $C_3 = 0$.

To find the constant B, you need to use the condition at the ends:

$$M = \int_0^{2\pi}\int_a^b \sigma_z \sin\theta \cdot r^2 dr d\theta;$$

$$\int \sin^2\theta \cdot d\theta = \int \left[\frac{1}{2} - \frac{1}{2}\cos 2\theta\right] \cdot d\theta = \frac{1}{2}\theta - \frac{1}{2}\cdot\frac{1}{2}\sin 2\theta + C.$$

$$\begin{aligned} M = \int_0^{2\pi}\int_a^b \sigma_z \sin\theta \cdot r^2 dr d\theta = \int_0^{2\pi}\int_a^b & \left[C_1 r^{\lambda-2}(\lambda-1)(\mu_{zr} + \lambda\mu_{zt})\right. \\ & + C_2 r^{k-2}(k-1)(\mu_{zr} + k\mu_{zt}) + C_3 r^{n-1} n(\mu_{zr} + \mu_{zt}(n+1)) \\ & \left. + Bl_n(n+2)(n+2)r^{n+1}(\mu_{zr} + \mu_{zt}(n+3))\right] \sin\theta \cdot \sin\theta \cdot r^2 dr d\theta \\ = \int_a^b & \left[C_1 r^{\lambda-2}(\lambda-1)(\mu_{zr} + \lambda\mu_{zt}) + C_2 r^{k-2}(k-1)(\mu_{zr} + k\mu_{zt})\right. \\ & \left. + C_3 r^{n-1} n(\mu_{zr} + \mu_{zt}(n+1)) + Bl_n(n+2)(n+2)r^{n+1}(\mu_{zr} + \mu_{zt}(n+3))\right] \\ & \left[\frac{1}{2}\theta - \frac{1}{4}\sin 2\theta\right]_0^{2\pi} r^2 dr = M \end{aligned}$$

$$\pi = \frac{1}{2}\theta - \frac{1}{4}\sin 2\theta$$

$$\begin{aligned} & \int_a^b C_1 r^{\lambda}(\lambda-1)(\mu_{zr} + \lambda\mu_{zt})\pi dr + \int_a^b C_2 r^k (k-1)(\mu_{zr} + k\mu_{zt})\pi dr \\ & + \int_a^b C_3 r^{n-1} n(\mu_{zr} + \mu_{zt}(n+1)) \cdot \pi dr \\ & + \int_a^b Bl_n(n+2)(n+2)r^{n+3}(\mu_{zr} + \mu_{zt}(n+3))\pi dr = M \end{aligned}$$

From the latter we find the constant of integration B.

Thus, for $m = 0$ we have:

$$f_0 = C_{10}r^{\lambda_0} + C_{20}r^{k_0} + C_{30}r^{1+0} + C_{40}r + 2B_0 l_n r^3;$$

$$\sigma_r = \left[C_{10}r^{\lambda_0-2} + C_{20}r^{k_0-2}(k_0 - 1) + 4B_0 l_n r\right] \cdot \sin\theta;$$

$$\sigma_\theta = \left[\lambda_0(\lambda_0 - 1)C_{10}r^{\lambda_0-2} + k_0(k_0 - 1)C_{20}r^{k_0-2} + 12B_0 l_n r\right] \cdot \sin\theta;$$

$$\sigma_z = \left[C_{10}r^{\lambda_0-2}(\lambda_0 - 1)\left(\mu_{zr} + \lambda_0\mu_{zt}\right) + C_{20}r^{k_0-2}(k_0 - 1)\left(\mu_{zr} + k_0\mu_{zt}\right) + 4B_0 l_n\left(\mu_{zr} + 3\mu_{zt}\right)\right] \sin\theta;$$

$$\tau_{rt} = -\left[C_{10}r^{\lambda_0-2} + C_{20}r^{k_0-2} + C_{30}r^{-1} + C_{40}\frac{1}{r} + 2B_0 l_n\right] \cdot \sin\theta - \left[\lambda_0 C_{10}r^{\lambda_0-2} + k_0 C_{20}r^{k_0-2} + C_{30}r^{-1} + C_{40}r\frac{1}{k} + 6B_0 l_n r\right] \cdot \cos\theta;$$

Condition: $r = a$, $\tau_{rt} = 0$, $\sigma_r = 0$.

$$C_{10}a^{\lambda_0-2} + C_{20}a^{k_0-2} + C_{30}a^{-1} + 2B_0 l_n a = 0;$$

$$\lambda_0 C_{10}a^{\lambda_0-2} + k_0 C_{20}a^{k_0-2} + C_{30}a^{-1} + C_{40}\frac{1}{a} + 6B_0 l_n a = 0.$$

The condition $r = a$, $\sigma_r = 0$ is fulfilled automatically. The second condition $r = b$, $\sigma_r = 0$, $\tau_{rt} = 0$:

$$C_{10}b^{\lambda_0-2} + C_{20}b^{k_0-2} + C_{30}b^{-1} + 2B_0 l_n b = 0;$$

From the condition at the ends

$$M = \int_0^{2\pi}\int_a^b \sigma_z \sin\theta \cdot r^2 dr d\theta;$$

We have:

$$C_{10}(\lambda_0 - 1)\left(\mu_{zr}^m + \lambda\mu_{zt}^m\right)\pi\frac{b^{\lambda_0+1} - a^{\lambda_0+1}}{\lambda_0 + 1} + C_{20}(k_0 - 1)\left(\mu_{zr}^m + k_0\mu_{zt}^m\right)\pi\frac{b^{k_0+1} - a^{k_0+1}}{k_0 + 1} + 4B_0 l_n\left(\mu_{zr}^m + \mu_{zt}^m \cdot 3\right)\pi\frac{b^4 - a^4}{4} = M_0,$$

where M_0—perceived (as yet unknown) part of the applied moment M; B_0—the constant of integration for this differential equation (with $m = 0$); μ_{zr}^m, μ_{zt}^m—average values of elasticity coefficients; by analogy with the modulus of elasticity, it can be written as follows:

$$\mu_{zr} = \mu_{zr}^{m} + \mu_{zr}^{a}\frac{\pi}{\delta}r - \mu_{zr}^{a}\left(\frac{\pi}{\delta}\right)^{3}\frac{r^{3}}{3!} + \mu_{zr}^{a}\left(\frac{\pi}{\delta}\right)^{5}\cdot\left(\frac{r^{5}}{5!}\right);$$

$$\mu_{zt} = \mu_{zt}^{m} + \mu_{zt}^{a}\frac{\pi}{\delta}r - \mu_{zt}^{a}\left(\frac{\pi}{\delta}\right)^{3}\frac{r^{3}}{3!} + \mu_{zt}^{a}\left(\frac{\pi}{\delta}\right)^{5}\cdot\left(\frac{r^{5}}{5!}\right);$$

In the calculations, it is necessary to substitute the values of μ_{zt} and $\mu_{zr}^{m}$from tables [20] instead of μ^{m}.

To find λ_0 and k_0, one should solve the differential equation

$$f_0 = C_{10}r^{\lambda_0} + C_{20}r^{k_0} + C_{30}r + C_{40}r + 2B_0 l_n r^3;$$

In which

$$\begin{Bmatrix}\lambda_0\\k_0\end{Bmatrix} = 1\sqrt{1+\alpha_0},$$

Further, the differential equations are solved in a similar way for $m = 1, 2, 3$. All stress equations will contain the constant $B_{1,\,2,\,3}$, determined from the conditions at the ends. The resulting equations for M_0, M_1, M_2, M_3 were summed up. On the left-hand side, we have obtained the sum of moments equal to the applied moment M. On the right-hand side, there will be constants C_i, B_i. Then they made and solved an equation with unknowns.

2nd option—express all constants C_{10}, C_{20}, C_{11}, C_{21} … C_{14}, C_{24} through one constant B, which we find using the 3rd condition.

$$f^{IV} - 4f^{III} + f^{II}\left[-\frac{1-\upsilon_{rz}\upsilon_{zr}}{E_{tm}}(5+\alpha^2) + 2\frac{\upsilon_{rt}+\upsilon_{rz}\upsilon_{rt}}{E_{rm}} - \frac{1}{G_{rtm}}\right]$$
$$+ f^{I}\left[6\frac{1-\upsilon_{tz}\upsilon_{zt}}{E_{tm}} - \frac{1}{G_{rtm}}\right] + f\left[-\frac{1-\upsilon_{rz}\upsilon_{zr}}{E_{rm}} + 2\frac{\upsilon_{rt}+\upsilon_{rz}\upsilon_{zt}}{E_{rm}}\right] = 0.$$

Characteristic equation:

$$k^4 - 4k^3 + k^2\left[-\frac{1-\upsilon_{rz}\upsilon_{zr}}{E_{tm}}(5+\alpha^2) + 2\frac{\upsilon_{rt}+\upsilon_{rz}\upsilon_{rt}}{E_{rm}} - \frac{1}{G_{rtm}}\right]$$
$$+ k\left[6\frac{1-\upsilon_{tz}\upsilon_{zt}}{E_{tm}} - \frac{1}{G_{rtm}}\right] + \left[-\frac{1-\upsilon_{rz}\upsilon_{zr}}{E_{rm}} + 2\frac{\upsilon_{rt}+\upsilon_{rz}\upsilon_{zt}}{E_{rm}}\right] = 0$$

The roots of this equation are k_1, k_2, k_3, k_4 for the corresponding values υ…, E_{rm}, E_{rt}, $\alpha^2 = 0.5$. Then the function

$$f(r) = C_1 r^{k_1} + C_2 r^{k_2} + C_3 r^{k_3} + C_4 r^{k_4}$$

This is one of the options when the roots are real, positive or negative.

$$\alpha_{23} = -\frac{\mu_{zta}}{E_{za}} = -\frac{0.17}{5533} = -0.0307 \cdot 10^{-3};$$

$$\alpha_{13} = -\frac{\mu_{zra}}{E_{za}} = -\frac{0.14}{5533} = -0.0253 \cdot 10^{-3};$$

$$\alpha_{33} = \frac{\delta}{E_{za} \cdot \pi} = \frac{4}{5533 \cdot 3.14} = 0.2302 \cdot 10^{-3};$$

$$\gamma_{66} = 2\frac{\mu_{tra}}{E_{ta}} = \frac{2 \cdot 0,1033}{187.3} = 1.10304 \cdot 10^{-3};$$

$$\gamma_{11} = \frac{\delta}{\pi E_{ra}} - \frac{(\mu_{zra})^2 \pi}{E_{za} \cdot \delta} = \frac{2}{3.14 \cdot 375} - \frac{0.14^2 \cdot 3.14}{5533 \cdot 2} = 1.69444 \cdot 10^{-3};$$

$$\gamma_{22} = \frac{\delta}{\pi E_{ta}} - \frac{(\mu_{zta})^2 \pi}{E_{za} \cdot \delta} = \frac{2}{3.14 \cdot 187.3} - \frac{0.17^2 \cdot 3.14}{5533 \cdot 2} = 3.3895 \cdot 10^{-3};$$

$$\gamma_{12} = -\frac{\mu_{tra}}{E_{ra}} - \frac{\mu_{zra}\mu_{zta}}{E_{ra}} \cdot \frac{\pi}{\delta} = -\frac{0.1033}{375} - \frac{0.14 \cdot 0.17 \cdot 3.14}{375 \cdot 2} = -0.37511 \cdot 10^{-3};$$

$$\alpha = \frac{1.69444 \cdot 10^{-3} - 3 \cdot 0.37511 \cdot 10^{-3} + 1.10304 \cdot 10^{-3}}{3.3895 \cdot 10^{-3}} = 0.4933.$$

$$E_{za} = E_{zm}\frac{1-\alpha_{02}^2}{1+\alpha_{02}^2} = 16600 \cdot \frac{1-0.5}{1+0.5} = 5533\,\text{МПа}; \left\{\frac{E_{z\Pi} = 5533 + 16600 = 22133\,\text{МПа}}{E_z = 16600 - 5533 = 11067\,\text{МПа}}\right\}$$

$$E_{ta} = E_{tm}\frac{1-\alpha_{01}^2}{1+\alpha_{01}^2} = 582 \cdot \frac{1-0.5}{1+0.5} = 194\,\text{МПа}; \left\{\frac{E_{t\Pi} = 194 + 582 = 776\,\text{МПа}}{E_t = 582 - 194 = 388\,\text{МПа}}\right\}$$

$$\frac{E_t}{E_r} = 0.5178; E_{ra} = \frac{E_{ta}}{\alpha_0^2} = \frac{194}{0.5178} = 375\,\text{МПа}; \left\{\frac{E_{r\Pi} = 375 + 1124 = 1499\,\text{МПа}}{E_r = 1124 - 375 = 749\,\text{МПа}}\right\}$$

$$\mu_{tra} = 0.31 \cdot \frac{1-0.5}{1+0.5} = 0.1033; \left\{\frac{\mu_{tr\Pi} = 0.31 + 0.1033 = 0.4133}{\mu_{tr} = 0.31 - 0.1033 = 0.2067}\right\}$$

$$E_a = 16600; E_t = 582; E_r = 1164; G_{rt} = 188; \frac{\mu_{tr}}{\mu_{rt}} = \frac{0.31}{0.68}; \frac{\mu_{at}}{\mu_{ta}} = \frac{0.51}{0.015}; \frac{\mu_{ra}}{\mu_{ar}} = \frac{0.038}{0.42};$$

$$\gamma_0^2 = \frac{E_t}{E_r} = \frac{582}{1124} = 0.5178.$$

Hypothetically, we accept

$\gamma_0^2 = 0.5178.$

$\mu_{tr} = 0.68 \cdot 0.5178 = 0.35; \mu_{ta} = 0.51 \cdot 0.036 = 0.01785; \mu_{ra} = 0.42 \cdot 0.07 = 0.0294;$

$\frac{E_t}{E_a} = \frac{582}{16600} = 0.035; \frac{E_r}{E_a} = \frac{1124}{16600} = 0.07; \mu_{zra} = 0.42 \cdot \frac{1-0.5}{1+0.5} = 0.14;$

$\mu_{zta} = 0.51 \cdot \frac{1-0.5}{1+0.5} = 0.1.$

The calculations took:

$$\mu_{rt} = 0.68; \mu_{at} = 0.51; \mu_{ar} = 0.42; \mu_{tr} = 0.35; \mu_{ta} = 0.018; \mu_{ra} = 0.03;$$
$$E_t = 583\,\text{МПа}; E_r = 1124\,\text{МПа}; E_a = 16600\,\text{МПа}; G_{rt} = 188\,\text{МПа}.$$
$$\delta = 4\text{ММ}.$$

See Fig. 4.

$$E_{tm} = \frac{E_{t\Pi} + E_{tp}}{2}; E_{ta} = \frac{E_{t\Pi} - E_{tp}}{2}; E_{t\Pi} = E_{tm} + E_{ta}; E_{tp} = E_{tm} - E_{ta}.$$

Normal stresses are determined by the formula:

$$\sigma_z = \frac{M}{J} \cdot r \cdot \sin\theta + \sigma_{z1} \cdot \sin\theta + \sigma_{z3} \cdot \sin\theta + \sigma_{z5} \cdot \sin\theta.$$

The values of this part of the formula $\frac{M}{J} \cdot r \cdot \sin\theta$ are known, and the values of the other part.

$\sigma_{z1} \cdot \sin\theta + \sigma_{z3} \cdot \sin\theta + \sigma_{z5} \cdot \sin\theta$ must be found. In terms 2, 3 and 4 it is necessary to substitute their values through M and the resulting component.

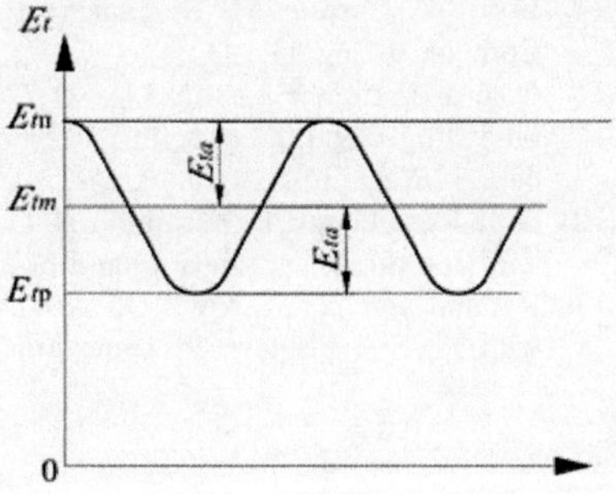

Fig. 4 Change in the coefficient of deformation E_{ta} by annual layers

4 Conclusion

- The problem of calculating the strength characteristics of structural sawn timber is solved, the macrostructure of the wood structure of which is a system of thin-walled shells coaxially strung on each other, simulating annual layers of wood.
- The dependences of the distribution of normal stresses in the cross section of the log are found.
- The dependence of the bending moment on the distribution of normal stresses along the radius of the cross-section of the log has been obtained.
- The obtained solution makes it possible to study the stress–strain state of the elements of load-bearing building structures with an annular structure of annual wood layers during transverse bending.

References

1. Vorontsov Y, Surovtseva L (2002) Efficiency of sawmills' specialization according to diameter groups of sawn raw material. Bulletin of Higher Educational Institutions. Russian For J 5:89–93
2. Chubinskii A (2014) Physical nondestructive methods for link he testing and evaluation of the structure of wood based materials. Russ J Nondestr Test 11:693–700
3. Wei Q, Leblon B, La Rocque A (2011) On the use of X-ray computed tomography for determining wood properties a revive. Can J For Res 41:2120–2140
4. Horvath B, Peszlen I, Peralta P, Horvath L, Kasal B, Li L (2010) Elastic modulus determination of transgenic aspen using a dynamic mechanical analyzer in static bending mode. For Prod J 60(3):296–300
5. Miller D (1962) Selective efficiencies of nondestructive strength test. For Prod J 5(8):87–92
6. Jlinen A (1952) Uber die mechanissche Schaftformtheorie der Baume. Techn. Hochschule in Finland. Wiss. Forschungen, 6, Helsinki
7. Byzov V, Melekhov V (2016) Structural sawn timber: resource enhancement. Mag Civil Eng 5:67–76
8. Byzov V, Melekhov V (2019) Manufacturing structural building components from round with hearwood rot. Mag Civil Eng 2:11–19
9. Byzov V, Melekhov V, Chernykh A, Mamedov Ch (2020) Strength properties of truss elements made of environmentally friendly structural lumber. Architec Eng 5(2):25–31
10. Borri A, Corradi M, Speranzini E (2013) Reinforcement of wood with natural fibers. Compos B Eng 53:1–8
11. Côté J-F, Luther J, van Lier O (2021) Assessing the impact of fine-scale structure on predicting wood fiber attributes of boreal conifer trees and forest plots. Forest Ecol Manage 479, 30 September, 1 January 2021, 118624
12. de la Rosa García P, Escamilla A, García M (2013) Bending reinforcement of timber beams with composite carbon fiber and basalt fiber materials. Compos B Eng 55:528–536
13. Karamisheva A, Yazyev S, Avakov A (2016) Calculation of plane bending stability of beams with variable stiffness. Procedia Eng 150:1872–1877

14. Mayencourt P, Mueller C (2020) Hybrid analytical and computational optimization methodology for structural shaping: Material-efficient mass timber beams. Eng Struct 215, 15 May (Cover date: 15 July 2020), art. 110532
15. Ramage MH, Burridge H, Busse-Wicher M, Fereday G (2017) The wood from the trees: the use of timber in construction. Renew Sustain Energy Rev 68:333–359
16. Sharaf H, Ishak M, Fattahi A (2020) Experimental and numerical investigation of the mechanical behavior of full-scale wooden cross arm in the transmission towers in terms of load-deflection test. J Market Res 5(June):7937–7946
17. Smardzewski J, Wojciechowski K (2019) Response of wood-based sandwich beams with three-dimensional lattice core. Compos Struct 216: 340–349, 2 March (Cover date: 15 May 2019)
18. Sorin E, Coureau J-L, Myriam Chaplain M (2021) Prediction of the ultimate load-carrying capacity of wooden notched beams with and without reinforcements using a splitting model. Constr Build Mater 271, 17 November (Cover date: 15 February 2021), art. 121518
19. Lekhnitskiy S (1977) The theory of elasticity of an anisotropic body. The science, Moscow
20. Ashkenazi Y (1977) Anisotropy of wood and wood materials. Timber industry, Moscow

Deformation and Failure of Prestressed Reinforced Concrete Frames in Ultimate States

Vitaly Kolchunov, Tatyana Iliushchenko, and Sergey Savin

Abstract The results of experimental and theoretical researches of crack resistance, development and growth of cracks of a prestressed reinforced concrete multi-storey frame of a monolithic building under a special impact caused by the sudden removal of the outside column of the frame are presented. The calculation of the frame was carried out by the decomposition method according to a two-level primary and secondary design models, taking into account the prestressing of the constructions framework beam. An analysis of the pattern cracking and destruction of frames is given in the frames with prestressed elements with a sudden change of flux of forces under the considered special impact. It was found that dynamic loading of frame elements and the failure pattern of the frame structures, arising in such structural systems, depend on the presence and level of prestress, the parameters of the deformation diagram of the sections of the frame elements and the place of application of the special emergency impact. The obtained research results can be used in the development of methodology and methods for protecting the frames of monolithic multi-storey buildings against progressive collapse under the considered influences.

Keywords Reinforced concrete structure · Prestressing · Crack resistance · Experimental research · Progressive collapse

1 Introduction

Due to the constant increase in the number and intensity of natural and technogenic impacts on buildings and structures, leading to a disproportionate failure of structures and in some cases to their complete failure, a new regulatory framework has

V. Kolchunov · T. Iliushchenko (✉)
South West State University, 50 let Oktyabrya street, 94, Kursk 305040, Russia

V. Kolchunov · S. Savin
Moscow State University of Civil Engineering, Yaroslavskoe shosse, 26, Moscow 129337, Russia

N. Vatin et al. (eds.), *Proceedings of MPCPE 2021*, Lecture Notes in Civil Engineering 182, https://doi.org/10.1007/978-3-030-85236-8_4

been introduced in Russia and in a number of foreign countries to protect buildings and structures from progressive collapse [1–4]. These standards consider the general provisions of the methodology and design calculation and don't specify the design requirements, taking into account the various design features of the supporting structures of buildings and structures. As a result, new research is needed to study the force resistance of various types of structures under such influences. In recent years, a number of theoretical [5–15] and experimental [16–21] studies of reinforced concrete structural systems have been carried out, aimed at solving this problem. However, there are practically no works on this topic devoted to the study of prestressed reinforced concrete structures. It is possible to note several experimental foreign works in which such structural systems are investigated [22, 23]. The work [24] presents the results of experimental studies of a symmetrical structure of a reinforced concrete frame, designed a fragment of a multi-storey reinforced concrete frame of a monolithic building with prestressed beams on a special impact caused by a sudden removal of the central column. In the development of this study, this work presents the results of an experimental and theoretical study of crack resistance, development and opening of cracks in a fragment of a prestressed reinforced concrete frame under the special impact caused by the sudden removal of the outside column of the frame, leading to asymmetric loading of the entire structural system under study. In a structural system, not only the level of additional loading of load-bearing structures changes, but also the nature of the application of additional actions in case of asymmetric additional loading of elements. In this regard, the exhaustion of the bearing capacity of the elements can be associated not only with the exhaustion of their strength, but also from the loss of stability.

2 Materials and Methods

An experimental study of frame structures was carried out on a specially designed stand, which includes a load-bearing frame made of metal structures and a lever-mechanical system of loading devices in the form of strands from channels to transfer the load to the experimental frame structure. The installation of the test frame structure was carried out at the stand by welding the outlets of the working reinforcement from the frame poles to the channels of the base of the test stand.

Before the design load, the frame beams were loaded with two concentrated forces in each beam span using a mechanical lever arrangement consisting of a lever and distribution beams that transfer the load to the frame structure (Fig. 1). The beyond design impact was simulated by the sudden removal of the outside column of the frame using a special device made according to the patent [25]. The diagram of this device is shown in Fig. 1.

The prestressing of the reinforcement in the beams above the first floor of the frame was carried out mechanically using a jack. At the same time, to the compression force reflection from prestressing, the timbering of the lower beams of the

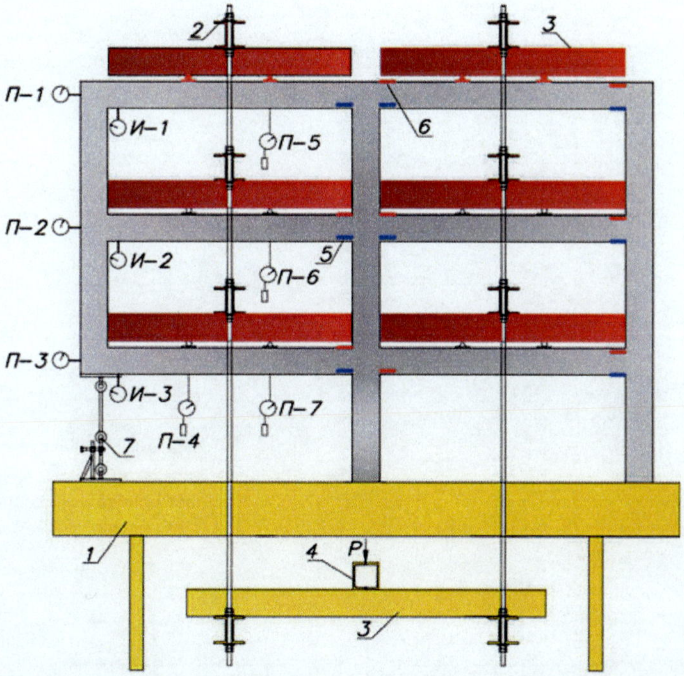

Fig. 1 Configuration diagram of tensoresistors and optical-mechanical devices: 1—power frame; 2—strands; 3—distribution beam; 4—lever; 5—tensoresistor on concrete; 6—tensoresistor on the reinforcement, 7—disconnecting device

frame was strengthen with a metal plate and a beam cage made of corners and two welded channels (Fig. 2). The design and materials of the FR-2 experimental frame tested when the outside column was removed were designed similarly to the design of the FR-1 experimental frame tested when the central column was removed [24]. The difference between the FR-2 experimental frame and the FR-1 frame was that there was no outside reinforced concrete column in it, which was replaced during testing, as already noted, with a special device that simulated the frame pole during the test. The prestressed reinforcement of the beams of the first floor of the frame was installed in the upper and lower cross-sectional zones of the beam from one rod of class A540, with a diameter of 8 mm. The value of the primary controlled stress is $\sigma_{sp} = 514MPa$. The first loss of prestress was 134 MPa, the second loss was 40 MPa, the stress in the prestressing reinforcement, taking into account all losses $\sigma_{sp2} = 340MPa$ (Fig. 3).

Before the start of the experiments, the calculation of the stress–strain state of the bearing capacity of the experimental frame structures was carried out. The calculation was carried out using the LIRA SAPR software package according to primary and secondary design models of different levels. The calculation according to the primary design model for the action of a given test load in the form of two

Fig. 2 Prestressing of reinforcement in the beam above the first floor of the experimental structure of the FR-2 frame

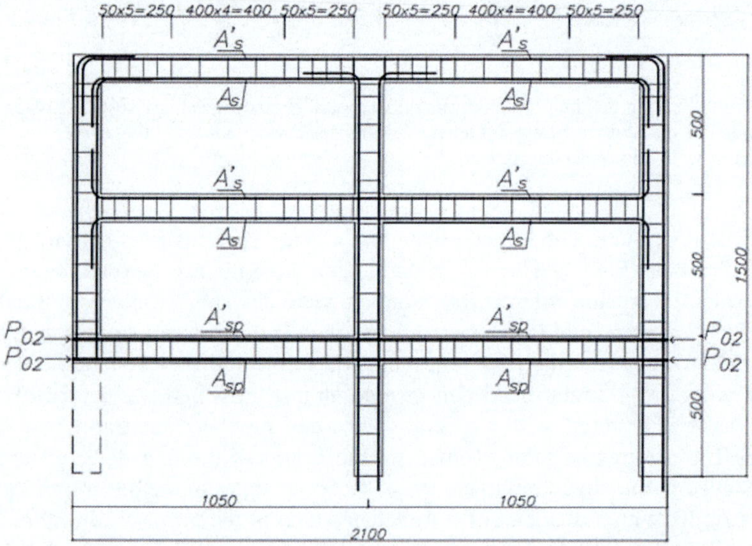

Fig. 3 Layout of reinforcement of the experimental structure of the FR-2 frame

concentrated forces in each span of the b was carried out in two variants: (1) using bar finite elements (Fig. 4a) and (2) using solid finite elements (Fig. 4b, c).

The prestress in the reinforcing rod was simulated in the first case by applying a longitudinal compressive force to the bar elements at the ends of the beams, equal

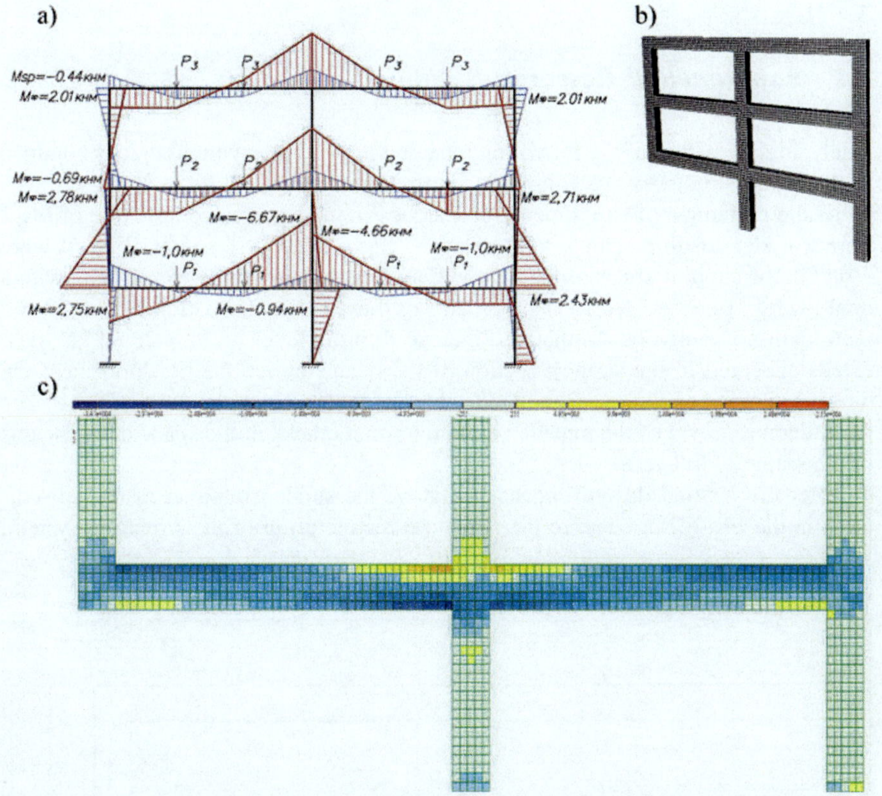

Fig. 4 Moment diagram for a frame with a prestressed beam (M_{sp}): before the sudden removal of the left column (blue) and after it is removal (red) (**a**); solid finite element model of the frame (**b**), mosaic of the distribution of longitudinal stresses (forces) in a prestressed beam above the first floor of the frame after removal the left column (**c**)

to the total compression force in the upper and lower reinforcing bars. In the second variant, when using solid finite elements, the prestress was taken into account by applying a longitudinal compression force in the center of the corresponding solid finite element.

The calculation according to the secondary design model is performed for the total action of the test load in the form of two concentrated forces in each span of the beam and a special effect caused by the removal of the outside column of the frame. Removal of the outside column of the frame was simulated by an application with the opposite sign of the reaction in this column, determined by the frame calculation according to the primary design model in accordance with the requirements [1] in the design model.

3 Results

3.1 Experimental Research Results

Analyzing the nature of the formation, opening and development of cracks obtained in the frame structures, the following can be noted.

At the first stage of experimental researches in the prestressed structure of FR-2 frames under loading with a total design $\sum P_{i,max}$, the first cracks (Fig. 5) were found in the support zones of the second floor beams near the central column with a total load $\sum$ Pi = 19.96 kN. The moments of these cracks formation were 0.59 and 0.56 kNm, respectively. As the load increased to the level $\sum$ Pi = 23.64 kN, new cracks appeared in the support section of the beams above the first floor near the outside columns and in the span. At the stage of loading with the design load before the sudden removal of the outside column, normal cracks in the spans of the beams of the frame didn't form.

After the beyond design impact caused by the sudden removal of the left column, in the FR-2 frame due to the structural restructuring of the structural system,

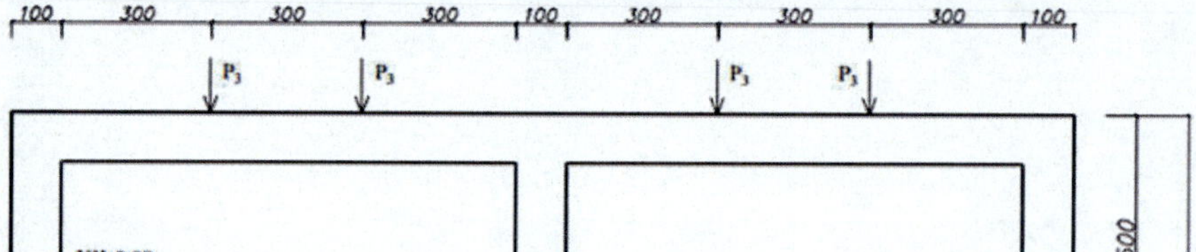

Fig. 5 Design of formation and opening width of cracks of all types in the structure of the FR-2 frame before the application of a special load

the previously formed cracks 1–6 opened (see Fig. 5), as well as their increment from the dynamic additional loading of the structural system elements.

New cracks were also formed in the support sections of the beam of the third floor of the frame, as well as normal cracks in the beams of all floors of the frame. The largest crack opening width after applying a special load was 0.70 and 0.65 mm in the support section of the central and right columns. At the same time, the experimental frame structure didn't receive visible damage (Fig. 6).

Analysis of the quantitative change in the width of the crack opening in the beams of the prestressed frame at the design load and after a special beyond design impact (Table 1) showed that the maximum increments were obtained by the cracks in the columns of the second floor (Cr-1, Cr-5). Using these data, the coefficient of dynamic loading of cross-sections of beams (θ_{ds}) from the considered beyond design impact was calculated as the ratio of the maximum crack opening width before and after the beyond design impact. Analyzing the obtained data, it can be noted that the structure of the prestressed reinforced concrete frame of the FR-2 didn't receive visible damage after the sudden removal of the left column. The criteria for the special limiting state for deformations of concrete and reinforcement [1] in the frame weren't met.

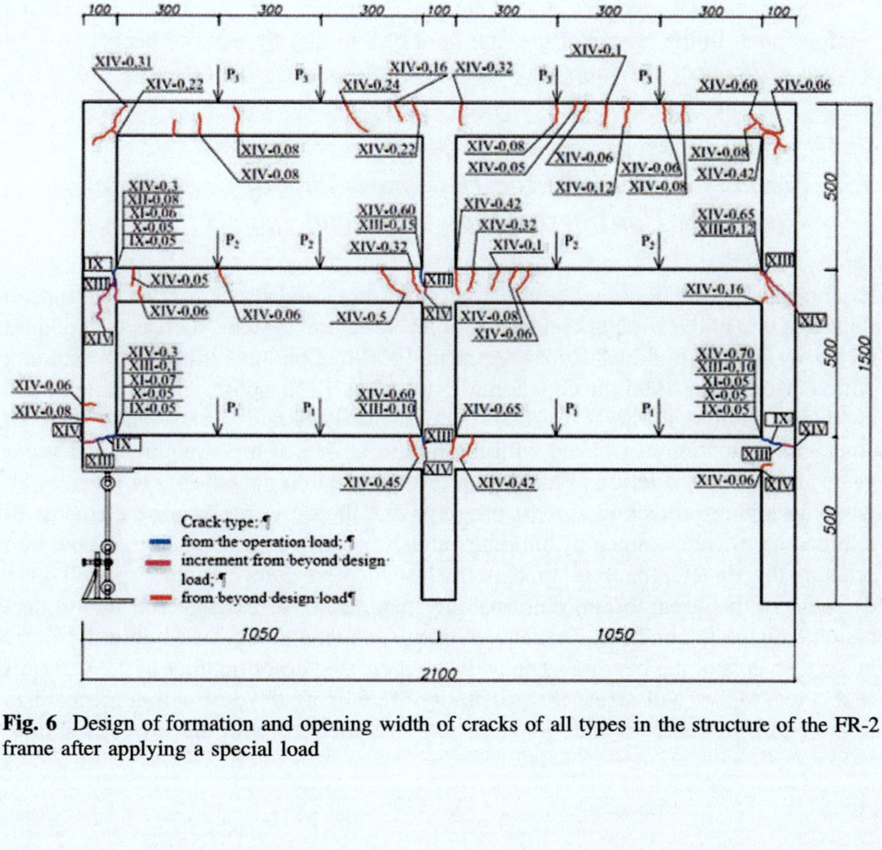

Fig. 6 Design of formation and opening width of cracks of all types in the structure of the FR-2 frame after applying a special load

Table 1 Width of crack opening of all types in the prestressed frame FR-2

Loading stage	Crack opening width of all types a_{crc} by crack number, mm						Note
	1	2	3	4	5	6	
Frame series FR-2							
IX	0.05	0.05	–	–	–	0.05	The moment of cracking of the experimental model
X	0.05	0.05	–	–	–	0.05	–
XI	0.06	0.07	–	–	–	0.05	–
XII	0.08	0.07	–	–	–	0.05	–
XIII	0.08	0.10	0.15	0.10	0.12	0.10	Before removing the column
XIV	0.30	0.30	0.60	0.60	0.65	0.70	After removing the column
θ_{ds}	3.75	3.00	4.00	6.00	5.42	7.00	–

When comparing the picture of cracking in the frame after the beyond design impact above the first and overlying floors, it can be seen that as the number of storeys increases, the coefficient of dynamic additional loading θ_{ds} decreases. So the maximum value θ_{ds} was recorded above the first floor in the right column of the frame and amounted to 7.00, while above the second floor this value was 5.42.

From the above analysis, it can be concluded that the placing of prestressed reinforcement in the beams of the first floor of building frames can become one of the ways to protect buildings and structures from progressive collapse.

3.2 *Stability Analysis of the Deformed State of the Frame Under the Considered Type of Special Impact*

Examples of partial or complete collapse of buildings and structures due to a sudden failure of one of the structural elements of the structural system, such as the collapse of the residential building Ronan Point in 1968 in London [26]; federal building Alfred Murrah in 1995 in Oklahoma City, USA [27]; sports and entertainment complex "Transvaal-Park" in 2004 in Moscow [28] and others, show that an emergency situation associated with the failure of one of the structures, can occur years after the introduction of a building or structure into operation. For this reason, when designing protection against progressive collapse in the bearing elements of reinforced concrete frames of buildings and structures, it is necessary to take into account the development in time of inelastic creep deformations, as well as a decrease in the strength and deformability parameters of the material due to corrosion damage to concrete. These phenomena can lead to a redistribution of forces in the elements of the bearing system and an increased deformability of the sections. In this regard, we will assess the possibility of realizing the destruction mechanism of the entire frame or its individual elements, associated with the loss of stability.

When analyzing the stability of the scale model of a reinforced concrete frame shown in Fig. 3, we will take into account creep deformations for girders by dividing the modulus of concrete deformation by (1 + φ_{cr}), where φ_{cr} is a coefficient that takes into account creep and is taken according to SP 63.13330.2018, depending on the class of concrete on the compression and humidity of the environment. The stiffness of the struts of the reinforced concrete frame under consideration under the action of only constant and long-term loads in the first approximation will be determined taking into account the eccentricities of the application of longitudinal forces to each of them according to the formula SP 63.13330.2018:

$$D = k_b E_b J + k_s E_s J_s \tag{1}$$

where E_b, E_s are initial modulus of elasticity of concrete and modulus of elasticity of steel reinforcement, respectively, J and J_s are moment of inertia of a concrete section and moment of inertia of reinforcement relative to the relative center of gravity of the concrete section;

$$k_b = 0.15/[\varphi_l(0.3 + \delta_e)], k_s = 0.7.$$

Here φ_l = 2 is a coefficient taking into account the effect of the duration of the load; δ_e = *e*/*h* is a relative eccentricity of application of longitudinal force.

Corrosion damage of the struts of all frames was taken into account by recalculating the stiffness using the corrosion damage function, for the approximation of which a polynomial of the second degree was adopted (see, for example, [29]). In this case, the depth of advancement of the corrosion front was taken conditionally equal to 20 mm, measured from the most compressed face, without evaluating the kinetics of the process. For the convenience of modeling the frame under consideration, the integral modulus of deformation of the sections was introduced into the calculation, which was taken to be constant along the length for each individual structural element in terms of the minimum stiffness of the section.

Additionally, in the cross-sections of the girders and frame struts, in which cracks formed at the stage of normal operation (Fig. 6), elastic-yielding fastenings were introduced between the FE to take into account the increased deformability. The compliance of such fixings was determined as an equivalent in deflection value based on preliminary numerical modeling of the corresponding structural elements using 8-node volumetric FE elements with joint joints at the location of the crack. Based on the results of a preliminary numerical simulation by FEM, the compliance of sections with a crack was: for girders—495 kNm, for racks—190 kNm.

According to the results of calculating the stability for the considered reinforced concrete frame in the absence of preliminary compression forces (Fig. 7a) and in the presence of preliminary compression (Fig. 7b), the safety factor for the first frame was—23.35, for the second—8.66, i.e., the stability of the two frames is ensured.

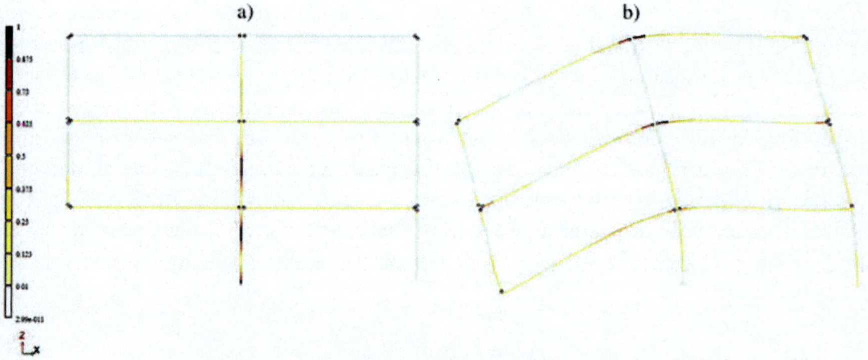

Fig. 7 Buckling modes of a model of a reinforced concrete frame under a special emergency impact: a frame without preliminary compression of the girders, the safety factor for stability is 23.35 (**a**); frame with a pre-compressed girder above the first floor, safety factor for stability 8.66 (**b**)

However, it should be noted that the preliminary compression of the girder led to more than a twofold decrease in the safety factor of the bearing system. In this case, in the model of the reinforced concrete frame considered above, the posts have flexibility $l_0/h = 1.2 \cdots 430/100 = 8.6$, and the cross-sectional height of the crossbars is 1/9 of the calculated length. Let us additionally evaluate the stability of a reinforced concrete frame with more graceful sections of structural elements: posts 50 × 50 mm ($l_0/h = 17.2$), crossbars 70 × 50 mm (section height is 1/13 of the span). Such an assessment is necessary to establish the limits of applicability of the proposed method of protection against progressive collapse.

The results of calculating the stability of the deformed state of the model of a reinforced concrete frame with changed parameters of sections of racks (50 × 50 mm) and girders (70 × 50 mm) with identical parameters of force and environmental influences, as for the original reinforced concrete frame, show: the safety factor for the frame without preliminary compression of the girder over on the first floor—14.44 (Fig. 8a), in the presence of a pre-compression of the girder—3.16 (Fig. 8b).

Thus, the stability of the frames is ensured in this case as well, however, a decrease in the safety factor of stability to 3.16 for a frame with a pre-compressed girder and reduced cross-sectional dimensions of structural elements shows that the use of pre-compression to increase the resistance of the structural system to progressive collapse with the flexibilities of the post elements close to $l_0/h = 20$ requires additional verification of the supporting system for stability criteria.

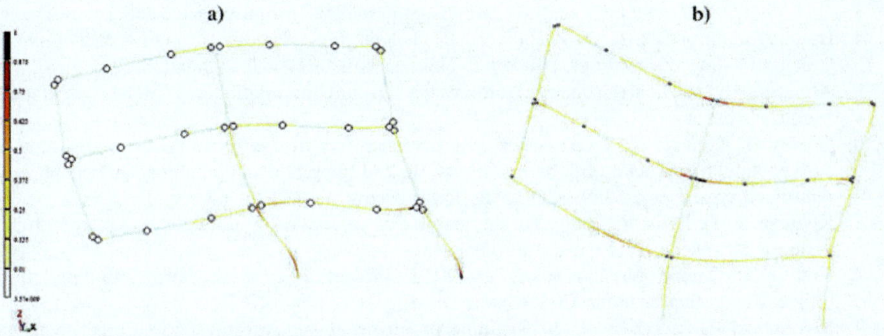

Fig. 8 Buckling modes of a model of a reinforced concrete frame with reduced cross-sectional dimensions in case of a special emergency impact: a frame without preliminary compression of the girders, the safety factor for stability is 14.44 (**a**); frame with a pre-compressed girder above the first floor, safety factor for stability 3.16 (**b**)

4 Conclusion

1. The parameters of the stress–strain state of crack resistance, development and opening of cracks were obtained from the results of experimental and theoretical studies of a fragment of a prestressed reinforced concrete frame simulating the frame of a multi-storey monolithic building under a special effect caused by the sudden removal of the outside frame column.
2. Based on the experimental data, the values of the coefficient of dynamic additional loading in the prestressed structure of the frame of the FR -2 were calculated, which was calculated as the ratio of the crack opening width after the sudden removal of the outside column to the crack opening width before the outside column floors and amounted to 7.0.
3. The obtained results of studies of crack resistance of prestressed frame structures can be used in the development of methodology and methods for protecting the frames of monolithic multi-storey buildings from progressive collapse, in particular, one of such methods can be prestressing the frame beams above the first floor.

Acknowledgements The reported study was funded by RFBR, project number 19-38-90111.

References

1. Building Code: SP 385.1325800.2018 (2019) Protection of buildings and structures against progressive collapse. Design code. Basic statements
2. Unified Facilities Criteria (2010) UFC 4-023-03. Design of buildings to resist progressive collapse text. department of defense

3. GSA (2016) Alternate path analysis and design guidelines for progressive collapse resistance
4. Building Code: DBN B.1.2-14-2009 (2009) General principles of ensuring the reliability and structural safety of buildings, structures, building structures and foundations
5. Almazov V (2005) Resistance to progressive fracture in multi-storey frame-type frames. MGSU, Moscow
6. Geniev G, Klyueva N (2005) Issues of structural safety of reinforced concrete structures in case of sudden beyond design basis impacts. In: 2nd International conference on concrete and reinforced concrete, pp 359–367. CARCRI, Moscow
7. Klyueva N, Fedorov V (2006) To the analysis of survivability of suddenly damaged frame systems. Str. Mech Anal Constr 3(205):7–13
8. Kodysh E, Trekin N, Chesnokov D (2016) Protection of multi-storey buildings from progressive collapse. Indus Civil Constr 6:8–13
9. Kolchunov V, Fedorova N (2018) Some problems of survivability of reinforced concrete structural systems under emergency impacts. Bull SRC Stroitelstvo 1:115–119
10. Rastorguev B, Plotnikov A (2008) Calculation of load-bearing structures of monolithic reinforced concrete buildings for progressive destruction, taking into account dynamic effects. In: Collection of scientific works of the Institute of Construction and Architecture MGSU, pp 65–72. MGSU, Moscow
11. Fedorova N, Korenkov P (2016) Static-dynamic deformation of monolithic reinforced concrete frames of buildings in limiting and out-of-limit states. Build Reconstr 6:90–100
12. Fedorova N, Ngoc V (2019) Deformation and failure of monolithic reinforced concrete frames under special actions. J Phys: Conf Ser 1425:012033
13. Hammad K, Lofty I, Naiem M (2021) Enhancing progressive collapse resistance in existing buildings. Des Constr Smart Cities 1:39–46
14. Alogla K, Weekes L, Levingshan A (2016) A new mitigation scheme to resist progressive collapse of RC structures. Constr Build Mater 125:533–545
15. Mohajeri N, Nima U, Abbasnia R (2018) Analytical investigation of reinforced concrete frames under middle column removal scenario. Adv Struct Eng 21(9):1388–1401
16. Klyueva N, Vetrova O (2006) Experimental and theoretical studies of survivability of operated reinforced concrete frames in case of sudden damage. Concr Reinforced Concr 6:12–15
17. Klyueva N, Korenkov P (2016) Method of experimental determination of parameters of survivability of reinforced concrete frame-bar structural systems. Indus Civil Constr 2:44–48
18. Kolchunov V, Osovskikh E, Alkadi S (2017) Deformation and destruction of reinforced concrete frame-bar spatial structural systems of multi-storey buildings in extreme states. Indus Civil Constr 8:73–77
19. Fedorova N, Fan D, Nguyen T (2020) Experimental studies of the survivability of reinforced concrete frames with crossbars reinforced with indirect reinforcement. Building Reconstr 1:92–100
20. Alshaikh I, Bakar B, Alwesabi E, Akil H (2020) Experimental investigation of the progressive collapse of reinforced concrete structures: an overview. Structures 25:881–900
21. Yu J, Gan Y, Liu J (2020) Numerical study of dynamic responses of reinforced concrete infilled frames subjected to progressive collapse. Adv Struct Eng 24(4):635–652. https://doi.org/10.1177/1369433220965273
22. Yang T, Chen W, Han Z (2020) Experimental investigation of progressive collapse of prestressed concrete frames after the loss of middle column. Adv Civil Eng 2020:1–12
23. Fan Y-L, Wang J, Wang H-L (2018) Experimental study on collapse performance of one-story reinforced concrete frames using external prestressing tendons. J Central South Univ 49:1244–1253
24. Ilyushchenko T, Kolchunov V, Fedorov S (2021) Crack resistance of prestressed reinforced concrete frame structures under special influences. Building Reconstr 1:74–84
25. Klueva N, Koenkov P (2016) Patent 2642542. Device for experimental determination of dynamic addition loading in frame-rod structural systems
26. Pearson C, Delatte N (2005) Ronan point apartment tower collapse and its effect on building codes. J Perform Constr Facil 19:172–177

27. Byfield M, Paramasivam S (2012) Murrah building collapse: reassessment of the transfer girder. J Perform Constr Facil 26:371–376
28. Belostotsky AM, Pavlov AS (2010) Long span buildings analysis under physical, geometric and structural nonlinearities consideration. Int J Comput Civ Struct Eng 6:80–86
29. Savin SY, Fedorov SS (2019) Stability analysis of reinforced concrete building frames damaged by corrosion under static-dynamic loading. J Phys Conf Ser 1425:012043

Engineering Calculations of Acidifier Retaining Walls During Water Treatment Facilities Designing

V. I. Rimshin, S. I. Roshchina, E. S. Ketsko, P. S. Truntov, and I. S. Kuzina

Abstract The water consumption growth by the population and industrial facilities leads to a constantly increasing amount of polluted wastewater discharged from the settlement, industrial enterprises and railway transport enterprises. The design of facilities for this wastewater treatment and disinfection is of great importance, since untreated wastewater ingress into water bodies leads to their pollution and exit from the water circulation system. This paper is devoted to the design and calculation process for the retaining walls of acidifier mixers. Acidification units are six interlocked rectangular basins measuring 20.0 × 20.0 m and 11.80 m deep. Each acidifier consists of horizontal and bunker parts. The designed number of six acidifiers ensures the normal acidification process of raw sludge. Retaining walls are designed taking into account horizontal and vertical external loads located on the collapse prism. Retaining walls are designed for two groups of limiting states: for bearing capacity (a wall stability against subgrade shear and strength, as well as for structural elements strength) and for serviceability (bases for permissible deformations and cracks). The soils analysis was carried out, the lateral pressure and the slope stability coefficient were determined. The position stability of the retaining walls against plane shear and two types of depth shear has been determined. The base calculation on deformations was performed. The normative soil pressure intensity on the walls is determined. The reinforcement area by forces for vertical walls and a horizontal foundation slab was selected.

Keywords Engineering calculations · Water treatment facilities · Retaining walls · Acidifiers · Limiting states · Bearing capacity · Serviceability · Slope stability coefficient

---

V. I. Rimshin · P. S. Truntov · I. S. Kuzina (✉)
Moscow State University of Civil Engineering (MGSU), Moscow, Russia

S. I. Roshchina
Vladimir State University Named After Alexander and Nikolay Stoletovs, Vladimir, Russia

E. S. Ketsko
Research Institute of Building Physics of the Russian Academy of Architecture and Building Sciences (NIISF RAASN), Moscow, Russia

N. Vatin et al. (eds.), *Proceedings of MPCPE 2021*, Lecture Notes in Civil Engineering 182, https://doi.org/10.1007/978-3-030-85236-8_5

1 Introduction

Geotechnical conditions are accepted in accordance with the data given in the results of geotechnical surveys. It is accepted to backfill the retaining wall sinuses with medium-grained sand with the following characteristics (natural soil): γ_n = 18.6 kN/m^3; γ_I = 18.2 kN/m^3; φ_n = 24°, φ_I = 23°; c_n = 1 kPa; c_I = 0; γ_{II} = 16.9 kN/m^3; φ_{II} = 24°; c_{II} = 1 kPa.

$$\lambda_a = tg^2(45° - \varphi/2) = tg^2(45° - 23°/2) = tg^2 33.5° = 0.6619^2 = 0.438$$

The ratio between the backfill soils and natural constitution soils characteristics, according to the regulations, are as follows: γ_I^1 = 18.2 · 0.95 = 17.3 kN/m^3 or $\gamma_I^1 = \gamma_n$ = 18.6 kN/m^3 → from the two obtained values, select the maximum value γ_I^1 = 18.6 kN/m^3; φ_I^1 = 23° · 0.9 = 20.7° or $\varphi_I^1 = \varphi_n$ · 0.82 = 24° · 0.82 = 19.7° → from the two obtained values, select the minimum value φ_I^1 = 19.7° = 19°42; γ_{II}^1 = 16.9 · 0.95 = 16.1 kN/m^3; φ_{II}^1 = 24° · 0.9 = 21.6°; c_I^1 = 0. For further calculation, we take the following characteristics of sandy backfill soil:

$$\gamma^1 = 18.6\,kN/m^3;\ \varphi^1 = 19°42^I;$$
$$\lambda_a = tg^2(45° - \varphi/2) = tg^2\left(45° - 19°42^I/2\right) = tg^2 35°09^I = 0.7041^2 = 0.496;$$
$$\lambda_{II} = tg^2(45° + \varphi/2) = tg^2\left(45° + 19°42^I/2\right) = tg^2 54°51^I = 1.4203^2 = 2.017$$

Due to the retaining wall location in the backfilling soil of acidificators under its foundation bottom, the documentation provides for a layer-by-layer compacted sand and gravel cushion 0.8 m high with the following characteristics: $\varphi_I = \varphi_{II}$ = 30°; $\gamma_I = \gamma_{II}$ = 18 kN/m^3; $c_I = c_{II}$ = 0. Retaining wall 1 arrangement along the A axis of the mixing chamber is shown in Fig. 1.

2 Lateral Pressure Determination

Lateral pressure determination circuit is presented in Fig. 2.

Random load on the ground surface is q_1 = 10 kN/m^2. Conditional angle of the collapse plane is tgƐ = 2.6/6.9 = 0.3768; Ɛ = 20°40^I ≈ 21°.

Lateral pressure ordinate from load at the chamber top: $p_{h1} = k \cdot q \cdot \lambda_a$ = 1.1 · 10 · 0.496 = 5.5 kN/m^2.

Lateral pressure ordinate from load and backfill soil at the bottom level: $p_{h2} = p_{h1} + k \cdot \gamma \cdot h \cdot \lambda a$ = 5.5 + 1.1 · 1.86 · 6.9 · 0.496 = 5.5 + 7.0 = 12.5 kN/m^2.

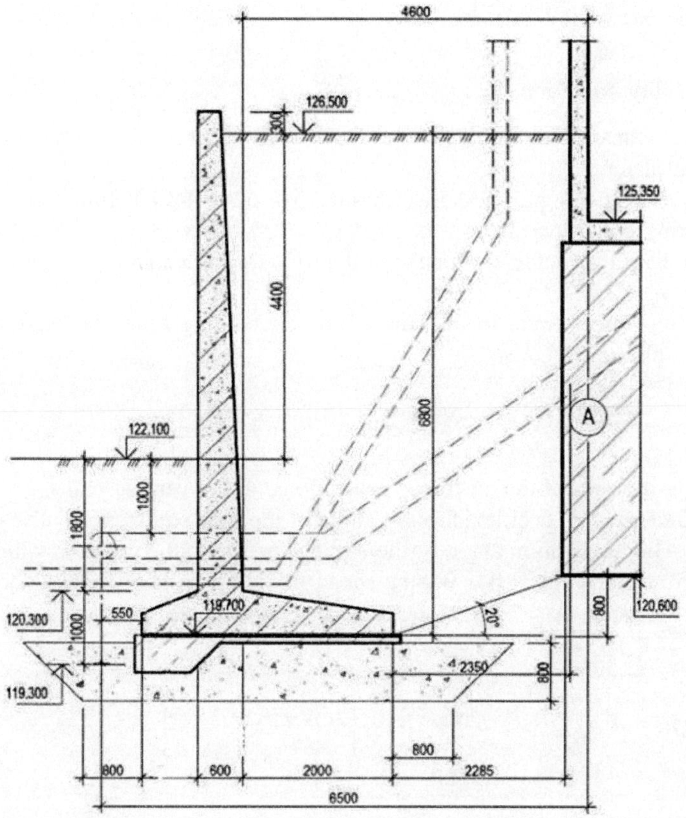

Fig. 1 Retaining wall 1 arrangement along the A axis of the mixing chamber

Fig. 2 Lateral pressure determination circuit

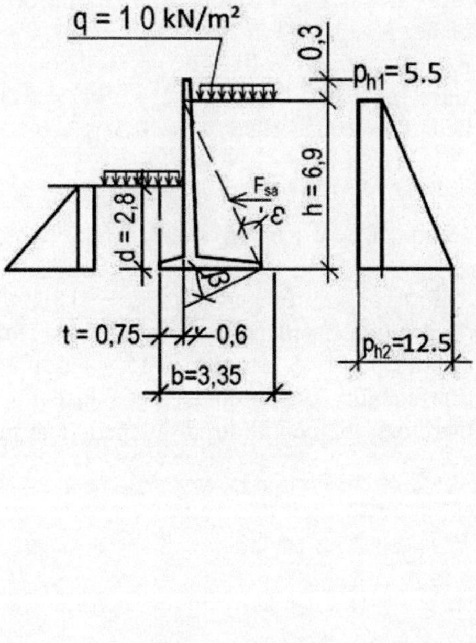

3 Wall Stability Against Shear

Shear stability condition: $F_{sa} \leq (\gamma_c/\gamma_n) \cdot F_{sr}$,

where F_{sa}—shear force equal to the sum of all shear forces projections on the horizontal plane:
$F_{sa} = 0.5 \cdot h \cdot (p_{h1} + p_{h2}) = 0.5 \cdot (5.5 + 12.5) \cdot 6.9 = 62.1$ kN/m;
γ_c—working condition factor—for sand $\gamma_c = 1$;
γ_n—the reliability coefficient for the structure purpose, taken as 1.15 for a class II structure: $\gamma_n = 1.15$;
F_{sr}—holding force, equal to the sum of the all holding forces projections on the horizontal plane:

$$F_{ar} = F_v \cdot tg(\varphi_1 - \beta) + A \cdot c_1 + E_{hr} \tag{1}$$

where F_v—the sum of the all forces projections on the vertical plane, φ_1 and c_1—internal friction and specific adhesion angle of the base soil respectively: $\varphi_I = 30°$; $c_I = 0$, β—the inclination angle of the sliding surface to the horizon, which should be taken for three values: $\beta = 0$—flat shear, $\beta = 0.5 \cdot \varphi_1$ and $\beta = \varphi_1$—depth shift, A—wall foot area, $A = 3.35 \cdot 1 = 3.35$ m^2, E_{hr}—passive soil resistance.

Define

$$F_v = F_{sa} \cdot tg(E + \varphi_I^1) + \gamma_I^1 \cdot \gamma_f \left[\frac{h \cdot (b - t)}{2} + t \cdot d\right] + \frac{\gamma_1 \cdot tg\beta \cdot b^2}{2}$$
$$= 487.4 + 10.1 \cdot tg\beta \tag{2}$$

Define E_{hr}, F_v and F_{sr} consistently for each inclination angle of the sliding surface β:

The first case is $\beta = 0$—flat shear; $\lambda_{II} = 2{,}017$, but $\lambda_{IImax} = 1.0$. Passive soil resistance is $E_{hr} = 61$ kN/m. The sum of the all forces projections on the vertical plane: $F_v = 487.4 + 10.1 \cdot tg\beta = 487.4 + 10.1 \cdot tg0° = 487.4$ kN. Holding force $F_{sr} = F_v \cdot tg(\varphi_1 - \beta) + A \cdot c_1 + E_{hr} = 342.4$ kN. The stability condition is met, since $F_{sa} = 280$ kN < $(1/1.15) \cdot 342.4 = 300$ kN, therefore, the condition is satisfied. The second case is $\beta = 0.5\varphi_1 = 0.5 \cdot 30° = 15°$—depth shift.

$$\lambda_{II} = tg^2(45° + \varphi/2) = tg^2(45° + 19°42^I/2) = 2.017$$
$$h_r = d + b \cdot tg\beta = 2.8 + 3.35 \cdot tg15° = 2.8 + 3.35 \cdot 0.2679 = 3.7\,m$$

Passive soil resistance is $E_{hr} = 215$ kN/m. The sum of the all forces projections on the vertical plane: $F_v = 514.5$ kN. Holding force $F_{sr} = Fv \cdot tg(\varphi_1 - \beta) + A \cdot c_1 + E_{hr} = 51.45 \cdot tg(30° - 15°) + 0 + 21.5 = 51.45 \cdot 0.2679 + 21.5 = 353$ kN. The stability condition is met, since $F_{sa} = 280$ kN < $(1/1.15) \cdot 353 = 307$ kN, therefore, the condition is satisfied. The third case is $\beta = \varphi_1 = 30°$—depth shift.

$$\lambda_{II} = tg^2(45^\circ + \varphi/2) = tg^2(45^\circ + 19^\circ 42^I/2) = 2.017$$
$$h_r = d + b \cdot tg\beta = 2.8 + 3.35 \cdot tg30^\circ = 2.8 + 3.35 \cdot 0.5774 = 4.73\,m;$$

Passive soil resistance is E_{hr} = 351 kN/m. The sum of the all forces projections on the vertical plane: F_v = 487.4 + 10.1 · tgβ = 487.4 + 10.1 · tg30° = 545.7 kN. Holding force F_{sr} = Fv · tg(φ_1 − β) + A · c_1 + E_{hr} = 0 + 0 + 351 = 351 kN. The stability condition is met, since F_{sa} = 280 kN < (1/1.15) · 351 = 305 kN, therefore, the condition is satisfied. The condition $F_{sa} < F_{sr}/1.15$ is satisfied in all three cases. Thus, the position stability of the retaining wall against shear is ensured in all three cases of plane shear.

4 Subgrade Strength Calculation

Subgrade strength calculation should be carried out at $tg\delta_1 < \sin\varphi_1$, where the tangent of the inclination angle to the resultant external load vertical on the base is determined from the condition $tg\delta_1 = F_v / F_{sa}$ = 280/487.4 = 0.57; $\sin\varphi_1$ = sin30° = 0.5. Due to the fact that the condition $tg\delta_1 < \sin\varphi_1$; 0.57 > 0.5 not completed, there is no need to calculate the subgrade strength [1–5].

5 Subgrade Calculation for Deformations

Due to the retaining wall location in the backfilling soil of acidificators under its foundation sole, the documentation provides for a layer-by-layer compacted sand and gravel cushion 0.8 m high with the following characteristics: $\varphi_I = \varphi_{II} = 30^\circ$; $\gamma_I = \gamma_{II} = 18$ kN/m^3; $c_I = c_{II} = 0$. Calculation of foundation soil resistance of the R:

$$R = \left(\frac{\gamma_{c1} \cdot \gamma_{c2}}{k}\right) \cdot \left[M_y \cdot k_z \cdot b \cdot \gamma_{II} + M_q \cdot d_I \cdot \gamma_{III} + (M_q - 1) \cdot d_b \cdot \gamma_{III} + M_c \cdot c_{II}\right], \tag{3}$$

where γ_{c1} and γ_{c2}—working condition factor: at the base—sand and gravel mixture: γ_{c1} = 1.4; γ_{c2} = 1.1; k = 1; at φ_{II} = 30°, therefore, M_γ = 1.15; M_q = 5.59; M_c = 7.95; k_z = 1; γ_{II}—the calculated specific gravity of the soil lying below the foundation base: at the base—sand and gravel mixture: γ_{II} = 18 kN/m^3.
γ_{III}—the calculated specific gravity of the soil lying above the foundation base: γ_{III} = 16.1 kN/m^3;
c_{II}—the calculated specific adhesion of the soil lying directly under the foundation sole: c_{II} = 0 kN/m^2 (in reserve);
b—foundation base width, b = 3.35 m.

d_I—the foundations depth—for slab foundations—the smallest depth from the slab base to the planning level: $d_I = 2.8$ m; $R = 495$ kN/m^2.

The normative soil pressure intensity on the wall: the random load on the soil surface are accepted $q_1 = 10$ kN/m^2. $\lambda_a = tg^2(45° - \varphi/2) = tg^2(45° - 21.6°/2) = tg^2\ 34°12^1 = 0.68^2 = 0.462$. Retaining wall diagram for calculating the subgrade strength is presented in Fig. 3 (the foundation slab front is t = 0.75).

From the side of the mixing chamber, the lateral pressure ordinate from the load at the wall top: $p_{h1} = k \cdot q \cdot \lambda_a = 1.0 \cdot 10 \cdot 0.462 = 4.6$ kN/m^2. Lateral pressure ordinate from load and soil at the bottom of the wall foundation: $p_{h2} = p_{h1} + k \cdot \gamma \cdot h \cdot \lambda_a = 4.6 + 1.0 \cdot 1.61 \cdot 6.9 \cdot 0.462 = 4.6 + 5.13 = 9.73 \approx 9.7$ kN/m^2.

M_o—sum of the all vertical and horizontal forces moments about an axis passing through the gravity center of the sole:

$$\begin{aligned} M_o = & F_{sa} \cdot \left[h_1 - tg(E + \varphi_1^1) \cdot \left(\frac{b}{2} - h_1 \cdot tgE\right)\right] \\ & + \gamma_1^1 \cdot \gamma_f \cdot (b - t) \cdot \frac{[h \cdot (b - 4 \cdot t) + 6 \cdot t \cdot d]}{12}, \end{aligned} \tag{4}$$

where h_1—distance from the resultant shear force to the wall base bottom, γ_f—load safety factor, taken equal to 1.0.

$F_{sa} = 209$ kN. $h_1 = 2.47$ m. $M_o = 429$ kN · m. $F_v = 371$ kN.

e—application eccentricity of all forces resultant relative to the axis passing through the gravity center of the wall sole. $e = F_v / M_o = 42.9/37.1 = 1.16$ m $> b/3 = 3.35/3 = 1.12$ m, that is, the condition $e \leq b/3$ is not completed and the retaining wall foundation slab dimensions should be changed [6–11]. The front part of the foundation slab is 0.9 m, that is, it should be increased by 15 cm. Retaining wall diagram for calculating the subgrade strength (t = 0.90) is presented in Fig. 4.

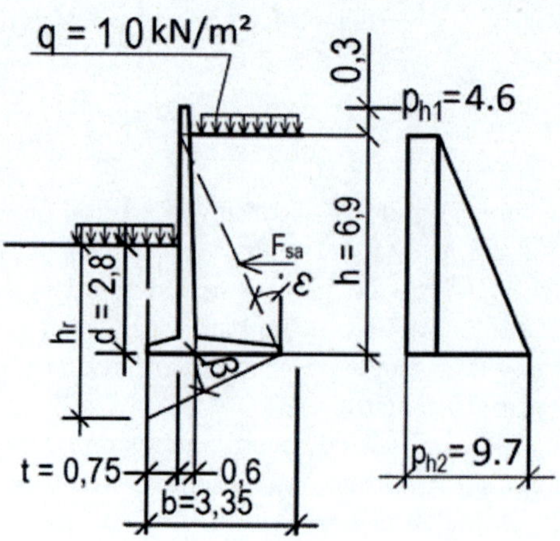

Fig. 3 Retaining wall diagram for calculating the subgrade strength (t = 0.75)

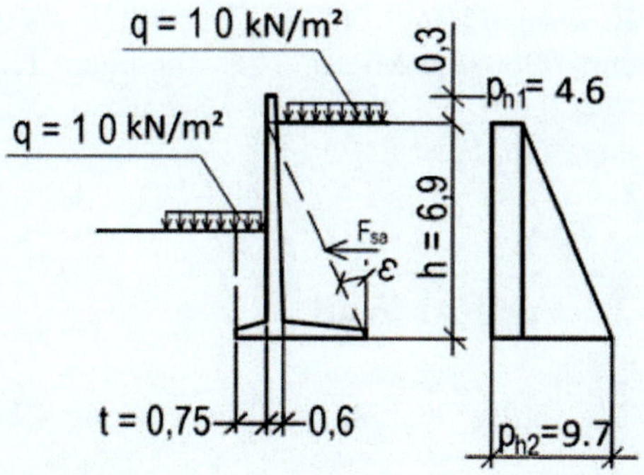

Fig. 4 Retaining wall diagram for calculating the subgrade strength (t = 0.90)

$M_o = 412$ kN · m. $F_v = 377$ kN.
e—application eccentricity of all forces resultant of relative to the axis passing through the gravity center of the wall sole. $e = F_v/M_o = 412/377 = 1.1$ m < b/3 = 3.5/3 = 1.17 m, that is $e \leq b/3$ is satisfied.

Boundary ground pressures: $p_{min}^{max} = F_v \cdot (1 \pm 6 \cdot e/b = 37.7 \cdot (1 \pm 6 \cdot 1.1/3.5)/3.5 = 10.77 \cdot (1 \pm 1.89)$.

$p_{max} = 311.2$ kN/m_2 < R = 495 · 1.2 = 594 kN/m². $p_{min} = -95.8$ kN/m².

Plot of edge pressures on soil is presented in Fig. 5.

Partial separation of the foundation from the ground is 0.825 m and does not exceed 1/4 of the foundation width: 3.5/4 = 0.875 m > 0.825 m, that is, the condition is met. The conditions for calculating the base for deformations are satisfied [12–15].

6 Reinforcement Area Selection Based on Efforts

The concrete class is B30: $R_b = 173$ kg/cm² (17 MPa); $R_{b,ser} = 224$ kg/cm² (22 MPa); $R_{bt} = 11.7$ kg/cm² (1.15 MPa); $R_{bt,ser} = 17.8$ kg/cm² (1.75 MPa); $E_b = 331{,}000$ kg/cm² (32.5 MPa). The reinforcement class is A500C:

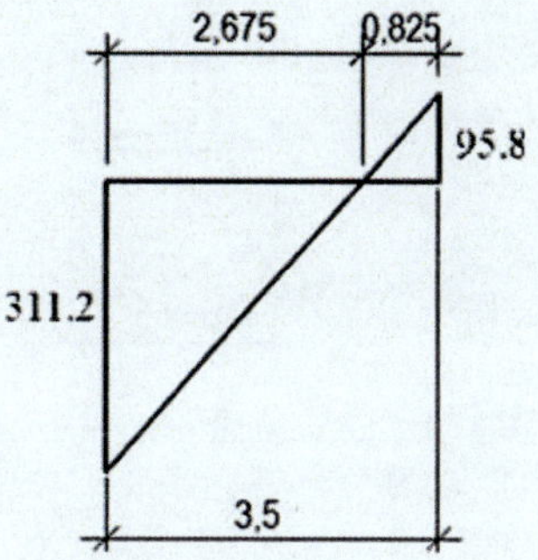

Fig. 5 Plot of edge pressures on soil

$R_s = 4430$ kg/cm^2 (435 MPa); $R_{sw} = 3060$ kg/cm^2 (300 MPa); $E_s = 2{,}000{,}000$ kg/cm^2 (200,000 MPa). The minimum reinforcement percentage at h = 60 cm: $F_a = 0.001 \cdot 100 \cdot 52 = 5.2$ cm^2, therefore, accept Ø18A500C, a step is 200, therefore, $F_a = 12.72$ cm^2.

7 Vertical Wall

The vertical retaining wall calculating scheme is shown in Fig. 6.

$M_{support} = 686$ kN · m. $Q_{fact} = 280$ kN.
$Q_{max} = 456$ kN > $Q_{fact} = 280$ kN.
M = 686 kN · m; $A_o = 6{,}860{,}000/(173 \cdot 100 \cdot 52^2) = 0.147$, therefore, $\vee = 0.92$; $F_a = 6{,}860{,}000/(4430 \cdot 0.92 \cdot 52) = 32.4$ cm^2; accept Ø28A500C, a step is 100, therefore, $F_a = 61.58$ cm^2. M = 454 kN · m; $A_o = 4{,}540{,}000/(173 \cdot 100 \cdot 45^2) = 0.13$, therefore, $\vee = 0.93$; $F_a = 454{,}000/(4430 \cdot 0.93 \cdot 45) = 24.5$ cm^2; accept Ø28A500C, a step is 200, therefore, $F_a = 30.79$ cm^2, +Ø18A500C, a step is 200, therefore, $F_a = 12.72$ cm^2 = $F_a = 43.51$ cm^2. M = 222 kN · m; $A_o = 2{,}220{,}000/(173 \cdot 100 \cdot 34^2) = 0.111$, therefore, $\vee = 0.944$; $F_a = 2{,}220{,}000/(4430 \cdot 0.944 \cdot 34) = 15.6$ cm^2; accept Ø28A500C, a step is 200, therefore, $F_a = 30.79$ cm^2.

8 Foundation Slab

Bending moments in wall elements is presented in Fig. 7.

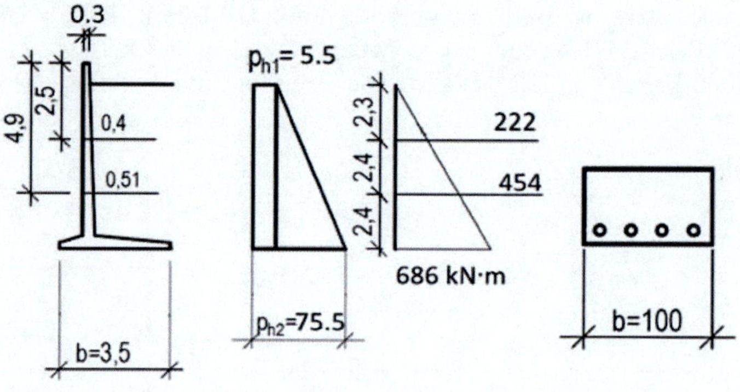

Fig. 6 The vertical retaining wall calculating scheme

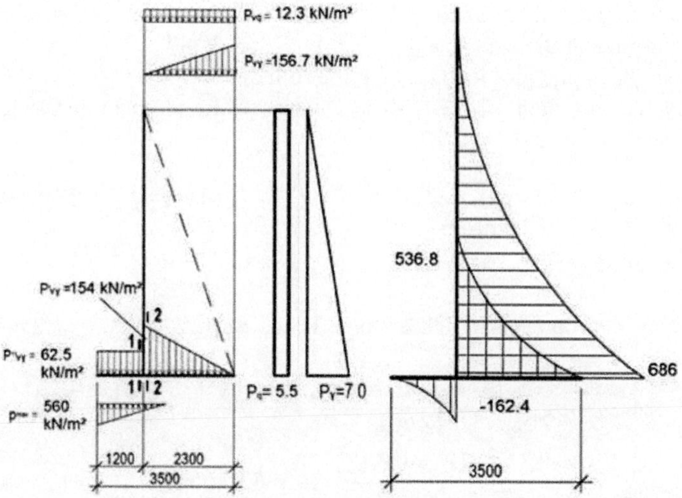

Fig. 7 Bending moments in wall elements, kN · m

9 Design Forces in the Foundation Slab

P_Y, P_q—horizontal pressure intensity from the ground and vertical pressure;

P_{vY}, P_{vq}—vertical pressure intensity from the own soil weight in the collapse prism and the vertical pressure of the temporary load:

$P_{vY} = P_Y \cdot tg(\mathcal{E} + \varphi_I^1) / tg\mathcal{E} = 7.0 \cdot tg(21° + 19.7°) / tg21° = 156.7\ kN/m^2$
$P_{vq} = P_q \cdot tg(\mathcal{E} + \varphi_I^1) / tg\mathcal{E} = 0.55 \cdot tg(21° + 19.7°) / tg21° = 12.3\ kN/m^2$

P^1_{vY}, P^{11}_{vY}—vertical pressure intensity from own soil weights in the collapse prism and above the front console respectively:

$P^1_{vY} = \gamma_I^1 \cdot \gamma_f \cdot h = 154\ kN/m^2$. $P^{11}_{vY} = \gamma_I^1 \cdot \gamma_f \cdot d = 62.5\ kN/m^2$

M_o—sum of the all vertical and horizontal forces moments about an axis passing through the gravity center of the sole from design loads:

$$M_o = F_{sa} \cdot \left[h_1 - tg(E + \varphi_1^1) \cdot \left(\frac{b}{2} - h_1 \cdot tgE \right) \right] + \gamma_1^1 \cdot \gamma_f \cdot (b - t) \cdot \frac{[h \cdot (b - 4 \cdot t) + 6 \cdot t \cdot d]}{12} = 568.2\ kNm \tag{5}$$

F_v—the sum of the all forces projections on the vertical plane: $F_v = 487.4$ kN.

e—application eccentricity of all forces resultant relative to the axis passing through the gravity center of the wall sole: $e = F_v/M_o = 568.2/487.4 = 1.17$ m > b/6 = 3.5/ 6 = 0.58 m. $3C_o$—the plot length along the foundation foot: $C_o = 0.5b - e = 0.5 \cdot 3.5–1.17 = 0.58$ m, then $3C_o = 1.74$ m. Determine the moment and shear force in the cross Sect. 1—1 at $x_1 = 0.9$ m:

$M_{1\text{-}1} = P^{11}{}_{v\gamma} \cdot x_1{}^2 / 2 - p_{max} \cdot x_1{}^2 \cdot (1 - x_1 / 9C_o) / 2$, where $p_{max} = 2F_v / 3C_o = 560$ kN/m², then: $M_{1\text{-}1} = -162.4$ kN·m

$Q_{1\text{-}1} = P^{11}{}_{v\gamma} \cdot x_1 - p_{max} \cdot x_1 \cdot (1 - x_1 / 6C_o) = -317.4 \text{ kN} < Q_{max} = 456 \text{ kN}$

Determine the moment and shear force in the Sect. 2—2 at $x_2 = 2.6$ m:

$$M_{2-2} = p_{max} \cdot \frac{(3 \cdot C_0 - b + x^2)^3}{18 \cdot C_0} - \frac{P_{v\gamma} \cdot x_2^2}{2} - \frac{P_{vq} \cdot (x^2 - \xi)^2}{2} - \frac{x_2^3 \cdot (P^1_{v\gamma} - P_{vq})}{6 \cdot (b - t)} = -536.8 \,\text{kNm}; \tag{6}$$

$$Q_{2-2} = p_{max} \cdot \frac{(3 \cdot C_0 - b + x^2)^2}{6 \cdot C_0} - P_{v\gamma} \cdot x^2 - \frac{P_{vq} \cdot (x^2 - \xi)^2}{2} - \frac{x_2^3 \cdot (P^1_{v\gamma} - P_{vq})}{2 \cdot (b - t)} = -322 \,\text{kN} < 456 \,\text{kN}; \tag{7}$$

M = 536.8 kN · m; $A_o = 5{,}368{,}000/(173 \cdot 100 \cdot 52^{2)} = 0.115$, therefore, $\vee = 0.939$; $F_a = 5{,}368{,}000/(4430 \cdot 0.939 \cdot 52) = 25.0$ cm²; accept Ø28A500C, a step is 100, therefore, $F_a = 61.58$ cm². Accept Ø28A500C, a step is 200, therefore, $F_a = 30.79$ cm², + Ø18A500C, a step is 200, therefore, $F_a = 12.72$ cm². $F_a = 43.51$ cm²; M = 162.4 kN · m; $A_o = 1{,}624{,}000/(173 \cdot 100 \cdot 52^2) = 0.035$, therefore, $\vee = 0.982$; $F_a = 1{,}624{,}000/(4430 \cdot 0.982 \cdot 52) = 7.2$ cm²; accept Ø18A500C, a step is 200, therefore, $F_a = 12.72$ cm². Retaining wall 2 scheme (along the A-axis of acidificators) is presented in Fig. 8.

10 Lateral Pressure Calculation

Lateral pressure determination circuit is shown in Fig. 9.

The random load on the soil surface is $q_1 = 10$ kN/m². Conditional angle of the collapse plane is $tg\mathcal{E} = 2.9/5.2 = 0.5577$; $\mathcal{E} = 29°\ 09I \approx 29°$. The lateral pressure ordinate at the bottom of the foundation from the soil layer of the backfill height 5.2 m is equal to 55.2 kN/m². The lateral pressure ordinate at the foundation bottom from the backfill soil layer located within the slope: b = 3.3 m; $h_1 = 1.9$ m; $\angle 45° + \varphi/2 = \angle 54°\ 51^I$; $h_2 = 3.3 \cdot tg54°\ 51^I = 4.7$ m; $h_3 = 0.5$ m; $\lambda = 0.496$; $\sqrt{\lambda} = \sqrt{0.496} = 0.704$; $e_\Delta = (h/\sqrt{\lambda} - 0.5 \cdot b) \cdot h_1\, \gamma^1 \sqrt{\lambda}/(h_3 + 0.5 \cdot h_2) = 50$ kN/m².

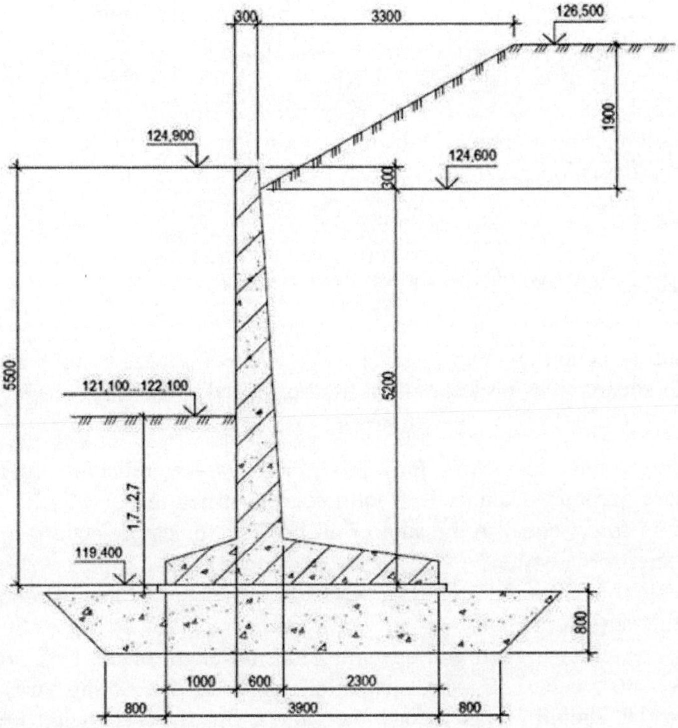

Fig. 8 Retaining wall 2 scheme (along the A-axis of acidificators)

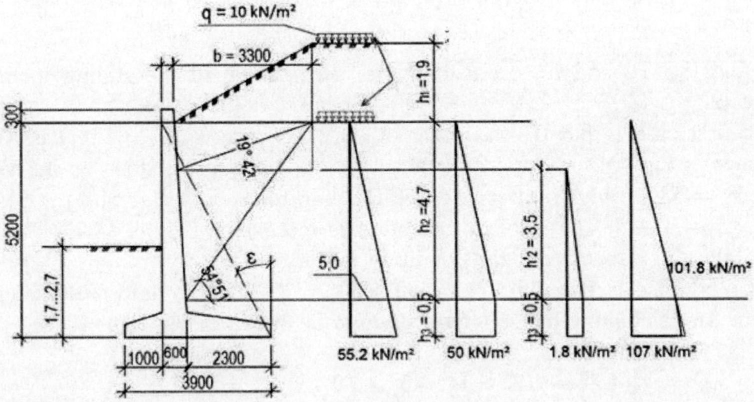

Fig. 9 Lateral pressure determination circuit

The lateral pressure ordinate at the foundation bottom from the uniformly distributed load at the slope top: $h_1' = b \cdot tg\varphi^1 = 3.3 \cdot tg19°42^I = 1.18$ m. $h_2' = h_2 - h_1' = 4.7 - 1.18 = 3.5$ m. $e_q = q \cdot (h\sqrt{\lambda} - b)/(h_3 + 0.5\ h_2') = 1.8\ kN/m^2$. The total lateral pressure ordinates: $e_{\Sigma} = 107\ kN/m^2$. $e_{\Sigma}' = 101.8\ kN/m^2$. Determine the maximum values to align the pressure plot at the foundation bottom: $P = 113\ kN/m^2$, we accept for further calculation.

11 Wall Stability Against Shear

Shear stability condition: $F_{sa} < (\gamma_c/\gamma_n) \cdot F_{sr}$, where F_{sa}—shear force equal to the sum of all shear forces projections on the horizontal plane: $F_{sa} = h/(2 \cdot P) = 294$ kN/m;

γ_c—working conditions factor—for sands $\gamma_c = 1$; γ_n—the reliability coefficient for the structure purpose, taken as 1.15 for a class II structure: $\gamma_n = 1.15$;
F_{sr}—holding force, equal to the sum of all holding forces projections on the horizontal plane:
$F_{sr} = F_v \cdot tg(\varphi_1 - \beta) + A \cdot c_1 + E_{hr}$, where F_v—the sum of the all forces projections on the vertical plane,
φ_1 and c_1—internal friction and specific adhesion angle of the base soil respectively: $\varphi_I = 30°$; $c_I = 0$, β—the inclination angle of the sliding surface to the horizon, which should be taken for three values: $\beta = 0$ − flat shear, $\beta = 0.5 \cdot \varphi_1$ and $\beta = \varphi_1$—depth shift, A—wall foot area, $A = 3.9 \cdot 1 = 3.9\ m^2$, E_{hr}—passive soil resistance.

$$F_v = F_{sa} \cdot tg\,(\mathcal{E} + \varphi^1{}_1) + \gamma_I{}^1 \cdot \gamma_f\,[h \cdot (b-t)/2 + t \cdot d] + \gamma_I \cdot tg\beta \cdot b^2/2 = 534 + 13.7 \cdot tg\beta$$

Define E_{hr}, F_v and F_{sr} consistently for each angle of the sliding inclination surface β:

The first case is $\beta = 0$—flat shear; $\lambda_{II} = 2.017$, but $\lambda_{IImax} = 1.0$. Passive soil resistance is $E_{hr} = 42$ kN/m. The sum of the all forces projections on the vertical plane: $F_v = 534 + 13.7 \cdot tg\beta = 534$ kN. Holding force $F_{sr} = F_v \cdot tg(\varphi_1 - \beta) + A \cdot c_1 + E_{hr} = 350$ kN. The stability condition is met, since $F_{sa} = 294\ kN < (1/1.15)$ $350 = 304$ kN, respectively, the condition is satisfied.

The 2nd case is $\beta = 0.5 \cdot \varphi_1 = 0.5 \cdot 30° = 15°$—depth shift. Retaining wall diagram for calculating the subgrade strength is shown in Fig. 10.

$$\lambda_{II} = tg^2(45° + \varphi/2) = tg^2(45° + 19°42^I/2) = 2.017$$
$$h_r = d + b \cdot tg = 2.2 + 3.9 \cdot tg15° = 2.2 + 3.9 \cdot 0.2679 = 3.24\,m;$$

Passive soil resistance is $E_{hr} = 183.2$ kN/m. The sum of the all forces projections on the vertical plane: $F_v = 534 + 13.7 \cdot tg\beta = 571$ kN. Holding force $F_{sr} = F_v \cdot tg$

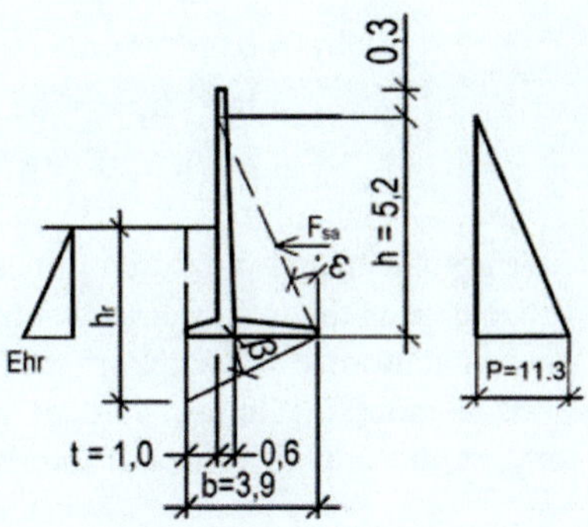

Fig. 10 Retaining wall diagram for calculating the subgrade strength

$(\varphi_1 - \beta) + A \cdot c_1 + E_{hr} = 336.2$ kN. The stability condition is met, since $F_{sa} = 294$ kN $\approx (1/1.15)336.2 = 292$ kN, respectively, the condition is satisfied.

3rd case $\beta = \varphi_1 = 30°$—depth shift

$$\lambda_{II} = tg^2(45° + \varphi/2) = tg^2(45° + 19°42^I/2) = 2.017$$
$$h_r = d + b \cdot tg\beta = 2.2 + 3.9 \cdot tg30° = 2.2 + 3.9 \cdot 0.5774 = 4.45\,m;$$

Passive soil resistance is $E_{hr} = 345$ kN/m. The sum of the all forces projections on the vertical plane: $F_v = 613$ kN. Holding force is $F_{sr} = F_v \cdot tg(\varphi_1 - \beta) + A \cdot c_1 + E_{hr} = 0 + 0 + 345 = 345$ kN. The stability condition is met, since $F_{sa} = 294$ kN < 300 kN, respectively, the condition is satisfied. The condition $F_{sa} \leq F_{sr}/1.15$ is satisfied in all three cases. Thus, the position stability of the retaining wall against shear is ensured in all three cases of plane shear [16–19].

12 Base Calculation on Deformations

b—foundation base width, b = 3.9 m. d_I—foundation depth for slab foundations—the smallest depth from the slab base to the planning level: $d_I = 2.2$ m; R = 429 kN/m^2.

The intensity of the normative soil pressure on the wall: the random load on the soil surface is $q_1 = 10$ kN/m^2. $\lambda_a = 0.462$. The ordinate of the lateral pressure at the level of the foundation bottom from the soil layer of the backfill height 5.2 m: $e = k \cdot \gamma \cdot h \cdot \lambda_a = 38.7$ kN/m^2. The ordinate of the lateral pressure at the level of the foundation bottom from the backfill soil layer located within the slope: b = 3.3 m; $h_1 = 1.9$ m; $\angle 45° + \varphi/2 = \angle 55°48^I$; $h_2 = 3.3tg\ 55°48^I = 3.3 \cdot 1.4715 = 4.86$ m; $h_3 = 0.34$ m; $\lambda = 0.462$; $\sqrt{\lambda} = \sqrt{0.462} = 0.68$; $e_\Delta = (h/\sqrt{\lambda} - 0.5 \cdot b) \cdot h_1 \cdot \gamma^1 \cdot \sqrt{\lambda}/(h_3 + 0.5 \cdot h_2) = 45$ kN/m^2.

The ordinate of the lateral pressure at the level of the foundation bottom from the uniformly distributed load at the top of the slope:

$$h_1' = b\,tg\varphi^1 = 3.3 \cdot tg21°10^I = 3.3 \cdot 0.3879 = 1.28\,m.$$
$$h_2' = h_2 - h_1' = 4.86 - 1.28 = 3.58\,m.$$
$$e_q = q \cdot (h \cdot \sqrt{}\lambda - b)/(h_3 + 0.5 \cdot h_2) = 1.1\,kN/m^2.$$

The total lateral pressure ordinates: $e_{\Sigma} = 85.2$ kN/m^2. $e'_{\Sigma} = 82.7$ kN/m^2.

Determine the maximum value for aligning the pressure diagram at the bottom of the foundation: P = 89 kN/m^2, we take it for further calculation.

M_o—moments in all vertical and horizontal forces about the axis passing through the gravity center of the sole:

$$M_o = F_{sa} \cdot \left[h_1 - tg(E + \varphi_1^1) \cdot \left(\frac{b}{2} - h_1 \cdot tgE\right)\right] + \gamma_1^1 \cdot \gamma_f \cdot (b - t) \cdot \frac{[h \cdot (b - 4 \cdot t) + 6 \cdot t \cdot d]}{12} \tag{8}$$

where h_1—distance from the resultant shear force to the bottom of the wall base, γ_f—load safety factor, taken equal to 1.0.

$F_{sa} = 0.5 \cdot h \cdot P = 231.4$ kN/m; $h_1 = 5.2/3 = 1.73$ m. $M_o = 170$ kN · m. $F_v = 439$ kN.

Eccentricity of all forces relative to the axis passing through the gravity center of the wall sole $e = F_v/M_o = 17.0/43.9 = 0.4$ m < 3b = 3.9/3 = 1.3 m, that is, the condition b/3 is satisfied. Boundary ground pressures: $p_{max}{}^{min} = F_v \cdot (1 \pm 6 \cdot e/b)/b = 112.6(1 \pm 0.62)$.

$p_{max} = 182.4$ kN/m^2 < R = 515 kN/m^2. $p_{min} = 70$ kN/m^2.

Boundary pressure determination diagram is shown in Fig. 11.

The conditions for calculating the base for deformations are satisfied. Reinforcement area selection is based on efforts. The concrete class is B30: $R_b = 173$ kg/ cm^2 (17 MPa); $R_{b,ser} = 224$ kg/ cm^2 (22 MPa); $R_{bt} = 11.7$ kg/cm^2 (1.15 MPa); $R_{bt,ser} = 17.8$ kg/cm^2 (1.75 MPa); $E_b = 331{,}000$ kg/cm^2 (32.5 MPa). The reinforcement class is A500C: $R_s = 4430$ kg/cm^2 (435 MPa); $R_{sw} = 3060$ kg/ cm^2 (300 MPa); $E_s = 2{,}000{,}000$ kg/cm^2 (200,000 MPa). The minimum reinforcement percentage at h = 60 cm: $F_a = 0.001 \cdot 100 \cdot 52 = 5.2$ cm^2, respectively, accepted Ø18A500C, a step is 200, $F_a = 12.72$ cm^2.

Fig. 11 Boundary pressure determination diagram

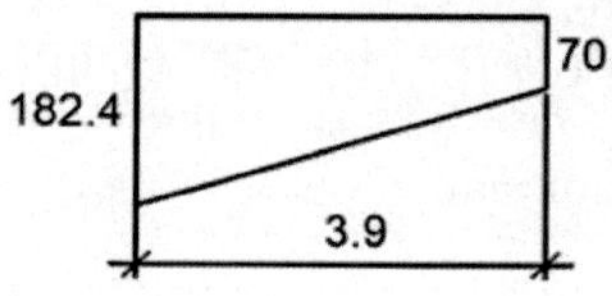

13 Vertical Wall

Calculating scheme for the vertical wall of the retaining wall is shown in Fig. 12.

$M_{support} = 11.3 \cdot 5.2^2/6 = 510$ kN · m.
$Q_{fact} = 29.4$ kN.
$Q_{max} = 456$ kN > $Q_{fact} = 294$ kN.
M = 510 kN · m; $A_o = 5{,}100{,}000/173 \cdot 100 \cdot 52^2 = 0.109$, respectively, $\gamma = 0.944$; $F_a = 5{,}100{,}000/4430 \cdot 0.944 \cdot 52 = 23.5$ cm²; accept Ø25A500C, a step is 100, respectively, $F_a = 49.09$ cm².
M = 333 kN · m; $A_o = 3{,}330{,}000/173 \cdot 100 \cdot 44^2 = 0.099$, respectively, $\gamma = 0.944$; $F_a = 3{,}330{,}000/4430 \cdot 0.944 \cdot 44 = 18.1$ cm²; accept Ø25A500C, a step is 200, respectively, $F_a = 24.54$ cm², +Ø18A500C a step is 200, respectively, $F_a = 12.72$ cm².
$F_a = 37.26$ cm². M = 157 kN · m; $A_o = 1{,}570{,}000/173 \cdot 100 \cdot 34^2 = 0.079$, respectively, $\gamma = 0.959$; $F_a = 1{,}570{,}000/4430 \cdot 0.959 \cdot 34 = 10.9$ cm²; accept Ø25A500C, a step is 200, respectively, $F_a = 24.54$ cm².

Bending moments in wall elements is presented in Fig. 13.

14 Foundation Slab: Design Forces in the Foundation Slab

P_Y—the horizontal pressure intensity from the ground. There is no uniformly distributed load q = 10 kN/m² on the collapse prism surface, therefore P_q is not taken into account in the further calculation.

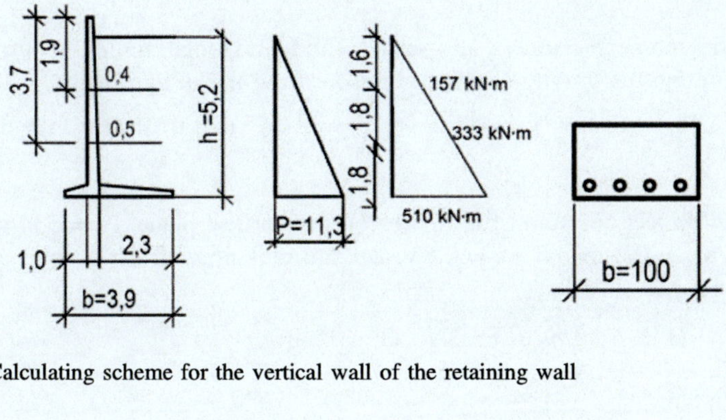

Fig. 12 Calculating scheme for the vertical wall of the retaining wall

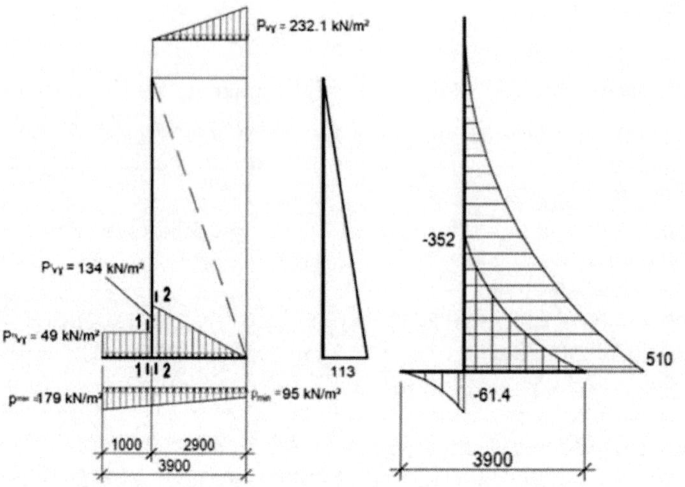

Fig. 13 Bending moments in wall elements

P_{vY}—the vertical pressure intensity from the own weight of the soil in the collapse prism:

$P_{vY} = P_Y \cdot tg(\mathcal{E} + \varphi_I^1) / tg\mathcal{E} = 232.1\ kN/m^2$

$P^1{}_{vY}$, $P^{11}{}_{vY}$—the vertical pressure intensity from the dead weight of the soil in the collapse prism and above the front console respectively:

$P^1{}_{vY} = \gamma_I{}^1 \cdot \gamma_f \cdot h = 18.6 \cdot 1.2 \cdot 6.0 = 134\ kN/m^2$ (in connection with the variable soil surface within the collapse prism, the height of the soil layer is determined as follows: $h = 5.2 + 1.9/3.3 \cdot [2.2/2 + 0.3] = 5.2 + 0.8 = 6.0$ m).

$P^{11}{}_{vY} = \gamma_I{}^1 \cdot \gamma_f \cdot d = 18.6 \cdot 1.2 \cdot 2.2 = 49\ kN/m^2$

M_o—the moments sum of all vertical and horizontal forces about the axis passing through the gravity center of the sole from the design loads.

$M_o = F_{sa} \cdot [h_l - tg(\mathcal{E} + \varphi_I{}^1) \cdot (b/2 - h_l \cdot tg\mathcal{E})] + \gamma_I{}^1 \cdot \gamma_f \cdot (b - t) \cdot [h \cdot (b - 4 \cdot t) + 6 \cdot t \cdot d] / 12 = 108.1\ kN \cdot m$

F_v—the projections sum of all forces on the vertical plane: $F_v = 534$ kN.

$e = F_v/M_o = 0.2\ m < 6 \cdot b = 0.65$ m. Edge pressures on the ground under the foundation base:

$p^{max}_{min} = F_v \cdot (1 \pm 6 \cdot e/b)/b = 13.7 \cdot (1 \pm 0.307)$.

$p_{max} = 179\ kN/m^2 < R = 515\ kN/m^2$.

$p_{min} = 95\ kN/m^2$.

Determine the moment and the shear force in the Sect. 1—1 at $x_1 = 1.0$ m:

$$M_{1-1} = \frac{P^{11}_{vy} \cdot x_1^2}{2} - p_{max} \cdot x_1^3 \cdot \frac{\left(\frac{p_{min}}{p_{max}} + \frac{3 \cdot b}{x_1} - 1\right)}{6 \cdot b} = -61.4\,\text{kNm} \tag{9}$$

$$Q_{1-1} = P^{11}_{vy} \cdot x_1^2 - p_{max} \cdot x_1^2 \cdot \frac{\left(\frac{p_{min}}{p_{max}} + \frac{3 \cdot b}{x_1} - 1\right)}{6 \cdot b} = -119\,kN < Q_{max} = 456\,kN \tag{10}$$

Determine the moment and shear force in Sect. 2—2 at $x_2 = 2.9$ m:

$$M_{2-2} = p_{min} \cdot x_2^3 \cdot \frac{\frac{p_{min}}{p_{max}} + \frac{3 \cdot b}{(X_2 - 1)}}{6 \cdot b} - \frac{P_{vy} \cdot x^2}{2} - \frac{x_2^3 \cdot (P^1_{vy} - P_{vy})}{6 \cdot (b - t)} = -352\,\text{kNm}; \tag{11}$$

$$Q_{2-2} = p_{min} \cdot x_2^2 \cdot \frac{\frac{p_{min}}{p_{max}} + \frac{2 \cdot b}{(x_2 - 1)}}{2 \cdot b} - P\gamma \cdot x^2 - \frac{x_2^2 \cdot (P\gamma - P\gamma)}{2 \cdot (b - t)} = -165\,kNm; \tag{12}$$

M = 352 kN · m; respectively, ⋎ = 0.96;

F_a = 3,520,000/4430 · 0.96 · 52 = 16.0 cm^2; accept Ø18A500C, a step is 100, respectively, F_a = 25.45 cm^2. M = 61.4 kN · m;

A_o = 614,000/173 · 100 · 52^2 = 0.013, respectively, ⋎ = 0.992; F_a = 614,000/4430 · 0.992 · 52 = 2.7 cm^2; accept Ø18A400, a step is 200, respectively, F_a = 12.72 cm^2.

15 Conclusions

The calculations were made in accordance with the current regulations. The calculation result is the calculated reinforcement area of monolithic reinforced concrete retaining walls located near the acidifiers [20, 21]. Retaining walls are designed for two groups of limit states:

(1) by bearing capacity: by the wall stability against shear and the soil base strength and by the structural elements strength;
(2) by suitability for service: bases on permissible deformations and cracks.

A cushion of sand and gravel mixture is provided under the retaining wall foundations foot. The soil resistance of the base does not exceed the maximum stress under the sole with the accepted foundation dimensions: for retaining wall 1—R = 495 kN/m^2 > σ_{max} = 311.2 kN/m^2,

for retaining wall 2—R = 429 kN/m^2 > σ_{max} = 182.4 kN/m^2, for retaining wall 3—R = 283 kN/m2 > σ_{max} = 101.4 kN/m^2.

The retaining walls stability against shear is ensured only at three angles values that determine the sliding plane. Vertical working walls reinforcement in the maximum moments zone: for retaining wall 1—Ø28 A500C with a step of 100, for retaining wall 2—Ø25 A500C with a step of 100. Horizontal working reinforcement in the foundation part of the retaining wall in the maximum moments zone: for retaining wall 1—Ø28 A500C with a step of 200 + Ø18 A500C with a step of 200 (total step is 100 mm), for retaining wall 2—Ø18 A500C with a step of 100 [2].

References

1. Rimshin VI, Kuzina ES, Shubin IL (2020) Analysis of the structures in water treatment and sanitation facilities for their strengthening. J Phys: Conf Ser 1425(1):012074. https://doi.org/10.1088/1742-6596/1425/1/012074
2. Kuzina ES, Rimshin VI (2020) Calculation method analysis for structure strengthening with external reinforcement. IOP Conf Ser: Mater Sci Eng 753(2):022004. https://doi.org/10.1088/1757-899X/753/2/022004
3. Rimshin V, Kuzina E, Neverov A (2019) Residual resource of force resistance to deformation. E3S Web Conf 135:01069. https://doi.org/10.1051/e3sconf/201913501069
4. Lisyatnikov MS, Roshchina SI, Chukhlanov VY, Ivaniuk AM (2020) Repair compositions based on methyl methacrylate modified with polyphenylsiloxane resin for concrete and reinforced concrete structures. IOP Conf Ser: Mater Sci Eng 896(1):012113. https://doi.org/10.1088/1757-899X/896/1/012113
5. Modin A, Lukin M, Vlasov A, Hisham E (2020) Energy-efficient indicators of panel housing mass construction in the climatic conditions of central Russia. IOP Conf Ser: Mater Sci Eng. https://doi.org/10.1088/1757-899X/896/1/012063
6. Lukin MV, Popov MV, Lisyatnikov MS (2020) Short-term and long-term deformations of the lightweight concrete. IOP Conf Ser: Mater Sci Eng. https://doi.org/10.1088/1757-899X/753/3/032071
7. Evseev V, Barkhi R, Pleshivtsev A, Scrynnik A (2019) Modeling the influence of weather and climatic conditions on the safety characteristics of the construction process. E3S Web Conf 97:03035. https://doi.org/10.1051/e3sconf/20199703035
8. Lisyatnikov MS, Shishov II, Sergeev MS, Hisham E (2020) Precast monolithic coating of an industrial building based on variable-height beam-slabs. IOP Conf Ser: Mater Sci Eng. https://doi.org/10.1088/1757-899X/896/1/012064
9. Telichenko VI, Benuzh AA, Rud NS (2021) Indoor air quality requirements for civil buildings in Russian regulations in comparison with international green standards. Int J Comput Civil Struct Eng 17(1). https://doi.org/10.22337/2587-9618-2021-17-1-98-107
10. Erofeev VT, Rodin AI, Kravchuk AS, Kaznacheev SV, Zaharova EA (2018) Biostable silicic rock-based glass ceramic foams. Mag Civil Eng 84(8):48–56. https://doi.org/10.18720/MCE.84.5
11. Telichenko V, Rimshin V, Ketsko E (2020) Reinforced concrete structures stress-strain state strengthen with composite materials. IOP Conf Ser: Mater Sci Eng 869(5):052003. https://doi.org/10.1088/1757-899X/869/5/052003
12. Erofeev VT, Rodin AI, Yakunin VV, Bochkin VS, Chegodajkin AM (2018) Alkali-activated slag binders from rock-wool production wastes. Mag Civil Eng 82(6):219–227. https://doi.org/10.18720/MCE.82.20
13. Erofeev V, Rodin A, Rodina N, Kalashnikov V, Irina E (2016) Biocidal binders for the concretes of underground constructions. Procedia Eng 165:1448–1454. https://doi.org/10.1016/j.proeng.2016.11.878
14. Gerasimov SI, Erofeev VI, Travov YF, Kalmykov AP, Lapichev NV (2021) Experimental determination of the drag coefficient of conical penetrators and a penetrator with a flat front end during supersonic motion in sandy soil. Tech Phys 66(3). https://doi.org/10.1134/S1063784221030087

15. Erofeev V, Dergunova A, Piksaikina A, Bogatov A, Kablov E, Startsev O, Matvievskiy A (2016) The effectiveness of materials different with regard to increasing the durability. MATEC Web Conf. https://doi.org/10.1051/matecconf/20167304021
16. Karpenko NI, Kolchunov VI, Kolchunov VI, Travush VI (2021) Calculation model of a complex-stressed reinforced concrete element under torsion with bending. Int J Comput Civil Struct Eng 17(1). https://doi.org/10.22337/2587-9618-2021-17-1-34-47
17. Travush V, Erofeev V, Bulgakov A, Kruglova T (2020) Operation mode optimization for monolithic construction mechatronic complex according technical condition of its servos. In: International multi-conference on industrial engineering and modern technologies FarEastCon 2020, 9271263. https://doi.org/10.1109/FarEastCon50210.2020.9271263
18. Travush VI, Gordon VA, Kolchunov VI, Leontiev EV (2020) Dynamic loading of the beam on the Pasternak base initiated by the sudden settlement of part of the base. IOP Conf Ser: Mater Sci Eng 896(1):012041. https://doi.org/10.1088/1757-899X/896/1/012041
19. Travush VI, Shulyat'ev OA, Shulyat'ev SO, Shakhraman'yan AM, Kolotovichev YA (2019) Analysis of the results of geotechnical monitoring of Lakhta center tower. Soil Mech Found Eng 56(2). https://doi.org/10.1007/s11204-019-09576-9
20. Travush VI, Karpenko NI, Erofeev VT, Smirnov VF, Rodina NG (2017) Development of biocidal cements for buildings and structures with biologically active environments. Power Technol Eng 51(4):377–384. https://doi.org/10.1007/s10749-017-0842-8
21. Roschina SI, Lisyatnikov MS, Koshcheev AA (2019) Technical- and economic efficiency of reinforced wooden structures. IOP Conf Ser: Mater Sci Eng 698(2):022005. https://doi.org/10.1088/1757-899X/698/2/022005

Determination of Deformability of LVL Structures with Toothed Plates Connectors

Aleksandr Chernykh, Stefania Mironova, Egor Danilov, Shirali Mamedov, Tatyana Kazakevich, and Pavel Koval

Abstract A method for calculating joints with toothed plates in LVL structures is proposed. The hypothesis about the possibility of using computational programs to determine the deformations of structures taking into account the movements in the node, based on empirical data, is tested. Full-scale experimental studies were carried out to determine the stiffness of the nodes of structures made of LVL with toothed plates connectors, the operation of such joints was simulated in a computational program. According to the results of experimental tests, data on the stiffness of the joints were obtained. Such information can be used to determine the deformability of a structure with such joints. This data was used to model individual assemblies and structures as a whole to determine deformations under load. Based on the conducted studies, when comparing the obtained values of displacements and the actual ones studied earlier, it was found that the proposed method for modeling the behavior of LVL structures with toothed connector plates can be applied for wide implementation.

Keywords Timber structures · LVL structures · Finite element model · Joint stiffness · Toothed plates connectors · Linear stiffness

1 Introduction

With the development of the construction industry, the use of LVL is widespread in timber construction, because this material has demonstrated high strength and stiffness. In LVL structures, the use of joints is justified, providing also high strength and stiffness characteristics. Dowels with claw washers (toothed plates «Bulldog») can be used as such joints [1–9].

A. Chernykh · S. Mironova · E. Danilov (✉) · S. Mamedov · T. Kazakevich · P. Koval
Saint Petersburg State University of Architecture and Civil Engineering, 4 Vtoraya Krasnoarmeiskaya ul., Saint Petersburg 190005, Russia
e-mail: mamedov_am@bk.ru

N. Vatin et al. (eds.), *Proceedings of MPCPE 2021*, Lecture Notes in Civil Engineering 182, https://doi.org/10.1007/978-3-030-85236-8_6

Nowadays joints with claw washers are quite popular abroad. However, in Russia, this type of connection is not widely used due to the lack of the sufficient theoretical and practical knowledge. Design of the structures made of LVL with claw washers implies, among other things, checking the second group of limit states. In this case, the calculations imply the use of the linear stiffness of the joint to determine the actual displacements. The stiffness can be calculated by formula 1:

$$k_z = N_z/\delta_z \tag{1}$$

where N_z is the value of the force applied to the joint;
δ_z—displacement of the joint elements.

Modern regulatory literature does not take a number of factors into consideration that affect the change in the stiffness of such joints and structures: moisture changes, the angle of application of force to the fibers, the size and shape of the dowels, the types and dimensions of claw washers and elements of wood-based joints, therefore, the definition of linear stiffness according to the existing methods [5, 10, 11] requires clarification.

In several works [1, 12, 13], the topic of research of structure joints made of unidirectional glued laminated timber with claw washers was highlighted. Earlier studies were carried out to determine the linear stiffness for a double-shear joint under the action of loads at different angles to the fibers of the joint elements as well [12].

The purpose of this work is to develop engineering methodology for calculation of building structures made of LVL with claw washers according to the second group of limit states, taking into account the linear stiffness of the joint, obtained experimentally and theoretically.

2 Methods

To describe the deformed state of some structures under load, it is common to use special types of finite elements of the "elastic type joint" in the design scheme, which allow modeling the nodes of structures with a certain flexibility to increase the accuracy of calculations for the second group of limit states. The characteristic of real structural units implies different patterns of the distribution of the joint stiffness, depending on the angle of the load applied. That is why for a joint with the same characteristics of linear stiffness for any directions of the load (in a plane coordinate system), the distribution of the stiffness characteristics is described in Fig. 1a. The general case with non-uniform stiffness distribution is described in Fig. 1b. According to the studies [12], the linear stiffness of joints in LVL structures is distributed as shown in Fig. 1c.

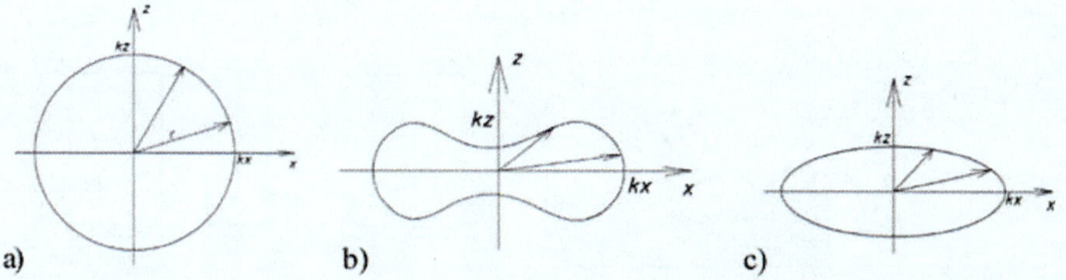

Fig. 1 An idealized picture of the distribution of the linear angle stiffness depending on the angle of application of force to the fibers: **a** uniform distribution of linear stiffness; **b** non-uniform distribution of the linear stiffness of the node; **c** linear stiffness of bolted joints in LVL

Let us consider the general case of a symmetrical double-shear joint of structures, shown in Fig. 2a. The arrangement of the fibers for the pad elements and the middle element is directed in different ways, while the external force is applied at an arbitrary angle γ to the main axes. The stiffness distribution is described by two ellipses that characterize the linear stiffness for each joint element. In this case, the shaded area reflects the area with the minimum distribution of stiffness, depending on the angle of load application. At a certain arbitrary angle of application of the force γ, the actual minimum linear stiffness of the joint can be calculated.

In LVL truss structures, there is often a case when it is required to design a symmetrical multi-cut joint with the elements placed at different angles to each other. The use of standard formulas of norms is possible if such a compound is considered as consisting of several groups. Figure 3 shows an example of such a 5-element joint. If we consider the bearing capacity of the group including elements 1, 2, 1, separately as well as groups with elements 2, 3, 2, then the use of standard

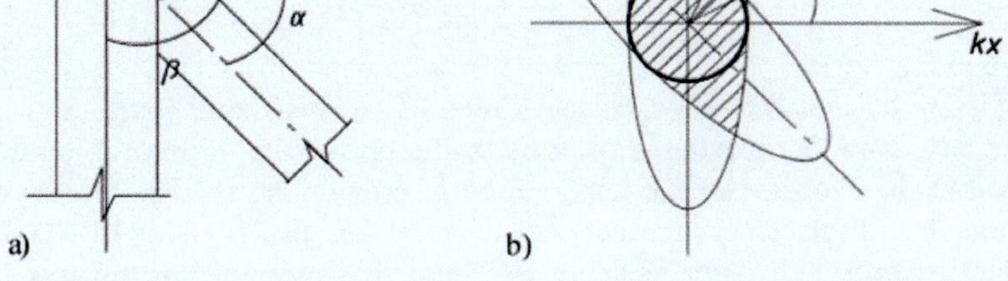

Fig. 2 Joint diagram: **a** general joint diagram; **b** a schematic diagram of the distribution of the linear stiffness of the joint in dependence to the angle of application of force to the fibers

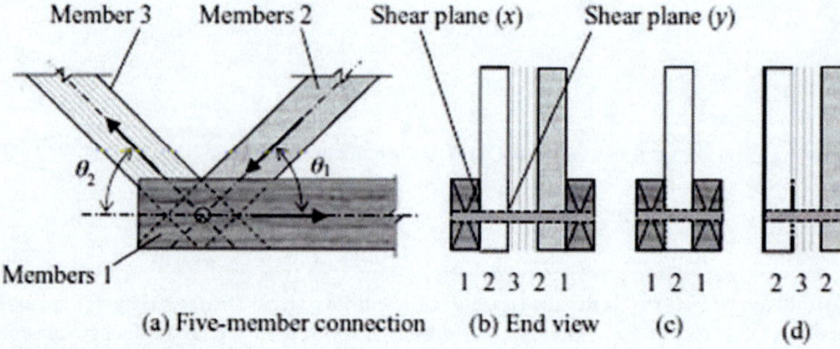

Fig. 3 Scheme for calculating a joint with several shear planes [14]

formulas of norms allows you to evaluate the bearing capacity right. Similarly, it is possible to evaluate the compliance of such joints and the structure as a whole, if we know the linear stiffness for a specific plane of joint between the elements (Fig. 3).

To determine the linear stiffness, experiments were previously made on samples with the application of a compressive load on the middle element, rotated relatively to the extreme ones by the angle $\alpha = 0°, 45°, 90°$ [12]. To determine the linear stiffness, the graphs of the displacements dependence in the joint on the live load were given, according to the data obtained, where the zone of elastic deformations was determined [13, 15]. According to the results of the experiment, the linear stiffness of the joints was determined in the elastic zone, compared with the experimental values. The difference between theoretical and experimental values averaged 15%. This indicates the right choice of the research tools used (Fig. 4).

To study the operation of LVL structures with claw washers, experimental studies of a truss with parallel belts with a span of 4.6 m were carried out [1, 16]. The experiment was made in order to calculate the deformed state of the trusses under the action of a static load applied at the nodes of the compressed belt. The manufacture of a pilot truss from LVL with claw washers was made in the mechanical laboratory of St. Petersburg State University of Architecture and Civil Engineering in St. Petersburg in accordance with the requirements [10, 11, 17].

An LVL Ultralam-R with a section of 100 × 45 mm was used as a material for the structural elements [18]. Steel cylindrical bolts made of C255 steel with a diameter of 16 mm with nuts and reinforced washers and claw washers of the "Bulldog" type C1 with a diameter of 75 mm manufactured by Simpson (USA) were used as joints. The truss was designed for the level of breaking load corresponding to $P_p = 6.18$ kN per node in accordance with [10, 11, 17]. In the manufacture of the structure, a universal press device was used [19].

The tests were done using Amsler hydraulic press. The horizontal joints were attached to the elements of the upper chord to provide the spatial stiffness of the structure. The displacement measurements were taken using Aistov 6PAO deflection meters with a division value of 0.01 mm, by installing instruments in the support zone of the truss and along the nodes of the upper and lower chords.

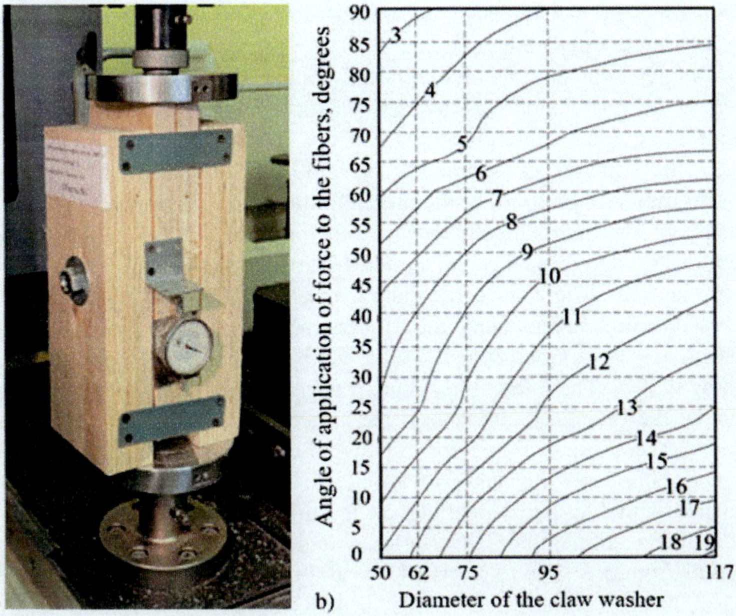

Fig. 4 Determination of the linear stiffness of the joints: **a** testing **b** theoretical distribution of the linear stiffness depending on the angle of application of force to the fibers and the diameter of the claw washer [12, 13]

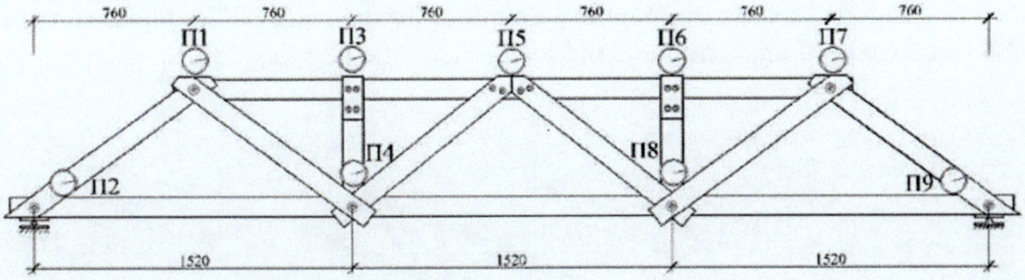

Fig. 5 Scheme of the tested LVL truss with claw washers

9 deflection meters were installed, the location of which is shown in Fig. 5. During the main test, the load was applied to the upper chord in steps, according to the recommendations, to the test of timber structures [15]. The load was applied for 5 min. Exposure of the structure to redistribute the forces in the structure was 25 min.

In the middle of the tests, the displacements of the nodes of the truss elements were recorded during the transfer of the load in the range from 2.16 to 10.31 kN per node with a loading step of 0.65 kN. After completing the main cycle of loading and unloading (up to 10.31 kN), the truss structure was brought to failure at a load level of 14.23 kN per node.

To simulate the experiment with the tested truss made of LVL with claw washers, a design scheme was created, shown in Fig. 6. To create a model of the structure, the graphic editor AutoCAD 2021 with the tools "3dface" was used. The length of the structure is assumed to be 4560 mm, the height of the elements is 45 × 100 mm. After exporting the model to the SCAD 21 computational complex, each element was automatically assigned the properties of a lamellar 4-angle finite element of the shell (type 44). Triangular finite elements of the shell (type 42) were used to "smooth" the non-rectilinear faces of the elements at the nodes of the elements docking. In this case, the properties for the plates were assigned as for orthotropic (E_x = 14 GPa, E_y = 0.5 GPa, ν_{xy} = 0.018, ν_{yx} = 0.45, G_{xy} = 0.7 GPa), according to [20]. The interconnections of the elements were modeled using a finite element of the "elastic joint" type (type 55), while the stiffness of such elements were taken according to Fig. 5 for joints with claw washers d 75 mm and a dowel 16 mm. In this case, the angle of application of loads to the fibers is taken in accordance with Fig. 6a, depending on the arrangement of the elements. This is how the linear stiffness for 38° angles was determined (k_x = 9.63 kN/mm, k_y = 10,000 kN/mm, k_z = 9.63 kN/mm, U_x = 10,000 kN * m/rad, U_y = 0 kN * m/rad, U_z = 10,000 kN * m/rad) and 75° (k_x = 4.58 kN/mm, k_y = 10,000 kN/mm, k_z = 4.58 kN/mm, U_x = 10,000 kN * m/rad, U_y = 0 kN * m/rad, U_z = 10,000 kN * m/rad). Since the computational algorithms of the SCAD 21 program do not allow applying the stiffness distribution for finite elements of type 55 with rotation of the main axes by an arbitrary angle, the same values are taken for the stiffness in the X and Z directions for a known angle of application of force to the fibers (that is, $k_x = k_z$). Figure 2b shows a graphical definition of the value of the linear stiffness at a fixed angle of application of the load.

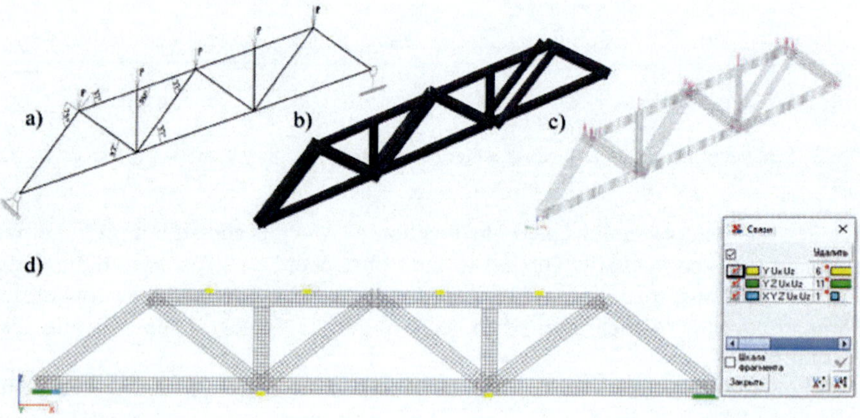

Fig. 6 Design model of the simulated structure: **a** general design model, **b** design in AutoCAD 2021 editor, **c** design model in SCAD 21, **d** fixing conditions

The boundary conditions for fixing the structure are shown in Fig. 6d. Loading was carried out by applying concentrated loads to the nodes, while the force was distributed equally to each branch of the joint element. In total, three load levels of 5.542, 6.867 and 8.927 kN were applied to the nodes per node.

3 Results and Discussion

As a result of the numerical and full-scale experiment, diagrams of the displacement of the truss nodes were formed depending on the load level (Table 1, Figs. 7 and 8). The analysis of the displacements of the results of the physical experiment corresponds to the elastic nature of the work of the joints up to the load level of 3.48 kN per truss unit; with an increase in the load to 7.55 kN per unit, a change in the slope of the load–displacement curve is noticeable, which is characterized by a proportional change in displacement from the load level.

The results of numerical studies in SCAD 21 are characterized by elastic work of joints for all load levels. The design was tested at three load levels: 5.542, 6.867 and 8.927 kN per node. In this case, the deformed load diagram of 5.542 kN for is shown in Fig. 7.

Table 1 and Fig. 8 show the difference between the theoretical and experimental values for a load level of 5.542 kN per node differs by 29%, which is caused by a peculiarity of the calculation method for the elastic deformation zone with a short loading time (up to 24 h). In the values from the deflection meters P3 and P6, there is a difference in the experimental data of 16% among themselves (an asymmetrical pattern of deflections), which may be a consequence of the uneven inclusion of the truss elements in the operation due to the insufficient load level to seal the connection of joints. For a load level of 6.687 kN per node, there is a noticeable difference in the results of experiments and calculations at the level of 9%, which is a satisfactory convergence of the results, there is a discrepancy with theoretical values of 1% for a load of 8.927 kN per node.

Table 1 Results of the numerical and physical experiment

Load on the node, kN	Vertical displacement in the node, mm				
	Physical experiment		Numerical experiment		Average convergance (%)
	П4	П5	П4	П5	
5.542	9.11	7.885	11.5	12.6	29
6.867	14.09	13.03	14.18	15.7	9
8.927	19.85	19.28	18.5	20.2	−1

Fig. 7 Deformed truss scheme at 5.542 kN load per node

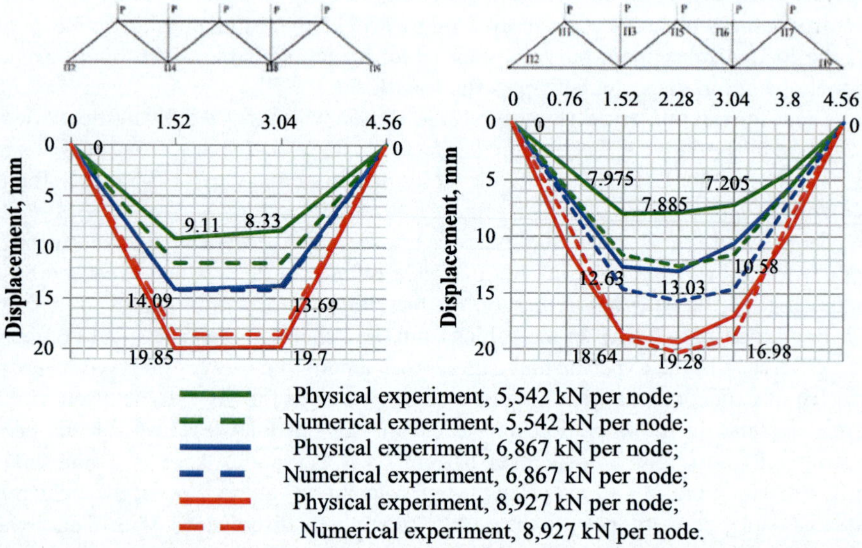

Fig. 8 The results of the data comparison of the numerical and physical experiment for the compressed and stretched chords of the truss

4 Conclusions

In the middle of experimental and theoretical studies, the deformations of the nodes of the LVL truss structure using claw washers were identified and compared. The accuracy of the calculation of the proposed method in the area of the projected design load on the truss structure is acceptable for use in the design of similar structures. The proposed method for calculating building structures made of LVL with claw washers according to the second group of limiting states can be used in engineering process.

References

1. Glukhikh VN, Chernykh AG, Danilov EV (2018) Wooden structures with the use of toothed plates and taking into account the initial stresses of the wood SPbGASU
2. Rimshin VI, Labudin BV, Melekhov VI, Popov EV, Roshchina SI (2016) Connections of elements of wooden structures on dowels and washers. Vestnik MGSU
3. Popov EV, Labudin BV, Melekhov VI (2019) Shear testing of elements of wooden structures connected with the use of toothed plates "Bulldog" Safety of the Construction Fund of Russia. Problems and solutions
4. Danilov EV, Chernykh AG (2013) Methods of research of joints of wooden structures on toothed plates. In: Modern problems of science and education
5. Blass HJ, Aune P, Choo BS, Gorlacher R, Griffiths DR, Hilson BO et al (1995) Timber engineering STEP 1: basis of design, material properties, structural components and joints almere, Netherlands, Centrum Hout
6. Blass HJ, Schlager M (1993) Trial calculations for determination of the load-carrying capacity of joints with bulldog connectors. Papers of the 23th meeting of the working commission W18
7. Blaß JH, Ehlbeck J, Schlager M (1993) Characteristic strength of tooth-plate connector joints Holz als Roh und Werkstoff
8. Mettem CJ, Page AV, Davis G (1995) Validatory tests and proposed design formulae for the load-carrying capacity of toothed-plate connectors joints. Papers of the 26th meeting of the working commission W18—timber structures
9. Standard NEN 3852:1973 (1972) Technische grondslagen voor de berekening van bouwconstructies Houtconstructies
10. Standard of the Association of Wooden Housing Construction StADD-3.2-2011 (2012) Wooden structures. Joints of wooden elements using toothed plates
11. Eurocode 5. Design of timber structures. Part: General rules and rules for buildings. ENV 1995-1-1. CEN, Brussels
12. Danilov EV (2017) Determination of the linear rigidity of joints with toothed plates in LVL beams. Vestnik grazhdanskikh inzhenerov
13. Chernykh AG, Danilov EV, Koval' PS (2020) Stiffness analysis of connections of LVL structures with claw washers. Lesnoi Zhurnal
14. Porteous J, Kermani A (2017) Structural timber design to Eurocode 5. Library of congress cataloging-in-publication data
15. TsNIISK (1976) Recommendations for testing wooden structures. Stroiizdat
16. Danilov EV (2018) Experimental study of the strength of a wooden truss made of unidirectional glued timber using toothed plates. Innovatsii v derevyannom stroitel'stve: materialy 9-i Mezhdunarodnoi nauchno-prakticheskoi konferentsii SPbGASU
17. SP 64.13330-2017 (2017) Wooden structures. Updated version of SNiP II-25-80. Stroyizdat
18. STO 36554501-021-2010 (2010) Wooden structures. Multi-layer glued veneer material Ultralam. General technical requirements. JSC «SIC» «Stroitelstvo»
19. Danilov EV, Chernykh AG, Gimmelfarb IA, Krupin EV (2014) Device for pressing claw washers. Utility model patent No. 146232
20. Tsiulin EY, Shmidt AB (2018) Numerical simulation of LVL elastic modulus with different combinations of mutually perpendicular veneer layers. Lesnoy Zhurnal

Bending Test of Stress-Laminated Timber Decks Using Laser Scanning

Aleksandr Chernykh, Pavel Koval, Shirali Mamedov, Egor Danilov, and Denis Nizhegorodtsev

Abstract The article presents the experimental investigation of bending stress-laminated timber decks (SLTD). The aim of the test was to study the effect of various combinations of design parameters of a deck on its deflections. The original method of experimental determination of vertical deformations of large-scale plate-type structures (timber decks) by continuous laser scanning of deflecting surface was developed and applied. The mathematical regression model describing the deflected surface of bending SLTD is obtained. It is revealed that the deformability properties of SLTD depend on a combination of such design parameters as its width, the pre-stressed rods spacing and the value of pre-stressing. The most significant of the considered variables is the width of the plate. With an increase or decrease in the width of SLTD, the deflections of the structure increase. The tensioned rods spacing has the greatest effect on the deflection at the minimum values, and with an increase in the distance between the rods, this effect is significantly reduced. The tension forces of the rods have a significant effect on the SLTD stiffness, with an increase in these forces the deflection of the structure linearly decreases.

Keywords Stress-laminated timber deck · Timber deck · Deformability · Deflected surface of plate · Laser scanning · Timber structure · Bending · Transversal plates

1 Introduction

Plotting the deflected surface of the plate allows determining its stress and strain state [1–5]. It means the internal forces and stresses calculation in a stress-laminated timber deck is possible if its deflections at each point are determined. Existing calculation methods of SLTD [6, 7] don't take into account the actual design model of the structure as they replace the slab with a fictive beam. Furthermore the results

A. Chernykh · P. Koval (✉) · S. Mamedov · E. Danilov · D. Nizhegorodtsev
Saint Petersburg State University of Architecture and Civil Engineering, 4 Vtoraya Krasnoarmeiskaya ul., Saint Petersburg 190005, Russian Federation

N. Vatin et al. (eds.), *Proceedings of MPCPE 2021*, Lecture Notes in Civil Engineering 182, https://doi.org/10.1007/978-3-030-85236-8_7

of calculating the bearing capacity and deflections of the plate according to various methods are different for the same conditions [8]. According to studies [8–11], there are significant deviations of the calculated deflections of the structure from the experimental ones and, in some cases, its values are underestimated.

The experimental determination of the deflected surface of the plate by conventional methods is a task with a high degree of laboriousness. For bending beams it is recommended to determine vertical deformations in the middle of the span, on the supports, and in the quarters of the span, but this is not enough for large-scale plate-type structures that are in complex stress and strain station. Bending of the SLTD which is hinged bearing on two sides and loaded with concentrated or irregularity distributed forces, occurs in two directions [12–14]. Thus there is a need to evaluate deflections in a large number of points. In addition, behaviour of the structure may be accompanied by a mutual shearing of the elements relative to each other [15, 16]. So there is a task of experimental investigations of the SLTD deformability and the determination of deflections at all points of the plate.

2 Methods

The original method of experimental determination of vertical deformations of large-scale plate-type structures (timber decks) by continuous laser scanning of deflecting surface was developed for minimizing the laboriousness of deflections measurement. The developed method has the following features:

- a laser scanner is used to measure the vertical deformations of the loaded structure, and specialized software is used to process the results obtained;
- there are no restrictions on the number of points at which vertical deformations will be determined;
- the scanning step of the point grid can be adjusted;
- it is possible to conduct the experiment before the specimen is destroyed;
- scanner measurement accuracy up to 10^{-5} m.

The algorithm for determining deformations is shown in the Fig. 1.

Note that it is not possible to move the scanner during scanning or between loading steps of one specimen.

3 Results and Discussions

The results of the laser scanning (experiment) include photofixation of test specimens at each loading stage, combined with clouds of points with determined coordinates. The deflected surface of the slab and the cross sections may be obtained from this data.

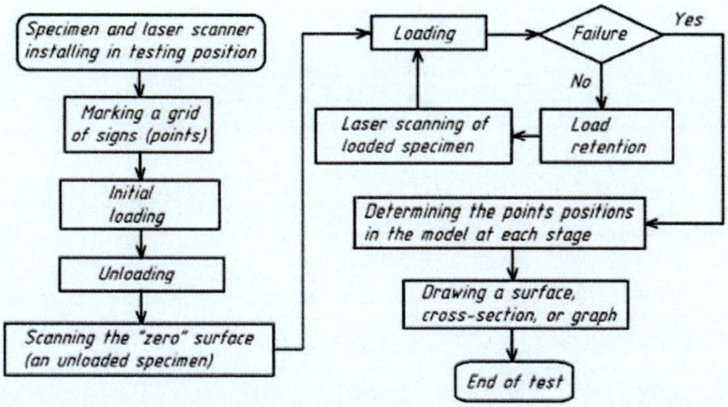

Fig. 1 Algorithm of experimental determination of vertical deformations by continuous laser scanning

The experimental study of deformability of SLTD was carried out as the complex experiment (multifactorial). The tests were conducted according to the Box's design B3 [17–19]. The width of the plate, the tensioned rods spacing (hence their number), and the tension force of the rods were chosen as variable factors. Each factor varied on three levels. The values of the variable factors are given in the Table 1.

The test conditions, material of specimen (pine), plate thickness (120 mm), plate span (2760 mm), and loading method remained unchanged. The deflection of the central point of the plate was taken as the main parameter of the experiment.

According to the Box's design (Table 1), 14 test specimens of SLTD with a different combination of variable factors were performed, 3 pieces of each type. The tension forces to the prestressed rods were applied manually using a torque wrench. The tests were carried out according to the static bending scheme, the plates were hinged bearing on two opposite sides, and the load was applied concentrated in the center of the plate. The structure and general view of the assembly testing set are shown in the Fig. 2.

Table 1 Variable factors of complex experiment by Box's design B3

#	Factor		Factor value		
			slab width	**rods spacing**	**tension**
1	Measure unit		m (number of planks)	m	N (Nm)
2	Designation	natural	b	s	F
3		code	x_1	x_2	x_3
4	Factor levels	low (−)	0.945 (21)	0.250	7650 (35)
5		middle (0)	1.395 (31)	0.500	15,300 (70)
6		high (+)	1.845 (41)	0.750	22,950 (105)
7	Range		0.450 (10)	0.250	7650 (35)

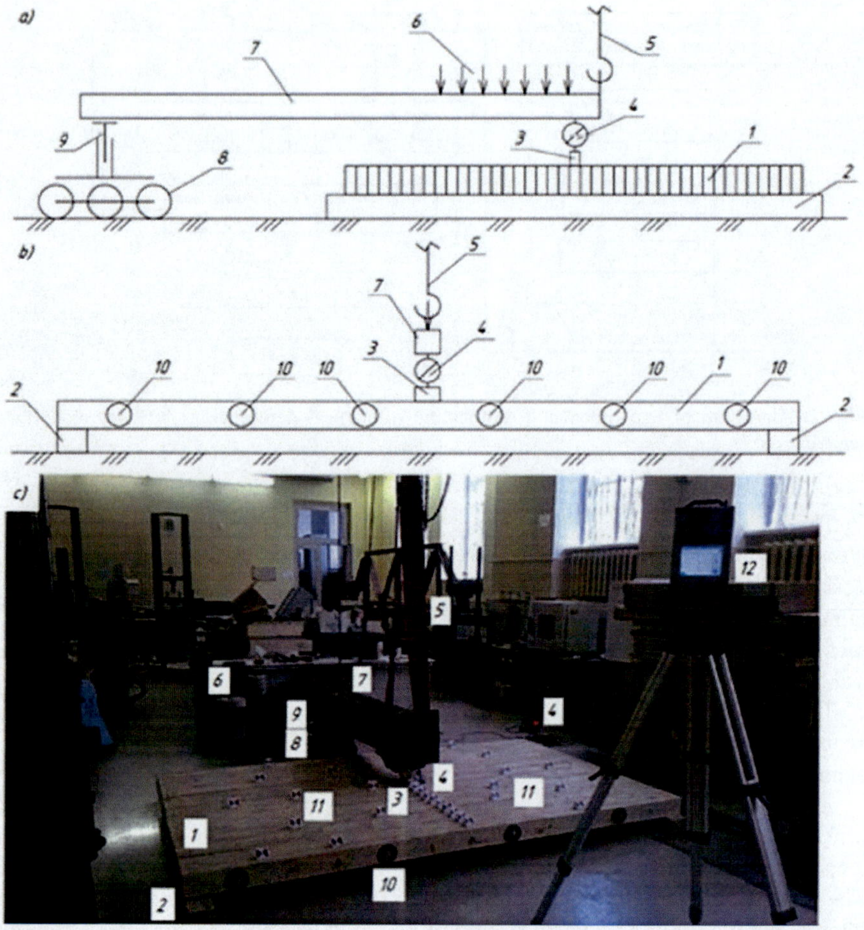

Fig. 2 Testing set: **a** side view; **b** front view; **c** assembled set, general view; 1—stress-laminated timber deck; 2—support beam; 3—steel element; 4—dynamometer; 5—hoist; 6—loading units; 7—loading beam; 8—horizontal adjustment; 9—vertical adjustment; 10—washer; 11—points (signs); 12—laser scanner

Loading was carried out stepwise with loading units (6) put up on the loading beam (7), with one edge supported on a vertically (9) and horizontally (8) adjustable support, and the other edge supported on a steel element (3) that transmits the load directly to the test specimen (1). The dynamometer (4) was installed between the loading beam (7) and the steel element (3) to control the load on each stage. To completely remove the load from the specimen, the loading beam (7) was lifted by hoist (5) by one edge.

Metal loading units put up manually, the load retention at each stage was 10 min, the deformations in the system were measured after retention. The destruction of the test specimen wasn't performed. Deflections at each loading stage were determined by the method of continuous laser scanning of deflected surface—see the Fig. 1. The laser scanner Z+F IMAGER 5010C was used for this purpose (Fig. 2).

Immediately before each test, the tension forces of each SLTD tensioned rod were monitored.

As a result of the tests, 3D models were obtained, including: photofixation of test specimens at each loading stage; clouds of points; information about the coordinates of each point; date and time of the study. These point clouds were processed in special laser scanning software. A mathematical model of SLTD bending is obtained in the form of a regression Eq. (1) with constraints (2):

$$w = 2.79375 + 0.1x_1 + 0.22x_2 - 0.17x_3 + \\ + 0.45625x_1^2 - 0.14375x_2^2 - 0.0875x_1x_2 - 0.0625x_2x_3. \quad (1)$$

$$\left.\begin{aligned} 945\ \text{mm} \le x_1 \le 1845\ \text{mm}, \\ 250\ \text{mm} \le x_2 \le 750\ \text{mm}, \\ 7650\ \text{N} \le x_3 \le 22950\ \text{N} \end{aligned}\right\}. \quad (2)$$

Dependence graphs of the deflections of the plate central point at the last stage of loading on a combination of variable factors (x_1, x_2, x_3) are made—see the Fig. 3.

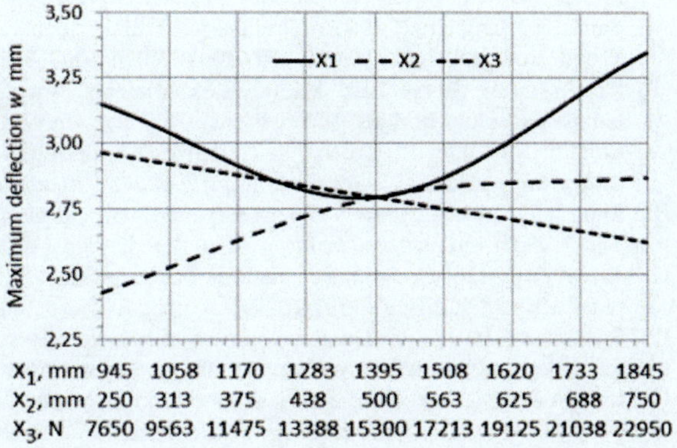

Fig. 3 Dependence $w(x_i)$ of deflections on variable factors

4 Conclusions

The analysis of the dependences in Fig. 3 shows that the deformability properties of SLTD depend on a combination of its design parameters: the width of the plate, the tensioned rods spacing, and the tension force of the rods. The most significant of the considered variables is the width of the plate: with an increase in the width, the deflections of the structure increase nonlinearly, and, with a decrease in this parameter, however the deflections also increase, which can probably be explained by a reduce in the moment of inertia of the section. The tensioned rods spacing has the greatest effect on the deflection at the minimum values, and with an increase in the distance between the rods, this effect is significantly reduced, and the curve character is also nonlinear. The tension forces of the rods have a significant effect on the SLTD stiffness, with an increase in these forces the deflection of the structure linearly decreases.

References

1. Whitney JM (1969) The effect of transverse shear deformation on the bending of laminated plates. J Compos Mater. https://doi.org/10.1177/002199836900300316
2. Carrera E (2002) Theories and finite elements for multilayered, anisotropic, composite plates and shells. https://doi.org/10.1007/BF02736649
3. Irgens F (2008) Continuum mechanics. https://doi.org/10.1007/978-3-540-74298-2
4. Goldenveizer AL, Kaplunov JD, Nolde EV (1993) On Timoshenko-Reissner type theories of plates and shells. Int J Solids Struct. https://doi.org/10.1016/0020-7683(93)90029-7
5. MacEri A (2010) Theory of elasticity. https://doi.org/10.1007/978-3-642-11392-5
6. Calil C (2010) Field performance after 5 years of a transverse cellular glulam box prestressed bridge. In: 11th World conference on timber engineering 2010, WCTE 2010
7. Sartori T, Crocetti R (2016) Prefabricated timber-concrete composite floors. Eur J Wood Wood Prod. https://doi.org/10.1007/s00107-016-1007-4
8. Andersson E, Bergendahl J (2009) Experimental and numerical investigations on stress laminated timber bridges. M.Sc. thesis. Chalmers University of Technology, Göteborg
9. Ekholm K, Kliger R, Crocetti R (2012) Full-scale ultimate-load test of a stress-laminated-timber bridge deck. J Bridg Eng. https://doi.org/10.1061/(asce)be.1943-5592.0000304
10. Idnurm J, Funk A, Salm S (2013) Experimental and numerical investigations of timber. In: The XXVIII international baltic road conference, pp 1–10. Baltic Road Association, Vilnius
11. Koval PS (2016) Comparative analysis of modern methods for calculating prestressed wood planks. Bull Civil Eng 6(59):97–101
12. Freedman GJH (2006) The development of transversely stress-laminated timber arch bridges for pedestrian and minor vehicle use. Ph.D. thesis. Napier University, Edinburgh
13. Lindquist M (2006) Confiabilidade estrutural de pontes laminadas protendidas de madeira: Tese de Doutorado. Universidade de São Paulo, São Paulo
14. Steige R, Schubert S, Gülzow A, Hugener M, Gsell D (2010) Vibration and damping behaviour of a cable-stayed timber deck bridge with asphalt pavement. In: 11th World conference on timber engineering 2010, WCTE 2010
15. Ekholm K, Ekevad M, Kliger IR (2014) Modeling slip in stress-laminated timber bridges: comparison of two finite-element-method approaches and test values. J Bridg Eng. https://doi.org/10.1061/(asce)be.1943-5592.0000595

16. Chernykh AG, Petrov AN, Koval PS, Danilov EV (2019) Experimental study of the deformability of prestressed multilayer wood-plates by the method of continuous laser scanning. Bull Civil Eng 5(76):97–104
17. Wohlin C, Runeson P, Höst M, Ohlsson MC, Regnell B, Wesslén A (2012) Experimentation in software engineering. Springer-Verlag Berlin Heidelberg. https://doi.org/10.1007/978-3-642-29044-2
18. Belytschko T, Lu YY, Gu L (1994) Element-free Galerkin methods. Int J Num Methods Eng 37:229–256. https://doi.org/10.1002/nme.1620370205
19. Chernykh AG (1998) Formation of protective and decorative coatings of wood by filling method. D.Sc. thesis. Saint Petersburg State Forest Technical University, Saint Petersburg

Bending of Multilayer Slabs Lying on Elastic Half-Space, Considering Shear Stresses

Mirziyod Mirsaidov, Kazokboy Mamasoliev, and Kubaymurat Ismayilov

Abstract This article is devoted to the development of models and methods for solving contact problems of a structure with a foundation. The study of internal force factors in multilayer slab-strips lying on an elastic half-space, taking into account shear stresses, is an urgent task. In this study, a mathematical model is developed and an analytical method for solving contact problems is proposed to determine the internal force factors of multilayer slab-strips interacting with a half-space, taking into account the shear stresses in the structure-foundation contact. The problem, using orthogonal Chebyshev polynomials, is reduced to solving infinite systems of algebraic equations. To obtain a result with satisfactory accuracy, the required number of terms of the Chebyshev polynomial is established. From the result of numerical examples, the influence of the filler on the internal forces of the slab-strip is determined. At the same time, the internal forces of layered slabs-strips are compared, corresponding to different stiffness characteristics of the aggregate. An increase in the numerical values of the stiffness coefficients of the filler leads to a convergence of the values of bending moments. Based on the comparison, theoretical conclusions about the effect of the aggregate on the force factor of layer slabs, which is important for the designer when calculating projects of multilayer slabs, are presented. An analytical method for solving the problem was proposed to assess the internal force factors of multilayer slab-strips, based on the approximation of the orthogonal Chebyshev polynomials. An account for the shear stresses arising in the contact of multilayer slabs with the foundation leads to a decrease in the force factors in the slab-strips.

Keywords Analytical solution · Multilayer slab-strip · Foundation · Interaction · Shear stress · Chebyshev polynomial

---

M. Mirsaidov (✉)
Tashkent Institute of Irrigation and Agricultural Mechanization Engineers, 39 Kori Niyoziy str., 100000 Tashkent, Uzbekistan

K. Mamasoliev · K. Ismayilov
Samarkand State Architectural and Civil Engineering Institute, 70 Lolazor str., 140147 Samarkand, Uzbekistan

N. Vatin et al. (eds.), *Proceedings of MPCPE 2021*, Lecture Notes in Civil Engineering 182, https://doi.org/10.1007/978-3-030-85236-8_8

1 Introduction

A sufficient number of scientific publications are devoted to the development of various methods for calculating structures interacting with elastic foundations. These publications consider various models and methods to solve specific problems. The foundations of structures and buildings, plates of hydraulic structures, airfield and road paving, rails and railway sleepers, etc. are the structures, for which the interaction with an elastic foundation should be taken into account. In the methods of calculating a structure with a foundation, the main attention is paid to the study of the relationship between the contacting elements of the structures [1–4].

At present, a sufficient number of scientific publications, which investigate various issues arising from the interaction of structural elements with an elastic foundation, are available.

For example:

- in [5] contact interactions of two plates made of materials with different mechanical properties were considered, taking into account the structural and physical nonlinearity. To study the stress-strain state of the plates, the method of variational iteration was applied. The problem considered in partial differential equations was reduced to the solution of ordinary differential equations;
- in [6] provided solutions to an axisymmetric displaced problem of elasticity theory for a cone, fixed along the lateral surface, with an attached spherical segment;
- an axisymmetric quasi-static contact problem on the thermal and mechanical interaction of a circular punch and an inhomogeneous elastic foundation was presented in [7]. Thermophysical and mechanical properties were described by arbitrary functional along the depth coordinate. Numerical calculations were implemented for various dependencies of material properties on the depth of the foundation;
- in [8] the finite difference method was used to calculate the bending of the plates. The given examples illustrate the high accuracy of calculation and the simplicity of the algorithm;
- in [9], the solution to the problem for multilayer slab-strips, lying on an elastic foundation under static loads, was given. The problem with the orthogonal Chebyshev polynomials was reduced to solving infinite systems of algebraic equations. The regularity of infinite systems of algebraic equations is proved and the corresponding estimates are obtained;
- in [10, 11], an overview of the development of mathematical modeling and methods of elastic analysis of inhomogeneous solids was given, as well as an overview of various theoretical models of elastic and viscoelastic foundations in vibrational systems;
- in [12], dynamic problems were considered for the structure with the foundation, using artificial non-reflecting boundary conditions on the boundary of the finite area of the foundation;

– a number of problems devoted to the study of internal force factors and the behavior of inhomogeneous elastic and viscoelastic systems in the presence of various influences with account for their features, were considered in [13–20]. The analysis of the power factors and the behavior of systems interacting with the external environment were given [21].

In the above publications, each approach and method has its own advantages and shortcomings. Despite this, they are taken into account when solving various practical problems.

This is an overview of just a few publications on the study and solution of various problems in the joint work of a structure and an elastic foundation. According to the formulation and the results obtained in the above studies, we can notice the incompleteness of research, especially in the field of obtaining analytical and refined solutions to the problem.

Therefore, this article is devoted to the development of mathematical models and analytical methods for studying the interaction of multilayer slab-strips lying on an elastic half-space, taking into account the shear stresses arising between the slab-strip and the foundation. This is an urgent problem in the mechanics of a deformable rigid body.

2 Methods

2.1 Mathematical Model

Consider n-layer slab-strips lying on an elastic half-space (see Fig. 1). It is assumed that each layer of the slab has different mechanical characteristics. The geometrical parameters of the slab are taken as different in height; equal in width; infinite in length.

External loads are applied to each layer of the strip-slab arbitrarily across the slab. It is assumed that the response of the elastic filler between the slabs is proportional to the difference in deflections of the connecting slabs. Here, the aspect ratio is understood as the stiffness coefficient of the filler. The bottom plate, which is in contact with the elastic half-space, in addition to the reactive response, fillers and external loads, is affected by the reactive normal and shear stresses on the contact with the foundation.

When modeling the stress-strain state, the n-layer strip-slabs, under the above assumptions, are considered as n-layer beam slabs cut out with a width equal to one (see Fig. 2). Now the problem under consideration for n layer strip-slabs is reduced to an assessment of the stress-strain state of an n layer beam slab of length 2 l, of thickness $h_i, (i = 1, 2, 3, ..., n)$, and a width equal to one. For convenience, the origin is set at the symmetrical center of the bottom beam slab. Then the solution of the problem along the abscissa axis is given in the interval $[-l; l]$, i.e. $-l \leq x \leq l$,

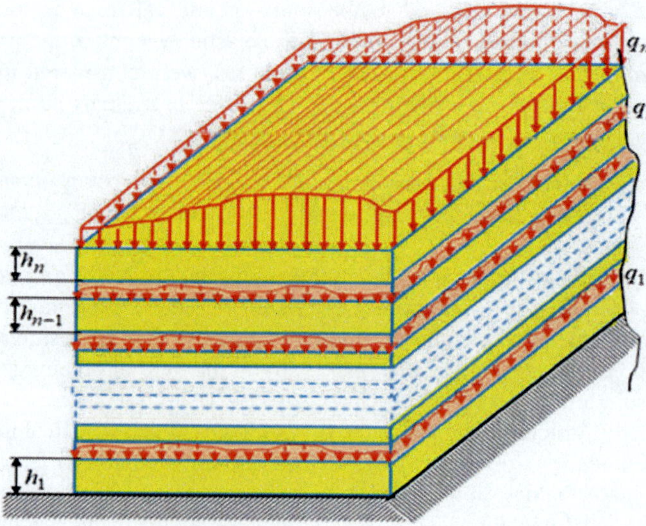

Fig. 1 Design scheme of n-layer slab-strips

Fig. 2 Design scheme of n-layer beam slab

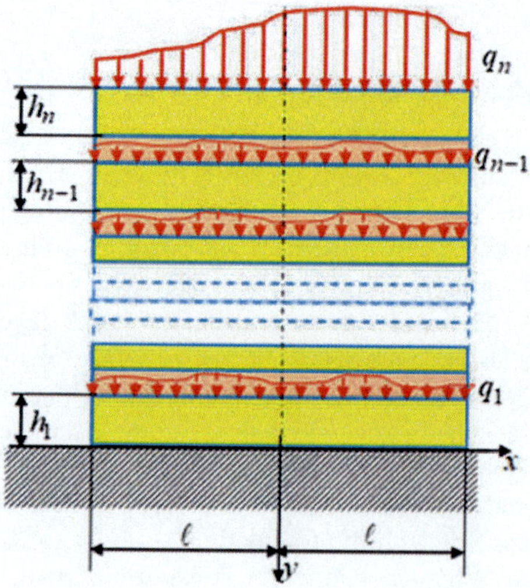

variable along the ordinate of the slab deflection as a function of the variable x, i.e. $y_i = y_i(x)$. Here y_i is the deflection of the i-th beam slab.

To model the deformation process of beam slabs, based on the above assumptions, the following systems of differential equations are used assumptions, the following systems of differential equations are used

$$D_i y_i^{IV} = q_i + k_i(y_{i+1} - y_i) - k_{i-1}(y_i - y_{i-1}) - \delta_i(\rho + 0,5 \cdot h_1 \cdot \tau'),\\ i = 1, 2, 3, \ldots, n \tag{1}$$

Here D_i is the cylindrical stiffness of the slab; k_i is the stiffness coefficients of the filler;

$q_i = q_i(x)$ is the intensity of external loads; $p = p(x)$ and $\tau = \tau(x)$ are the reactive normal and shear stresses of the foundation, respectively;

$$k_n = k_0 = 0; \quad \delta_i = \begin{cases} 1 & at \quad i = 1; \\ 0 & at \quad i = 2, 3, \ldots, n. \end{cases} \tag{2}$$

The horizontal displacement of the sole of the bottom beam slab is determined by the formulas given in [3, 4].

$$u_\tau = \frac{1 - \nu_1}{h_1 E_1} \int_0^x N(s)ds - 0,\ 5hy_1' + \tilde{B}_{1\tau}, \tag{3}$$

Here ν_1, E_1 are the Poisson's ratio and the modulus of elasticity of the first slab, respectively; $N(s)$ are the internal longitudinal forces related to the tangential forces, determined by the following equation

$$N(x) = \int \tau(x)dx - \tilde{B}_{2\tau}, \tag{4}$$

where $\tilde{B}_{1\tau}$, $\tilde{B}_{2\tau}$ are the unknown constants, determined from the boundary conditions of the problem.

As is well known, vertical and horizontal displacements of points of the foundation surface in the theory of elasticity are determined by formulas [3, 4]

$$V = \alpha_1 \cdot \int_{-l}^{l} p(s) \cdot \ln\frac{1}{|x - s|}ds + \alpha_2 \cdot \left[\int_{-l}^{x} \tau(s)ds - \int_{x}^{l} \tau(s)ds\right]; \tag{5}$$

$$U = \alpha_1 \cdot \int_{-l}^{l} \tau(s) \cdot \ln\frac{1}{|x - s|}ds - \alpha_2 \cdot \left[\int_{-l}^{x} p(s)ds - \int_{x}^{l} p(s)ds\right], \tag{6}$$

where $\alpha_1 = \frac{2 \cdot (1 - \nu_0^2)}{\pi E_0}$, $\alpha_2 = \frac{(1 + \nu_0) \cdot (1 - 2\nu_0)}{2E_0}$.

ν_0, E_0 are the Poisson's ratio and the elastic modulus of the foundation, respectively.

If the sums of vertical, horizontal forces and the moment of all external loads relative to the middle of the slab are denoted by *R, T* and *M* respectively, then the equilibrium equations for the slabs are written in the following form:

$$\int_{-l}^{l} p(x)dx = R, \quad \int_{-l}^{l} \tau(x)dx = T, \quad \int_{-l}^{l} xp(x)dx = M. \tag{7}$$

It is assumed that there is a two-way relation between the surface of the bottom plate and the elastic foundation, i.e., the separation of the slabs from the foundation is not supposed.

In this case, the tight fit of the surface of the bottom slab and the elastic foundation, taken as contact conditions interconnecting slabs with the foundation, are written in the form

$$y_1(x) = V(x), \; u_\tau(x) = U(x), \; -l \le x \le l. \tag{8}$$

Thus, the bending problem of multilayer slab-strips interacting with a linearly deformable half-space, considering the reactive shear stresses between the surface of the bottom slab-strip and the elastic half-space, is reduced to searching a solution to Eqs. (1) and (2) the system of differential equations, and Eqs. (5) and (6) integral equations, simultaneously satisfying conditions Eq. (8).

2.2 Solution Method

For convenience, we proceed to the dimensionless coordinate along *x* equal to the ratio of the absolute coordinate to the half-length of the beam slabs *l*.

The distribution patterns of reactive normal and shear stresses of an elastic foundation are presented in the form of a series of orthogonal Chebyshev polynomials of the first kind with weight functions

$$p(x) = \rho(x) \sum_{n=0}^{\infty} A_n T_n(x); \tag{9}$$

$$\tau(x) = \rho(x) \sum_{n=0}^{\infty} B_n T_n(x), \tag{10}$$

Here A_n, B_n are the unknown constant coefficients; $p(x)$, $T_n(x)$ are the weight function and the Chebyshev polynomial, respectively [22], defined by

$$p(x) = \left(1 - x^2\right)^{-\frac{1}{2}}, \; T_n(x) = \cos(n \arccos x). \tag{11}$$

The norm of orthogonal Chebyshev polynomials has the form

$$\|T_0(x)\| = \pi, \quad \|T_n(x)\| = \frac{\pi}{2}, \quad n = 1, 2, 3, \ldots \tag{12}$$

$$A_0 = \frac{R}{\pi l}, \; A_1 = \frac{2M}{\pi l^2}, \; B_0 = \frac{T}{2l}. \tag{13}$$

Based on the known dependencies [22]

$$\int_{-1}^{1} \frac{T_k(s)}{\sqrt{1 - s^2}} \ln|x - s| ds = \begin{cases} \pi \ln 2, & \text{at } k = 0; \\ -\frac{\pi}{k} T_k(x), & \text{at } k = 1, 2, 3, \ldots \end{cases} \tag{14}$$

$$\int_{-1}^{x} (1 - s^2)^{-\frac{1}{2}} T_k(s) ds - \int_{x}^{1} (1 - s^2)^{-\frac{1}{2}} T_k(s) ds = -\frac{2}{k} U_k(x). \tag{15}$$

The expressions for vertical Eq. (5) and horizontal Eq. (6) displacements of the points of foundation surface take the form

$$V = \pi\alpha_1 \left[-A_0 \ln 2 + \sum_{n=1}^{\infty} A_n \frac{T_n(x)}{n} \right] + 2\alpha_2 \left[B_0 \arcsin x - \sum_{n=1}^{\infty} B_n \frac{U_n(x)}{n} \right], \tag{16}$$

$$U = \pi\alpha_1 \left[-B_0 \ln 2 + \sum_{n=1}^{\infty} B_n \frac{T_n(x)}{n} \right] - 2\alpha_2 \left[A_0 \arcsin x - \sum_{n=1}^{\infty} A_n \frac{U_n(x)}{n} \right], \tag{17}$$

Here $U_n(x) = \sin(n \arccos x)$.

To illustrate the effectiveness of the proposed method, a two-layer beam slab lying on an elastic half-space is considered. Then Eq. (1) the system of differential equations for vertical displacements of beam slabs takes the following form

$$\begin{cases} D_2 l^{-4} y_2^{IV} = q_2 - k_1(y_2 - y_1), \\ D_1 l^{-4} y_1^{IV} = q_1 + k_1(y_2 - y_1) - p - 0,5 h_1 l^{-1} \tau. \end{cases} \tag{18}$$

Taking into account Eqs. (9) and (10), the general solution of Eq. (18) the system of differential equations is given in the following form

$$y_1 = \frac{l^4}{D_1 + D_2}\left\{\sum_{i=1}^{4} C_i x^{4-i} + f_q(x) - \sum_{n=0}^{\infty} A_n f_n(x) - \frac{h_1}{2l}\sum_{n=0}^{\infty} B_n f_n'(x) - D_2\left[\sum_{i=1}^{4} \tilde{C}_i u_i(\alpha x) + \psi_q(x) + \frac{1}{D_1}\sum_{n=0}^{\infty} A_n \varphi_n(x) + \frac{1}{D_1}\frac{h_1}{2l}\sum_{n=0}^{\infty} B_n \omega_n(x)\right]\right\}; \tag{19}$$

$$y_2 = \frac{l^4}{D_1 + D_2}\left\{\sum_{i=1}^{4} C_i x^{4-i} + f_q(x) - \sum_{n=0}^{\infty} A_n f_n(x) - \frac{h_1}{2l}\sum_{n=0}^{\infty} B_n f_n'(x) - D_1\left[\sum_{i=1}^{4} \tilde{C}_i u_i(\alpha x) + \psi_q(x) + \frac{1}{D_1}\sum_{n=0}^{\infty} A_n \varphi_n(x) + \frac{1}{D_1}\frac{h_1}{2l}\sum_{n=0}^{\infty} B_n \omega_n(x)\right]\right\}, \tag{20}$$

Here C_i, $\tilde{C}_i$ are the arbitrary constants of integration determined from the boundary conditions of the problem under consideration

$$\left.\begin{aligned} u_1(x) &= chx\cos x; & u_2(x) &= chx\sin x + shx\cos x; \\ u_3(x) &= shx\sin x; & u_4(x) &= chx\sin x - shx\cos x. \end{aligned}\right\} \tag{21}$$

$$f_q^{IV}(x) = q_1(x) + q_2(x), \tag{22}$$

$$\psi_q(x) = \frac{1}{4\alpha^3}\int_0^x \left[\frac{q_1(z)}{D_2} - \frac{q_2(z)}{D_1}\right] u_4[\alpha(x-z)]dz, \tag{23}$$

$$f_n(x) = \frac{1}{2^4 n(n-1)(n-2)(n-3)}(1-x^2)^{\frac{7}{2}} P_{n-4}^{\left(\frac{7}{2},\frac{7}{2}\right)}(x), \quad n > 3 \tag{24}$$

$$f_n'(x) = -\frac{1}{2^3 n(n-1)(n-2)}(1-x^2)^{\frac{5}{2}} P_{n-4}^{\left(\frac{5}{2},\frac{5}{2}\right)}(x), \quad n > 2 \tag{25}$$

$$f_n''(x) = \frac{1}{2^2 n(n-1)}(1-x^2)^{\frac{3}{2}} P_{n-4}^{\left(\frac{3}{2},\frac{3}{2}\right)}(x), \quad n > 1 \tag{26}$$

$$f_n'''(x) = -\frac{1}{2n}(1-x^2)^{\frac{1}{2}} P_{n-4}^{\left(\frac{1}{2},\frac{1}{2}\right)}(x), \quad n > 0 \tag{27}$$

$$\varphi_n(x) = \frac{1}{4\alpha^3}\int_0^x u_4[\alpha(x-z)](1-z^2)^{-\frac{1}{2}} T_n(z)dz, \tag{28}$$

$$\omega_n = \varphi_n'(x) - T_0(0)U_n(\alpha x), \tag{29}$$

Here $P_n^{(\alpha,\beta)}(x)$ are the Jacobi polynomials [22].

Using the Chebyshev polynomial [22, 23], it is possible to obtain formulas for determining function $f_n(x)$, for $n<4$.

Horizontal displacements of the sole points of the slabs Eq. (3), with Eqs. (10) and (20), take the following form:

$$u_t = \frac{(1-\nu_1^2)l}{h_1 E_1}\left[\tilde{B}_1 + \tilde{B}_2 x + \sum_{n=0}^{\infty} A_n f_n''(x)\right] - \frac{h_1}{2l}\frac{l^4}{D_1 + D_2}\left\{\sum_{i=1}^{3} C_i(4-x)x^{3-i} + f_q'(x)\right.$$
$$- \sum_{n=0}^{\infty} A_n f_n'(x) - \frac{h_1}{2l}\sum_{n=0}^{\infty} B_n f_n''(x)$$
$$\left. - D_2\left[\sum_{i=1}^{4} \tilde{C}_i u_i'(\alpha x) + \omega_q'(x) + \frac{1}{D_1}\sum_{n=0}^{\infty} A_n \varphi_n'(x) + \frac{1}{D_1}\frac{h_1}{2l}\sum_{n=0}^{\infty} B_n \omega_n'(x)\right]\right\} \tag{30}$$

Equations (19), (20), and (30), which determine the vertical and horizontal displacements of the beam slabs are of a general nature because the patterns of distribution of external loads are still arbitrary.

When solving specific problems corresponding to the given patterns of distribution of external loads, the vertical and horizontal displacements of the slab are determined, satisfying the boundary conditions of the problems under consideration, (i.e., constant integration coefficients are determined from the boundary conditions of the corresponding problems).

Unknown coefficients A_n , B_n are determined from the condition of tight fit of the slab to the elastic half-space, i.e. from the contact condition Eq. (8). Substituting defined coefficients A_n and B_n into Eqs. (9), (10), (16), (17), (19), (20), (30), we obtain a solution to the considered problems; i.e., it is possible to study the patterns of distribution of the reactive pressure of the foundation and the internal forces of the slab under various external factors acting on the two-layer slabs.

3 Results and Discussion

3.1 *Solving the Problem of Bending of Two-Layer Beam Slabs Under Uniformly Distributed Loads*

When solving this problem, it is assumed that

$$q_1(x) = q_2(x) = q = const:$$

Due to the symmetry of the load, Eqs. (9) and (10) take the form

$$p(x) = (1 - x^2)^{-1/2} \sum_{n=0}^{\infty} A_{2n} T_{2n}(x), \tag{31}$$

$$\tau(x) = (1 - x^2)^{-1/2} \sum_{n=0}^{\infty} B_{2n+1} T_{2n+1}(x). \tag{32}$$

From the equilibrium Eq. (13), we obtain

$$A_0 = \frac{4q}{\pi}. \tag{33}$$

Expressions for vertical Eq. (16) and horizontal Eq. (17) displacements of the points of foundation surface are:

$$V = \pi\alpha_1 \left[-A_0 \ln 2 + \sum_{n=1}^{\infty} A_{2n} \frac{T_{2n}(x)}{2n} \right] + 2\alpha_2 \sum_{n=0}^{\infty} B_{2n+1} \frac{U_{2n+1}(x)}{2n+1}, \tag{34}$$

$$U = \pi\alpha_1 \sum_{n=0}^{\infty} B_{2n+1} \frac{T_{2n+1}(x)}{2n+1} - 2\alpha_1 \left[A_0 \arcsin x - \sum_{n=1}^{\infty} A_{2n} \frac{U_{2n}(x)}{2n} \right]. \tag{35}$$

Equations (11) and (12) take the form

$$f_q(x) = \frac{q}{12} x^4; \tag{36}$$

$$\varphi_q(x) = \frac{D_1 - D_2}{D_1 D_2} \frac{q}{4\alpha^4} [1 - u_1(\alpha x)]. \tag{37}$$

Satisfying the boundary conditions, the following expressions can be obtained to determine the slab deflections

$$\begin{aligned} y_1 = & \frac{e^4}{D_1 + D_2} \left\{ \left(\frac{x^2}{2} + \frac{x^4}{24} \right) q + C_4 - \frac{D_2}{D_1} \sum_{n=0}^{\infty} A_{2n} \left[F_{1,2n}(x) + \frac{D_1}{D_2} f_{2n}(x) \right] \right. \\ & \left. - \frac{h_1}{2l} \frac{D_2}{D_1} \sum_{n=0}^{\infty} B_{2n+1} \left[F_{1,2n}(x) + \frac{D_1}{D_2} f'_{2n+1}(x) \right] \right\}, \end{aligned} \tag{38}$$

$$\begin{aligned} y_2 = & \frac{e^4}{D_1 + D_2} \left\{ \left(\frac{x^2}{2} + \frac{x^4}{24} \right) q + C_4 - \sum_{n=0}^{\infty} A_{2n} \left[F_{1,2n}(x) - f_{2n}(x) \right] \right. \\ & \left. + \frac{h_1}{2l} \sum_{n=0}^{\infty} B_{2n+1} \left[F_{1,2n+1}(x) - f'_{2n+1}(x) \right] \right\}. \end{aligned} \tag{39}$$

Here

$$F_{1,2n}(x) = \varphi_{1,2n}u_1(\alpha x) + \varphi_{2,2n}u_3(\alpha x) + \varphi_{2n}(x), \tag{40}$$

$$F_{2,2n+1}(x) = \omega_{1,2n+1}u_1(\alpha x) + \omega_{2,2n+1}u_3(\alpha x) + \omega_{2n+1}(x), \tag{41}$$

$$\begin{aligned}\varphi_{1,2n} =& (8b\alpha^3)^{-1}\int_0^1 \{2u_1(\alpha)u_1[\alpha(1-z)] + u_4(\alpha)u_2[\alpha(1-z)]\}\\ &(1-z^2)^{-1/2}T_{2n}(z)dz.\end{aligned} \tag{42}$$

$$\begin{aligned}\varphi_{2,2n} =& (8b\alpha^3)^{-1}\int_0^1 \{2u_3(\alpha)u_1[\alpha(1-z)] - u_2(\alpha)u_2[\alpha(1-z)]\}\\ &(1-z^2)^{-1/2}T_{2n}(z)dz.\end{aligned} \tag{43}$$

$$\begin{aligned}\omega_{1,2n+1} =& -(b\alpha^2)^{-1}\int_0^1 \{u_1(\alpha)u_4[\alpha(1-z)] - u_4(\alpha)u_1[\alpha(1-z)]\}\\ &(1-z^2)^{-1/2}T_{2n+1}(z)dz.\end{aligned} \tag{44}$$

$$\begin{aligned}\omega_{2,2n+1} =& -(b\alpha^2)^{-1}\int_0^1 \{u_3(\alpha)u_4[\alpha(1-z)] + u_2(\alpha)u_1[\alpha(1-z)]\}\\ &(1-z^2)^{-1/2}T_{2n+1}(z)dz.\end{aligned} \tag{45}$$

$$b = u_1(\alpha)u_2(\alpha) + u_3(\alpha)u_4(\alpha). \tag{46}$$

Integration constants C_4 can be eliminated by passing to the relative deflection of the slabs. Horizontal displacements of the sole of the beam slabs, that satisfy the boundary conditions of the problem, take the following form

$$\begin{aligned}u_\tau =& \frac{h_1}{2l}\frac{l}{D_1+D_2}\left\{-\left(x+\frac{x^3}{6}\right)q + \frac{D_2}{D_1}\sum_{n=0}^{\infty}A_{2n}\left[F'_{1,2n}(x) + \frac{D_1}{D_2}f'_{2n}(x)\right]\right.\\ &\left.+\frac{h_1}{2l}\frac{D_2}{D_1}\sum_{n=0}^{\infty}B_{2n+1}\left[F'_{2,2n+1}(x) + \frac{4D_1+D_2}{3D_2}f''_{2n+1}(x)\right]\right\}.\end{aligned} \tag{47}$$

Substituting Eqs. (34), (35), (38) and (47) into Eq. (8), which are the contact conditions of the problem, multiplying the obtained equations by

$$(1 - x^2)^{-1/2} T_{2k}(x) \quad \text{and} \quad (1 - x^2)^{-1/2} T_{2k-1}(x), \quad k = 1, 2, 3, \ldots$$

then integrating them over the variable x ranging from −1 to 1, and considering the orthogonality of the Chebyshev polynomials, we obtain the following infinite system of algebraic equations with infinite unknowns relative to the unknown coefficients

$$A_{2n}, B_{2n+1} : a_{2k} + \sum_{n=0}^{\infty} A_{2n} a_{2n,2k} + \sum_{n=0}^{\infty} B_{2n+1} a_{2n+1,2k} = \frac{\pi^2 \alpha_1}{4k} A_{2k}, \quad k = 1, 2, 3, \ldots \tag{48}$$

$$b_{2k-1} + \sum_{n=0}^{\infty} A_{2n} b_{2n,2k-1} + \sum_{n=0}^{\infty} B_{2n+1} b_{2n+1,2k-1} = \frac{\pi^2 \alpha_1}{2(2k-1)} B_{2k-1}, \tag{49}$$
$$k = 1, 2, 3, \ldots$$

The resulting Eqs. (48) and (49) are considered together, and the following notation is introduced

$$a_{2k} = \frac{l^4 q}{D_1 + D_2} \int_{-1}^{1} \left(\frac{x^2}{2} + \frac{x^4}{24} \right) (1 - x^2)^{-\frac{1}{2}} T_{2k}(x) dx.$$

$$a_{2n,2k} = -\frac{l^4}{D_1 + D_2} \frac{D_2}{D_1} \int_{-1}^{1} \left[F_{1,2n}(x) + \frac{D_1}{D_2} f'(x) \right] (1 + x^2)^{-\frac{1}{2}} T_{2k}(x) dx$$

$$a_{2n+1,2k} = -\frac{l^4}{D_1 + D_2} \frac{D_2}{D_1} \frac{h_1}{2l} \int_{-1}^{1} \left[F_{2,2n+1}(x) + \frac{D_1}{D_2} f_{2n+1}(x) \right] (1 - x^2)^{-\frac{1}{2}} T_{2k}(x) dx.$$

$$b_{2k-1} = \frac{l^4 q}{D_1 + D_2} \frac{h_1}{2l} \int_{-1}^{1} \left(x + \frac{x^3}{3} \right) (1 - x^2)^{-\frac{1}{2}} T_{2k-1}(x) dx.$$

$$b_{2n,2n-1} = \frac{l^4}{D_1 + D_2} \frac{D_2}{D_1} \frac{h_1}{2l} \int_{-1}^{1} \left[F'_{1,2n}(x) + \frac{D_1}{D_2} f'_{2n}(x) \right] (1 - x^2)^{-\frac{1}{2}} T_{2k-1}(x) dx.$$

$$b_{2n+1,2k-1} = \frac{l^4}{D_1 + D_2} \frac{D_2}{D_1} \left(\frac{h_1}{2l} \right)^2 \int_{-1}^{1} \left[F'_{2,2n+1}(x) + \frac{4D_1 + D_2}{3D_2} f''_{2n+1}(x) \right] (1 - x^2)^{-\frac{1}{2}} T_{2k-1}(x) dx.$$

Applying the method of integration by parts, it is possible to eliminate the singularities of the integrals, which acquire a convenient form for the implementation on a computer. Thus, the solution of the problem of bending of two-layer beam slabs lying on a linearly deformed half-space, in the presence of a friction force in the contact between the layers of the slab and the foundation, is reduced to solving an infinite system of algebraic equations with infinite unknowns.

3.2 Numerical Example

The following characteristics for the slab and foundation are taken:

$$l = 500\,sm,\ h_1 = h_2 = 45\,sm,\ \nu_1 = \nu_2 = 0.167,\ E_1 = E_2 = 1.25 \cdot 10^5 kg/sm^2,$$
$$\nu_0 = 0.3,\ E_0 = 5 \cdot 10^2 kg/sm^2$$

We restrict ourselves to the first four terms in Eq. (31) since the coefficient of the first term A_0 is known. In Eq. (32), we restrict ourselves to the first three terms. Then, under such assumptions, the system of algebraic Eqs. (48) and (49) is reduced to a system of six equations with six unknowns $A_2, A_4, A_6, B_1, B_3, B_5$. The numerical values of these coefficients corresponding to different values of the filler coefficient "*k*" are shown in Table 1.

Analysis of the results given in Table 1 shows that:

- changes in the numerical values of the stiffness coefficients of the filler, does not lead to a significant change in the values of the solution of algebraic equations;
- changes in the numerical values of the stiffness coefficients of the filler do not significantly affect the distribution of reactive pressures of the foundation;
- retention of 3 terms in the Eqs. (31) and (32) when searching for a solution to the problem of assessing the internal force factors of slab-strips provides sufficient accuracy of solutions.

Table 2 shows the maximum values of bending moments (M_1 and M_2) in the middle of the slab for different values of the filler coefficient "*k*". The maximum values of the bending moments formed in the middle of the slab that correspond to different values of the filler coefficient are given in Table 2.

Analysis of the change in the values of bending moments (Table 2) corresponding to the section $x = 0$, at different values of the filler coefficient "*k*" shows that:

- an increase in the numerical values of the stiffness coefficients of the filler leads to a convergence of the values of bending moments.
- a consideration of the shear stress between the slab and the foundation leads to a significant (up to 24%) reduction in the bending moments of a slab.

Table 1 The value of the unknown coefficients at different values of the filler coefficient

$k\ (kg/sm^3)$	0.25	1.00	2.50	10.00	25.00
A_0/q	1.273239	1.273239	1.273239	1.273239	1.273239
A_2/q	−0.308264	−0.303361	−0.300591	−0.300161	−0.209873
A_4/q	−0.004023	−0.004161	−0.004219	−0.004288	−0.004307
B_1/q	0.817629	0.803259	0.765981	0.726177	0.709615
B_3/q	−0.427618	−0.421134	−0.407619	−0.361219	−0.290124
B_5/q	−0.044363	−0.041242	−0.038764	−0.035947	−0.030967

Table 2 The maximum values of bending moments at different values of the filler coefficient "*k*"

$k\ (kg/sm^3)$	The maximum values of bending moments of a slab	
	$M_1(0)/(ql^2)$	$M_2(0)/(ql^2)$
0.25	0.025764	0.016191
1.00	0.24519	0.017284
2.5	0.022071	0.018969
10.0	0.021105	0.019703
25.0	0.020387	0.019986

4 Conclusions

1. A mathematical model was developed to study the internal force factors of multilayer slab-strips interacting with an elastic foundation, considering the shear stresses arising in contact with the foundation.
2. An analytical method for solving the problem was proposed to assess the internal force factors of multilayer slab-strips, based on the approximation of the orthogonal Chebyshev polynomials.
3. The sufficiency of the number of terms of the Chebyshev polynomial to obtain a result with satisfactory accuracy was established with test examples.
4. It was determined that an account of the shear stresses arising in the contact of multilayer slabs with the foundation leads to a decrease in force factors in the slab-strip.

References

1. Butenko Y (2002) Perturbation method in integrating the equations of bending of multilayer structures. Bull High Educ Institutions 2(5):3–6
2. Starovoitov EI, Yarovaya AV, Leonenko DV (2006) Deformation of three-layer structural elements on an elastic foundation. Fizmatlit, Moscow
3. Shirinkulov TSh, Zaretsky YK (1986) Creep and soil consolidation. Fan, Tashkent
4. Popov GY (1982) Concentration of elastic stresses near stamps of sections of thin inclusions and reinforcements. Nauka, Moscow
5. Awrejcewicz J, Krysko VA, Zhigalov MV, Krysko AV (2018) Contact interaction of two rectangular plates made from different materials with an account of physical nonlinearity. Nonlinear Dyn (91):1191–1211
6. Vaisfel'D ND, Popov GY, Reut VV (2013) The axisymmetric mixed problem of elasticity theory for a cone clamped along its side surface with an attached spherical segment. J Appl Math Mech 1(77):70–78
7. Krenev LI, Aizikovich SM, Tokovyy YV, Wang YC (2015) Axisymmetric problem on the indentation of a hot circular punch into an arbitrarily nonhomogeneous half-space. Int J Solids Struct (59):1–9
8. Gabbasov RF, Uvarova NB (2012) Application of the generalized equations of the finite difference method to the calculation of plates on an elastic foundation. Vestnik FSBEI HPE (4):12–20

9. Mirsaidov MM, Mamasoliev K (2020) Contact problems of slabs interaction on an elastic foundation. In: ICECAE 2020, pp 1–12. IOP Conference Series: Materials Science and Engineering, Tashkent
10. Younesian D, Hosseinkhani A, Askari H, Esmailzadeh E (2019) Elastic and viscoelastic foundations: a review on linear and nonlinear vibration modeling and applications. Nonlinear Dyn (97):853–895
11. Tokovyy Y, Ma CC (2019) Elastic analysis of inhomogeneous solids: history and development in brief. J Mech 35(5):1–14
12. Takekawa J, Mikada H (2016) An absorbing boundary condition for acoustic-wave propagation using a mesh-free method. Geophysics (2017):4231–4235
13. Usarov M, Salokhiddinov A, Usarov DM, Khazratkulov I, Dremova N (2020) To the theory of bending and oscillations of three-layered plates with a compressible filler. In: FORM-2020, pp. 1–13, IOP Conference Series: Materials Science and Engineering
14. Sultanov KS, Kumakov JX, Loginov PV, Rikhsieva BB (2020) Strength of underground pipelines under seismic effects. Mag Civil Eng (9):97–120
15. Mavlonov T, Yuldoshev B, Ismayilov K, Toshev S (2020) Compressed rectangular plates stability beyond the elastic limit. In: CONMECHYDRO-2020, pp 1–8. IOP Conference Series: Materials Science and Engineering
16. Indiaminov R, Butaev R, Isayev N, Ismayilov K, Yuldoshev B, Numonov A (2020) Nonlinear integro-differential equations of bending of physically nonlinear viscoelastic plates. In: FORM-2020, pp 1–9, IOP Conference Series: Materials Science and Engineering
17. Mirsaidov M (2019) An account of the foundation in assessment of earth structure dynamics. In: Form-2019, pp 1–11, E3S Web of Conferences
18. Mirsaidov MM, Toshmatov ES (2019) Spatial stress state and dynamic characteristics of earth dams. Mag Civil Eng 89(5):3–15
19. Bakhodirov AA, Ismailova SI, Sultanov KS (2015) Dynamic deformation of the contact layer when there is shear interaction between a body and the soil. J Appl Math Mech 79(6):587–595
20. Mirsaidov MM, Dusmatov OM, Khodjabekov MU (2020) The problem of mathematical modeling of a vibration protected rod under kinematic excitations. In: Proceedings of VII International Scientific Conference Integration, pp 1–7. Tashkent
21. Sultanov KS, Vatin NI (2021) Wave theory of seismic resistance of underground pipelines. Appl Sci 11:1785–1797
22. Suetin PK (2007) Classical orthogonal polynomials. Fizmatlit, Moscow
23. Abramovitsa M, Stigan I (1979) Handbook of special functions. Nauka, Moscow

The New Linear Deformations Hypothesis of Reinforced Concrete Under Combined Torsion and Bending

Vladimir Kolchunov, Alexey Demyanov, Vyacheslav Shankov, and Sergey Grichishnikov

Abstract The authors found a simple method from a family of mesh methods for developing linear deformation functionals by approximating rectangular sections in compressed and stretched regions. The general undefined functional is a complex function containing three function. The displacement is denoted as absolute deformations under the action of the deplanation of the cross section. The new linear deformation hypothesis was proposed. The new hypothesis of linear deformations and its filling of the diagram under combined torsion and bending was presented. The proposed new hypothesis is the kinematics between the fibers for the relative longitudinal fibers of the upper and lower deformations of concrete and reinforcement for determination the ratios at a distance from the neutral axis. The new hypothesis is similar to the modernized Bernoulli hypothesis, but has a special geometric figure for the summed function of additional deformation, as well as a parameter between the elastic and plastic regions of concrete to obtain the elastic deformation equation. The analysis of the new functional with the approximation of the Timoshenko-Goodyear function is carried out. The error in finding the value of the functional is 2% at the points considered and 7% at any points of the cross section. The authors recorded the bending moment, defined for any square or undefined for any point. The fillings of the diagram are in the form of expressions using functionals. A complex analysis was carried out with approximation, functionals, graphs.

Keywords Beams · Reinforced concrete · Bending moment · Torsion · Relative strains

---

V. Kolchunov (✉) · A. Demyanov · S. Grichishnikov
South West State University, 50 let Oktyabrya Street, 94, Kursk 305040, Russian Federation

V. Shankov
Bryansk State Engineering and Technological University, Stanke Dimitrov prospectus, 3, Bryansk 241037, Russian Federation

N. Vatin et al. (eds.), *Proceedings of MPCPE 2021*, Lecture Notes in Civil Engineering 182, https://doi.org/10.1007/978-3-030-85236-8_9

1 Introduction

In widespread practice in the design and construction of reinforced concrete structures experiencing a complex stress state-torsion with bending, the methods of resistance calculation of such structures, considered in scientific publications [1–9] and used in the regulatory documents of different countries, remain not sufficiently strict. Truss analogy models are used up to the present time.

The problem of defining deformations in torsion with bending requires research. Experimental research is being developed [10–24], modeling is carried out taking into account the stress state of the process of deformation of reinforced concrete structures with a combination of bending and torsion. This is due to different strength classes of concrete, different rectangular cross-sections and the need to develop a design scheme and take into account a number of new effects of deformation of reinforced concrete with spatial cracks.

The purpose of this research is to develop a new hypothesis of linear deformations in reinforced concrete and to determine the filling of the diagram in bending with torsion.

2 Methods

The authors considered the approach of constructing a model for cross-sections of reinforced concrete structures in the new hypothesis of linear deformations and its filling of the diagram in bending with torsion. A complex analysis with approximation, functionals, graphs has been carried out. The authors have found a simple new way from the families of the method of meshes for the development of functionals.

New first and second functionals were considered (see formulas (1) and (4)) for approximating any rectangular mean sections in compressed and stretched zones using squares. For the second iteration and correction of the values of the obtained function, the steps taken were repeated again using a more frequent division of the cross section using other points (Fig. 1b).

For this, the **first function** was previously recorded as a horizontal parabola relative to the y-axis—$f_{5,n,*}(y)$, and the second function as a parabola relative to the z-axis—$f_2(z)$.

We got a new first functional from both functions of the system after two iterations. Received the analytical first functional $f_{5,*}(y,z)$ using a horizontal parabola about the y-axis $f_{1,*}(y)$ and vertical parabola about the z-axis: $f_{2,*}(z)$:

$$\begin{aligned} f_{5,*}(y,z) &= f_{1,*}(y)\blacklozenge f_{2,*}(z) \\ &= \pm\left[-\frac{3(47b^2-200y^2)}{25b^2h^2}\cdot z^2+\frac{487b^2-2280y^2}{500b^2h}\cdot z+0.923-\frac{93y^2}{25b^2}\right] \end{aligned} \tag{1}$$

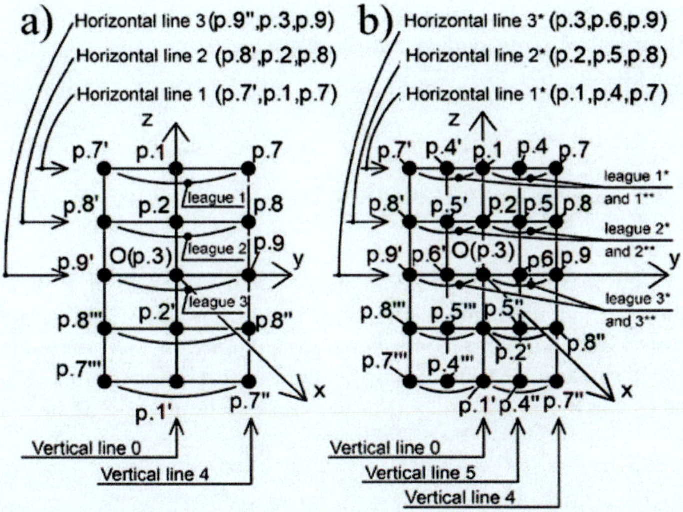

Fig. 1 Functionals for the primary coarse mesh partition (**a**) and for the division into smaller mesh parts (**b**) lines—horizontal and vertical; leagues (i.e. functions along the y or z axis)—between points

Here ♦—a transition operation between functions; for quadrants I, III and II, IV,—signs «+» and «−» are taken respectively.

$$A = -\frac{3(47b^2 - 200y^2)}{25b^2h^2}; \quad B = \frac{487b^2 - 2280y^2}{500b^2h}; \quad C = 0.923 - \frac{93y^2}{25b^2}$$

Then we got an analytic indefinite second functional (a function of three functions).

The elementary exponential first function has the form:

$$f_1(x) = \mathrm{e}^{-\lambda\left(\frac{x}{l}+A\right)} + C \tag{2}$$

Here A, λ, C are the coefficients of the function in formula (2).

The second function is a straight line from z (see Fig. 1,—points 9):

$$f_2(z) = \frac{A_{line1} \cdot z - A_{line2} \cdot z + A_{line2} \cdot z_{line1} - A_{line1} \cdot z_{line2}}{z_{line1} - z_{line2}} \tag{3}$$

A_{line1} and A_{line2} are the value of the function at the points of one vertical column (Fig. 1), but in different lines 1 and 2, for different values of the z coordinate (z_{line1} and z_{line2} respectively).

The third function y has the form of a curve (parabola y,—Fig. 1).

Substituting the corresponding values, we have defined the analytic undefined second functional:

$$f_{***}(x, y, z) = \mathrm{e}^{-3.19\left(\frac{x}{l} + \frac{3.04h - 22.96z}{b^2 h}\cdot (y)^2 - \frac{2.00h - 10.3z}{bh}\cdot y - \frac{0.31h - 0.36z}{h}\right)} - \frac{9.39h - 27.02z}{b^2 h}\cdot (y)^2 - \frac{7.16h - 17.39z}{bh}\cdot y - \frac{-0.306h + 0.232z}{h} \quad (4)$$

Here are the first and second iterations (see Fig. 1) and so on n iterations. The first function is $f_1(x)$, the second is $f_2(z)$ and the third is $f_3(y)$. The first iteration has an error of 15% and the second iteration has 2%. We denoted the absolute deplanation located at a distance x from the support as w (from [25]). The displacement U due to the deplanation of the cross section was written in the form

$$\varepsilon_{x,d} = \frac{\partial U}{\partial x} = D_1 \cdot y \cdot z\left[-D_2 \cdot x \cdot \mathrm{e}^{-\lambda_{***}\left(\frac{x}{l} + A_{***}\right)} + D_3 \cdot \mathrm{e}^{-\lambda_{***}\left(\frac{x}{l} + A_{***}\right)} + D_4\right] \quad (5)$$

Here $D_1 = \frac{M_t}{G_{rec} \cdot I_{rec}} \cdot \frac{a_*^2 - b_*^2}{a_*^2 + b_*^2}$; $D_2 = \frac{\lambda_{***}}{l^2}$; $D_3 = \frac{1}{l}$; $D_4 = \frac{C_{***}}{l}$; a_* and b_*—parameters equal to half the height and width of the section, respectively (see [25]) and the second functional; $C_{***}(y, z) = \frac{9.39h - 27.02z}{b^2 h} \cdot (y)^2 - \frac{7.16h - 17.39z}{bh} \cdot y - \frac{-0.306h + 0.232z}{h}$.

We have signs («+» and «−») taken for the corresponding quadrants. For example, the functional is accepted with the "−" sign in section 1–1 at $z = h/4$, in section 3–3 at $z = h/2$, etc.

3 Results

The authors developed a new hypothesis and gave its definition as a result of the analysis of functionals and the use of functionals in determining deformations. The **new hypothesis of the linear deformation**—the kinematics between fibers for the relative longitudinal fiber upper and lower deformations of concrete and reinforcement ($\varepsilon_{x,b}$ and $\varepsilon_{x,s}$) for their ratios in distances from the neutral axis, similar to the modernized Bernoulli hypothesis, but with special geometric shapes for functions $f_{sum,\Delta-d}$ and compressed concrete between plastic and elastic regions to obtain an equation with deformation $\varepsilon_{b,el}$.

A deformation curve was constructed (deplanation subtraction from the Bernoulli hypothesis from triangles). For this, we added the values of the relative longitudinal deformations $\varepsilon_{b,x}$ found from the new first hypothesis and the relative longitudinal deformations of the deplanation $\varepsilon_{x,d}$ found from the functional:

$$\varepsilon_{b,x,sum} = \varepsilon_{b,x} \pm \varepsilon_{x,d} \tag{6}$$

$$\frac{\varepsilon_{b,x}}{(\varepsilon_s - \varepsilon_0) \cdot \psi_s} = \frac{x}{h_0 - x} \tag{7}$$

Knowing the point $A_{1,u}(z_c = 0.5h;\ \varepsilon_{b,u})$ and $A_{2,s}(-[h_0 - z_c]; -\varepsilon_{s,m})$ we got:

$$\pm\varepsilon_x = \frac{\varepsilon_{s,m}(z - z_c) + \varepsilon_{b,u}(h_0 + z - z_c)}{h_0} \cdot b|_{y=const} \tag{8}$$

Here $y = const$ (i.e., discrete for its required point); $\varepsilon_{s,m} = (\varepsilon_s - \varepsilon_0) \cdot \psi_s$—average deformation of reinforcement, ψ_s—tensile strength of concrete through its parameter; ε_0—deformation of prestress.

We have a bending moment in the cross section (section 1–1 and section 3–3 for a block, a compartment with a spatial crack). In compressed concrete, the filling of the diagram was obtained in the form of a function (undefined) or an integral of the lower and upper limits (definite).

An indefinite bending moment $M_{bend,x}$ from compressed concrete to fill the diagram ω_ε (or ω_σ) or a certain (in the form of a number) moment $M_{bend,def,x}$ were determined in the cross section for external moments $M_{1-1} = R_{\text{sup}} \cdot a_1$, $M_{3-3} = R_{\text{sup}} \cdot a_3$ (i.e. $R_{\text{sup}} \cdot x$) and internal moments (from the diagram):

$$M_{bend,x} = \sigma_{b,x} \cdot \omega_\sigma(x,y,z) \cdot A_b(z) \cdot z_{b,\sigma}(z) = \varepsilon_{b,x} \cdot E_b(\lambda) \cdot \omega_\varepsilon(x,y,z) \cdot A_b(z) \cdot z_{b,\varepsilon}(z) \tag{9}$$

or

$$M_{bend,def,x} = \sigma_{b,x,u} \cdot \omega_\sigma \cdot A_b \cdot z_{c,\sigma} = \varepsilon_{b,x,u} \cdot E_b(\lambda) \cdot A_b \cdot z_{c,\varepsilon} = \varepsilon_{b,x,u} \cdot E_b \cdot \omega_\varepsilon \cdot A_b \cdot z_{c,\varepsilon} \tag{9*}$$

a_1 and a_3—the distances from the support R_{sup} to the cross-section 1–1 and 3–3; $\sigma_{b,x}$, $\omega_\sigma(x,y,z)$, $A_b(z)$, $z_{b,\sigma}(z)$, $\varepsilon_{b,x}$, $E_b(\lambda)$, $\omega_\varepsilon(x,y,z)$, $z_{b,\varepsilon}(z)$—parameters for an indefinite moment: stress, filling of the stress diagram, area of compressed concrete, shoulder for the i-th elementary stress from a point A_i to the neutral axis, deformations, secant modulus, filling of the deformation diagram, shoulder for the i-th elementary deformation from a point A_i to the neutral axis; $\sigma_{b,x,u}$, ω_σ, A_b, $z_{c,\sigma}$—parameters (numbers) for a certain moment: stress, filling of the deformation diagram, area of compressed concrete, shoulder from the center of the general diagram to the neutral axis.

The spatial new first deformation hypothesis is:

$$\begin{aligned} \pm\, \varepsilon_{x,*}(z,y,x) &= \frac{\varepsilon_{s,m}(z-z_1)+\varepsilon_{b,u}(h_0+z-z_1)}{h_0}\cdot b|_{y=const}\cdot\left[\varepsilon_{b,x}\right]_1\cdot x \\ &= \frac{\varepsilon_{s,m}(z-z_1)+\varepsilon_{b,u}(h_0+z-z_1)}{h_0}\cdot b|_{y=const}\cdot\frac{R_{\text{sup}}\cdot 1}{E_b\cdot\nu_b\cdot\omega_\varepsilon\cdot z_c}\cdot x \end{aligned} \quad (10)$$

Here $\left[\varepsilon_{b,x}\right]_1 = \frac{R_{\text{sup}}\cdot 1}{E_b\cdot\nu_b\cdot\omega_\varepsilon\cdot z_c}$; instead of $z_1 = z_c = z_u$, i.e. ∾—analogue of the distance x_u for compressed concrete to the neutral axis.

Adding $\varepsilon_{b,x}(z,y,x)$ i.e. formula (10) and $\varepsilon_{x,d}$:

$$\varepsilon_{b,x}(z,y,x) \pm \varepsilon_{x,d}(x,y,z) = \varepsilon_{b,x,sum}(x,y,z) \quad (11)$$

Quadrants I, II, III, IV with signs «+» or «−» were designated as: $B_1 = \frac{\varepsilon_{s,m}}{h_0}$; $B_2 = \frac{\varepsilon_{b,u}}{h_0}$; $B_3 = b|_{y=const} = b$; $B_4 = \left[\varepsilon_{b,x}\right]_1 = \frac{R_{\text{sup}}\cdot 1}{E_b\cdot\nu_b\cdot\omega_\varepsilon\cdot z_c}$.

Then:

$$\pm\varepsilon_{b,x}(z,y,x) = \pm[B_1\cdot(z-z_c)+B_2\cdot(h_0+z-z_c)]\cdot B_3\cdot B_4\cdot x \quad (12)$$

Subtraction of a geometric figure for function (13) and from a triangle This special function is obtained (13) $f_{sum,\Delta-d}$:

$$\begin{aligned} f_{sum,\Delta-d} &= \varepsilon_{x,sum}(x,y,z) = \pm[B_1\cdot(z-z_c)+B_2\cdot(h_0+z-z_c)]\cdot B_3\cdot B_4\cdot x \\ &\pm D_1\cdot y\cdot z\left[-D_2\cdot x\cdot \mathrm{e}^{-\lambda_{***}\left(\frac{x}{l}+A_{***}\right)}+D_3\cdot \mathrm{e}^{-\lambda_{***}\left(\frac{x}{l}+A_{***}\right)}+D_4\right] \end{aligned} \quad (13)$$

Here $B_1 = \frac{\varepsilon_{s,m}}{h_0}$; $B_2 = \frac{M_{bend}}{\nu_b\cdot E_b\cdot A_b\cdot z_b}$—**II group**, through bending indefinite moment M_{bend}, (strength—$B_2 = \frac{\varepsilon_{b,u}}{h_0}$); $B_3 = b|_{y=const} = b$;$B_4 = \left[\varepsilon_{b,x}\right]_1 = \frac{R_{\text{sup}}\cdot 1}{E_b\cdot\nu_b\cdot\omega_\varepsilon\cdot z_c}$; $D_1 = \frac{M_t}{G_{rec}\cdot I_{rec}}\cdot\frac{a_*^2-b_*^2}{a_*^2+b_*^2}$; $D_2 = \frac{\lambda_{***}}{l^2}$; $D_3 = \frac{1}{l}$; $D_4 = \frac{C_{***}}{l}$; $A_{***}(y,z) = \frac{3.84h-22.96z}{b^2h}\cdot(y)^2 - \frac{2.88h-12.3z}{bh}\cdot y - \frac{0.34h-0.36z}{h}$; $C_{***}(y,z) = \frac{9.39h-27.02z}{b^2h}\cdot(y)^2 - \frac{7.16h-17.39z}{bh}\cdot y - \frac{-0.306h+0.232z}{h}$.

We get the proportion for the kinematic upper and lower fibers:

$$\frac{f_{sum,\Delta-d,up}}{z_c} = \frac{f_{sum,\Delta-d,d}}{h_0-z_c} \quad (14)$$

The equation for finding λ_* was used from (13). For point 2 (x = a_i—distance from support to cross-section; y; $z_c - \lambda_*\cdot z_c$):

$$
\begin{aligned}
f_B(p2) &= \pm[B_1 \cdot (z - z_c) + B_2 \cdot (h_0 + z - z_c)] \cdot B_3 \cdot B_4 \cdot x \pm D_1 \cdot y \cdot z \\
&\left[-D_2 \cdot x \cdot \mathrm{e}^{-\lambda_{***}\left(\frac{x}{l}+A_{***}\right)} + D_3 \cdot \mathrm{e}^{-\lambda_{***}\left(\frac{x}{l}+A_{***}\right)} + D_4\right] \\
&= \pm[B_1 \cdot (-\lambda_* z_c) + B_2 \cdot (h_0 - \lambda_* z_c)] \cdot B_3 \cdot B_4 \cdot a_i \pm D_1 \cdot y \cdot (z_c - \lambda_* z_c) \\
&\left[-D_2 \cdot a_i \cdot \mathrm{e}^{-\lambda_{***}\left(\frac{x}{l}+A_{***}\right)} + D_3 \cdot \mathrm{e}^{-\lambda_{***}\left(\frac{x}{l}+A_{***}\right)} + D_4\right]
\end{aligned}
\tag{15}
$$

Took $f_B(p2) = \varepsilon_{x,el}$:

$$
\begin{aligned}
\varepsilon_{x,el} = &\pm [B_1 \cdot (-\lambda_* z_c) + B_2 \cdot (h_0 - \lambda_* z_c)] \cdot B_3 \cdot B_4 \cdot a_i \pm D_1 \cdot y \cdot (z_c - \lambda_* z_c) \\
&\left[-D_2 \cdot a_i \cdot \mathrm{e}^{-\lambda_{***}\left(\frac{x}{l}+A_{***}\right)} + D_3 \cdot \mathrm{e}^{-\lambda_{***}\left(\frac{x}{l}+A_{***}\right)} + D_4\right]
\end{aligned}
\tag{16}
$$

Here $B_2 = \frac{\varepsilon_{b,u}}{h_0}$ and $B_1, B_2, B_3, B_4, D_1, D_2, D_3, D_4, C_{***}(y, z)$—from (13).

λ_* follows from (16) (the elastic-plastic area, see Fig. 2b, d).

We received the first jump (Fig. 2c), downward from the upper face of the structure, from the spatial crack of the upper compressed area for x_b. First jump is function $f_{sum,\Delta 1,\Delta-d,crc}$, $\Delta 1$—first jump function, Δ—deformations between a curved line and a straight line:

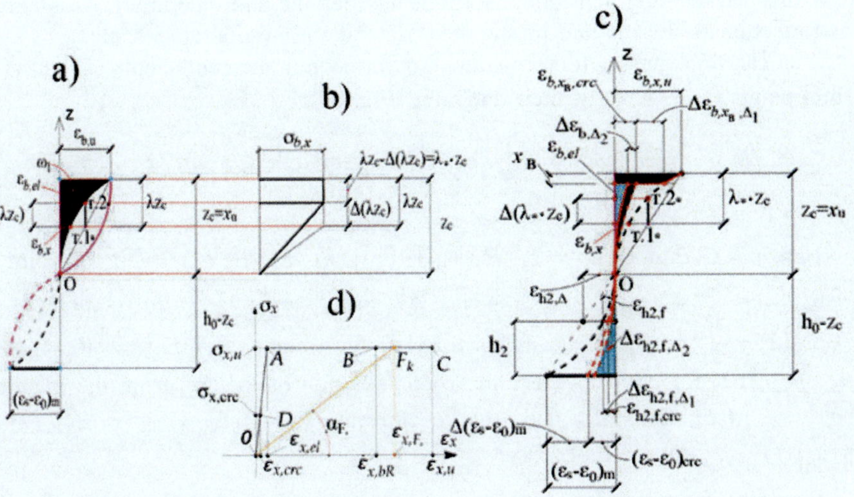

Fig. 2 Diagram of deformation of longitudinal concrete and reinforcement for stage I—(**a**); elastoplastic stress diagram of compressed concrete—(**b**); stages II and III—(**c**); diagram«$\varepsilon_x - \sigma_x$»—(**d**); $\Delta\varepsilon_{b,x_b,\Delta_1}$—the first leap; $\Delta\varepsilon_{b,\Delta_2}$—the second leap

$$f_{sum,\Delta_1,\Delta-d,crc} = \varepsilon_{x,sum,crc} = \pm\left[B_{1,crc}\cdot(z-(z_c-x_B)) + B_{2,crc}\cdot(h_0+z-z_c)\right]\cdot B_{3,crc}\cdot B_{4,crc}\cdot x \pm D_{1,crc}\cdot y\cdot z\left[-D_{2,crc}\cdot x\cdot \mathrm{e}^{-\lambda_{***}\left(\frac{x}{l}+A_{***}\right)} + D_{3,crc}\cdot \mathrm{e}^{-\lambda_{***}\left(\frac{x}{l}+A_{***}\right)} + D_{4,crc}\right] \quad (17)$$

Here $B_{1,crc} = \frac{\varepsilon_{s,m,crc}}{h_0 - x_B}$; $B_{2,crc} = \frac{\varepsilon_{b,x_B,crc}}{h_0 - x_B}$; $B_{3,crc} = y$; $B_{4,crc} = \left[\varepsilon_{b,x,crc}\right]_1 = \frac{R_{\sup,crc}\cdot 1}{E_{b,crc}\cdot\omega_\varepsilon\cdot(z_c - x_B)}$; $D_{1,crc} = \frac{M_{t,crc}}{G_{rec}\cdot I_{rec}}\cdot\frac{a_*^2 - b_*^2}{a_*^2 + b_*^2}$; $D_{2,crc} = D_2 = \frac{\lambda_{****}}{l^2}$; $D_{3,crc} = D_3 = \frac{1}{l}$; $D_{4,crc} = D_4 = \frac{C_{***}}{l}$.

We received the second jump from the extreme lateral border of the structure to the inner area and from the spatial crack (Fig. 2c). The second jump is a function $f_{sum,\Delta_2,\Delta-d,tr}$.

During the formation of a crack the function of longitudinal deformations from the second jump has the form:

$$f_{\Delta_2} = f_{sum,\Delta-d} - f_{sum,\Delta_1,\Delta-d,crc} \quad (18)$$

Here $f_{sum,\Delta_1,\Delta-d,crc}$—first fump (17), $f_{sum,\Delta-d}$—function (13).

The function of deformations when a second lateral, second normal crack occurs (a second jump also):

$$\begin{aligned} f_{sum,\Delta_2,\Delta-d,tr} &= f_{sum,\Delta_1,\Delta-d,crc} - f_{\Delta_2} \\ &= f_{sum,\Delta_1,\Delta-d,crc} - f_{sum,\Delta-d} + f_{sum,\Delta_1,\Delta-d,crc} = 2f_{sum,\Delta_1,\Delta-d,crc} - f_{sum,\Delta-d} \end{aligned} \quad (19)$$

We also obtained the coefficients φ_{ij} for the projection of the principal stresses and strains for compressed concrete according to normal and tangential diagrams. We obtained bending and torsional moments (definite and indefinite), its integrals and functionals. Elastic and plastic areas, lateral and normal cracks, etc.

You can find stresses (deformations), by projecting the coefficients φ_{ij} and $\nu(\lambda)$ from points C, B, A, D of their diagrams (Fig. 2d):

$$\varphi_{\sigma_x j} = \frac{\sigma_{x,j}}{\sigma_{1,j}} = \frac{\sigma_{x_1,j}\cos^2\alpha + \sigma_{z_1,j}\sin^2\alpha + \tau_{z_1,x_1j}\sin 2\alpha}{1,149R_b} \quad (20)$$

Here $j = C, B, A, D$; $\sigma_x = \varepsilon_x\cdot E_b(\lambda) = \varepsilon_x\cdot\nu_b(\lambda)\cdot E_b$; $E_{b,j}(\lambda) = \frac{\sigma_{b,i}}{\varepsilon_{b,i}}$; for point C—$E_{b,u}(\lambda) = \frac{\sigma_{i,u}}{\varepsilon_{i,u}} = \frac{R_b}{0.0035}$, $\mu_{b,u}(\lambda) = 0.357$, $k_{*,u} = \frac{M_{bend,u}}{M_{bend,crc}}$; for point B—$E_{bR}(\lambda) = \frac{\sigma_{i,bR}}{\varepsilon_{i,bR}} = \frac{R_b}{0.0020}$, $\mu_{bR}(\lambda) = 0.251$; for point A—$E_{b,el} = \frac{\sigma_{i,el}}{\varepsilon_{i,el}} = \frac{R_b}{0.0015}$, $\mu_{b,el} = 0.167$; for point F_k (second group,—section of the AC from the diagram "$\sigma_x - \varepsilon_x$" (see Fig. 2d) and the coefficient φ_{ij})—$tg\alpha_{F_k} = \frac{\sigma_{x,u}}{\varepsilon_{x,F_k}} = E_{b,F_k}(\lambda) = E_b\nu_{b,F_k}(\lambda)$, $\nu_{b,F_k}(\lambda) = \frac{\sigma_{x,u}}{\varepsilon_{x,F_k}\cdot E_b}$.

Similarly, for any point, we got ε_x, ε_{x1}, ε_z, ε_{z1}, γ_{sum}, $\gamma_{sum,\alpha}$ and etc. We also got ν_{ij}, coefficients $\nu_{F_k,i}(\lambda)$ (for points C, B, A, D).

When calculating the first group (according to the proposed model, for the criterion of strength by ultimate shortening of deformations), longitudinal deformations are calculated proportionally in the transition from section k–k to section I–I.

In the second group, the stresses can be found by carrying out a proportional transition to the shortening deformations from section I–I to section k–k with known R_{sup} and moment M_{bend}.

Relative mean deformations of reinforcement between a spatial crack is $\varepsilon_{s,m,x}$, $\varepsilon_{s,m,x,lef}$, $\varepsilon_{s,m,x,rig}$, $\varepsilon_{s,m,x,i}$; coefficients for the resistance of tensile concrete between spatial cracks of reinforced concrete structures in bending with torsion is ψ_s, $\psi_{s,lef}$, $\psi_{s,rig}$ for left or right reinforcement or $\psi_{s,k}$ for any k-th reinforcement.

We have definite and indefinite bending (or twisting) moments for its deformation (linear and shears) or stresses (normal and tangential), as well as filling diagrams of their deformations ω_ε, $\omega_{\gamma,sum}$ and stresses ω_σ, $\omega_{\tau,sum}$.

The bending moment to determine the deformation and stress is calculated by (9*). λ_* for compressed concrete is calculated by (16).

We get the shoulder $z_{b,*j}$:

$$z_{b,*} = z_c - \frac{\sum_{j=1}^{j} S_j}{\sum_{j=1}^{j} A_j} = z_c - \frac{\sum_{j=1}^{j} \left(\lambda_* \cdot z_c \cdot b \cdot (z_c - 0.5 \cdot \lambda_* \cdot z_c) + 0.5 \cdot (z_c - \lambda_* \cdot z_c) \cdot b \cdot \frac{2}{3}(z_c - \lambda_* \cdot z_c)\right)_j}{\sum_{j=1}^{j} (\lambda_* \cdot z_c \cdot b + 0.5 \cdot (z_c - \lambda_* \cdot z_c) \cdot b)_j} \tag{21}$$

Here A_j and S_j—j-th areas and static moments for compressed concrete (elastic, plastic deformation and crack,—lateral, normal, etc.); $\omega_{\varepsilon,def}$—defined filling factor for the deformation diagram; $\omega_{\sigma,def}$—same for stress diagram (using diagram «$\sigma_x - \varepsilon_x$»).

Indefinite bending moment for its deformation or stress from (9) and using the new functional:

$$\begin{aligned} M_{bend,i} &= v_b(\lambda) \cdot E_b \cdot z_{b,i} \cdot \iiint f_{sum,\Delta-d} dxdydz \\ &= v_b(\lambda) \cdot E_b \cdot z_{b,\varepsilon,i} \cdot f_{\varepsilon,\mu ltimap\ \text{int,vol}}(x, y, z) \end{aligned} \tag{22}$$

Here function $f_{sum,\Delta-d}$ (17) is calculated over the volume. We got another function $f_{\varepsilon,\mu ltimap\ \text{int,vol}}(x, y, z)$ after integration.

We can also find the filling of the diagram $\omega_{\varepsilon,i}(x, y, z)$ from the undefined functional for any small squares ($z_{b,\varepsilon,i}(x, y, z)$(or $z_{b,\sigma,i}(x, y, z)$) the shoulder from the point Aj to the neutral axis). The filling of the diagram was obtained from its function after some algebraic transformations for any squares, i.e. field of squares

$$\omega_{0,i}(x, y, z) = \frac{\int f_{sum,\Delta-d} dxdydz}{\varepsilon_{b,x} \cdot A_{b,c}} \tag{23}$$

Here $A_{b,j} = \Delta h_j \cdot \Delta b_j$—area of one square; $\omega_\sigma(x, y, z)$—filling the indefinite diagrams of normal stresses from functional ($\int f_{sum,\Delta-d} dxdydz$).

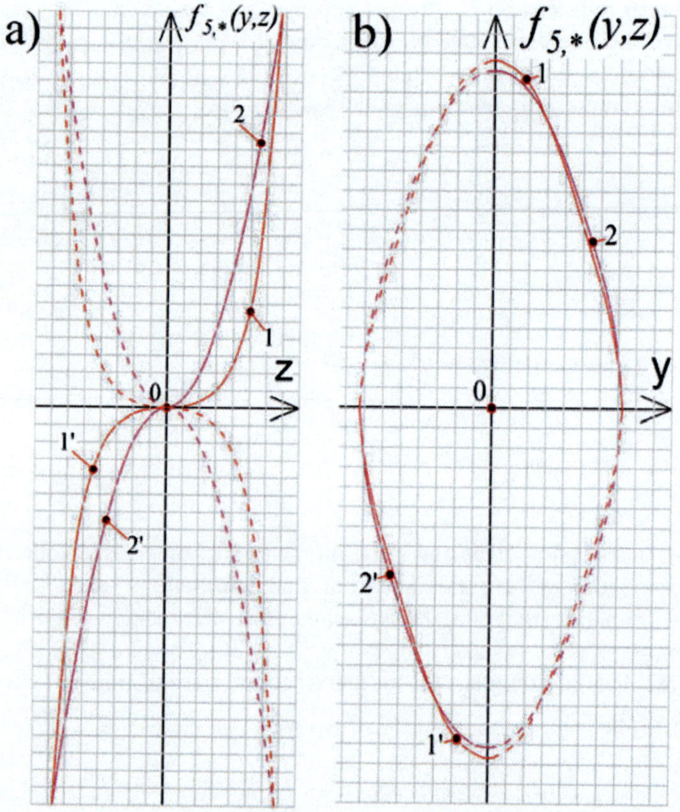

Fig. 3 Approximation of branches of graphs of Timoshenko-Goodyère functions [25] (1) and our new first functional (2) in the axes «$z - f_{5,*}(y, z)$» (**a**), in the axes «$y - f_{5,*}(y, z)$» (**b**); quadrants I–IV

4 Discussion

The error is up to 2% of the considered points of the section (Fig. 2b) and up to 7% at any points of the section when using functional (1) to find the values of functions and determine the deformations. After approximation using the graph of the theory of elasticity S.P. Timoshenko, Goodier [25] the proposed new functional received an error: 7% in the axes "$f_{5,*}(y,z) - z$" (Fig. 3a) and less than 1% in the axes "$f_{5,n,*}(y,z) - y$" (Fig. 3b).

Since the obtained error is small, the main directions of future research are the development of a new other hypothesis of angular deformations and the determination of the total relative deformations of structures in bending with torsion. The hypothesis of angular deformations should have ratios between the fibers of the relative transverse total shear strains of concrete and reinforcement ($\gamma_{sum,b}$ and $\gamma_{sum,s}$) and a special figure to determine the equation with relative deformation.

5 Conclusions

A simple new method from the families of the mesh method is found for the development of deformation functionals when approximating rectangular mean sections in compressed and stretched zones using their special squares.

1. The new hypothesis is proposed for linear deformations and its filling diagrams in bending with torsion. The analytical second functional is a general indefinite function of three functions—exponent, straight line and parabola.
2. The new hypothesis is proposed for linear deformations and its filling diagrams in bending with torsion. The analytical second functional is a general indefinite function of three functions—exponent, straight line and parabola. The new hypothesis is the kinematics between fibers for the relative longitudinal fiber upper and lower deformations of concrete and reinforcement for their ratios in distances from the neutral axis, similar to the modernized Bernoulli hypothesis but having a special geometric figure for the summed function of additional deplanation and parameter compressed concrete between plastic and elastic regions for constructing the equation.
3. Authors obtained stresses (deformations) by projecting the coefficients φ_{ij} and $\nu(\lambda)$ from points C, B, A, D of diagrams (Fig. 3d).
4. The filling coefficients of the diagrams were obtained for the spatial triple integral from the longitudinal deformations for the first hypothesis. Jumps appear (lateral cracks, normal cracks, etc.) for stages I–III of average deformations of longitudinal concrete and reinforcement.
5. A complex analysis of the proposed new functional from the Timoshenko theory of elasticity with its approximation, functionals and graphs is carried out. Functional have a grid (the first or second iteration) for the horizontal and

vertical line and leagues between the points (i.e., functions along the y or z axis) to find the values of the functions. The error is up to 2% at the considered points of the cross-section or up to 7% at any points.

References

1. Lin W (2021) Experimental investigation on composite beams under combined negative bending and torsional moments. Adv Struct Eng 24:1456–1465. https://doi.org/10.1177/1369433220981660
2. Kim C, Kim S, Kim K-H, Shin D, Haroon M, Lee J-Y (2019) Torsional behavior of reinforced concrete beams with high-strength steel bars. Struct J 116:233–251
3. Kandekar SB, Talikoti RS (2018) Study of torsional behavior of reinforced concrete beams strengthened with aramid fiber strips. Int J Adv Struct Eng 10:465–474. https://doi.org/10.1007/s40091-018-0208-y
4. Jan L, Vítek Jaroslav Průša, Vladimír Křístek LB (2020) Torsion of rectangular concrete sections. ACI Symp Publ 344:111–130
5. Křístek V, Průša J, Vítek JL (2018) Torsion of reinforced concrete structural members. Solid State Phenom 272:178–184. https://doi.org/10.4028/www.scientific.net/SSP.272.178
6. Karpyuk VM, Kostyuk AI, Semina YA (2018) General case of nonlinear deformation-strength model of reinforced concrete structures. Strength Mater 50:453–454. https://doi.org/10.1007/s11223-018-9990-9
7. Bernardo L (2019) Modeling the full behavior of reinforced concrete flanged beams under torsion. Appl Sci 9. https://doi.org/10.3390/app9132730
8. Rahal KN (2000) Torsional strength of reinforced concrete beams. Can J Civ Eng 27:445–453. https://doi.org/10.1139/cjce-27-3-445
9. Tsai H-C, Liao M-C (2019) Modeling torsional strength of reinforced concrete beams using genetic programming polynomials with building codes. KSCE J Civ Eng 23:3464–3475. https://doi.org/10.1007/s12205-019-1292-7
10. Karpenko NI (1996) General models of reinforced concrete mechanics. Stroyizdat, Moscow
11. Karpenko NI (1976) The theory of deformation of reinforced concrete with cracks. Stroyizdat, Moscow
12. Karpenko NI (1972) To the calculation of deformations of reinforced concrete rods with cracks in bending with torsion. Theory Reinf Concr 10:50–59
13. Karpenko NI, Elagin EG (1970) Deformations of reinforced concrete tubular elements subjected to torsion after cracking. Concr Reinf Concr 3:3–12
14. Kalkan I, Kartal S (2017) Torsional rigidities of reinforced concrete beams subjected to elastic lateral torsional buckling. World Acad Sci Eng Technol Int J Civ Environ Eng 11:969–972
15. Arzamastsev SA, Rodevich VV (2015) To the calculation of reinforced concrete elements for bending with torsion. Proc High Educ institutions Constr 9(681):99–199
16. Zalesov AS, Khozyainov BP (1991) Strength of reinforced concrete elements in torsion and bending. Proc Univ Constr Archit 1:1–4
17. Lessing NN, Rullay LK (1972) General principles for calculating the torsional flexural strength of reinforced. Theory Reinf Concr 43–49
18. Lessing NN (1959) Determination of the bearing capacity of reinforced concrete elements of rectangular cross-section, working in bending with torsion. Investig Strength Elem Reinf Concr Struct 5:3–28
19. Demyanov A, Kolchunov I (2017) The dynamic loading in longitudinal and transverse reinforcement at instant emergence of the spatial srack in reinforced concrete element under the action of a torsion with bending. Istraz i Proj za privredu 15:381–386. https://doi.org/10.5937/jaes15-14663

20. Kolchunov VI, Dem'yanov AI, Naumov NV, Mikhaylov MM (2019) Calculation of the stiffness of reinforced concrete structures under the action of torsion and bending. J Phys Conf Ser 1425:012077. https://doi.org/10.1088/1742-6596/1425/1/012077
21. Demyanov AI, Kolchunov VI, Pokusaev AA (2017) Experimental studies of deformation of reinforced concrete structures subjected torsion and bending. Struct Mech Eng Constr Build 6:37–44. https://doi.org/10.22363/1815-5235-2017-6-37-44
22. Travush VI, Karpenko NI, Kolchunov VI, Kaprielov SS, Dem'yanov AI, Konorev AV (2019) Main results of experimental studies of reinforced concrete structures of high-strength concrete B100 round and circular cross sections in torsion with bending. Struct Mech Eng Constr Build 15:51–61. https://doi.org/10.22363/1815-5235-2019-15-1-51-61
23. Travush VI, Karpenko NI, Kolchunov VI, Kaprielov SS, Demyanov AI, Bulkin SA, Moskovtseva VS (2020) Results of experimental studies of high-strength fiber reinforced concrete beams with round cross-sections under combined bending and torsion. Struct Mech Eng Constr Build 16:290–297. https://doi.org/10.22363/1815-5235-2020-16-4-290-297
24. Travush VI, Karpenko NI, Kolchunov VI, Kaprielov SS, Konorev AV (2018) The results of experimental studies of structures square and box sections in torsion with bending. Build Reconstr 6:32–43
25. Timoshenko SP, Goodyer J (1972) Theory of elasticity. Nauka, Moscow

Crack Resistance of Tension Reinforced Concrete Members with Bond Failure Areas

Igor Rudniy, Natalia Vorontsova, Dmitry Korolkov, and Pavel Pachulia

Abstract The paper presents a method for estimating the crack formation force and crack width for centrically tension reinforced concrete members with bond failure areas between reinforcement and concrete. The method is based on the theory of compound bars and allows to consider an unlimited number of bond failure areas and their unspecified location. The concrete behavior is presented in the form of an idealized Prandtl diagram. The behavior of the reinforcement is considered only on the elastic section up to the yield stress. The article presents a theoretical assessment of the influence of bond failure areas on the process of formation and development of cracks. The crack width is considered as the difference between the strains of concrete and rebar, taking into account the influence of the strain capacity of their bonding in the area between two adjacent cracks. Experimental investigations have been carried out, which confirm the correctness of the theoretical model and good agreement with the experimental data. The test program included 3 series of specimens with different lengths of the bond failure area. The first two series consisted of 3 specimens, with the length of the bond failure area equal to 300 mm and 600 mm of the total length of the member. The third series consisted of 2 specimens with ensured bonding along the entire length of the members. A total of 8 specimens were made. Experimental data on the influence of bond failure areas on the process of crack formation in tension members are presented.

Keywords Crack resistance · Bond failure · Tension members · Crack width · Crack spacing · Experimental investigations · Theory of compound bars

I. Rudniy (✉) · N. Vorontsova · D. Korolkov · P. Pachulia
Saint Petersburg State University of Architecture and Civil Engineering,
4 Vtoraya Krasnoarmeiskaya, Saint Petersburg 190005, Russian Federation

N. Vatin et al. (eds.), *Proceedings of MPCPE 2021*, Lecture Notes in Civil Engineering 182, https://doi.org/10.1007/978-3-030-85236-8_10

1 Introduction

The features inherent in reinforced concrete structures, such as the interaction of different modulus materials, the discrete location of cracks, the inelastic behavior of the reinforcement, the bond performance between reinforcement and concrete, cause the complexity of assessing the stress-strain state and developing calculation methods. The mechanics of the interaction of reinforcement and concrete have been studied in many articles [1–4], and the processes of formation and development of cracks in the contact zone, as well as the stages of the stress-strain state of the contact zone, have been described for elements with ensured bonding.

The presence of discretely located bond failure areas between reinforcement and concrete changes the stress-strain state of the contact zone and the member itself and makes it difficult to calculate such structures [5, 6] and, consequently, to make reasonable design decisions. The presence of bond failure areas makes it necessary to consider the strain capacity of the bonding of reinforcement with concrete and use models that describe the sequence of formation and growth of discretely located cracks.

The experience of the survey of reinforced concrete structures shows that one of the most common defects is a bond failure of reinforcement with concrete, up to 35% of the total number of damaged structures. Defect of the bond failure between reinforcement and concrete occur both at the construction phase of structures—this is not high-quality concrete work and non-compliance with technologies, and during their operation: from the impact of various types of petroleum products, temperature deformations, corrosion of reinforcement and mechanical damage to the concrete cover [7, 8].

However, in addition to the fact that the bond failure in any reinforced concrete structures [9, 10] can be accidental, prestressed structures with unbonded reinforcement are widely used in construction practice [11, 12].

In structures with bond failure areas between reinforcement and concrete, the crack formation force decreases, and with a large ratio of the reinforcement, the crack spacing, the crack width, and the deformations of the members increase [13–15].

To study the process of crack formation and the parameters of bond between reinforcement and concrete, uniaxial tests of prisms for central tension reinforced with a single rebar are most popular [4, 16–19].

In [18], analytical dependences are proposed for determining the stress in the reinforcement bar, as well as the bond stress between the reinforcement and the concrete. For the investigation, prisms with a cross-section size of 150 × 150 mm and a length of 300 mm were made of concrete with a compressive strength of 20, 40 and 60 MPa, reinforced with a central reinforcing bar Ø12. It is noted that this test method more accurately reflects the stress-strain state at the bond failure area than the tests for pulling the rebar out of the prism.

In the paper [15], a "b-ellipse model" is proposed that considers the tension stiffening and tension softening effects, as well as the effect of deterioration of the bond. This model is a logical development of the traditional model of reinforcement bond with concrete. This model was implemented in the finite element analysis (FEA) for calculating the crack width when using fiber elements of columns and beams.

In the article [19], the behavior of the bond of reinforcement and concrete is investigated, considering the effect of the influence of transverse tensile forces. It is shown that the bond strength is a function of the transverse tensile forces. When the transverse tensile stresses exceed 80% of the concrete's ultimate tensile strength, the bond strength is reduced by 54%.

To uniformly distribute the tensile force over the prism cross-section, specimens with a cross-section size of 150 × 150 mm, a length of 500 or 1210 mm, reinforced with not one, but several reinforcement bars 4Ø10, 4Ø12, 8Ø8 or 16Ø5 were also tested [20]. The influence of the ratio of reinforcement, the diameter of the reinforcing bar, as well as the size of the concrete cover, varying from 15 to 50 mm, was studied. The results obtained are compared with the provisions of the CEB-FIP Model Code.

Even though many scientists have been engaged in the study of crack formation since the appearance of reinforced concrete, the interest of scientists in this topic is still observed to this day. Moreover, the issues related to the violation of the bond failure of reinforcement with concrete remain underinvestigated.

The paper examines the influence of defects of bond failure on the formation and propagation of cracks in tension reinforced concrete members at all stages of the stress-strain state. The calculation method is based on the theory of compound bars.

2 Methods

The initial equation of the strain compatibility at the level of the contact layer:

$$\varepsilon_{sv} = \varepsilon_s - \varepsilon_{bt}, \tag{1}$$

where ε_{sv}—strain of bond; ε_s—strain of reinforcement; ε_{bt}—strain of concrete at the level of reinforcement.

When an external load acts on a reinforced concrete member due to the bonding forces of the reinforcement with the concrete, the stresses between the reinforcement and the concrete are redistributed. The general case of calculating a compound bar is considered. The shear relations by the contact of the reinforcement with concrete are replaced by reactions (S).

Taking the shear force (S) as external to the parts of the rod, ε_s, ε_{sv} and ε_{bt} is expressed as

$$\varepsilon_s = \frac{S}{E_s A_s}; \quad \varepsilon_{sv} = \frac{S''}{G};$$
$$\varepsilon_{bt} = -\left(\frac{S}{E_b A_b} + \frac{S m_0^2}{B} \pm \frac{N}{E_b A_b} - \frac{M \pm Ne}{B} m_0\right) \tag{2}$$

Substituting (2) in (1) is obtained

$$\frac{S''}{G} = \gamma S + \Delta, \tag{3}$$

where $\gamma = \frac{1}{E_s A_s} + \frac{1}{E_b A_b} + \frac{m_0^2}{B}$; $\Delta = \pm \frac{N}{E_b A_b} - \frac{M - Ne}{B} \cdot m_0$; E_s—modulus of elasticity of reinforcement; A_s—area of reinforcement; E_b—modulus of elasticity of concrete; A_b—area of concrete section; B—flexural rigidity of concrete branch; G—modulus of deformation of bonding; m_0—distance between the centroids of the branches; e—eccentricity of normal force N.

The solution of the differential Eq. (3)

$$\begin{cases} S = C_1 sh\lambda x + C_2 ch\lambda x + \frac{G}{\lambda}\int\limits_0^x \Delta(\xi) sh\lambda(x-\xi)d\xi; \\ S' = \lambda\left[C_1 ch\lambda x + C_2 sh\lambda x + \frac{G}{\lambda}\int\limits_0^x \Delta(\xi) ch\lambda(x-\xi)d\xi\right]; \\ S'' = \lambda^2\left[C_1 sh\lambda x + C_2 ch\lambda x + \frac{G}{\lambda}\int\limits_0^x \Delta(\xi) sh\lambda(x-\xi)d\xi\right], \end{cases} \tag{4}$$

where $\lambda = \sqrt{G\gamma}$; ξ—additional variable.

The pattern of the distribution of the strains of the tension concrete, depending on the method of applying the load, allows to find the crack formation force for members with bond failure. The concrete work is presented in the form of an idealized Prandtl diagram (see Fig. 1).

When a crack occurs, there is a redistribution of forces between the concrete and the reinforcement: near the crack face, the stresses in the concrete drop to zero, and

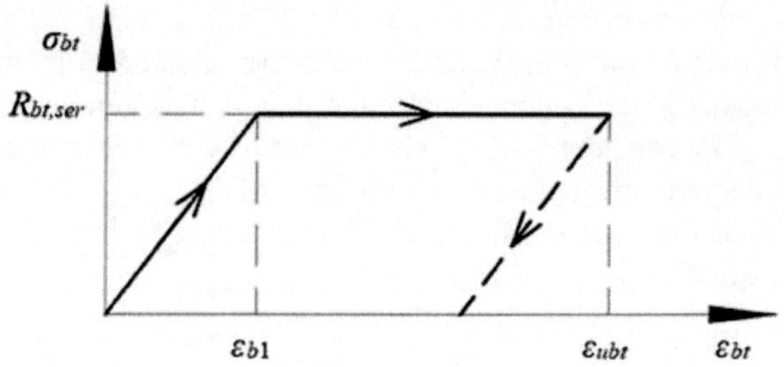

Fig. 1 Idealized diagram $\sigma_{bt} - \varepsilon_{bt}$

the strains to the value of $\varepsilon_{bt} - \varepsilon_{b1}$; the stresses in the reinforcement increase by a certain amount. Taking into account the above

$$\varepsilon_{bt(x)} = \varepsilon_{ubt} - \varepsilon_{b1} + \varepsilon_{bt,el(x)}. \tag{5}$$

The crack formation force is determined from the condition $\varepsilon_{b(x)} = \varepsilon_{ubt}$, therefore

$$\varepsilon_{b1} = \varepsilon_{bt,el(x)}, \tag{6}$$

where ε_{b1} defined according to the diagram (Fig. 1) $\varepsilon_{b1} = \frac{R_{bt,ser}}{E_b}$.

The forces of successive crack formation are determined step by step, first considering the total length of the member to find $N_{crc,1}$, then evaluating each block of the member into which each subsequent crack divides it. At the same time, the length of each block is determined—the crack spacing. The calculation is carried out until the maximum strains in the valve or in the contact layer of stresses equal to τ_u are reached.

Figure 2 shows relationships of the first crack formation force on the length of sections with different bonding characteristics λ_1 and λ_2, at different ratios ($\lambda_1/\lambda_2 \rightarrow 1$—member with ensured bonding; $0 \leq \lambda_1/\lambda_2 < 1$—member with bond failure areas). On the graph along the abscissa axis, the ratio of the length of the bond failure area to the total length of the specimen is plotted. The graphs of the propagation of the bond failure area from the center of the member (l_1) and from the edge of the member (l_2) are combined, since the development of the bond failure area from the edge or center of the member does not significantly affect the force of the formation of the first crack.

The graphs show that the lower the ratio of the bonding characteristics and the longer the length of the bond failure area, the higher the crack formation force. A small ratio of the bonding characteristics ($\lambda_1/\lambda_2 = 0.17$ or less) (Fig. 2 (1)) has a significant effect on the value of the crack formation force (22%), when the length of the bond failure area is more than 80% of the total length of the member.

When the ratio λ_1/λ_2 is from 0.33 to 0.5, the effect of the length of the bond failure area on the crack formation force is minimal (less than 5%). When the ratio of the bonding characteristics is 0.5 or more, the influence of the length of the bond failure area is practically absent.

A decrease in the stress transfer zone from l_3 to l_1 (Fig. 3) leads to a decrease in the value of strains in the tension concrete from ε_{bt3} to ε_{bt1}, as a result of which a greater load must be applied to achieve the maximum strains of ε_{ubt}, so there is an increase in the crack formation force. The bond failure between the reinforcement and the concrete in the stress transfer zone affects the value of strains in the tension concrete in a similar way. Bond failure leads to an increase in the load range at which active crack formation occurs, and to an increase in the load of stabilizing the crack formation process.

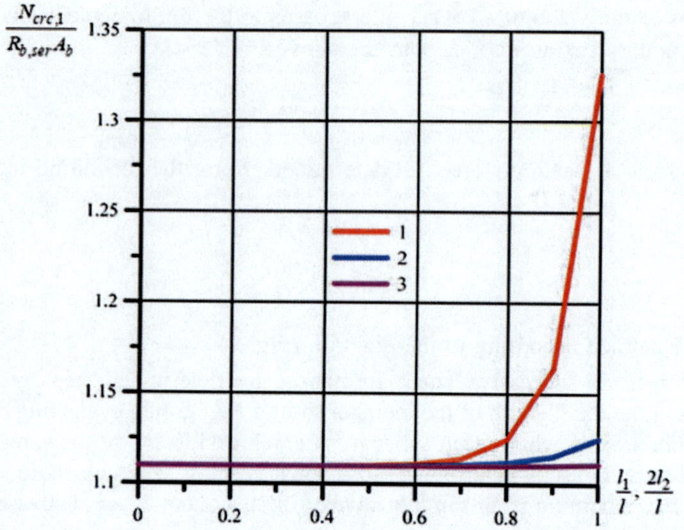

Fig. 2 Relationship $\frac{N_{crc,1}}{R_{b,ser}A_b}$ on the length of bond failure area at different ratios λ_1/λ_2: 1 – 0.17; 2 – 0.33; 3 – 0.5

Fig. 3 Diagram of the change in the value of strain ε_{bt} with a decrease in the stress transfer zone

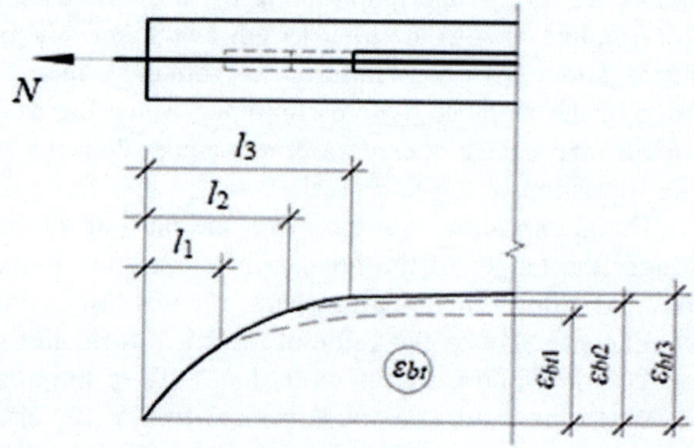

The influence of the bond failure areas on the process of crack formation is illustrated by the example of the two members shown in Fig. 4. The location of the bond failure area in the central zone of the member (Fig. 4 a) leads to an increase in the zone, with stresses in the range $0.95\varepsilon_{ubt} \leq \varepsilon_{bt} \leq \varepsilon_{ubt}$, in which the formation of cracks occurs. The location of the bond failure area in the stress transfer zone (Fig. 4 b) leads to a significant reduction in this zone and brings the crack formation process closer to the idealized scheme.

Nonhomogenity of concrete and the presence of hidden defects cause the deviation of the crack formation process from the idealized scheme. It is known that in the initial period, the formation of cracks is quite chaotic, and with an increase in

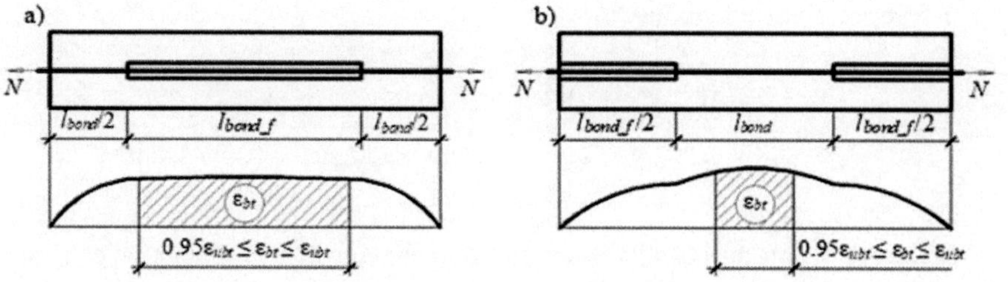

Fig. 4 Distribution of strains of concrete when the bond failure areas are located in the center (**a**) or at the ends of the members (**b**)

the load and a decrease in the crack spacing, the process approaches the idealized one. The presence of bond failure areas, as shown above, can significantly affect this process in some cases, or have almost no effect in others.

In general, the crack width is determined from the following equation

$$a_{crc(n)} = (\delta_{c_{n-1}} + \delta_{c_n}) - \frac{R_{bt,ser}}{E_b} \cdot (c_n + c_{n-1}), \tag{7}$$

where $\delta_{cn\text{-}1}$, δ_{cn}—relative movement of reinforcement and concrete during the elastic behavior of the contact layer from sections of length $c_{n\text{-}1}$ and c_n, respectively. The second term in Eq. (7) is the proper elongation of the concrete in the area between the cracks.

In the work of A V Trofimov, it was shown that when the process of crack formation is stabilized, the zone of active bonding must decrease in order to meet the condition $\sigma_{bt} \leq R_{bt,ser}$, thus, in a section of some length near the crack faces, there is a bond failure area of reinforcement with concrete. Based on this, in the area between the cracks, 3 characteristic zones can be distinguished: the zone of bond failure, the zone of nonelastic strains and the zone of elastic strains of the contact layer. The equations for determining the length of the sections of these zones from, taking into account the irregular step of the cracks when calculating the tangential stress plot from the zero point, are given below:

(1) the zone of bond failure

$$c_0 = c_k \left[1 - a_0 - \frac{N}{\tau_u S_p c_k (1+\alpha)} + \frac{\sqrt{t_1^2+1}}{\lambda t_1 c_k}\right], \tag{8}$$

(2) the zone of nonelastic strains

$$b_0 = \frac{N}{(1+\alpha)\tau_u S_p} - \frac{\sqrt{t_1^2+1}}{\lambda t_1}, \tag{9}$$

(3) the zone of elastic strains

$$a_0 = \frac{2\ln\left(t_1 + \sqrt{t_1^2 + 1}\right)}{\lambda}, \quad t_1 = \frac{\tau_u S_p (1+\alpha)}{\lambda \sigma_{s0} A_s \left(1 - \frac{R_{bt,ser}(1+\alpha)}{\sigma_{s0}\mu}\right)}. \tag{10}$$

The crack width during the elastic-plastic behavior of the contact layer is also determined by the Eq. (7), but the relative movements between the reinforcement and the concrete are found by the Eq. (4), and δ_{cn} and $\delta_{cn\text{-}1}$ are determined by the Eq. (11).

$$\delta_{c_n} = \delta_{c0(n)} + \delta_{b0(n)} + \delta_{a0(n)}; \quad \delta_{c_{n-1}} = \delta_{c0(n-1)} + \delta_{b0(n-1)} + \delta_{a0(n-1)}. \tag{11}$$

The proposed method for calculating the tension members allows to determine the crack width, taking into account their irregular step during the elastic and elastic-plastic behavior of the contact layer.

The test program included 3 series of specimens with different lengths of the bond failure area. The first two series consisted of 3 specimens, with the length of the bond failure area equal to 300 mm and 600 mm of the total length of the member (Fig. 5). The third series consisted of 2 specimens with ensured bonding along the entire length of the members. A total of 8 specimens were made.

Crack resistance studies were carried out on the INSTRON 5982 floor-type test system with a maximum force of 100 kN (Fig. 6). The built-in gauge of the test system tracked the total elongations of the specimen and plotted the "force—absolute strains" relationship. During the tests, the relative movements between the reinforcement and the concrete at both ends of the member were measured using four dial gauges with a scale division of 0.001 mm. The gauges were attached to the reinforcement bar with the help of special equipment.

Figures 7, 8 and 9 show the relationship "$\sigma_{s0} - \Delta l$" for three series of specimens with different lengths of the bond failure area, obtained on the INSTRON 5982 test

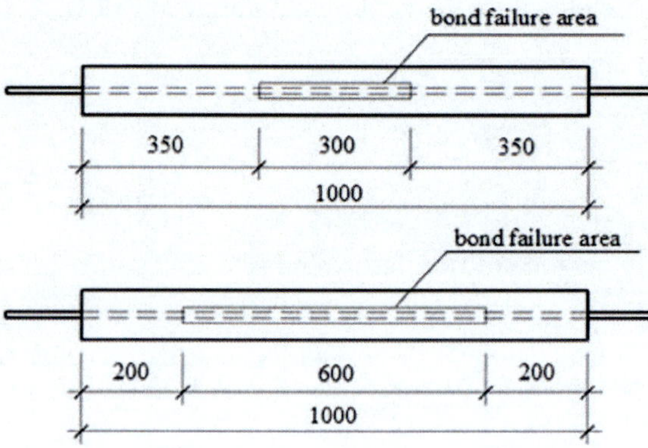

Fig. 5 Location of the bond failure areas in specimens 1 and 2 of the series

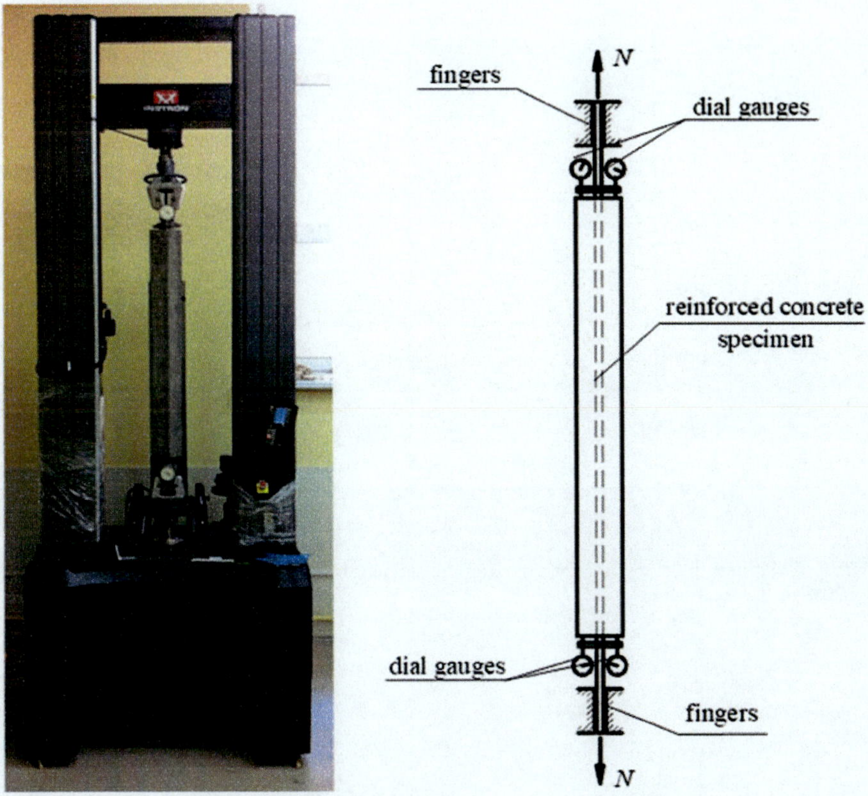

Fig. 6 Test system INSTRON 5982 with specimen and test scheme of the reinforced concrete specimen

system. Figures 7 and 8 show a spike resulting from the formation of the first crack in the members with bond failure. In addition to a sudden drop in load in the member, significant longitudinal deformations of the sample occurred, caused by the elongation of the reinforcement in the bond failure area after the formation of the first crack.

Failure of the specimens is shown in the Fig. 10.

3 Results and Discussions

The force of the first crack formation ($N_{crc,1}$) was determined by the load drop on Figs. 8 and 9. A comparison of the experimental and theoretical values of the first crack formation force is given in Table 1. The value of the coefficient χ in the areas covered with plasticine was assumed to be 0.

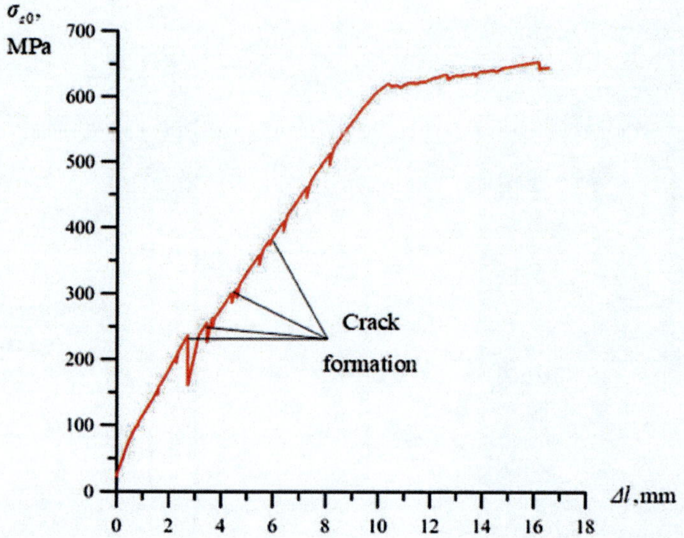

Fig. 7 Relationship "$\sigma_{s0} - \Delta l$" for centrically tensioned members with bond failure areas length of 30% of the total length

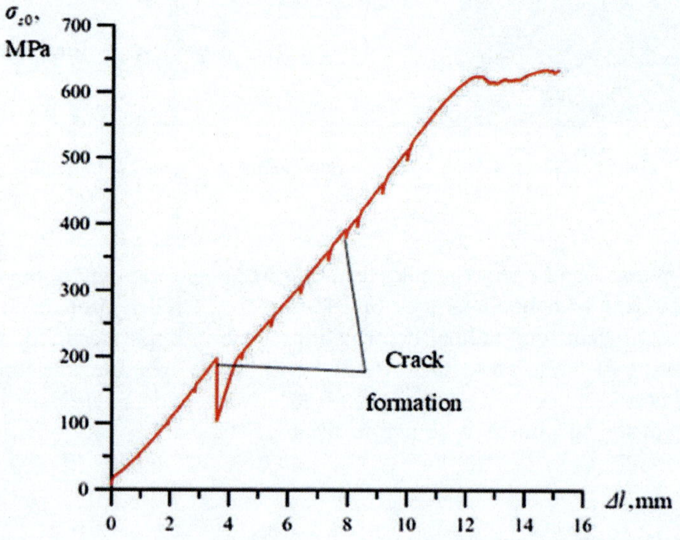

Fig. 8 Relationship "$\sigma_{s0} - \Delta l$" for centrically tensioned members with bond failure areas length of 60% of the total length

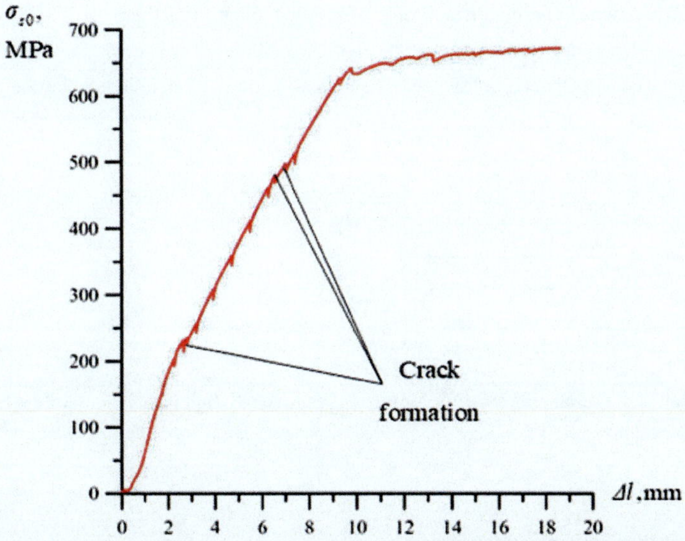

Fig. 9 Relationship "$\sigma_{s0} - \Delta l$" for centrically tensioned members with ensured bonding of the total length

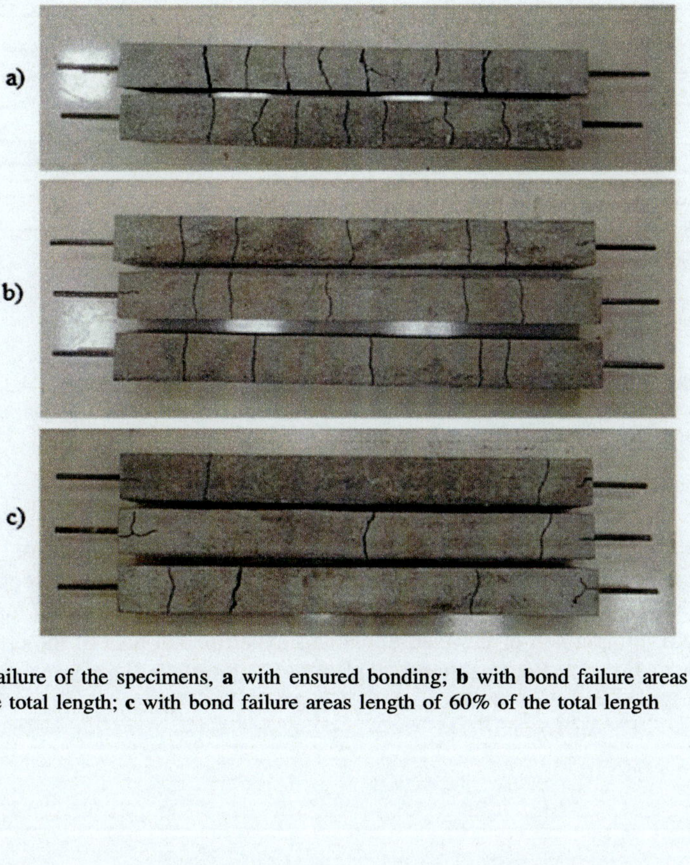

Fig. 10 Failure of the specimens, **a** with ensured bonding; **b** with bond failure areas length of 30% of the total length; **c** with bond failure areas length of 60% of the total length

Table 1 Experimental and theoretical force of the first crack formation

Specimen no	N_{crc}1, kN		
	Experimental value	Theoretical value	Building code requirements in the Russian Federation
1–0	26.13	27.38	27.77
2–0	25.92		
1–30	26.77	27.68	
2–30	26.68		
3–30	25.78		
1–60	28.43	30.02	
2–60	26.86		
3–60	25.72		

The first digit is the serial number of the specimen, the second digit is the length of the bond failure areas in %

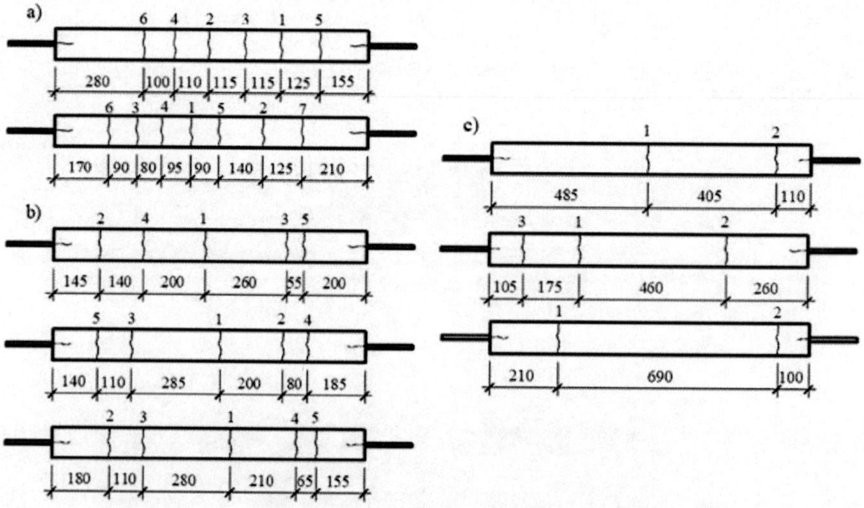

Fig. 11 Scheme of the location of cracks and the sequence of their appearance in the specimens, **a** with ensured bonding; **b** with bond failure areas length of 30% of the total length; **c** with bond failure areas length of 60% of the total length

There is a decrease in the extreme blocks of the specimens (Fig. 11), which confirms the influence of the bond failure areas on the areas with ensured bonding.

The crack width in the specimens was measured using a microscope with a 24-fold increase and a scale division of 0.05 mm at each stage of loading the member. A comparison of the theoretical and experimental data of the crack width is shown in Fig. 12. For specimens with bond failure areas, a comparison of the crack width located only in the zone of bond failure is given.

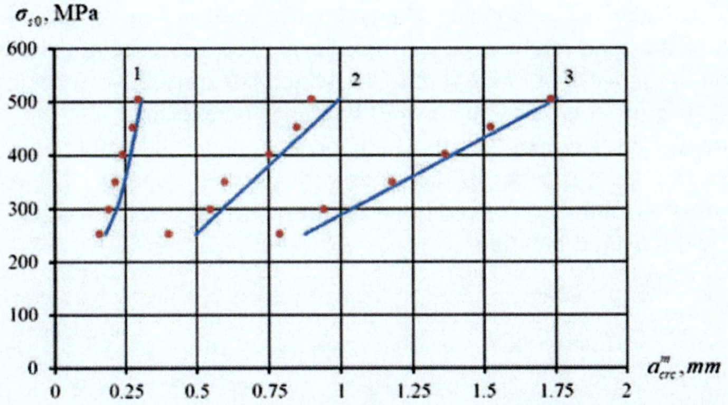

Fig. 12 Relationship «$\sigma_{s0} - a^{m}_{crc}$»: 1—specimens with ensured bonding; 2—specimens with bond failure areas length of 30% of the total length; 3—specimens with bond failure areas length of 60% of the total length

For members with bond failure areas length of 30% of the total length, the crack opening width increased by 3.3 times compared to the check specimens, and for specimens with bond failure areas length of 60% of the total length by 6.5 times. The divergence between the experimental a_{crc} values and the theoretical ones is from 3 to 13%. The average square deviation of the experimental data of the average crack width from the theoretical one is 9.31%.

4 Conclusions

According to the results of experimental investigations, the following conclusions are made:

1. The analytical equations obtained as a result of the performed investigation allow to evaluate the stress-strain state of tension reinforced concrete members with discretely located bond failure areas, taking into account the elastic and elastic-plastic behavior of the contact layer.
2. The cases of fully ensured and bonding failure are special cases of the developed calculation method.
3. An increase in the crack formation force in the centrally loaded members with bond failure area was found in comparison with the members with ensured bonding. The assessment of the influence of different lengths of bond failure area on the crack formation forces in tension members revealed that when the length of the bond failure areas is more than 80% of the total length of the member, the ratio of the bonding characteristics ($\lambda_1/\lambda_2 = 0.17$ or less) affects the crack formation force.

4. When the ratio of the bonding characteristics is 0.5 or more, the effect of the length of the bond failure area is insignificant. The influence of the bond failure area on the process of crack formation, which is expressed in a decrease in the cracks spacing in areas with ensured bonding, is revealed.
5. The smaller the value of the ratio λ_1/λ_2 or, the greater the bond failure area, the smaller the crack spacing in the areas with ensured bonding. The experiment confirmed the influence of the bond failure areas on the cracks spacing in the areas with ensured bonding.

References

1. Broms BB (1965) Stress distribution in reinforced concrete members with tension cracks. ACI J Proc 62:1095–1108. https://doi.org/10.14359/7740
2. Broms BB (1964) Stress distribution, crack patterns, and failure mechanisms of reinforced concrete members. ACI J Proc 61:1535–1558. https://doi.org/10.14359/7838
3. Goto Y (1971) Cracks formed in concrete around deformed tension bars. ACI J Proc 68:244–251. https://doi.org/10.14359/11325
4. Gerstle W, Ingraffea AR, Gergely P (1984) Fracture mechanics of bond in reinforced concrete. In: Proceedings of the U.S. national congress of applied mechanics, pp. 477–478. ASME. https://doi.org/10.1061/(asce)0733-9445(1984)110:4(871)
5. Mousa MI (2016) Effect of bond loss of tension reinforcement on the flexural behaviour of reinforced concrete beams. HBRC J 12:235–241. https://doi.org/10.1016/j.hbrcj.2015.01.003
6. Islam MS, Ahmed SJU, Alam FM (2019) The failure bond force of untreated and epoxy-treated reinforcement: an analytical study. Constr Build Mater 201:81–89. https://doi.org/10.1016/j.conbuildmat.2018.12.142
7. Fang C, Lundgren K, Chen L, Zhu C (2004) Corrosion influence on bond in reinforced concrete. Cem Concr Res 34:2159–2167. https://doi.org/10.1016/j.cemconres.2004.04.006
8. Li F, Yuan Y (2013) Effects of corrosion on bond behavior between steel strand and concrete. Constr Build Mater 38:413–422. https://doi.org/10.1016/j.conbuildmat.2012.08.008
9. Lisyatnikov MS, Shishov II, Sergeev MS, Hisham E. Precast monolithic coating of an industrial building based on variable-height beam-slabs recent citations precast monolithic coating of an industrial building based on variable-height beam-slabs. https://doi.org/10.1088/1757-899X/896/1/012064
10. Lukin MV, Popov MV, Lisyatnikov MS (2020) Short-term and long-term deformations of the lightweight concrete. In: IOP conference series: materials science and engineering, p. 032071. Institute of Physics Publishing. https://doi.org/10.1088/1757-899X/753/3/032071
11. Ralls ML, Carrasquillo RL, Burns NH (1998) Shoshov. https://doi.org/10.3141/1624-16
12. Alharbi YR, Galal M, Abadel AA, Kohail M (2020) Bond behavior between concrete and steel rebars for stressed elements. Ain Shams Eng J. https://doi.org/10.1016/j.asej.2020.10.001
13. Chen Y, Yu J, Younas H, Leung CK (2020) Experimental and numerical investigation on bond between steel rebar and high-strength Strain-Hardening Cementitious Composite (SHCC) under direct tension. Cement Concr Compos 112:103666. https://doi.org/10.1016/j.cemconcomp.2020.103666
14. Kaklauskas G, Ramanauskas R, Ng PL (2019) Predicting crack spacing of reinforced concrete tension members using strain compliance approach with debonding. J Civ Eng Manag 25:422–430. https://doi.org/10.3846/jcem.2019.9871

15. Xu LY, Nie X, Zhou M, Tao MX (2017) Whole-process crack width prediction of reinforced concrete structures considering bonding deterioration. Eng Struct 142:240–254. https://doi.org/10.1016/j.engstruct.2017.03.060
16. Brown CB (1966) Bond failure between steel and concrete. J Franklin Inst 282:271–290. https://doi.org/10.1016/0016-0032(66)90256-0
17. Hooshmand A, Kianoush R, Siad H, Lachemi M, Moslemi M (2021) Investigation of leakage and self-healing of direct tension cracks under sustained loading and high-water pressure. Constr Build Mater 267:120879. https://doi.org/10.1016/j.conbuildmat.2020.120879
18. Tang CW (2021) Modeling uniaxial bond stress–slip behavior of reinforcing bars embedded in concrete with different strengths. Materials 14:1–21. https://doi.org/10.3390/ma14040783
19. Zhang X, Dong W, Zheng JJ, Wu ZM, Hu Y, Li Q (2014) bin: Bond behavior of plain round bars embedded in concrete subjected to lateral tension. Constr Build Mater 54:17–26. https://doi.org/10.1016/j.conbuildmat.2013.12.031
20. Gribniak V, Rimkus A, Pérez Caldentey A, Sokolov A (2020) Cracking of concrete prisms reinforced with multiple bars in tension–the cover effect. Eng Struct 220. https://doi.org/10.1016/j.engstruct.2020.110979

Nonlinear Vibrations of an Orthotropic Viscoelastic Rectangular Plate Under Periodic Loads

Mirziyod Mirsaidov, Rustamkhan Abdikarimov, Dadakhan Khodzhaev, Bakhodir Normuminov, Svetlana Roshchina, and Nikolai Vatin

Abstract Modern methods and technologies for the manufacture of structures make it possible to obtain structures of various shapes and sizes. This, in turn, determines the possibility of using structures of variable thickness in modern technology and engineering. During operation, they are often subjected to various loads. Among the loads, periodic loads are of particular interest. On the basis of the Kirchhoff-Love theory, nonlinear parametric vibrations of an orthotropic viscoelastic rectangular plate of variable thickness are investigated without considering the elastic wave propagation. The mathematical model of the problem is described by a system of nonlinear integrodifferential equations, where the weakly singular Koltunov-Rzhanitsyn kernel is used as the relaxation kernel. The resolving equations of the problem are obtained by the Bubnov-Galerkin method and by a numerical method based on the use of quadrature formulas. The behavior of an orthotropic viscoelastic rectangular plate under the action of an external periodic load is investigated. The graphs obtained with the developed computer program show the effect on the amplitude-frequency response of the plate on various physical, mechanical, and geometrical parameters.

Keywords Rectangular plate · Viscoelasticity · Orthotropy · Variable thickness · Periodic load · Nonlinear parametric oscillations

M. Mirsaidov · D. Khodzhaev (✉) · B. Normuminov
Tashkent Institute of Irrigation and Agricultural Mechanization Engineers, Tashkent, Uzbekistan

R. Abdikarimov
Tashkent Institute of Finance, Tashkent, Uzbekistan

S. Roshchina
Vladimir State University Named After Alexander and Nikolay Stoletovs, Vladimir, Russia

N. Vatin
Peter the Great St. Petersburg Polytechnic University, St. Petersburg, Russia

N. Vatin et al. (eds.), *Proceedings of MPCPE 2021*, Lecture Notes in Civil Engineering 182, https://doi.org/10.1007/978-3-030-85236-8_11

1 Introduction

Thin-walled shell structures of variable thickness made of composite materials are among the most widespread structures used in many fields of modern technology. This is due to the great functionality of such structures and the successful combination of their properties of lightness and strength. Modern technological progress makes it possible to manufacture structures of various shapes and sizes made of various materials.

At the same time, during operation, such structures are subjected to various dynamic loads. Among such loads, periodic loads are of particular interest for research. On the other hand, difficulties arise in the calculation and design of such structures, and the determination of their stress-strain state causes both computational and principal difficulties. Therefore, at present, the development of new mathematical models, improvement of calculation methods and algorithms is one of the urgent tasks.

Currently, there are many publications devoted to the construction of various theories, models and methods to assess the stress-strain state of thin-walled shell structures of constant and variable thickness under the action of various static and dynamic loads. A significant contribution to the study of such problems was made by Bolotin [1], Volmir [2], and over the past decades by Awrejcewicz and Krys'ko [3], Amabili [4], Grigorenko and Grigorenko [5].

There are a number of articles devoted to the study of vibrations and stability of plates and shells of variable thickness.

The study in [6] is devoted to the parametric vibrations of plates under the action of static and periodic loads. A numerical-analytical method for solving the problem using the Bolotin method was proposed. In that, the plates can have an arbitrary geometric shape.

In [7], the problem of parametric vibrations of an isotropic cylindrical shell of variable thickness under the action of a load along its generatrix is considered. An exact solution is obtained for different ratios of the parameters.

The study in [8] is devoted to the parametric vibrations of conical shells of variable thickness under static and periodic loads. Using the Galerkin method, the problem is reduced to solving an equation of the Mathieu type. The influence of various parameters on the region of dynamic instability was studied.

In [9], free vibrations of composite shells and plates of variable thickness are investigated. The solutions obtained are compared with analytical and numerical solutions known in the literature.

The development of a technique for solving the three-dimensional problem of bending of orthotropic plates of variable thickness is given in [10]. The problem is reduced to solving two independent problems, described by two independent systems of two-dimensional infinite equations.

Loja et al. [11] is devoted to the determination of the dynamic instability of composite plates of variable thickness. To solve the problem, the Bolotin method was used. The influence of various geometric parameters, as well as the properties of the material on the stability of the plate, was studied.

In [12], parametric vibrations of composite plates of variable thickness under periodic loads were investigated. The mathematical model of the problem is described by an equation of the Karman type. The finite element method is used to solve the problem.

The study of free vibrations of shells with a linear change in thickness under various boundary conditions is given in [13]. The finite element method is used to solve the problem. The influence of the variability of the thickness, dimensions of the shell and other parameters on the amplitude-frequency response of its vibrations was studied.

In [14], the finite element method is used to solve the problem of dynamic stability of rectangular panels of variable thickness under compressive loads. The problem is reduced to solving the system of Mathieu-Hill equations.

In [15], on the basis of the classical theory of shells in a nonlinear formulation, forced vibrations and dynamic stability of cylindrical shells of variable thickness subjected to mechanical stress are investigated. To derive the resolving equations, the methods of Galerkin and Runge-Kutta were used.

Analysis of published works shows that insufficient attention was paid to nonlinear parametric vibrations of nonhomogeneous viscoelastic plates and shells of variable thickness [16–18].

The article investigates nonlinear parametric vibrations of an orthotropic viscoelastic rectangular plate of variable thickness without considering elastic wave propagation.

2 Materials and Methods

Consider an orthotropic viscoelastic plate, rectangular in plan, with sides a and b, variable thickness $h = h(x, y)$, and with initial imperfections $w_0 = w_0(x, y)$; the plate is subjected to acting periodic load $P(t) = P_0 + P_1 \cos \Theta t$ ($P_0, P_1 = const$; Θ is the frequency of an external periodic load) along side a.

We will assume that there are no tangential inertial forces. Then the system of three equations [16] can be reduced to a system of two equations with two unknowns.

Following the results of [2], we introduce the stress function Φ in the middle surface in accordance with the following formulas:

$$\sigma_x = \frac{N_x}{h} = \frac{\partial^2 \Phi}{\partial y^2}, \quad \sigma_y = \frac{N_y}{h} = \frac{\partial^2 \Phi}{\partial x^2}, \quad \tau_{xy} = \frac{N_{xy}}{h} = -\frac{\partial^2 \Phi}{\partial x \partial y} \tag{1}$$

The following equation of the Karman type is obtained:

$$
\begin{aligned}
&(1-\Gamma^*)\left\{\frac{h^3}{12}\left[B_{11}\frac{\partial^4(w-w_0)}{\partial x^4}+(8B+B_{12}+B_{21})\frac{\partial^4(w-w_0)}{\partial x^2\partial y^2}+B_{22}\frac{\partial^4(w-w_0)}{\partial y^4}\right]\right.\\
&+\frac{1}{4}\left[2h\left(\frac{\partial h}{\partial x}\right)^2+h^2\frac{\partial^2 h}{\partial x^2}\right]\left(B_{11}\frac{\partial^2(w-w_0)}{\partial x^2}+B_{12}\frac{\partial^2(w-w_0)}{\partial y^2}\right)\\
&+\frac{1}{2}h^2\frac{\partial h}{\partial x}\left[B_{11}\frac{\partial^3(w-w_0)}{\partial x^3}+(B_{12}+4B)\frac{\partial^3(w-w_0)}{\partial x\partial y^2}\right]\\
&+\frac{1}{2}h^2\frac{\partial h}{\partial y}\left[B_{22}\frac{\partial^3(w-w_0)}{\partial y^3}+(B_{21}+4B)\frac{\partial^3(w-w_0)}{\partial x^2\partial y}\right]\\
&+\frac{1}{4}\left[2h\left(\frac{\partial h}{\partial y}\right)^2+h^2\frac{\partial^2 h}{\partial y^2}\right]\left(B_{22}\frac{\partial^2(w-w_0)}{\partial y^2}+B_{21}\frac{\partial^2(w-w_0)}{\partial x^2}\right)\\
&\left.+\left[2h\frac{\partial h}{\partial x}\frac{\partial h}{\partial y}+h^2\frac{\partial^2 h}{\partial x\partial y}\right]2B\frac{\partial^2(w-w_0)}{\partial x\partial y}\right\}\\
&=\frac{12(1-\mu_1\mu_2)}{\sqrt{E_1E_2}}\left[h\frac{\partial^2 w}{\partial x^2}\frac{\partial^2\Phi}{\partial y^2}+h\frac{\partial^2 w}{\partial y^2}\frac{\partial^2\Phi}{\partial x^2}-2h\frac{\partial^2 w}{\partial x\partial y}\frac{\partial^2\Phi}{\partial x\partial y}\right]\\
&+\frac{12(1-\mu_1\mu_2)}{\sqrt{E_1E_2}}q-P(t)\frac{\partial^2 w}{\partial x^2}-\frac{12(1-\mu_1\mu_2)}{\sqrt{E_1E_2}}\rho h\frac{\partial^2 w}{\partial t^2},\\
&\delta_2\frac{\partial^4\Phi}{\partial x^4}+2\delta_3\frac{\partial^4\Phi}{\partial x^2\partial y^2}+\delta_1\frac{\partial^4\Phi}{\partial y^4}=(1-\Gamma^*)\left\{\left(\frac{\partial^2 w}{\partial x\partial y}\right)^2-\frac{\partial^2 w}{\partial x^2}\frac{\partial^2 w}{\partial y^2}\right\},
\end{aligned}
\tag{2}
$$

where $\delta_1=\frac{1}{E_1}$, $\delta_2=\frac{1}{E_2}$, $2\delta_3=\frac{1}{G}-\frac{\mu_1}{E_1}-\frac{\mu_2}{E_2}=\frac{1}{G}-\frac{2\mu_1}{E_1}$.

The solution of the system (2) with respect to the deflection w and the stress function Φ is found in the form

$$
w(x,y,t)=\sum_{n=1}^{N}\sum_{m=1}^{M}w_{nm}(t)\psi_{nm}(x,y),\quad \Phi(x,y,t)=\sum_{n=1}^{N}\sum_{m=1}^{M}\Phi_{nm}(t)\chi_{nm}(x,y) \tag{3}
$$

where $w_{nm}=w_{nm}(t)$ and $\Phi_{nm}=\Phi_{nm}(t)$—are the unknown time functions; $\psi_{nm}(x,y)$, $\chi_{nm}(x,y)$, $n=1,2,\ldots,N$; $m=1,2,\ldots,M$ are the coordinate functions that satisfy the boundary conditions of the problem.

Substituting (3) to (2) and introducing the following dimensionless quantities

$$
\begin{aligned}
&\overline{w}=\frac{w}{h_0};\ \overline{w}_0=\frac{w_0}{h_0};\ \overline{x}=\frac{x}{a};\ \overline{y}=\frac{y}{b};\ \overline{t}=\omega t;\ \overline{h}=\frac{h}{h_0};\ \lambda=\frac{a}{b};\ \delta=\frac{b}{h_0};\\
&\overline{q}=\frac{q}{\sqrt{E_1E_2}}\left(\frac{b}{h_0}\right)^4;\ \overline{\theta}=\frac{\theta}{\omega};\ \omega t;\ \frac{\Gamma(t)}{\omega};\ \delta_0=\frac{P_0}{P_{cr}};\ \delta_1=\frac{P_t}{P_{cr}};\ \Delta=\sqrt{E_1/E_2};\ g=\frac{G_{12}}{\sqrt{E_1E_2}}
\end{aligned}
$$

while retaining the previous notation with respect to the unknowns $w_{nm}=w_{nm}(t)$, $\Phi_{nm}=\Phi_{nm}(t)$, we obtain the following system

$$2\pi^4\lambda^4[1+\Delta\mu_2+2(1-\mu_1\mu_2)g]\sum_{n=1}^{N}\sum_{m=1}^{M}a_{k\ln m}\ddot{w}_{nm}$$

$$+\sum_{n=1}^{N}\sum_{m=1}^{M}p_{k\ln m}^2(1-2\mu_{k\ln m}\cos\Theta t)w_{nm}-(1-\Gamma^*)\sum_{n=1}^{N}\sum_{m=1}^{M}a_{1k\ln m}w_{nm}$$

$$=12(1-\mu_1\mu_2)\lambda^2\sum_{n,i=1}^{N}\sum_{m,j=1}^{M}a_{2k\ln mij}w_{nm}\Phi_{ij}+12(1-\mu_1\mu_2)\lambda^4q_{kl}, \tag{4}$$

$$\sum_{n=1}^{N}\sum_{m=1}^{M}b_{k\ln mij}\Phi_{nm}=\lambda^2\sum_{n,i=1}^{N}\sum_{m,j=1}^{M}b_{1k\ln mij}w_{nm}w_{ij},$$

$$w_{nm}(0)=w_{0nm},\ \dot{w}_{nm}(0)=\dot{w}_{0nm}$$

The solution of the system (3) is found by the numerical method proposed in [19]. This method is based on the use of quadrature formulas and eliminates the singularity in the relaxation kernel. In this case, a weakly singular Koltunov-Rzhanitsyn kernel of the following form is used as a relaxation kernel [20]:

$$\Gamma(t)=Ae^{-\beta t}\cdot t^{\alpha-1},\quad A>0,\quad \beta>0,\quad 0<\alpha<1 \tag{5}$$

An efficient computational algorithm was developed and a program in the Delphi algorithmic language was developed and implemented on a computer. The results of the study of parametric vibrations of orthotropic viscoelastic plates of variable thickness without taking into account the propagation of elastic waves at various physical and geometric parameters are given in the form of graphs.

The thickness variation law is set analytically and, in the general case, can be of any form. To obtain numerical results, the law of thickness variation is chosen in the form:

$$h=1-\alpha^*x, \tag{6}$$

where α^* is the parameter of the change in thickness.

3 Results and Discussion

The influence of the viscoelastic properties of the material on the behavior of the plate is shown in Fig. 1. The results of the study show that an account for the viscosity of the plate material leads to a decrease in the vibration amplitude.

Figure 2 shows the influence of the nonhomogeneity parameter Δ on the behavior of the viscoelastic plate. The results show that an account for the nonhomogeneous properties of the plate material leads to a decrease in the vibration amplitude.

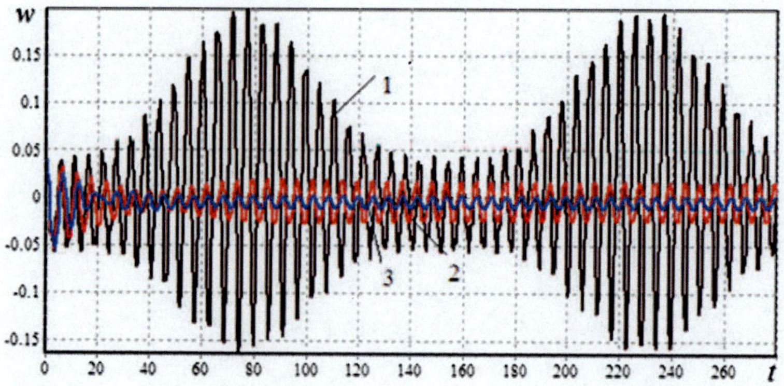

Fig. 1 The dependence of the deflection on time: $A_{ij} = 0$ (1); 2, $A_{ij} = 0.05$ (2); 3, $A_{ij} = 0.1$ (3)

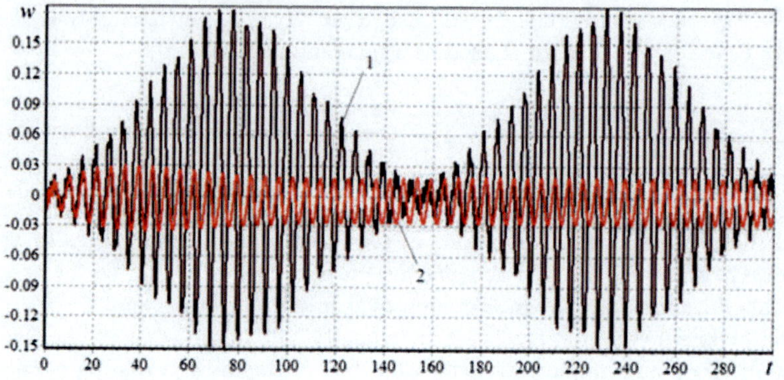

Fig. 2 The dependence of the deflection on time: $A_{ij} = 0.02$, $\Delta = 1.1$ (1); $A_{ij} = 0.02$, $\Delta = 1.3$ (2)

The change in the deflection of a viscoelastic plate as a function of time for different values of the thickness variability parameter α^* is shown in Fig. 3. It is seen that an increase in this parameter leads to a decrease in the vibration amplitude.

Figure 4 shows the results of studying the parametric vibrations of an orthotropic viscoelastic rectangular plate of variable thickness under various boundary conditions. The results obtained show that with an increase in the number of fixed sides of the plate, the vibration amplitude decreases, and the vibration frequency increases.

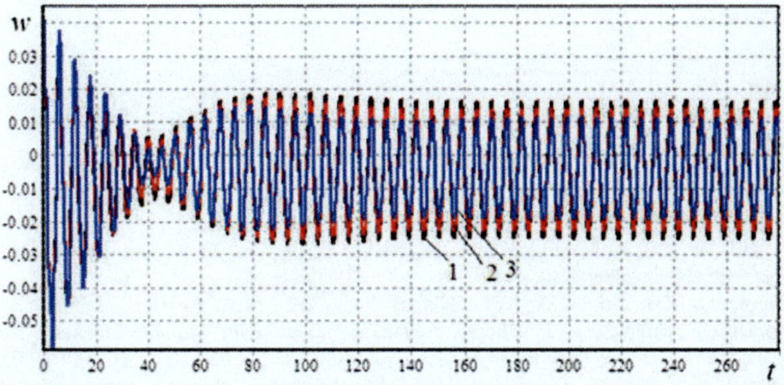

Fig. 3 The dependence of the deflection on time: α* = 0 (1); α* = 0.3 (2); α* = 0.5 (3)

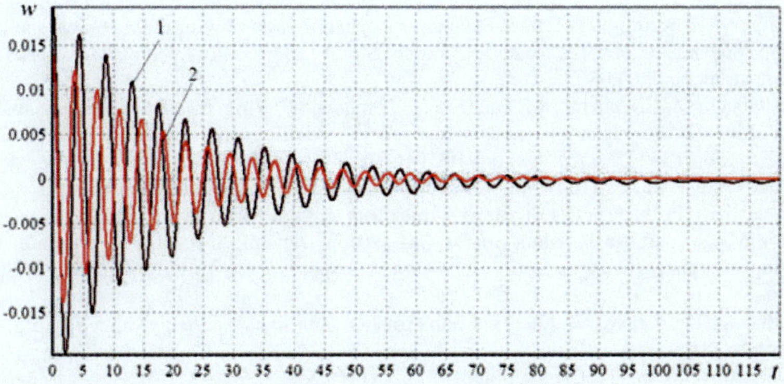

Fig. 4 The dependence of the deflection on time: two opposite sides are simply supported, the other two are clamped (1); all sides are clamped (2)

4 Conclusions

The study of parametric vibrations of orthotropic viscoelastic rectangular plates of variable thickness gave the following results:

1. A mathematical model, a method and an algorithm for solving the problem of parametric vibrations of orthotropic viscoelastic rectangular plates of variable thickness without considering the propagation of elastic waves were developed.
2. The proposed method can be used for various viscoelastic thin-walled structures such as plates, panels and shells of variable thickness.
3. The developed technique allows obtaining the results of the study of parametric vibrations of an orthotropic viscoelastic plate of variable thickness and for other laws of thickness variation.

References

1. Bolotin VV (1964) The dynamic stability of elastic systems. Holden-Day, San Francisco
2. Volmir AS (1974) The nonlinear dynamics of plates and shells. Foreign Technology Division Wright-Patterson Air Force, USA, Ohio
3. Awrejcewicz J, Krys'ko AV (2003) Analysis of complex parametric vibrations of plates and shells using Bubnov-Galerkin approach. Arch Appl Mech 73:495–504. https://doi.org/10.1007/s00419-003-0303-8
4. Amabili M (2018) Nonlinear vibrations and stability of shells and plates. Cambridge University Press, New York
5. Grigorenko YM, Grigorenko AY, Zakhariichenko LI (2018) Analysis of influence of the geometrical parameters of elliptic cylindrical shells with variable thickness on their stress-strain state. Int Appl Mech 54:155–162. https://doi.org/10.1007/s10778-018-0867-1
6. Kurpa LV, Mazur OS, Tkachenko VV (2014) Parametric vibration of multilayer plates of complex shape. J Math Sci 203:165–184. https://doi.org/10.1007/s10958-014-2098-2
7. Mochalin AA (2015) Parametric oscillations of a non-uniform circular cylindrical shell of variable density at different boundary conditions. News Saratov Univ Ser Math Mech Comput Sci 15:210–215
8. Sofiyev AH, Kuruoglu N (2016) Domains of dynamic instability of FGM conical shells under time dependent periodic loads. Compos Struct 136:139–148. https://doi.org/10.1016/j.compstruct.2015.09.060
9. Bacciocchi M, Eisenberger M, Fantuzzi N, Tornabene F, Viola E (2016) Vibration analysis of variable thickness plates and shells by the generalized differential quadrature method. Compos Struct 156:218–237. https://doi.org/10.1016/j.compstruct.2015.12.004
10. Usarov MK (2017) To the solution of three-dimensional problems of oscillations in the theory of elasticity in thick plates of variable thickness. J Appl Math Phys 5:79987
11. Loja MAR, Barbosa JI, Mota Soares CM (2017) Dynamic instability of variable stiffness composite plates. Compos Struct 182:402–411. https://doi.org/10.1016/j.compstruct.2017.09.046
12. Akhavan H, Ribeiro P (2017) Geometrically non-linear periodic forced vibrations of imperfect laminates with curved fibres by the shooting method. Compos Part B Eng 109:286–296. https://doi.org/10.1016/j.compositesb.2016.10.059
13. Golpayegani IF, Arani EM, Foroughifar AA (2019) Finite element vibration analysis of variable thickness thin cylindrical fgm shells under various boundary conditions. Mater Perform Charact 8:20180148. https://doi.org/10.1520/mpc20180148
14. Rasool M, Singha MK (2019) Stability of variable stiffness composite laminates under compressive and shearing follower forces. Compos Struct 225. https://doi.org/10.1016/j.compstruct.2019.111003
15. Phu KV, Bich DH, Doan LX (2021) Nonlinear forced vibration and dynamic buckling analysis for functionally graded cylindrical shells with variable thickness subjected to mechanical load. Iran J Sci Technol Trans Mech Eng. https://doi.org/10.1007/s40997-021-00429-1
16. Abdikarimov RA, Zhgutov VM (2010) Mathematical models of the problems of nonlinear dynamics of viscoelastic orthotropic plates and shells of variable thickness. Mag Civ Eng 16:38–47
17. Abdikarimov R, Khodzhaev D, Vatin N (2018) To calculation of rectangular plates on periodic oscillations. MATEC Web Conf. https://doi.org/10.1051/matecconf/201824501003
18. Normuminov B, Abdikarimov R, Khodzhaev D, Khafizova Z (2020) Parametric oscillations of viscoelastic orthotropic plates of variable thickness. IOP Conf Ser Mater Sci Eng 896. https://doi.org/10.1088/1757-899X/896/1/012029

19. Abdikarimov RA, Khodzhaev DA (2014) Computer modeling of tasks in dynamics of viscoelastic thinwalled elements in structures of variable thickness. Mag Civ Eng 49:83–94. https://doi.org/10.5862/MCE.49.9
20. Mal'tsev LE (1979) The analytical determination of the Rzhanitsyn-Koltunov nucleus. Mech Compos Mater 15:131–133

Method of Calculation and Placement of Spring Force Compensators in Log Structures of Wooden Housing Construction

A. Galyautdinov, A. Chernykh, V. Glukhikh, E. Furman, and V. Polozhencev

Abstract By their nature, all trees initially have internal tensions. They depend on the type of wood, the place of growth, the impact of external factors. Internal stresses of growth arising in a growing tree are the reason for the appearance of deformations in wooden housing objects. The main object of this article is a wooden house made of rounded logs. To exclude the effect of bulging of logs of wooden structures and to prevent the appearance of gaps between them, a solution was proposed to compensate for such effects. For this purpose, the use of spring power compensators is proposed. This paper we analyzes the results of theoretical studies on the effect of internal growth stresses, distributed over the section of the trunk, on the stress-strain state of a rounded log. A theoretical assumption is made about the influence of prestressed devices on the compensation of deformations of log structures. As a result, a method for determining the effective location of compensation devices along the length of the log has been developed. The most rational technical characteristics of Spring Force Compensators are revealed. The graphs of a compensating deformation versus location in the logs were obtained for different technical characteristics of the devices.

Keywords Timber structures · Wood · Spring force compensators · Log house · Wooden structures · Timber log-walls · Growth stresses · Wooden housing objects · Bulging effect

---

A. Galyautdinov · A. Chernykh (✉) · V. Glukhikh · V. Polozhencev
Saint Petersburg State University of Architecture and Civil Engineering, 4 Vtoraya Krasnoarmeiskaya ul., Saint Petersburg 190005, Russian Federation

E. Furman
Bolsheprolet, Fermskoe Sh., 30 l. A, office f/1-2, Saint-Petersburg 197341, Russian Federation

N. Vatin et al. (eds.), *Proceedings of MPCPE 2021*, Lecture Notes in Civil Engineering 182, https://doi.org/10.1007/978-3-030-85236-8_12

1 Introduction

1.1 General Information

Timber structures is now immensely popular because wood has a beneficial effect on human health. An increase in the share of wood use in construction will have a positive effect on the environmental situation in the world [1, 2]. But people living in wooden houses often face the problem of various kinds of deformations of the logs in the crowns. Often, mechanically processed logs, such as rounded logs, are subject to such deformations [3]. Some of these deformations include fissures, cracking, and buckling [4]. All this has a negative impact on the quality of the house itself. Special Spring Force Compensators can be used to reduce the deformations of the logs from the longitudinal axis.

One of the reasons for the appearance of log deformation is internal growth stresses, which are described [5–8]. Formulas were obtained for the resultant force of internal forces and bending moment because of a theoretical study of internal growth stresses in a tree trunk, which appear during cylindering [9].

1.2 Design of Spring Force Compensator

The design of the spring force compensator was based on the "spring unit force" device from the company "Sila-Center" (Fig. 1).

A spring force compensator pulls two logs together. For this, a landing nest with a diameter of 5 mm larger than the diameter of the spring is drilled in the upper log. The depth of the hole depends on the diameter of the logs used (Fig. 2).

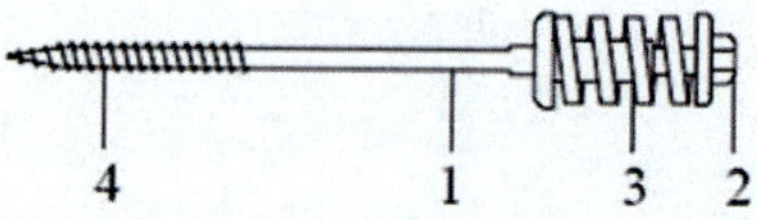

Fig. 1 Spring device diagram: 1-rod; 2-rod head; 3-spring; 4-thread

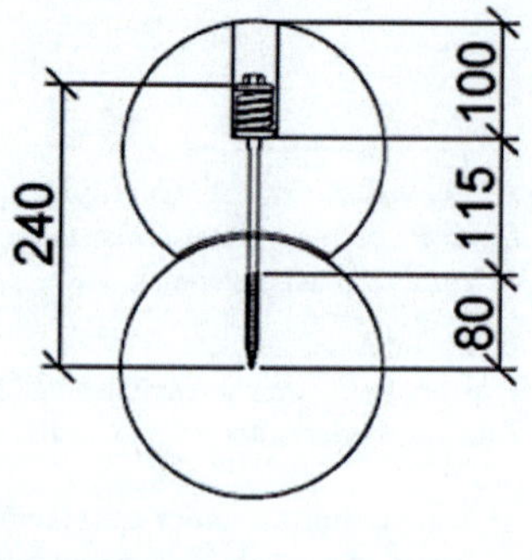

Fig. 2 Arrangement of the devices

In further calculations, we assume that the spring is compressed until the coils touch.

2 Methods and Materials

This section presents a methodology for calculating and placing spring force compensators, it should become universal in the design of wooden houses. The algorithm of actions, based on which the methodology was developed, is presented in (Fig. 3).

2.1 Collection of Loads

As a load, we consider only the constant that is in effect during the whole year. It can include only the dead weight of structures—the weight of the roof, the weight of the ceiling, the weight of the wall logs.

Logs located at different levels relative to the height of the wall will perceive different loads. The upper log will be the least loaded, at the same time the lower log, which is mated with the foundation, will be the most loaded of all. This affects the rate of the deformation that must be compensated by using Spring force compensators. Therefore, at the end of this paragraph, we will make a table indicating the value of the distributed load on each log in the house (Table 1).

To determine the load, a general formula has been drawn up:

$$Q_i = Q_{dead\ weight} \cdot (i - 1) + Q_{roof} + Q_{ceiling} + Q_{dead\ weight},\ N/m \qquad (1)$$

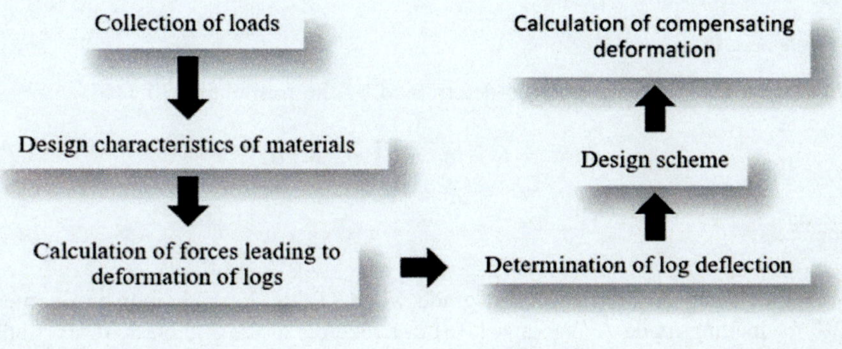

Fig. 3 Algorithm of actions

Table 1 Example of load information on a log in a wall

Wall in axes A–B				
Log number	Formula	Load name	Lad value	Unit of measure
1	$Q_i = Q_{dead\ \mathrm{weight}} \cdot (i-1)$	Dead weight of log	250	N/m
…	…	…	…	…

where:

$Q_{dead\ \mathrm{weight}}$—dead weight of logs in the wall structure, N/m;
Q_{roof}—dead weight of all elements of the roof structure, N/m;
$Q_{ceiling}$—dead weight of all elements of the ceiling structure, N/m;
i—sequential number of the log from less to more loaded (top-down).

2.2 Design Characteristics of Materials

At the beginning, it is necessary to determine the operating conditions of the structures:

- Purpose of the object;
- Class of functional purpose (Table A1, [10]);
- Class of operating conditions (table. 1, [10]);
- Approximate service life (Table A3, [10]);
- Class of structures (Table 2, [11]);
- Loading mode (tab. 4, [10]);
- Wood grade;
- Strength class (p. 6.2.2, [12]).

Then, the design characteristics are determined according to SP 64.13330.2017 [10]. It is also possible to determine the design characteristics according to EN 1995-1-1—Eurocode 5 [13].

Design resistance

The design resistance of wood is determined by the formula p. 6.1 [10]:

$$R^p = R^A \cdot m_{\partial\pi} \cdot \prod m_i,\ MPa \tag{2}$$

where:

R^A, MPa—design resistance of 1st grade wood (Table 3, [10]), moisture content 12% for loading mode A (Table 4, [10]) in structures of the 2nd class of functional purpose (Appendix B, [10]) with a service life of no more than 50 years;

$m_{\partial\pi}$—coefficient of long-term strength corresponding to the mode loading duration (Table 4, [10]).

$\prod m_i = m_\beta \cdot m_T \cdot m_б \cdot m_o \cdot m_a \cdot m_{c\pi} \cdot m_{cH} \cdot m_{c.c} \cdot m_{cM}$—product of operating conditions coefficients (6.9, [10]).

Calculated modulus of elasticity

This calculation is carried out according to the second limiting state, therefore, it is possible to restrict oneself to determining the calculated elastic modulus (shear modulus) of wood and wood materials according to the limiting states of the 2nd group according to the formula of clause 6.11 [10]:

$$E^{II} = E_{cp} \cdot m_{\partial\pi \cdot E} \cdot \prod m_i, \text{ GPa} \tag{3}$$

where:

E_{cp}, GPa—average modulus of elasticity in bending (adj. B, [10]);
$m_{\partial\pi \cdot E}$—duration factor for elastic characteristics (Table 4 [10]);
$\prod m_i = m_\beta \cdot m_T \cdot m_{c \cdot c}$—product of operating conditions coefficients (п.6.9a, 6.9б, 6.9и, [10]).

2.3 Calculation of Forces Leading to Deformation of Logs

The design scheme that must be used is shown in Fig. 4.

The design formulas for the bending moment and the resultant force exerted by the action of internal growth stresses were obtained in the article [9].

Formula for determining the resultant force:

$$N = \int_{-R_2}^{+R_2} dx \int_{-\sqrt{R_2^2 - x^2}}^{\sqrt{R_2^2 - x^2}} k_4 \cdot \left((x+a)^2 + (y+b)^2 \right)^4 + b_0 \, dy, \; kN \tag{4}$$

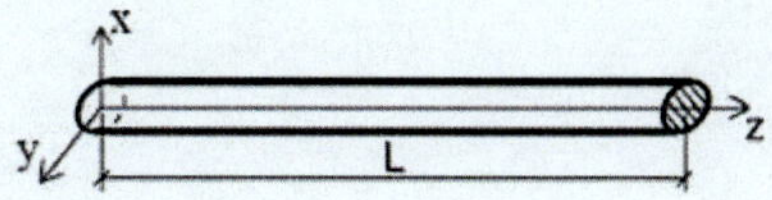

Fig. 4 A design scheme

Formula for determining the bending moment:

$$M_{x0} = \int_{-R_2}^{+R_2} dx \int_{-\sqrt{R_2^2-x^2}}^{\sqrt{R_2^2-x^2}} \left(k_4 \cdot \left((x+a)^2 + (y+b)^2\right)^4 + b_0\right) \cdot y\,dy, \text{ kNm} \quad (5)$$

where:

R_2, m—the radius of the rounded log;
a, m—displacement of the axis when rounding the log;
$k_4 = \frac{\sigma_R + \sigma_0}{R^4}$—coefficient;
σ_R, MPa—stresses on the tree trunk contour;
$\sigma_0 = \frac{\sigma_{BC}}{1{,}8255}$, MPa—stresses in the center of the tree trunk.

2.4 Determination of Log Deflection

As a design scheme, we assume that the log is symmetrically bent on both sides, and clamped in the middle (Fig. 5).

Knowing the maximum bending moment from the action of internal growth stresses, it is possible to obtain the maximum deflection that will appear in this case. General formula for determining the displacement of the end of a beam [14–16]:

$$E \cdot I \cdot y_z = \int dz \int M_z dz + C_1 \cdot z + C_2 \quad (6)$$

In this case, the formula will take the form:

$$y = \frac{M \cdot z^2}{2 \cdot E \cdot I_x} + C_1 \cdot z + C_2 \quad (7)$$

The values of the integration constants C_1 and C_2 are determined according to the boundary conditions. In the clamping point of the cantilever beam, the angle of rotation and movement are zero, therefore C_1 and C_2 are also equal to zero. Based on this, formula (7) will take the form:

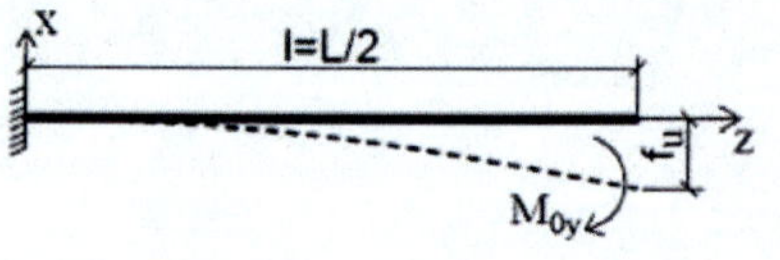

Fig. 5 A design scheme for determining the deflection of a log

$$f = \frac{M \cdot (l/2)^2}{2 \cdot E \cdot I_x} \tag{8}$$

2.5 Design Scheme

For further calculations, it is necessary to draw up a design scheme. As an example, let's take a low-rise wooden house, the walls of which are made of rounded logs. The log in the crowns of the log structure is acted upon by pressure from the overlying structures. We take this impact as a uniformly distributed load. In those crowns where the external load cannot compensate for the deformation of the logs, it is advisable to use spring force compensators, which serve as an additional source of impact.

The design scheme of the installation of spring force compensators is shown in Fig. 6.

Where:

f_1,f_2,f_n—part of the deflection eliminated by the spring force compensators, the corresponding amount on the console;
a_1, a_2, a_n—distance of installation of the spring force compensators;
F_1, F_2, F_n—forces from spring force compensators;
$q_{external}$—external uniformly distributed load from overlying structures.

The general deformation condition must be met:

$$f_1 + f_n + f_{external} = f_u \tag{9}$$

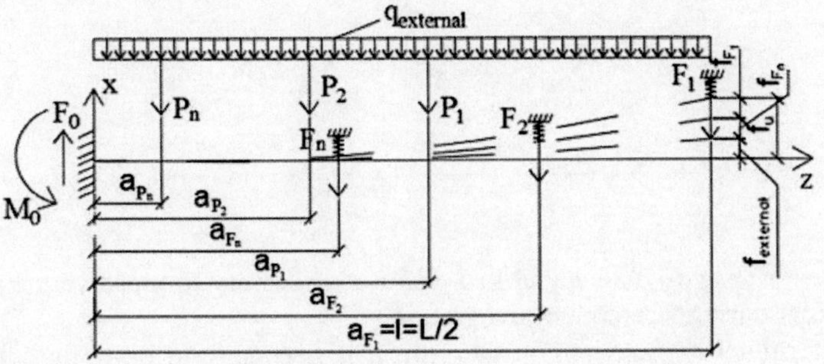

Fig. 6 The design scheme of the installation

2.6 Calculation of Compensating Deformation

The calculation is carried out in four stages:

- Determination of compensating deformation from external load;
- Revealing according to the formula (9) permanent deformation, which must be compensated by spring force compensators;
- Determination of deformation from spring force compensators at the end of the log. Consider that this clamping device is installed in each row;
- Determination of deformation from the required number of intermediate spring force compensators.

Below is a general calculation suitable for calculation at each stage.

For the calculation, we will use the method of initial parameters [14, 17], which allows for any number of sections to reduce the solution to finding only two constants—the deflection and the angle of rotation at the origin.

To implement the method of initial parameters, it is necessary to adhere to the following rules when drawing up the equation of moments for sections and integrating this equation:

(1) the origin of coordinates must be chosen common for all sections at the extreme left (or right) point of the beam;
(2) all components of the equation of moments in the previous section should remain unchanged in the equation of moments of subsequent sections;
(3) in the event of a break in the distributed load, it is extended to the end of the beam, and to restore the actual loading conditions, a "compensating" load of the opposite direction is introduced;
(4) the equations should be integrated in all sections without opening the brackets.
(5) the factor (x–a) must always be positive, terms with negative values (x–a) are discarded.

Let us define the internal bending moment in general form for an arbitrary section:

$$M_x = \sum_{n=0\ldots} F_n(x - a_{F_n}) + \sum_{n=1\ldots} P_n(x - a_n) + \\ + \sum_{n=1\ldots} \frac{q_n \cdot (x - a_{q_n})^2}{2} + \sum_{n=1\ldots} M_n \cdot (x - a_{M_n})^0 \tag{10}$$

The factor $(x - a_{M_n})^0$ is equal to 1, but it is necessary to preserve the similarity of the terms during subsequent integration.

To determine the reactions in the seal, it is necessary to draw up a system of equations:

$$\sum OY = \sum_{n=0...} F_n + \sum_{n=1...} P_n + \sum_{n=1...} q_n \cdot l_{q_n} \tag{11}$$

$$\sum M_A = \sum_{n=0...} M_n + \sum_{n=0...} F_n \cdot a_{F_n} + \sum_{n=1...} P_n \cdot a_{P_n} + \sum_{n=1...} q_n \cdot l_{q_n} \cdot (a_{q_n} + l_{q_n}/2) \tag{12}$$

The sign of the force is determined depending on its direction relative to the OY axis. A positive value is an upward direction.

The sign of the moment is determined depending on its direction of rotation of the OX axis. A positive value is clockwise twist.

Where:

P_n—concentrated effort from external load;
F_n—concentrated effort from the spring force compensators;
M_n—bending moments;
M_0—bending moment in the clamping point of the cantilever beam;
F_0—force in the clamping point of the cantilever beam;
q_n—distributed external load;
l_{q_n}—length of the load distribution area;
a_{F_n}—is the distance from point A to the place of application of the concentrated load from the spring force compensators;
a_{P_n}—is the distance from point A to the place of application of a concentrated external load;
a_{q_n}—is the distance from point A to the place of application of the extreme left point of the distributed load (in this case $a_{q_n} = 0$).

After drawing up Eq. (10) for a specific case, it is necessary to determine the compensating deformation provided in this or that case.

To do this, we will use the universal equation to determine the deflection in a given section:

$$y(x) = \frac{1}{E \cdot I_x} \cdot \iint M(x)dx \tag{13}$$

The boundary conditions for determining the integration constants will be as follows:

- In the clamping point of the cantilever beam: $\varphi(x) = 0,\ x = 0$;
- In the clamping point of the cantilever beam: $y(x) = 0,\ x = 0$.

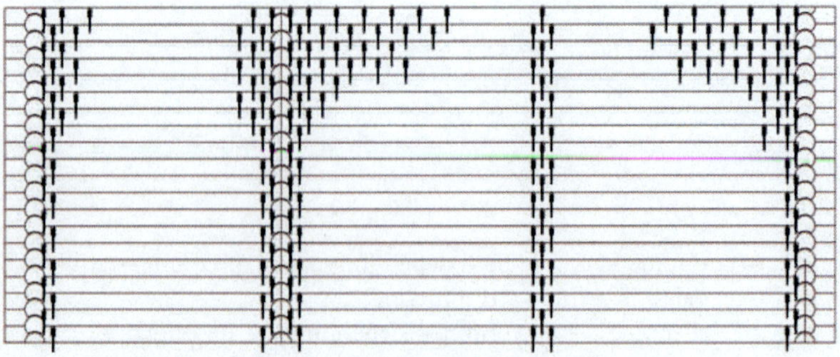

Fig. 7 Scheme of the arrangement of devices along the length of the log

2.7 *Drawing up a Scheme of the Arrangement of Devices Along the Length of the Log*

After finding the required number of devices in each crown of the log house, we draw up a diagram of the arrangement of spring force compensators along the length of the log for each wall of the house, as shown in Fig. 7.

3 Results

To find the optimal combination of the installation site of the spring force compensators and the force exerted by them, it is necessary to find the dependence of the compensating deformation on the load exerted by the spring force compensators.

If you initially set certain parameters:

- take several points on the log with a step of 0.1 m;
- split the forces that can be exerted by spring force compensators into several groups with a step of 0.1 kN.

As a result, using the formulas presented above, it is possible to plot the dependence of the compensating deformation versus the force of the spring force compensators for different positions of these devices along the length of the log (Fig. 8). To plot the graph, the software package Microsoft EXCEL was used.

The choice of rational technical characteristics of spring force compensators depends on the choice of the spring, that is, on the force that it can exert. To determine the optimal spring force depending on the installation location along the log, the graph in Fig. 8.

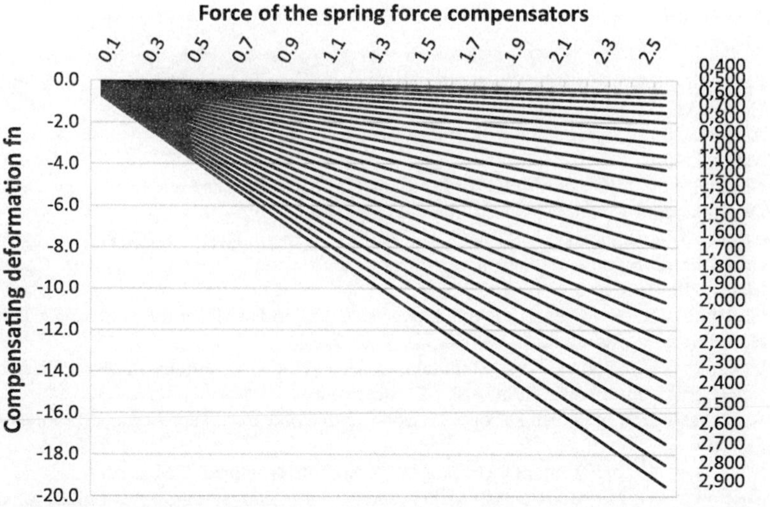

Fig. 8 Dependence diagram of compensating deformation versus the forces of spring force compensators

4 Discussion

Research in the field of application of spring power expansion joints has not been carried out at the moment. Usually, construction organizations that are engaged in the construction of wooden houses using log structures technology used these devices according to the general schemes provided by the manufacturers. This technique is the beginning of a further study of the work of log structures.

5 Conclusions

As a result of theoretical research, a "Methodology for calculating and placing Spring Force Compensators in log structures of wooden housing construction" was developed, which makes it possible for a scientific and substantiated approach to the use of spring force compensators in the design and construction of low-rise buildings.

With the help of the graph of the dependence of the compensating deformation versus the force of the spring force compensators, it is possible to determine the most effective location of the spring force compensators and select the most rational technical characteristics of the spring devices.

References

1. Serov EN, Sannikov YD, Serov AE (2011) Designing wooden structures. ASV Publishing House, M., St. Petersburg
2. Asif M (2009) Sustainability of timber, wood and bamboo in construction. Sustain Constr Mater 31–54
3. Khetsuriani ED, Kvashilava IG, Khetsuriani TE (2016) Building a house from a rounded log. Development of modern science: theoretical and applied aspects, pp 45–47
4. Cabaleiro M, Lindenbergh R, Gard WF, Arias P, Van De Kuilen JWG (2017) Algorithm for automatic detection and analysis of cracks in timber beams from LiDAR data. Constr Build Mater 130:41–53
5. Glukhikh VN, Akopyan AL, Okhlopkova AYu (2018) Natural features of wood. Publishing House of the Polytechnic University, St. Petersburg
6. Glukhikh VN, Chernykh AG, Okhlopkova AYu (2013) Anisotropy of wood Technological aspect. Publishing house SPbGASU, St. Petersburg
7. Ashkenazi EK (1978) Anisotropy of wood and wood materials. Lesnaya promyshlennost, Moscow
8. Kuznetsov AI (1950) Internal stresses in wood. Goslesbumizdat, Moscow
9. Chernykh AG, Mamedov ShM, Nizhegorodtsev DV, Gulyautdinov AV (2020) Analysis of efforts leading to the curvature and buckling of logs in wooden housing construction. Scientific Aspect 3(S):841–850
10. SP 64.13330.2017 (2017) Wooden structures [Text]: normative and technical material, p 90. Standardinform, Moscow
11. GOST 27751-2014 (2019) Reliability of building structures and foundations. Basic provisions. Rosstandart
12. GOST 33080-2014 (2019) Wooden structures. Strength classes of structural sawn timber and methods for their determination. Rosstandart
13. EN 1995-1-1—Eurocode 5—Design of timber structures—Part 1-1: general-common rules and rules for buildings—European Committee for Standardisation (CEN), Brussels, Belgium (2009)
14. Moutee M, Fafard M, Fortin Y, Laghdir A (2005) Modeling the creep behavior of wood cantilever loaded at free end during drying 37(3):521-534
15. Carlsen GG, Slitskoukhova YuV (1986) Structures made of wood and plastics, 4th edn. Stroyizdat, Moscow
16. Ross Carl TF, Chilver A (1999) Strength of materials and structures, 4nd edn. Butterworth-Heinemann, Great Britain
17. Kimball C, Tsai L-W (2002) Modeling of flexural beams subjected to arbitrary end loads. J Mech Des Trans ASME 124(2):223–235

Finite Element Modeling of Complexly Stressed Reinforced Concrete Structures

V. P. Pochinok, E. V. Greshkina, and M. M. Tamov

Abstract Paper presents application of the nonlinear finite element model Cyclic Softened Membrane Model (CSSM) to analyze the plane-stressed reinforced concrete elements, such as shear walls, partitions, lintels, wall panels, deep beams, and others. This constitutive model developed at the University of Houston and implemented in the software complex OpenSees (Open System for Earthquake Engineering Simulation) was successfully used in our previous analyses of shear critical zones of the experimental thin-webbed reinforced concrete beams with varied types of reinfocemenr (bars, mesh) and basic characteristics of material and shear span. The following results have been acquired: the appropriate choice of mesh size for finite elements results in best proximity of strength and deformation analysis of the reference and the experimental data for deep beams as well as composite constructions and strengthened and plane stressed elements that vary in sizes, types of reinforcement and load schemes. Here are given results of the calculation model practical usage that were obtained on the basis of the experimental researches published earlier.

Keywords Experimental research · Stress-strain behaviour · Plane stress reinforced composite constructions

1 Introduction

The stress-strain behaviour defines the bearing capacity, rigidity, and crack resistance of the reinforced concrete structures. Thus, the structural engineering and the design of structural members demand qualified evaluation of a stress-strain behaviour, that is considered to be quite a complex and time-consuming multi-foci task. In constructing of buildings and related structures on a large scale one can identify

V. P. Pochinok · E. V. Greshkina (✉) · M. M. Tamov
Kuban State Technological University, Moskovskaya str., Krasnodar 350072, Russian Federation

N. Vatin et al. (eds.), *Proceedings of MPCPE 2021*, Lecture Notes in Civil Engineering 182, https://doi.org/10.1007/978-3-030-85236-8_13

three groups of the reinforced concrete elements subjected to complex non-uniform loads.

They are as follows:

- plane stress structures of the constant section—webs and walls of the structural core in multistorey buildings that are characterized by different stress-related properties, lintels, precast panels of the bearing walls, deep beams in various structures, towers, and others;
- plane stress elements with composite cross-section that are formed in the process of restoration and rehabilitation of reinforced concrete structures and have different stress-related properties in different parts and layers of the concrete;
- short elements—cantilevers, pile caps, compound frame wall elements, short beams and other similar structures with a small shear span.

Review of the simulation models of the plane stress structures shows that despite the research rationale and a considerable amount of the experimental work no unifying computational theory has been evolved so far and the matter is still under discussion. The complexity of consideration of the nonlinear physical factors and a wide range of the analytical models result in the absence of the unifying basis for reliable and economical design of the constructions operating in the conditions of the complex stress state. In particular, existing building codes regulating the design and construction are lacking a recommended guideline on the reinforcement for this type of constructions [1–6]. For instance, a design codes for reinforced concrete frame constructions for industrial building establishes requirements for the wall panels design, referring to the general design code for concrete and reinforced concrete structures, which is not enough itself, since the design dependencies of this document do not take into account the effect on the bearing capacity of such structures, the nature of their reinforcement, the peculiarities of the loading patterns, the shapes and geometric characteristics of the sections, as well as the size of the load and support areas and a number of other factors. Leading scientists around the world are engaged in the issue of rational reinforcement of complexly stressed reinforced concrete structures, however, a single solution on the design provisions and principles of designing such elements has not been developed yet [7–23].

2 Possibility of Finite Element Modelling of Deep Beams or Wall Panels

The modern development of computing technology and programming tools allows the most accurate estimation of bearing capacity of the engineering constructions of any type of geometry under the combined stresses. For nonlinear modeling the finite element method (FEM) which has profound theoretical underpinning is widely used in engineering practice. This method is also utilized by many researchers who work in the field of plane stress calculations. Review of such works reveals that in many cases the major problem is the discrepancy between laboratory tests and analytical solutions.

This is because the insufficiency of necessary multivariable tests leads to a wide range of empirical allowances. Performing of non-linear calculations is complicated by the necessity of a greater number of iterations when applying the general simulation programs based on FEM and is considered to be difficult in the process of planning and design. In this case engineers tend to simplify and sometimes oversimplify computational models that in its turn results in loss of reliability and validity of the gained results. Now researchers from different countries are studying the issue of developing the general nonlinear plane stress computation model that allows performing highly accurate analysis of stress-strain state at all stages of loading till failure alongside taking into consideration all essential distinctive features of their structures and loading patterns. Within the framework of research it is needed to perform a numerical analysis as well as full-scale testing, to make a comparison between the newly offered plane stress computation models and the ones that have been being used so far.

The analysis of the plane stress-strain state of the reinforced concrete deep beams and bearing wall panels and its problems are close to the problems studied in the research of the effect of transverse compressive stresses on the web-crushing strength of beams. The theoretical studies on finite element modeling of nonlinear simulation of a thin web of I-shaped reinforced beam at all stages till the failure were fulfilled at Kuban State Technological University (Krasnodar) employing nonlinear finite element constitutive model in Cyclic Softened Membrane Model (CSMM) [24]. This model was developed at the University of Houston and implemented into OpenSees (Open System for Earthquake Engineering Simulation [25]) that in its turn was developed at Pacific Earthquake Engineering Research Center (PEER), UC Berkeley, USA. The successful results of these experiments are described in our works [26, 27].

To check the sensitivity of the simulation model to the extremely different parameters of the concrete strength in plane stress structures it was decided to make calculations of bearing capacity of deep beams mentioned in the work [28] of the following researchers: PhD in Technical Science N.I. Karpenko, C.N. Karpenko, A. N. Petrova. They presented the numerical simulation test of deep beams BS-1 and BS-2 with similar geometry parameters such as sizes and reinforcement but differing in concrete strength: 20 MPa and 100 MPa for the samples BS-1 and BS-2 respectively.

The results of our calculations received through CSMM were analyzed alongside with the mentioned above research results presented in the work [28] where the method based on physical laws in incremental (step-by-step evolved) form with the usage of material stress-strain diagrams in the form of broken lines was tested. The method of physically nonlinear analysis in the study [28] was performed as a computer simulation program for reinforced concrete deep beams with cracks that was used to conduct a comparative analysis of the stress-strain state of two structures characterized by the same reinforcement while concrete used for them is significantly different in strength.

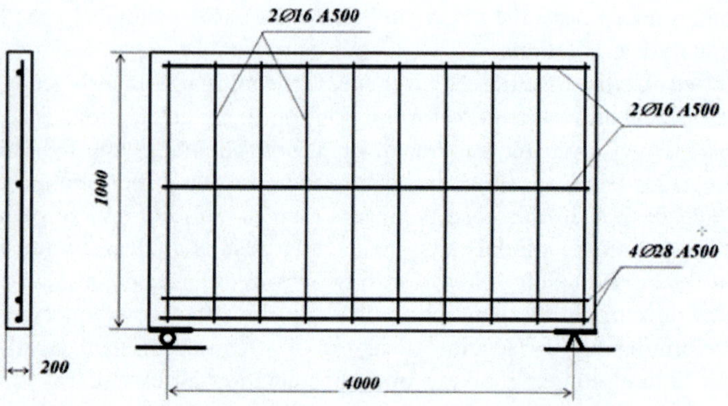

Fig. 1 Dimensions and reinforcement of deep beams BS-1 and BS-2 [12]

Deep beams BS-1 and BS-2 with the height of 1000 mm, thickness 200 mm and span of 4 m were reinforced by orthogonal mesh with bars Ø16 A500 at 200 mm. Four bars Ø28A500 were installed as longitudinal reinforcement in the tension zone (Fig. 1). The support of the deep beams was provided at the ends by steel plates 300 × 300 mm with a thickness of 30 mm. Their loading was modeled as a stepwise procedure with the transmission of a uniformly distributed load to the upper side face. Increasing the load to failure included 14 steps for the BS-1 deep beam and 24 steps for the BS-2 deep beam. The load at failure for the BS-1 deep beam was 280 kN/m and for the BS-2 deep beam, it was 960 kN/m. With a fivefold difference in the concrete strength, the failure load differed only by 3.43 times which is explained by the difference in the mechanisms for the exhaustion of the material's strength.

Sample BS-1 was failed due to concrete crushing in the compressed zone accompanied by the yield of compressed reinforcement. The tensile strength of the reinforcement was underutilized. The structural failure mechanism of the BS-2 deep beam was fundamentally different. It was caused by the yield of the longitudinal tension reinforcement while the stresses in the concrete of the compressed zone did not achieve the prism strength.

The obtained analysis results gained by the nonlinear computational model CSMM for plane-stressed concrete showed a fairly good agreement of the values of the failure loads: the best (960 kN/m) for the deep beam BS-2 and slightly more divergent (250 kN/m) for the deep beam BS-1. It was possible to achieve a better approximation of the analysis results for a beam made of concrete with lower strength by adjusting the sizes of finite elements, however, at this initial stage of trial calculations, it is not considered as the main issue.

The second group of deep beams for our analysis verification of the computational model of the University of Houston included more complex samples tested at the Kyiv National Aviation University [29–31]. T.V. Tugay under the supervision of a PhD in Technical Science Vl. I. Kolchunov conducted an experimental study

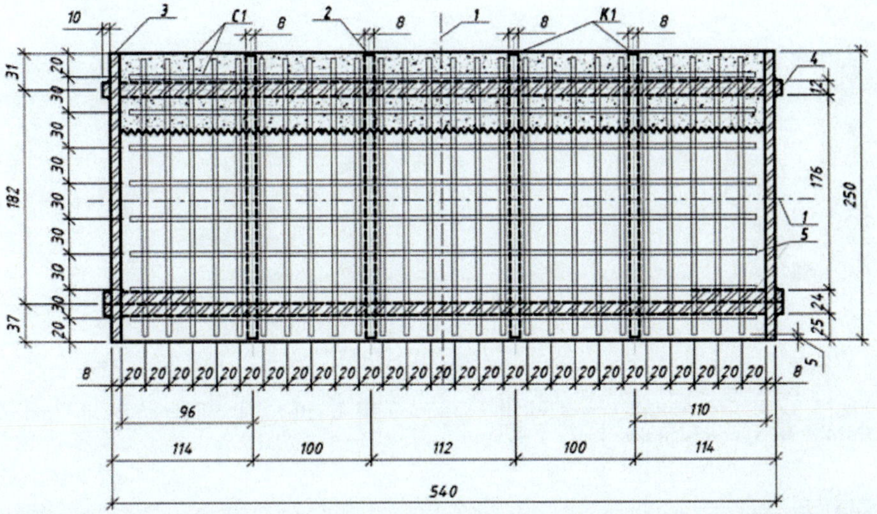

Fig. 2 Experimental reinforced concrete composite deep beams [13], 1—center line; 2—transverse reinforcement Ø8 A240C; 3—plate 8 mm thick, welded to the main longitudinal reinforcement; 4—longitudinal reinforcement Ø12 A500C; 5—reinforcing mesh C1 Ø3 mm 30 × 30 mm; 6—longitudinal reinforcement Ø12 A500C

and design of a method for determining the stiffness of composite plane-stressed reinforced concrete structures with different types of reinforcement and strength grade of concrete. The research program included laboratory tests of six series of deep beams that differed in reinforcement ratios, height and strength characteristics of the concrete in the main lower part and cast upper part of the strip simulating the structural strengthening (Fig. 2). The test program and the main parameters of the experimental samples as well as the testing procedure are given in the paper [29].

The samples varied in the reinforcement of the lower "main" part of the deep beam: for samples of series I and III it was taken as 2Ø12 A500C, for samples of the remaining series—as 4Ø12 A500C. The longitudinal reinforcement of the upper "cast" part of the deep beams of all series was the same and consisted of 2Ø12 A500C (Fig. 2).

The height of the main lower concrete layer for different specimens was taken as 180 or 210 mm, for the reference beams concrete without strengthening it was equal to 250 mm. The height of the concrete reinforcement layer was taken as 40 or 70 mm. The actual concrete strength was 16.48 MPa for specimens of the IV series; the strength of the primary concrete of all series and strengthening concrete of the I-IV series was equal to 32.18 MPa; for strengthening concrete of the V series is equal to 42.54 MPa.

To verify the CSMM model we performed calculations of displacements and failure loads for plane-stressed reinforced concrete composite deep beams of the III-VI series that were presented in the experiments [29–31]. The experimental [31]

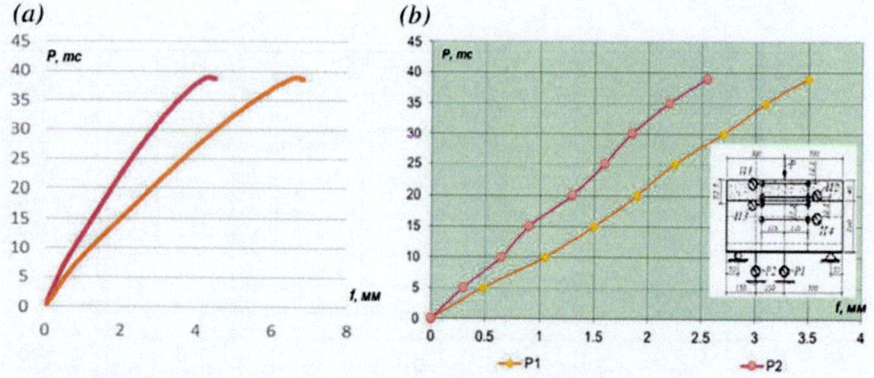

Fig. 3 Load-displacement curves for the specimen III BS30-30, **a** calculated by the model CSMM; **b** experimental data [31]

and obtained simulated "vertical load-displacement" diagrams shown in Figs. 3, 4, 5 and 6 demonstrate that the results of computer modeling were very close to experimental data up to the failure of these structural members. It was also confirmed that there was an increase in the displacements of strengthened specimens with an increase of the height of the concrete strengthening layer and the percentage of reinforcement.

The discrepancy between the simulated and experimental curves in the middle of the span before the failure of the deep beams presented in Figs. 3, 4, 5 and 6 does not exceed 10% in three out of four cases and the only twofold discrepancy was obtained for beam III BS 30–30. It should be noted that T.V. Tugai when calculating this beam in his thesis stated a significant discrepancy between the results of simulation and experiment as well [31, p. 151].

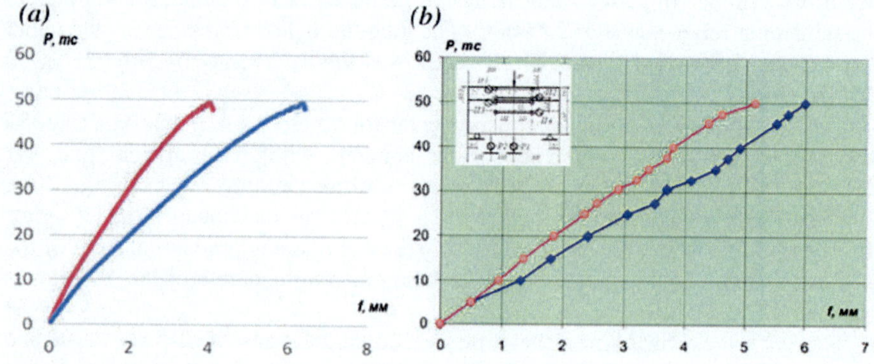

Fig. 4 Load-displacement curves for the specimen IV BS30, **a** calculated by the model CSMM; **b** experimental data [31]

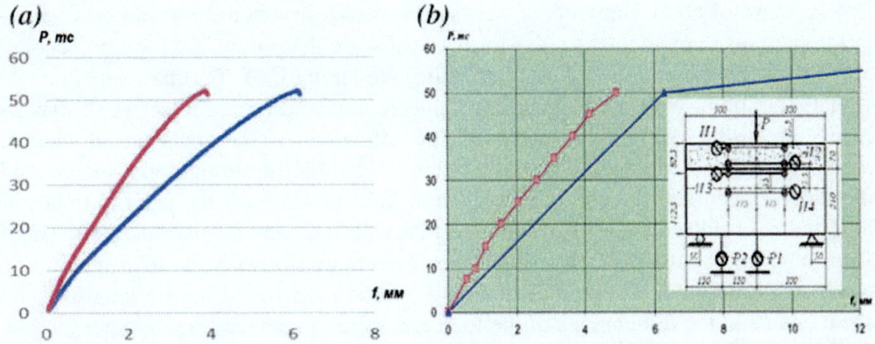

Fig. 5 Load-displacement curves for the specimen V BS30-40, **a** calculated by the model CSMM; **b** experimental data [31]

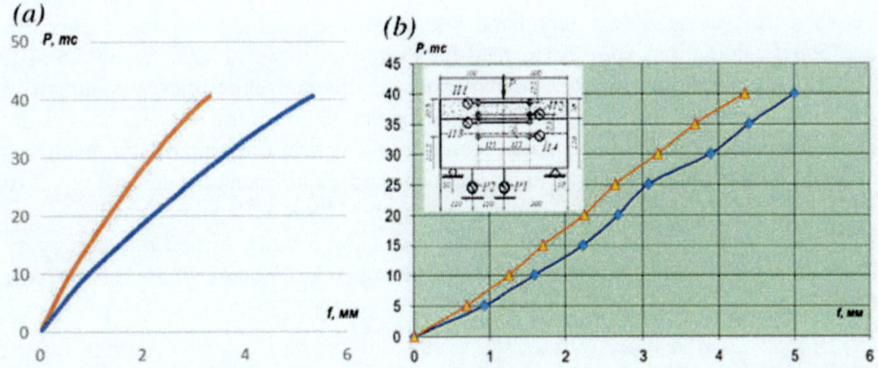

Fig. 6 Load-displacement curves the specimen III BS30-20, **a** calculated by the model CSMM; **b** experimental data [31]

The methods for analysis of short elements ($a/h_0 < 1.5$) under the action of transverse forces were developed by PhD in Sciences A.S. Zalesov and T.I. Baranova. Together they presented the concept for failure of short elements and deep beams along inclined compressed struts having created a simulation model [32] which described two types of failure: the loss of strength of an inclined compressed concrete strut and the rupture of a tensile reinforcing tie. In this case, the crucial issue is the analysis of principal compressive and tensile stresses. Experiments showed that the failure of a compressed struts has two features: in some cases, failure occurred as a result of an oblique shear, in others as a result of oblique compression. As a result, at the design stage, there is a problem of assessing the strength of a compressed strut located in the zone of maximum shear stresses. The solution to this problem depends on the degree of knowledge of the stress-strain state of short elements and the factors that determine it.

The class of short elements is vast so it is necessary to have a universal simulation method implemented in the form of software that allows taking into account all of the above-mentioned factors affecting the strength of short elements.

T.I. Baranova with a group of colleagues conducted a unique series of experiments including complex research of the influence of the main listed factors. Authors [33] studied dependence of the failure load on the strength of concrete and the value of the short shear span a/h_0 from 0.25 to 1.5 with the same number of longitudinal reinforcement. It was found that with the increase of concrete strength R from 30 to 50 MPa the failure loads increased more slowly than the growth of the concrete strength. The degree of this disproportion depended on the length of the shear span and the dimensions of the load and support areas and some other factors.

In the study [34] simulation model of short elements with a/h_0 from 1 to 1.5 from the experiments of T.I. Baranova in OpenSees software was made and researchers could compare the experimental failure loads with the predicted ones. After some adjustments of the analytical model, namely after correction of the FE mesh size towards a decrease in the node size we achieved a better convergence between the values of the simulated and experimental failure loads. Moreover, additional calculations were performed for short elements with a/h_0 from 0.25 to 1. The visual comparison of these calculations results and experiments is shown in Fig. 6. One can see that the peculiar experimental nature of the dependences of the experimental failure loads on shear span is replicated in our calculation results.

The average ratios value of the analytical and experimental data was equal to 1.007 for concrete beams with concrete strength of 29.8 MPa and 1.058 for beams with a strength equal to R = 51,9 coefficient of these ratios turned out to be very low and is almost same for both groups of beams: 0.0266 and 0.0255, respectively (Fig. 7).

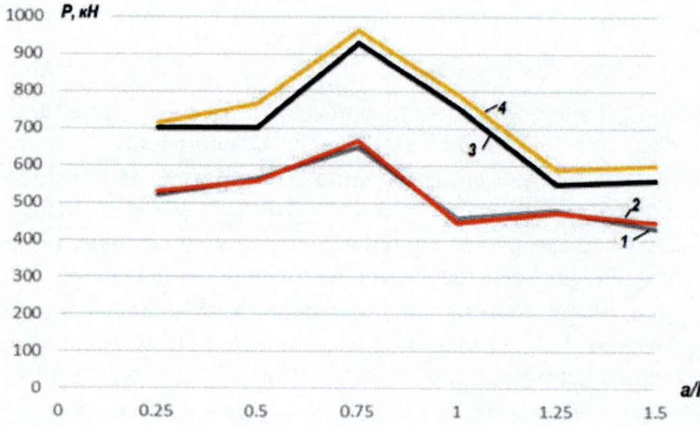

Fig. 7 Diagrams of failure load versus strength of concrete R and shear span a/h_0. 1 and 2—experimental and analytical values at R = 29.8 MPa; 3 and 4—experimental and simulated analytical values at R = 51.9 MPa

3 Conclusions

Presented study shows that the constitutive fixed angle model CSMM which was developed at the University of Houston, USA and incorporated into the OpenSees software package allows to perform precise calculations of the bearing capacity and displacements of plane-stressed reinforced concrete structures. The simulated structures were represented by composite elements and short beams with different shear spans, various reinforcement ratios, concrete grades and other design factors.

The comparative analysis allowed us to conclude that taking into account the actual properties of concrete under a plane stress state in the form of a nonlinear model with a fixed angle model CSMM [24] provided a sufficiently high accuracy of computer modeling of the deep beams.

The most versatile and accurate tool in the design process for assessing the stress state and serviceability of plane-stressed reinforced concrete structures is computer analysis based on nonlinear constitutive models where the actual strength and stress-related properties of concrete and reinforcement are taken into account.

Acknowledgements The innovation project was carried out with the financial support of the Kuban Science Foundation in the framework of the Commercializable scientific and innovation projects competition № NIP-20.01/27.

References

1. ACI Committee 318–95 (1995) Building code requirement of reinforced concrete. American concrete Institute, Farmington Hill, MI
2. ACI Committee 318–08 (2008) Building code requirements for structural concrete. American concrete Institute, Farmington Hill, MI
3. ACI Committee 318–14 (2014) Building code requirements for structural concrete. American concrete Institute, Farmington Hill, MI
4. IS 456:2000 (2000) Indian standard plain and reinforced concrete code of practice, 4th Revision. Bureau of Indian Standards, New Delhi
5. ACI Committee 318–11 (2011) Building code requirements for structural concrete. American Concrete Institute, Farmington Hill, MI
6. British Standards Institution (BSI) BS 8110–1 (1997) Structural use of concrete—Part 1: code of practice for design and construction. BSI, London, UK
7. Hong Kong Buildings Department (HKBD) (2013) Code of practice for structural use of concrete
8. Tan KH, Lu HY (1999) Shear behavior of large reinforced concrete deep beams and code comparisons. TACI Struct J 96(5):836–846
9. Yang K-H, Chung H-S, Lee E-T, Eun H-C (2003) Shear characteristics of high-strength concrete deep beams without shear reinforcements. Eng Struct 25(10):1343–1352
10. Tanimura Y, Sato T (2005) Evaluation of shear strength of deep beams with stirrups. Q Report RTRI 46(1):3–58
11. Salamy MR, Kobayashi H, Unjoh S (2005) Experimental and analytical study on RC deep beams. Asian J Civil En (AJCE) 6(5):409–422
12. Zhang N, Tan K-H (2007) Size effect in RC deep beams: experimental investigation and STM verification. Eng Struct 29(12):3241–3254

13. Garay JD, Lubell AS (2008) Behavior of concrete deep beams with high strength reinforcement. In: Structures congress—crossing borders, Vancouver, Canada, p 10
14. Lu WY, Lin IJ, Yu HW (2013) Shear strength of reinforced concrete deep beams. ACI Struct J 110(4):671–680
15. El-Sayed AK, Shuraim AB (2015) Size effect on shear resistance of high strength concrete deep beams. Mater Struct 49(5):1871–1882
16. Modin A, Lukin M, Vlasov A, Hisham E (2020) Energy-efficient indicators of panel housing mass construction in the climatic conditions of central Russia. In: IOP conference series: materials science and engineering
17. Lukin MV, Popov MV, Lisyatnikov MS (2020) Short-term and long-term deformations of the lightweight concrete. In: IOP conference series: materials science and engineering
18. Lisyatnikov MS, Shishov II, Sergeev MS, Hisham E (2020) Precast monolithic coating of an industrial building based on variable-height beam-slabs. In: IOP conference series: materials science and engineering
19. Andermatt MF, Lubell AS (2013) Behavior of concrete deep beams reinforced with internal fiber-reinforced polymer-experimental study. ACI Struct J 47(110):585–594
20. Birrcher DB, Tuchscherer RG, Huizinga M, Bayrak O (2013) Minimum web reinforcement in deep beams. ACI Struct J 26(110):297–306
21. Gara F, Ragni L, Roia D, Dezi L (2012) Experimental behaviour and numerical analysis of floor sandwich panels. Eng Struct, Elsevier Ltd 36:258–269
22. Raj JL, Rao GA (2013) Performance of RC deep beams with different combinations of web reinforcement. Appl Mech Mater 343:21–26
23. Tan KH, Cheng GH (2006) Size effect on shear strength of deep beams: investigating with strut-and-Tie model. J Struct Eng 132(5):673–685
24. Mansour M, Hsu TTC (2005) Behavior of reinforced concrete elements under cyclic shear: part 2—theoretical model. J Struct Eng/ASCE 131:54–65
25. McKenna F (2011) OpenSees: a framework for earthquake engineering simulation. Comput Sci Eng 58–66
26. Pochinok VP, Greshkina EV, Tamov MA, Tamov MM (2018) Effect of transverse compression on web-crushing strength of reinforced concrete i-shaped beams. Adv Eng Res (AER) 157:595–600
27. Greshkina EV, Pochinok VP, Popov VA (2020) Issledovaniye prochnosti priopornykh uchastkov tonkostennykh zhelezobetonnykh balok metodom konechnykh elementov [Study of the strength of the support and beam joints of thin-walled reinforced concrete beams by the finite element method]. In: Development i innovatsii v stroitel'stve, pp 627–640
28. Karpenko NI, Karpenko SN, Petrov AN (2018) Sovershenstvovaniye metodov rascheta ploskostnykh zhelezobetonnykh konstruktsiy s uchetom deystvitel'nykh svoystv vysokoprochnykh betonov [Advancing in calculating methods for plane reinforced concrete structures with considering the actual properties of high-performance concrete]. Int J Comput Civil Struct Eng 14:78–89
29. Kolchunov VI, Yakovenko IA, Tugay TV (2014) Metodika eksperimental'nykh issledovaniy zhestkosti ploskonapryazhennykh sostavnykh konstruktsiy [Experimental methods of the research of plane-stressed composite structures stiffness]. In: Místobuduvannya ta teritoríal'ne planuvannya: nauk.-tekhn. 52:178–185
30. Tugay TV (2014) Osnovnyye rezul'taty eksperimental'nykh issledovaniy zhestkosti ploskonapryazhennykh zhelezobetonnykh sostavnykh konstruktsiy [The main results of experimental studies of the stiffness of plane-stressed reinforced concrete composite structures]. In: Resursoekonomni materialy, konstruktsiyi, budivli ta sporudy: zb. nauk. prats' 29:369–375
31. Tugay TV (2015) Metodika rascheta zhestkosti ploskonapryazhennykh zhelezobetonnykh sostavnykh konstruktsiy. [Methods of calculating the stiffness of plane-stressed reinforced concrete composite structures] Dis… kand. tekhn nauk, p 210

32. Baranova TI, Zalesov AS (2003) Karkasno-sterzhnevyye raschetnyye modeli i inzhenernyye metody rascheta zhelezobetonnykh konstruktsiy [Frame-rod simulating models and engineering methods for calculating of reinforced concrete structures]: ucheb. posobiye dlya studentov, obuchayushchikhsya po napravleniyu 653500 "Str-vo". Izd. ASV, (PPP Tip. Nauka), Moscow
33. Baranova TI, Skachkov YP, Snezhkina OV, Ladin RA (2014) Modelirovaniye raboty korotkikh zhelezobetonnykh balok [Simulation model for sthin reinforced concrete beams]. Vestnik SibADI 2(36):54–60
34. Greshkina EV, Pochinok VP, Nelineynaya konechno-elementnaya model' dlya rascheta zhelezobetonnykh balok-stenok [Nonlinear finite element model for reinforced concrete deep beams simulation]. In: Materialy XI Mezhdunarodnoy nauchno-prakticheskoy konferentsii «Stroitel'stvo v pribrezhnykh kurortnykh regionakh», pp. 10–16. Sochi, SGU
35. Baranova TI (2006)Raschetnyye modeli soprotivleniya srezu szhatykh zon zhelezobetonnykh konstruktsiy [Simulation models for shear resistance of compressed zones in reinforced concrete structures]: uch. posobiye. Sarat. gos. tekhn. un-t, Saratov

Calculating the Strengthening of Construction Structures Before the Reconstruction of the Building

A. N. Neverov, P. S. Truntov, E. S. Ketsko, and V. I. Rimshin

Abstract At present, are many buildings of various purposes are in operation. Many of these buildings are in unsatisfactory condition after 15–20 years of use and need to be repaired. In addition, due to changes in process flow and structural load, it is also necessary to strengthen the building structure during renovation. It is of great significance to strengthen of the building structure, because the further technical condition of the building or building depends on timely repair. In the design of reinforced building structure, an important stage is to evaluate the technical condition and suitability of each component of the building. The emphasis this work is devoted to the process of designing and calculating the foundation slab, reinforced using the method of one-way build-up. The building has six floors, with an overall dimension of 63 × 66 m. The spatial rigidity of the building is provided longitudinal and cross walls, columns, slabs and roofs, as well as the foundation. The «LIRA-SAPR» and «STATICA» software complexes were used for constructive calculations (checking the carrying capacity of the design elements). This article introduces the calculation scheme of strengthening structure and reviews the calculation method. The results are analyzed, and the general conclusions about the calculation and design of building structural strengthening are obtained.

Keywords Gain calculation · Bearing capacity · Reconstruction · Limiting states · Serviceability

A. N. Neverov · E. S. Ketsko
Research Institute of Building Physics of the Russian Academy of Architecture and Building Sciences, Locomotive passage 21, Moscow 127238, Russian Federation

P. S. Truntov (✉) · V. I. Rimshin
National Research Moscow State University of Civil Engineering, Yaroslavskoe shosse, 26, Moscow 129337, Russian Federation

N. Vatin et al. (eds.), *Proceedings of MPCPE 2021*, Lecture Notes in Civil Engineering 182, https://doi.org/10.1007/978-3-030-85236-8_14

1 Introduction

The completion and usage of buildings and structures is always accompanied by the occurrence of various kinds of defects and damage to building structures. To ensure mechanical safety and durability of structures, it is necessary to repair and restore load-bearing structures, which in turn can significantly increase the service life of buildings and structures as a whole [1–5].

However, the need to improve the reliability of buildings and structures arises not only when various defects and damage to structures appear. This work considers the adjustment of design solutions caused by a change in planning solutions, a change in the installation scheme of technological equipment, as well as the device of pits for collecting water in the foundation slab [6–9].

Taking into account the changes introduced for new planning solutions, calculations of load-bearing structures were performed, and the existing design solutions were also corrected.

Within the framework of the study, a six-storey building with dimensions in plan 63 × 66 m was considered. The maximum height is 23.8 m. The building is divided into three blocks. The building frame is made of monolithic reinforced concrete with a slab foundation on a natural base, consisting of a grid of columns with a pitch of 4.7 ÷ 6.3 m and floor slabs [10–15].

Floor slabs are made with a thickness of 0.25 m. Foundation slab is made of monolithic reinforced concrete of class B30, F150, W6 and 700 mm thick. The columns are rigidly embedded in the foundation and interfloor ceilings, made with sections of 0.5 × 0.5 m and 0.4 × 0.4 m.

2 Materials and Methods

The calculation of the building together with the base was carried out in the LIRA-SAPR software package. The modeling of the structural elements of the building was carried out as follows:

- walls, floors, foundation slab are modeled using lamellar finite elements;
- columns and beams are modeled with bar finite elements;
- the decrease in the modulus of elasticity of reinforced concrete structures is taken into account;
- the stiffness characteristics of the foundation (variable in terms of the bed coefficient $C1$) were determined using the Soil module, taking into account the determination of settlement by the method of linearly deformable half-space. The results of engineering and geological surveys were taken as the initial data for determining the stiffness characteristics of the foundation;
- the reinforcement that directly affects the distribution of forces in the design scheme is taken into account explicitly, meaning that it is performed either by

modeling the corresponding added structural elements, or by changing the sections of existing elements;

– for structural calculations (checking the bearing capacity of structural elements) the software complexes LIRA-SAPR and STATICA were used.

To strengthen the foundation slab, two calculations were performed: the calculation of the reinforced foundation slab for punching and the calculation of the reinforced foundation slab along normal sections.

The STATICA software package was used when calculating a reinforced foundation slab for punching. Within the framework of the calculation, the most loaded (one which had the greatest deficit in bearing capacity according to the results of the verification calculation) support zone of the foundation slab was considered [16–22].

The software package performs the calculation taking into account the action of a concentrated force and bending moments in two mutually perpendicular planes. The calculation results must satisfy the condition:

$$\frac{F}{F_{ult}} + \frac{M_x}{M_{x,ult}} + \frac{M_y}{M_{y,ult}} \leq 1, \tag{1}$$

where F, M_x, M_y are the concentrated force and bending moments in the directions of the X and Y axes, taken into account when calculating the punching shear from an external load.

Where F_{ult}, $M_{x,ult}$, $M_{y,ult}$ are the limiting concentrated forces and bending moments in the directions of the X and Y axes, which can be perceived by concrete in the design cross section when they act separately.

The limiting concentrated force F_{ult} is determined by the formula:

$$F_{ult} = R_{bt} \cdot A_b, \tag{2}$$

where A_b is the area of the calculated cross-section, R_{bt} is the design resistance of concrete to axial tension.

$M_{x,ult}$ and $M_{y,ult}$ are calculated based on the following formulas:

$$\begin{aligned} M_{x,ult} &= R_{bt} \cdot W_x \cdot h_0, \\ M_{y,ult} &= R_{bt} \cdot W_y \cdot h_0, \end{aligned} \tag{3}$$

where W_x and W_y are the values of the moments of resistance of the design contour of concrete when punching in the directions of mutually perpendicular axes X and Y,

h_0 is the reduced working height of the section equal to $0.5(h_{0x} + h_{0y})$ where h_{0x} and h_{0y} are the working height of the section for longitudinal reinforcement located in the corresponding axes.

When calculating the reinforced foundation slab along normal sections, the LIRA-SAPR software package was used.

3 Calculations and Results

When calculating the punching shear reinforcement of the foundation slab, the following provisions and assumptions were taken:

- the calculation takes into account the build-up layer above the foundation slab;
- an increase in the load transfer circuit due to the supporting elements of the metal cage is not taken into account in the margin (given the complex nature of the pressure transfer through the corners, reinforced with ribs);
- soil rebound when calculating for punching into the stock is not taken into account (considering the possible variability of characteristics);
- concrete reinforcement class is taken over the concrete of the existing foundation slab.

Calculation scheme. Column section dimensions—0.5 × 0.5 m, slab thickness —1.0 m, working height of the slab section $h_{0x} = h_{0y} = 0.94$ m.

Punching force from the acting load—F = 3127 kN, moment relative to the X-axis—$M_x = 12.0$ kN·m.

The calculation uses heavy concrete of class B30, transverse reinforcement A500.

The calculation result is shown in Fig. 1.

Considering the increase in the thickness of the slab the results of the calculation of the most loaded support zone of the foundation slab for punching show that the adopted reinforcement provides the bearing capacity of the support zone without transverse reinforcement (which would be impossible to install in the actually existing foundation slab).

The deficiency of reinforcement revealed during the verification calculation was primarily related to the lower longitudinal reinforcement of the foundation slab. It is technically impossible to increase the bottom reinforcement of the foundation slab in the existing building; as a reinforcement, a build-up layer of concrete was adopted, which is involved in joint work with the existing foundation slab. By significantly increasing the working section height, the amount of reinforcement required to absorb the bending moment is reduced.

Figure 2 shows a fragment of the performed verification calculation.

The change in the required reinforcement in the areas with a deficit was analyzed in the process of selecting the thickness of the concrete layer to be built up. In this

Ultimate forces in concrete	Fb,ult [кН]	Mbx,ult [кНм]	Mby,ult [кНм]
	5603.9	2689.9	2689.9

Strength state condition $F / F_{ult} + M_x / M_{x,ult} + M_y / M_{y,ult} =$

0.5580 + 0.0045 + 0.0000 = 0.562 <= 1

Fig. 1 The result of the punching shear design for the support zone of the foundation slab. Fragment of the protocol from the PC "STATIC"

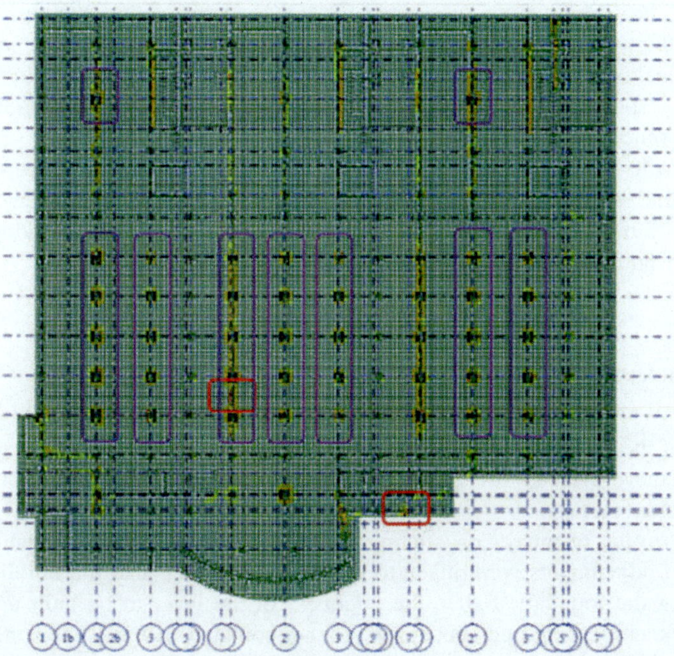

Fig. 2 Lower reinforcement of the foundation slab along the X-axis, cm^2/p.m. Areas with a bearing capacity deficit (up to 50%) are marked with red frames

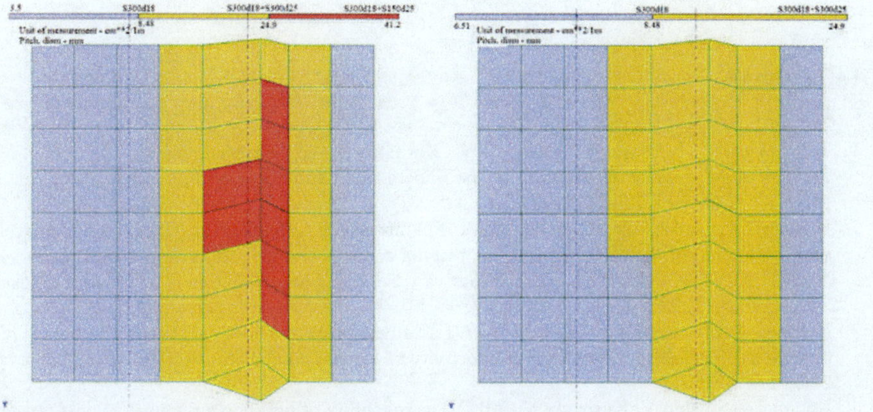

Fig. 3 Section of the foundation slab before and after reinforcement. The area with the required reinforcement exceeding the design reinforcement is highlighted in red on the reinforcement mosaic

case, the class of concrete is adopted for the concrete of the existing foundation slab. Figure 3 shows the calculation results for the most loaded area with the existing thickness of the foundation slab and after building up an additional layer of concrete, respectively. The required reinforcement does not exceed the design one in the presence of reinforcement.

The results of calculating the normal sections of the foundation slab, taking into account the reinforcement in the form of a built-up concrete layer, using the example of the most loaded sections, show the efficiency of the adopted reinforcement option.

4 Conclusions

The use of the STATICA and LIRA-SAPR software complexes made it possible to complete the project to strengthen the foundation slab and to provide it with the required characteristics.

Based on the results of calculating the reinforcement of the foundation slab, a number of adjustments were made to the structural scheme of the building. When reinforcing the foundation slab, the thickness of the layer to be built up is 0.3 m (total thickness, taking into account the foundation slab, 1.0 m). The concrete class of the reinforcement elements is accepted as B40, which is higher than the concrete class used in the reinforcement calculations. Reinforcement of the built-up concrete layer was made with reinforcement Ø18 mm with a step of 0.3 × 0.3 m. To ensure the joint work of the reinforcement with the existing foundation slab, it is necessary to prepare the surface of the slab in advance, to perform the gluing of transverse reinforcement.

References

1. Telichenko V, Rimshin V, Eremeev V, Kurbatov V (2018) Mathematical modeling of groundwaters pressure distribution in the underground structures by cylindrical form zone. MATEC Web Conf 196:02025
2. Krishan AL, Rimshin VI, Telichenko VI, Rakhmanov VA, Narkevich MY (2017) Practical implementation of the calculation of the bearing capacity trumpet-concrete column. Izvestiya Vysshikh Uchebnykh Zavedenii, Seriya Teknologiya Tekstil'noi Promyshlennosti 2017-January(2):227–232
3. Korotaev SA, Kalashnikov VI, Rimshin VI, Erofeeva IV, Kurbatov VL (2016) The impact of mineral aggregates on the thermal conductivity of cement composites. Ecol Environ Conserv 22(3):1159–1164
4. Kuzina E, Rimshin V, Kurbatov V (2018) The reliability of building structures against power and environmental degradation effects. IOP Conf Series: Mater Sci Eng 463(4):042009
5. Rimshin VI, Varlamov AA (2018) Three-dimensional model of elastic behavior of the composite. Izvestiya Vysshikh Uchebnykh Zavedenii, Seriya Teknologiya Tekstil'noi Promyshlennosti, 2018-January(3):63–68

6. Kuzina E, Rimshin V (2019) Strengthening of concrete beams with the use of carbon fiber. Adv Intell Syst Comput 983:911–919
7. Rimshin VI, Pudova AA, Shubin LI (2018) Evaluation of efficiency of use of photoelectric systems at operation of a residential house. Izvestiya Vysshikh Uchebnykh Zavedenii, Seriya Teknologiya Tekstil'noi Promyshlennosti, 2018-January(3):287–293
8. Travush VI, Karpenko NI, Erofeev VT, Smirnov VF, Rodina NG (2017) Development of biocidal cements for buildings and structures with biologically active environments. Power Technol Eng 51(4):377–384
9. Roschina SI, Lisyatnikov MS, Koshcheev AA (2019) Technical- and- economic efficiency of reinforced wooden structures. IOP Conf Series: Mater Sci Eng 698(2):022005
10. Erofeev V, Rodin A, Rodina N, Kalashnikov V, Irina E (2016) Biocidal binders for the concretes of unerground constructions. Procedia Eng 165:1448–1454
11. Rimshin V, Aralov R (2019) Sustainable regeneration of urban areas (using the example of Moscow renovation program). E3S Web of Conf 110:01011
12. Kuzina E, Rimshin V (2019) Experimental and calculated evaluation of carbon fiber reinforcing for increasing concrete columns carrying capacity. E3S Web Conf 97:04007
13. Evseev V, Barkhi R, Pleshivtsev A, Scrynnik A (2019) Modeling the influence of weather and climatic conditions on the safety characteristics of the construction process. E3S Web Conf 97:03035
14. Erofeev VT, Rodin AI, Kravchuk AS, Kaznacheev SV, Zaharova EA (2018) Biostable silicic rock-based glass ceramic foams. Mag Civil Eng 84(8):48–56
15. Erofeev VT, Rodin AI, Yakunin VV, Bochkin VS, Chegodajkin AM (2018) Alkali-activated slag binders from rock-wool production wastes. Mag Civil Eng 82(6):219–227
16. Telichenko V, Rimshin V, Kuzina E (2018) Methods for calculating the reinforcement of concrete slabs with carbon composite materials based on the finite element model. MATEC Web Conf 251:04061
17. Lisyatnikov MS, Roshchina SI, Chukhlanov VY, Ivaniuk AM (2020) Repair compositions based on methyl methacrylate modified with polyphenylsiloxane resin for concrete and reinforced concrete structures. IOP Conf Series: Mater Sci Eng 896(1):012113
18. Merkulov S, Rimshin V, Akimov E, Kurbatov V, Roschina S (2020) Regulatory support for the use of composite rod reinforcement in concrete structures. IOP Conf Series: Mater Sci Eng 896(1):012022
19. Sergeev MS, Gribanov AS, Roschina SI (2020) The stress STR in state of composite. Conf Series: Mater Sci Eng 896(1):012058
20. Rimshin V, Truntov P (2019) An integrated approach to the use of composite materials for the restoration of reinforced concrete structures. E3S Web Conf 135:03068
21. Roshchina S, Ezzi H, Shishov I, Lukin M, Sergeev M (2017) Evaluation of the deflected mode of the monolithic span pieces and preassembled slabs combined action. IOP Conf Series: Earth Environ Sci 012075
22. Shubin LI, Rimshin VI, Truntov PS, Suleymanova LA (2020) Evaluation method of technical condition façade systems. IOP Conf Series: Mater Sci Eng 896(1):012023

Method for Determining the Design Resistance of a Glued-In Twisted Elliptical Bar for Pulling Out in Elements of Wooden Structures

Boris Labudin, Olga Tyurina, Dmitriy Mavrin, and Wasim Hasan

Abstract Reinforcement is the most effective way to reinforce timber elements and transfer forces to nodal joints. In this regard, the search for rational parameters of rod elements for reinforcement is relevant. The article investigates a twisted rod of elliptical cross-section (the material of the rod is steel or composite). The main feature of the twisted rods is the high design characteristics of the pull-out resistance from the elements of wooden structures. This is because the holding force in the design pull-out resistance includes the friction force at the contact boundary and the crushing force of the wood under the surfaces of the spiral line. For this purpose, this paper proposes a method for determining the characteristics of the pulling strength of a twisted rod of elliptical cross-section, which includes these parameters.

Keywords Twisted rod of elliptical cross-section · Timber structures · Pull-out test · Pull-out resistance · Spiral line

1 Introduction

Reinforcement of structures allows you to significantly increase the bearing capacity of structures. The most interesting issue is the adhesion of the reinforcing bar. This characteristic depends on the longitudinal section of the rod and the periodicity of the profile [1–3].

Reinforcement of wood has been known since the beginning of the twentieth century and was proposed by scientists (A. Klaitila, A. Fischer). Developments in this area began to be implemented in practice together with the emergence of a variety of synthetic adhesives and fillers, which solved the problem of the joint work of the core material and wood. The most common type of reinforcement is structural reinforcement by gluing metal rods into it at an angle. This method was

B. Labudin · O. Tyurina (✉) · D. Mavrin · W. Hasan
Northern (Arctic) Federal University Named After M.V. Lomonosov, Arkhangelsk, Russian Federation

N. Vatin et al. (eds.), *Proceedings of MPCPE 2021*, Lecture Notes in Civil Engineering 182, https://doi.org/10.1007/978-3-030-85236-8_15

first proposed by Turkovsky [4] and was named "Sistema CNIISK". There are also researches in the field of longitudinal reinforcement of wooden structures at the stage of their manufacture [5, 6].

Despite the economic benefits and technological advantages, the gluing of rods has a number of problems that prevent its widespread introduction into the production process of construction. This is due to gaps in the regulatory framework. At the moment, the question of the stress-deformed state of non-standard rods (such as a twisted elliptical rod) has been tiny studied. This article describes the solution to the issue of determining the design resistance of a twisted elliptical rod to pull-out from wooden structures. The material of the twisted rods can be either steel or composite.

2 Methods

Figure 1 shows a general view of a twisted elliptical rod (a) and a cross-section with geometric characteristics (b). A twisted elliptical rod is characterized by the following parameters: L—rod length; D—diameter of the rod along the outer contour (D = 2a); d—diameter of the bar along the inner contour (d = 2b); α—the angle of inclination of the spiral lines to the plane perpendicular to the longitudinal axis of the bar. Characteristics narrowing elliptical cross section: $k = a/b \rightarrow a = kb$.

Pull-out strength of rods bonded in timber is determined by the pull-out tests (Fig. 2).

According to the Russian design code [7] the calculated value of the pulling capacity of rods glued-in parallel to the grain can be obtained from:

$$T_{s.h} = R_{s.h} d_0 \pi L k_s \tag{1}$$

where $R_{s.h}$—design resistance to pulling glued along the grain of wood, MPa; d_0—diameter of the (drilled) hole (for an ellipse $d_0 = 2 + 0.005$), м; L—length of the embedded part of the rod; k_s—coefficient that takes into account the uneven distribution of shear stresses depending on the length of the embedded part of the rod.

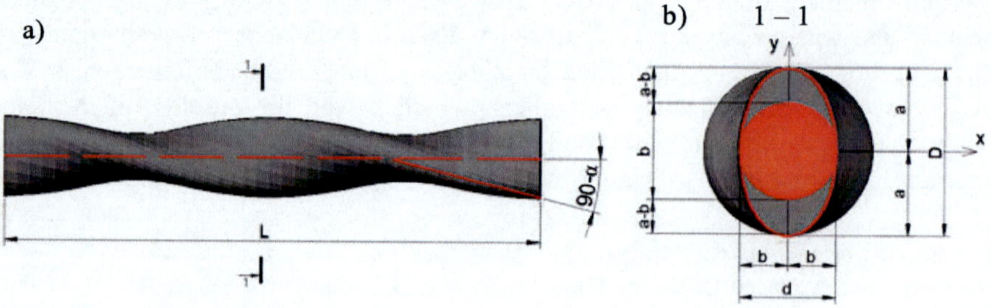

Fig. 1 Geometric properties of rod

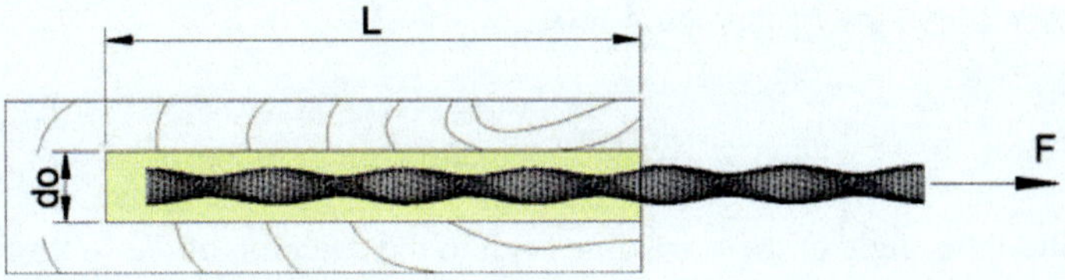

Fig. 2 Scheme of the pull-out test

For us, in Formula (1), the sought value is the calculated resistance to pulling out of an elliptical twisted rod. Let us take into account the calculation of the strength of the joints on screw nails under the action of a pull-out force [8]. According to this calculation [9], the holding force of the design pull-out resistance must take into account both the friction force at the contact boundary and the crushing force of the wood [10, 11] under the surfaces of the spiral lines. So, the design resistance of a helical elliptical rod to pulling out can be determined according to the formula:

$$R_{s.h} = R_0 + k_e \frac{R_c S \delta_e m}{\pi d_0} \quad (2)$$

where R_0—design resistance of a smooth rod to pulling out, which takes into account the friction force at the interface of the contact surfaces; R_c—resistance of wood [12] and adhesive to crushing over the entire surface, obtained from the results of experiments; S—the ratio of the lengths of the spiral line and its projection onto the longitudinal axis of the rod; δ_e—spiral lines protrusion width (for ellipse $\delta_e = a - b$); m—number of spiral lines in the cross-section of the bar (for elliptical twisted bar m = 2); k_e—coefficient derived from experimental studies.

The length of the spiral line is determined by (Fig. 3):

$$L_{s.1} = n\sqrt{(2\pi R)^2 + \lambda^2} \quad (3)$$

where R—the radius of the spiral line, m; λ—the pitch of the spiral line, m; n—number of turns in the spiral line.

Fig. 3 Length of the spiral line

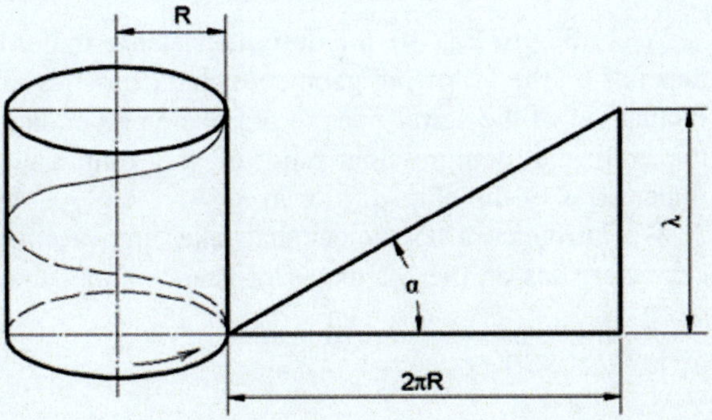

Introduce the angle of the spiral line α:

$$\tan\alpha = \frac{\lambda}{2\pi R} \tag{4}$$

Substitute the angle of the spiral line (4) into the equation of the spiral line length (3) and obtain the formula:

$$L_{s.l.} = n\sqrt{(2\pi R)^2 + \lambda^2} = n\sqrt{(2\pi R)^2\left(1 + \frac{\lambda^2}{(2\pi R)^2}\right)} = 2\pi Rn\sqrt{1+\tan^2\alpha} \tag{5}$$

The length of the projection of the spiral line to the longitudinal axis of the bar is determined according to the formula:

$$L_{p.s.l} = n\lambda \tag{6}$$

We substitute the equation for the angle of the spiral line (4) into the equation of the projection of the spiral line (6) and obtain the following equality:

$$L_{p.s.l} = n\lambda = n2\pi R\tan\alpha \tag{7}$$

Then the ratio of the lengths of the spiral line and its projection onto the longitudinal axis of the rod is calculated by the formula:

$$S = \frac{L_{s.l.}}{L_{p.s.l}} = \frac{2\pi R\sqrt{(1+\tan^2\alpha)}n}{2\pi Rn\tan\alpha}n = \frac{\sqrt{(1+\tan^2\alpha)}}{\tan\alpha} \tag{8}$$

In the formula for determining the design resistance of the spiral line elliptical rod to pulling out, we substitute the obtained values of S, $\delta_в$ and d_0 and obtain the dependence for the design resistance of the spiral line elliptical rod:

$$R_{s.h} = R_0 + k_e\frac{2R_c\frac{\sqrt{(1+\tan^2\alpha)}}{\tan\alpha}b(k-1)}{\pi(2bk+0.005)} \tag{9}$$

According to Eq. (9), the design resistance under the surfaces of the helical line depends on the following geometric characteristics of the rod [13–15]: the angle of inclination of the spiral lines to the plane perpendicular to the longitudinal axis of the rod; the compression ratio of the elliptical section; accepted geometric dimensions of the ellipse (b or a).

We introduce a coefficient that takes into account the influence of geometric characteristics on the calculated resistance under the spiral line surfaces:

$$k_{geom} = \frac{2b(k-1)}{\pi(2bk+0.005)} \frac{\sqrt{(1+\tan^2\alpha)}}{\tan\alpha} \tag{10}$$

Figure 4 shows the dependences of k_{geom} on the angle of inclination of the spiral line, taking into account the variation in the values of the semi-minor axis b and the compression ratio k of the elliptical section.

Combining Formulas (9) and (10), we obtain the following dependence for the calculated resistance of a spiral elliptical rod:

$$R_{s.h} = R_0 + k_e k_{geom} R_c, \tag{11}$$

k_{geom} is an integral indicator that takes into account the influence of the wood crumpling factor and the adhesive under the spiral line surfaces on the pull-out resistance $R_{s.h}$.

From Formulas (10) and (11), we can draw the following conclusions about the influence of the geometric characteristics of the rod on the $R_{s.h}$:

- increasing the angle of inclination of the spiral line α leads to a decrease in $R_{s.h}$;
- an increase in the compression ratio [16, 17] of the ellipse k leads to an increase in $R_{s.h}$;
- an increase in the size of the small semi-axis of the ellipse b leads to an increase in $R_{s.h}$.

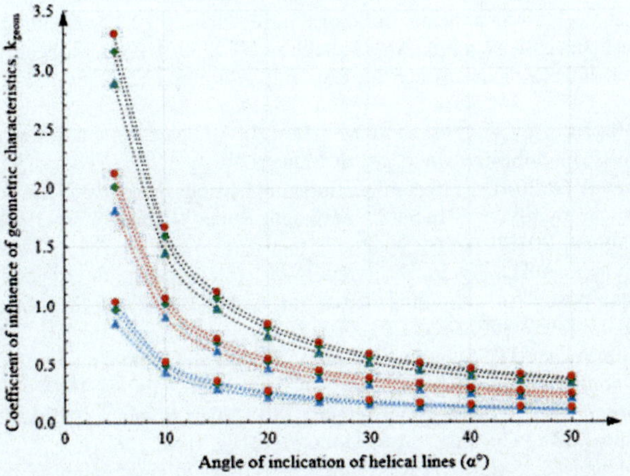

Fig. 4 Dependence "load—normal tensile stresses" during tensile tests at an angle to the fibers

3 Concluding Remarks and Further Research

This article proposes a method for determining the design resistance of a glued-in twisted rod of elliptical section for pulling out. Its main feature is taking into account both the friction force at the contact boundary and the crushing force of wood under the surfaces of the spiral line. It should be noted that there are a number of studies that determine additional parameters of influence on the process of pulling wood out of structures [18–22]. It is necessary to carry out experimental studies to determine the influence of unaccounted factors: temperature and humidity conditions, type of loading, type of wood and adhesive.

References

1. Karpenko NI, Kolchunov VI, Kolchunov VI, Travush VI (2021) Calculation model of a complex-stressed reinforced concrete element under torsion with bending. Int J Comput Civ Struct Eng 17:34–47. https://doi.org/10.22337/2587-9618-2021-17-1-34-47
2. Eryshev VA, Karpenko NI, Rimshin VI (2020) The parameters ratio in the strength of bent elements calculations by the deformation model and the ultimate limit state method. IOP Conf Ser Mater Sci Eng. https://doi.org/10.1088/1757-899X/753/2/022076
3. Karpenko N, Yarmakovsky V, Kadiev D (2020) Basis of diagrammatic deformation model for core reinforced concrete structures calculating. IOP Conf Ser Mater Sci Eng. https://doi.org/10.1088/1757-899X/953/1/012005
4. Turkovskij SB, Pogorel'tsev AA (2001) Wooden structures with rigid joints in structures with corrosive medium. Promyshlennoe i Grazhdanskoe Stroit 10–13
5. Lukin M, Prusov E, Roshchina S, Karelina M, Vatin N (2021) Multi-span composite timber beams with rational steel reinforcements. Buildings. https://doi.org/10.3390/buildings11020046
6. Gribanov AS, Roshchina SI, Naichuk AY, Melekhov VI (2020) Wooden beams with local wood modification. IOP Conf Ser Mater Sci Eng. https://doi.org/10.1088/1757-899X/896/1/012067
7. Turkovskij SB, Pogorel'tsev AA, Eknados'yan IL (2003) Selection of design scheme of lens-shaped trusses from adhesive wood. Stroit Mater 18–20
8. Somerville M, Deev A (2020) The effect of heating rate, particle size and gas flow on the yield of charcoal during the pyrolysis of radiata pine wood. Renew Energy 151:419–425. https://doi.org/10.1016/j.renene.2019.11.036
9. Lisyatnikov MS, Glebova TO, Ageev SP, Ivaniuk AM (2020) Strength of wood reinforced with a polymer composite for crumpling across the fibers. IOP Conf Ser Mater Sci Eng. https://doi.org/10.1088/1757-899X/896/1/012062
10. Karelskiy AV, Zhuravleva TP, Labudin BV (2015) Load-to-failure bending test of wood composite beams connected by gang nail. Mag Civ Eng. https://doi.org/10.5862/MCE.54.9
11. Gribanov AS, Roshchina SI, Popova MV, Sergeev MS (2018) Laminar polymer composites for wooden structures. Mag Civ Eng. https://doi.org/10.18720/MCE.83.1
12. Lukin MV, Roshchina SI, Smirnov EA, Shunqi M (2020) Strengthening of the operated wooden floor beams with external rigid reinforcement. IOP Conf Ser Mater Sci Eng. https://doi.org/10.1088/1757-899X/896/1/012065
13. Nagruzova L, Saznov K, Aytbu KK (2019) Thermal efficient panels on a wooden frame for quickly erectable low-rise buildings. E3S Web Conf. https://doi.org/10.1051/e3sconf/201911001023

14. Koshcheev AA, Roshchina SI, Naichuk AY, Vatin NI (2020) The effect of eccentricity on the strength characteristics of glued rods made of steel cable reinforcement in solid wood. IOP Conf Ser Mater Sci Eng. https://doi.org/10.1088/1757-899X/896/1/012059
15. Roschina SI, Lisyatnikov MS, Lukin MV, Popova MV (2018) Technology of strengthening the supporting zones of the glued-wood beaming structure with the application of nanomodified prepregs. Mater Sci Forum. https://doi.org/10.4028/www.scientific.net/MSF.931.226
16. Naychuk AY (2013) Estimation of load-bearing capacity and stiffness of timber beams with through-thickness cracks. Adv Mater Res. https://doi.org/10.4028/www.scientific.net/AMR.778.361
17. Roshchina S, Lukin M, Lisyatnikov M (2020) Compressed-bent reinforced wooden elements with long-term load. Lect Notes Civil Eng. https://doi.org/10.1007/978-3-030-42351-3_7
18. Madhoushi M, Ansell MP (2017) Effect of glue-line thickness on pull-out behavior of glued-in GFRP rods in LVL: finite element analysis. Polym Test 62:196–202. https://doi.org/10.1016/j.polymertesting.2017.06.029
19. Serrano E (2001) Glued-in rods for timber structures—a 3D model and finite element parameter studies. Int J Adhes Adhes 21:115–127. https://doi.org/10.1016/S0143-7496(00)00043-9
20. Azinović B, Serrano E, Kramar M, Pazlar T (2018) Experimental investigation of the axial strength of glued-in rods in cross laminated timber. Mater Struct Constr 51. https://doi.org/10.1617/s11527-018-1268-y
21. Lukina A, Roshchina S, Gribanov A (2021) Method for restoring destructed wooden structures with polymer composites. Presented at the 2021. https://doi.org/10.1007/978-3-030-72404-7_45
22. Lukin M, Sergeev M, Lisyatnikov M (2021) Non split wooden beam reinforced with composite reinforcement. Presented at the 2021. https://doi.org/10.1007/978-3-030-72404-7_12

Features of Preliminary Stresses in Wooden Structures

Mihail Moskalev

Abstract Stress adjustment while in service is the next step in the building structure development after the pre-stressing usage until the service. However, during the operation of real structures there were found the following features which must be applied when designing such structures: extremely high variability of properties along and across the fibers, leakage in knots which leads to deformation capacity of knot connections, influence of wood humidity, the nonlinear dependence of the elastic modulus on stresses during deformation of wood, rheological properties of wood. However, during the operation of real structures there were found the following features which must be applied when designing such structures: maximally increase the area of contact surfaces in the nodes to prevent "trampling down" of wood in them, with this nodal structure as shown by the experiments given in (Sanaev et al. in Prospects for using unique physicochemical properties of plants and their ability for self-regeneration in space, 2019) the effect of knot compliance and anisotropy of wood did not exceed 5%, maintain strict humidity control, regulation of stresses according to a special algorithm that takes into account the variable modulus of elasticity and also allows to compensate for other problems.

Keywords Timber · Active control structures · Actuators and smart structures · Active · Active control · Control algorithms · Structural control

1 Introduction

Stress adjustment while in service is the next step in the building structure development after the pre-stressing usage until the service [1]. However, the main features of its usage coincide with the features of pre-stressing. Pre-stressing is especially widespread in reinforced concrete structures due to the fact that it

M. Moskalev (✉)
Saint-Petersburg State University of Architecture and Civil Engineering,
Saint Petersburg, Russia

N. Vatin et al. (eds.), *Proceedings of MPCPE 2021*, Lecture Notes in Civil Engineering 182, https://doi.org/10.1007/978-3-030-85236-8_16

minimizes concrete creep [2]. By this time most of the plant-manufactured ferro-concrete has pre-stress. By the same token strut beam are widely used in metal structures [3].

In 1860, a wooden bridge (Elden Bridge) was built by using Long's trusses (Fig. 1). In Long's original drawings structural units and additional braces were installed in the truss and held by contraction forces. This was a result of using wedges located at different points in the truss panels [5].

Long described wedges as "3/4 inches thick at the end, 1/4 inches at the back, one-foot long". The pre-compression of the braces caused appearing the corresponding pre-stress states in the struts and chords. In 2002, there was the bridge structural calculations with modeling of pre-stress from wedges, which showed that pre-stressing allowed to redistribute the bridge structure forces increasing the structure safety and reducing the bridge deflection by 81% [6].

Zhadanov V. I., Dmitriev P. A. and Tisevich E. V. developed [7] an improved glu-lam. Its advantage is in the fact that rods equal in length to the package width are glued into the drilled holes of the pressed board set with reference to its reduction because of the pressure from compact. Uniform pre-stressing along the entire length of the set is ensured by the rod step from 300 to 500 mm.

Under the leadership of V. N. Golovach (Belarusian Polytechnic Institute), a pre-stressed frame was developed (Fig. 2). This frame is shown like an additional beam supported by braces which is laid under the crossbar supported by the posts [8].

Fig. 1 Elden Bridge, built in 1860 [4]

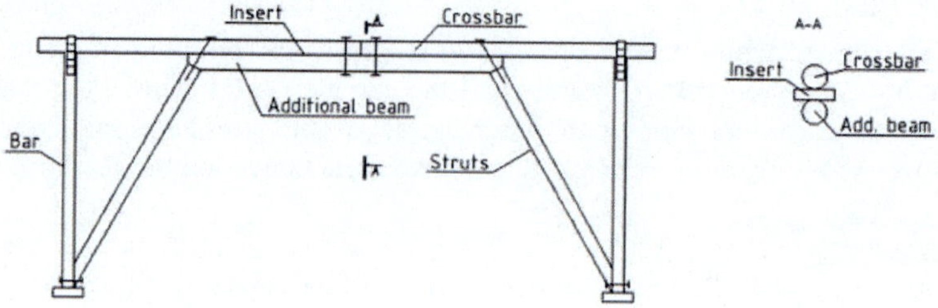

Fig. 2 Pre-stressed wooden frame

Structural stress control is a new approach compared to the "traditional" approach, which allows to save material by regulating stresses using torqueing during the operation of structures. A similar approach is used in metal structures however stress control in wooden structures faced the following problems [9].

2 Methods

However, during the operation of real structures there were found the following features which must be applied when designing such structures:

- Extremely high variability of properties along and across the fibers [10]. The strength and deformability of wood along and across the fibers differ 5–10 times (according to). In this regard, when stresses along the fibers are much lower than the ultimate strength, the wood across the fibers begins to sustain plastic flow. This phenomenon is known as "trampling down" of wood in the knots [11];
- Leakage in knots which leads to deformation capacity of knot connections. The deformation capacity of nodal joints under the action of deformations from compression-tension is characterized by the C-toughness factor, which is calculated by the following formula:

$$C = P/\delta, \quad (1)$$

where P is the compressive-tensile force; δ—linear deformation of compliance [12, 13]. Calculation of the average coefficient of C-toughness factor in bending, when the angle α (the angle between the direction of the force and the axis of symmetry of the cross section) varies from 0 to 90; is produced by the transition from linear to angular deformation:

$$C = M/\theta \quad (2)$$

where M is the bending moment; θ is the angular deformation of the compliance. In this case, the tolerance caused by the edge deformations is not taken into account due to the small edge deformations in comparison with the deformations of the bond compliance due to bending and trampling down of the wood.

$$\delta = \delta_{heat} + \delta_y \quad (3)$$

where δ_{nepl}—compliance due to the looseness of the nodal connection; δ_y—compliance of elastic properties of the nodal element;

- Influence of wood humidity [14, 15]. An increase of humidity in wooden structures leads to decreasing elastic characteristics (Young's modulus, shear modulus, etc.) in accordance to the following formula [2]:

$$E_{12} = \frac{E_W}{1 - \alpha(W - 12)} \tag{4}$$

When the humidity changes the elastic modulus changes up to 2 or more times. The strength characteristics of wood depend on humidity according to the following formula:

$$B_{12} = B_W(1 + a(W - 12)) \tag{5}$$

where E_{12} is the elastic modulus at 12% humidity, E_W is the elastic modulus at W humidity, B_{12} is the strength characteristic at 12% humidity, BW is the strength characteristic at W humidity, a is the correction factor for various types of stress state, a = 0.05 at compression, a = 0.04 for bending and a = 0.03 for shearing [16]. This formula is valid in the sample humidity range from 8 to 23%. Thus, with an increase in moisture content, wood fluidity and loss of the effect of pre-stressing are also observed; Wood in structures is dried to a moisture content of 12%. It is then exposed to short-term (daily) and long-term (seasonal) changes in relative humidity during operation. But at the same time, short-term changes in humidity affect only the surface layers with a thickness of no more than 10–15 mm. To protect against long-term changes, it is recommended to cover the wood with any protective coat of paints and varnishes, and also to prevent significant long-term changes in humidity by ventilation and heating [17].

- Relatively recently, another factor was revealed—the nonlinear dependence of the elastic modulus on stresses during deformation of wood [18, 19].
- Rheological properties of wood [20, 21]. Wood consists of components with different elastic properties, what results in redistribution of internal stresses between the component parts of the material. As a result, the stress state and deformation of structural elements change in some time. On the basis of the regression analysis, it was established that at a constant loading time the deformation process obeys a single regularity:

$$\delta(t) = \delta_1[1 + \varphi(t)] \tag{6}$$

where δ_1 is the deformation of the joint at the initial moment of loading, determined experimentally and depending on the type and material of the joint, the level of loading, temperature and humidity operating conditions and technological factors; $\varphi(t)$ is the creep characteristic of the joint, which is the ratio of creep to δ1. The $\varphi(t)$ value is determined by the expression

$$\varphi(t) = \varphi_\infty(1 - e^{-\gamma t}) \tag{7}$$

where φ_∞ is the limiting characteristic of the joint creep at the moment of stabilization of its deformation, γ—is the coefficient characterizing the rate of development of creep deformation in time t, days. The strength of wood with

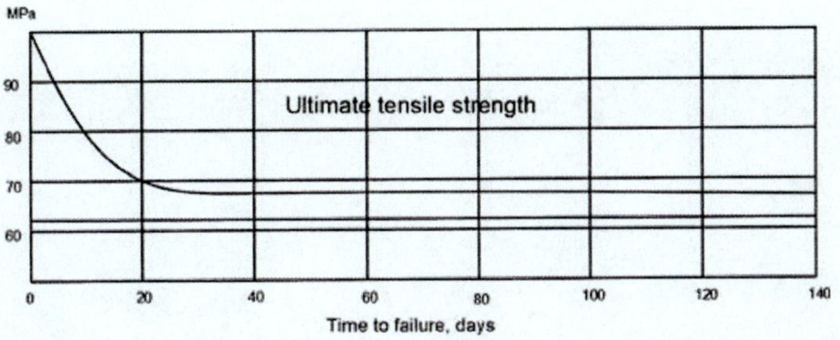

Fig. 3 Changes in the strength of wood over time

prolonged application of the load decreases to 1.5 times according to the following graph (Fig. 3). Deformations of wood also increase during time. This influences the fact that the maximum pre-stressing value must be assigned taking into account the maximum long-term deformations [19].

As a result of the influence of the above mentioned factors designing of structures with the regulation of stresses in wooden structures is complicated.

3 Results and Discussion

However, during the operation of real structures there were found the following features which must be applied when designing such structures:

1. Maximally increase the area of contact surfaces in the nodes to prevent "trampling down" of wood in them. An example of such a node is shown in Fig. 4 from work. With this nodal structure as shown by the experiments given in the effect of knot compliance and anisotropy of wood did not exceed 5%.
2. Maintain strict humidity control. Avoid changes in humidity during operation of the structure, or correct pre-stressing taking into account changes in humidity.
3. Regulation of stresses according to a special algorithm that takes into account the variable modulus of elasticity and also allows to compensate for other problems.

According to the static scheme this node is hinged since in its composition the "sliders" can rotate in a "hemisphere" around the center of the node. The elements of the upper chord or braces transfer forces directly to the sliders. The elements of the lower chord are directly attached to the hemisphere by means of welding or end elements of the "Markhi" knot type. The strength of the knot connection is ensured by the spherical shape of the knot element, since the ball works well in compression. The connection of the sliders with the elements of the braces is carried out using end elements, also of the "Markha" type. If plywood pipes are used as braces,

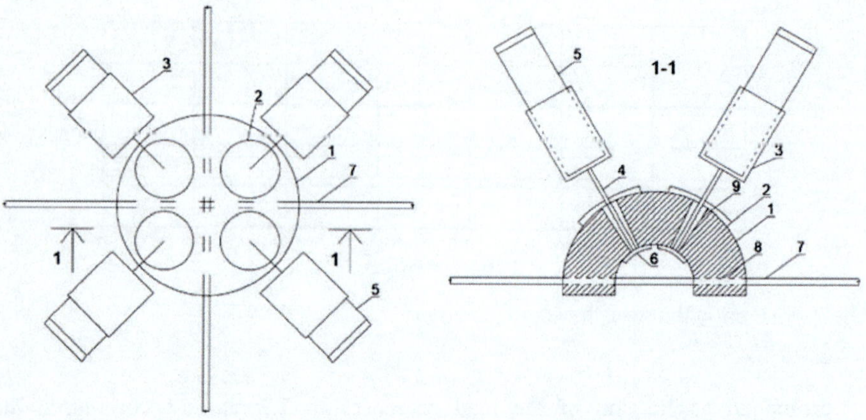

Fig. 4 Elden Bridge, built in 1860. 1—node spherical element, 2, 6—spherical washers, 3—lugs in the form of clips for rod elements of braces, 4—connectors transmitting forces from braces to the bearing node element, 5—metal-wood braces PSC, 7—metal belt element, 8, 9—tapered holes allowing rotation of the brace elements around the center of the node element

then the connection with the sliders is carried out using the thread onto which the pipes are screwed [14].

The use of a similar unit (see Fig. 4) is also rational when connecting the wooden elements of the upper chord and the wooden elements of the lattice of the braces of the metal-wood PSC.

The support unit of the metal-wood PSC in which the puffs are attached to it is shown in Fig. 5. This node based on the Kislovodsk system is made with some

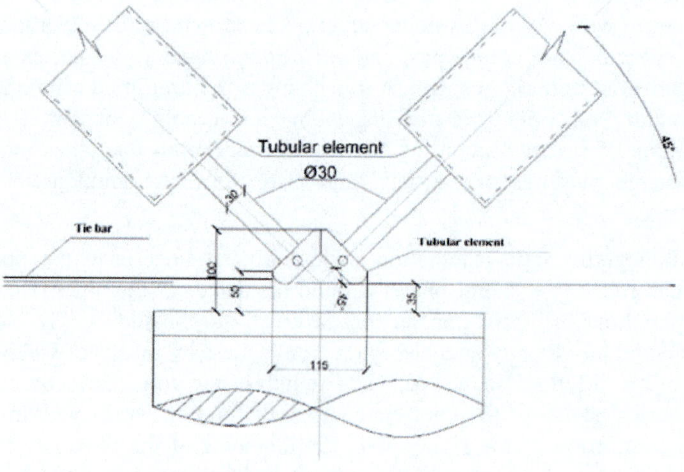

Fig. 5 The support unit of the metal-wood PSC in which the puffs are attached

changes. The brace and tie axes converge at one point. The node element is a truncated polyhedron.

4 Conclusion

Considering the above recommendations, the following should be noted:

• regulation of stresses in timber structures is quite possible;
• one of the peculiarities of stress regulation in wooden structures is a high dependence on humidity;
• at the same time, it is extremely important to avoid crushing the wood in the knots;
• the author proposed a model and received a patent for a nodal connection with the exclusion of hugging in the nodes.

References

1. Abovskiy NP, Maximova OM (2007) Neuro-prognosis based on step model with teaching for natural tests results of building structures. Opt Mem Neural Netw. https://doi.org/10.3103/s1060992x07010055
2. Li H, Huo L (2010) Advances in structural control in civil engineering in China. Math Probl Eng 2010. https://doi.org/10.1155/2010/936081
3. Sanaev VG, Gorbacheva GA, Galkin VP, Vinokurov SA, Kumaniaev SE, Klassen NV (2019) Prospects for using unique physicochemical properties of plants and their ability for self-regeneration in space. In: AIP conference proceedings. https://doi.org/10.1063/1.5133343
4. da Porta F (2002) Eldean bridge (Allen's Mill Bridge). Hist Am Eng Rec 8–38
5. Smith T, Pampanin S, Fragiacomo M, Buchanan A (2008) Design and construction of prestressed timber buildings for seismic areas. In: 10th world conference on timber engineering 2008, pp 1746–1753
6. Datta T (2003) A state-of-the-art review on active control of structures. ISET J Earthq Technol 40:1–17
7. Luo HY, Fan RL, Wang HJ, Zhang LM (2020) Physics of building vulnerability to debris flows, floods and earth flows. Eng Geol 271:105611. https://doi.org/10.1016/j.enggeo.2020.105611
8. Soong TT, Manolis GD (1987) Active structures. J Struct Eng 113:2290–2302. https://doi.org/10.1061/(ASCE)0733-9445(1987)113:11(2290)
9. Zhu X, Jing X, Cheng L (2011) A magnetorheological fluid embedded pneumatic vibration isolator allowing independently adjustable stiffness and damping. Smart Mater Struct 20. https://doi.org/10.1088/0964-1726/20/8/085025
10. Pavlenko AD, Rybakov VA, Pikht AV, Mikhailov ES (2016) Non-uniform torsion of thin-walled open-section multi-span beams. Mag Civ Eng 67:55–69. https://doi.org/10.5862/MCE.67.6
11. Cheng F, Jiang H, Lou H (2008) Smart structures: innovative systems for seismic response control. https://doi.org/10.1201/9781420008173

12. Casciati S, Chassiakos AG, Masri SF (2014) Toward a paradigm for civil structural control. Smart Struct Syst 14:981–1004. https://doi.org/10.12989/sss.2014.14.5.981
13. Kobori T (1996) Future direction on research and development of seismic-response-controlled structures. Comput Civ Infrastruct Eng. https://doi.org/10.1111/j.1467-8667.1996.tb00444.x
14. Smith T (2014) Post-tensioned timber frames with supplemental damping devices
15. Shan S, Li S (2020) Fire-induced progressive collapse mechanisms of steel frames with partial infill walls. Structures 25:347–359. https://doi.org/10.1016/j.istruc.2020.03.023
16. Wang X, Lu H, Gorbacheva G, Hossain M, Fu YQ (2021) Multi-modal commutative dynamics in semi-crystalline polymers undergoing multiple shape memory behavior. Smart Mater Struct 30. https://doi.org/10.1088/1361-665X/abe4e5
17. Ugolev BN, Skuratov NV, Gorbacheva GA (1998) On differential wood shrinkage coefficient. Drev Vysk Res 43:1–11
18. Pásztory Z, Gorbacheva G, Czimondor D, Rébék-Nagy P, Sanaev V, Rykunin S, Czupy I, Börcsök Z (2018) The effects of windows and reduced night temperature on the heating energy demand in different regions of Russia. IOP Conf Ser Mater Sci Eng. https://doi.org/10.1088/1757-899X/371/1/012050
19. Gorbacheva GA, Ivankin AN, Sanaev VG, Ageev AK, Kiryukhin DP, Kichigina GA, Kushch PP, Badamshina ER (2017) Surface modification of cellulose-containing materials with solutions of tetrafluoroethylene telomers. Russ J Appl Chem 90:1365–1371. https://doi.org/10.1134/S1070427217080286
20. Ugolev BN (1976) General laws of wood deformation and rheological propertions of hardwood. Wood Sci Technol 10:169–181. https://doi.org/10.1007/BF00355738
21. Ugolev BN, Skuratov NV (1992) Stress-strain state of wood at kiln drying. Wood Sci Technol 26:209–217. https://doi.org/10.1007/BF00224293

Regulation of Stresses in Structures of Buildings Located in Extreme Wind Conditions

Mihail Moskalev and Dmitriy Charnik

Abstract The Arctic and the Antarctic areas are characterized by extreme wind loads that exceed the standardized by SP. These areas are characterized by a significant distance from material and raw material bases and road transport arteries. Under these circumstances, the problem of perception of extreme wind loads must be solved with a minimum consumption of building material. One of the ways to minimize the consumption of building materials is to use an innovative approach to regulate stress in building structures The software package SCAD++ is used for the calculation. 1. Metal columns—column I-beam 23K1 GOST 26020-83, I-beam wide-flange 26SH1 GOST 26020-83, I-beam wide-flange 23SH1, I-beam norma 23B1. Covering load due to its own weight—0.15 t/m^2. Snow load—0.19 t/m^2. Wind load: upwind side 0.87 t/m^2, leeward side—0.55 t/m^2. Change in the value of displacement along the x-axis in the first version of the wind load and an increase in the tension of steel ropes from 1 to 25 t. Due to the tension of the ropes, it is possible to reduce the horizontal movement from 24.171 to 20.84 mm, the difference was 3.31, the movement decreased by 14%.

Keywords Active control structures · Actuators and smart structures · Active · Active control · Control algorithms · Structural control · Steel · Steel constructions

1 Introduction

The Arctic and the Antarctic areas are characterized by extreme wind loads that exceed the standardized by SP [1, 2]. These areas are characterized by a significant distance from material and raw material bases and road transport arteries [3]. Under these circumstances, the problem of perception of extreme wind loads must [3–6]

M. Moskalev (✉)
Saint-Petersburg State University of Architecture and Civil Engineering, Saint Petersburg, Russia

D. Charnik
Emperor Alexander I St. Petersburg State Transport University, Saint Petersburg, Russia

N. Vatin et al. (eds.), *Proceedings of MPCPE 2021*, Lecture Notes in Civil Engineering 182, https://doi.org/10.1007/978-3-030-85236-8_17

be solved with a minimum consumption of building material [7–9]. One of the ways to minimize the consumption of building materials [10] is to use an innovative approach to regulate stress in building structures [11–13]. In this work the survey of the effectiveness of applying this approach to the design of a modular building located in the Antarctic conditions [14] with the following characteristics was conducted [15]:

- Length—54 m.
- Width—from 18 to 30 m.
- Number of storeys—2, the height of the first and second floor is 3390 and 3110 mm respectively.
- The height of the building is 8.31 m.
- The building is installed on piles buried below the freezing depth.
- According to the constructive solution, the frame of the building was adopted as a frame-braced one [16].
- The appointment of the building is a scientific laboratory, with office, communal and residential premises.

The solution to the problem of voltage regulation is assumed in the nodes of building structures using puffs [5, 13, 17–20].

2 Methods

The software package SCAD++ is used for the calculation [21].

Structural elements:

1. Metal columns—column I-beam 23K1 GOST 26020-83 (Figs. 1, 2, 3, 4, 5, 6 and 7).

Fig. 1 Column I-beam 23K1

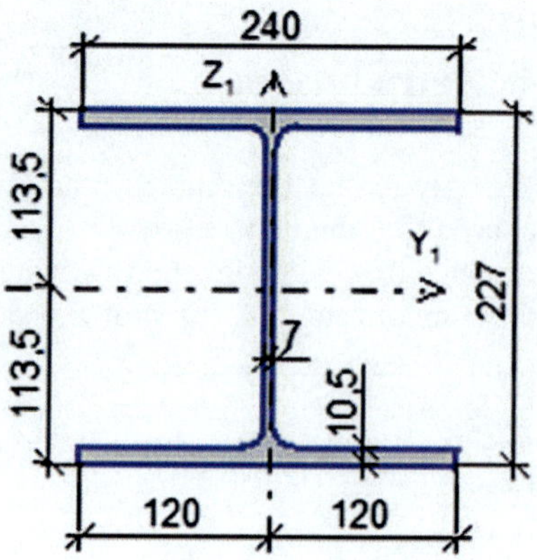

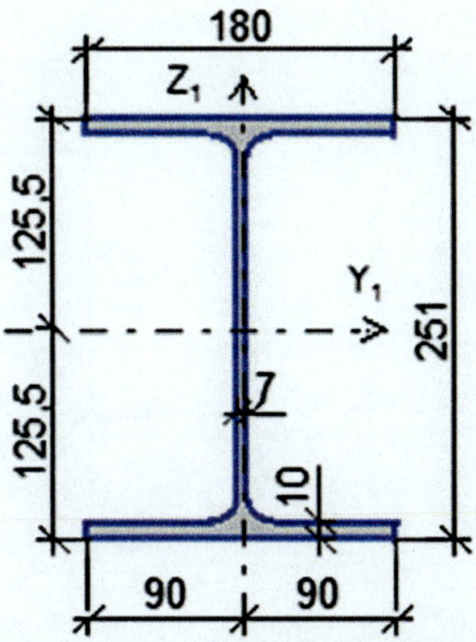

Fig. 2 I-beam wide-flange 26SH1 GOST 26020-83

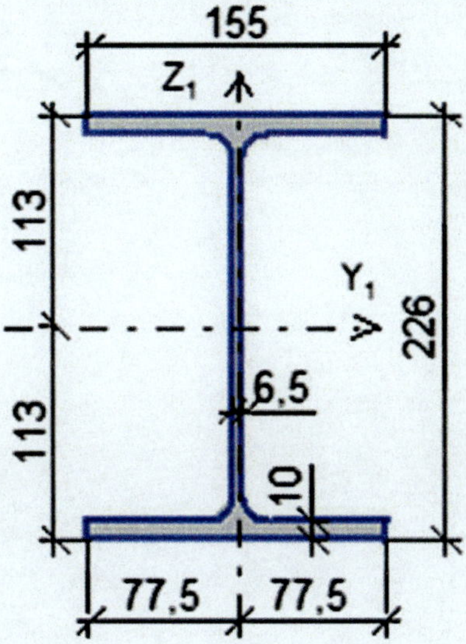

Fig. 3 I-beam wide-flange 23SH1

Covering load due to its own weight—0.15 t/m^2. Snow load—0.19 t/m^2. Wind load: upwind side 0.87 t/m^2, leeward side—0.55 t/m^2.

Before the calculation, the following load combinations cases were compiled (Fig. 8).

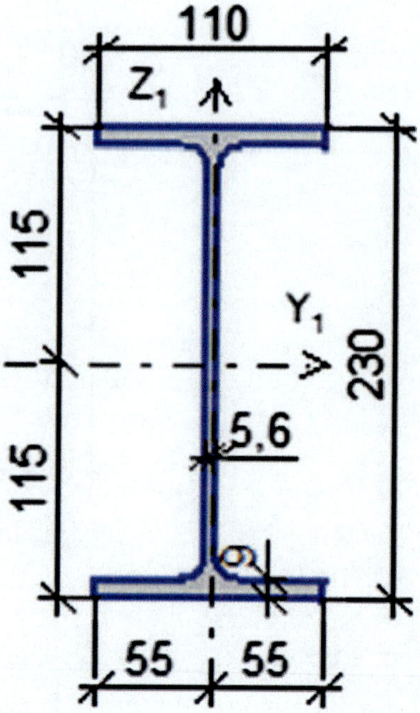

Fig. 4 I-beam norma 23B1

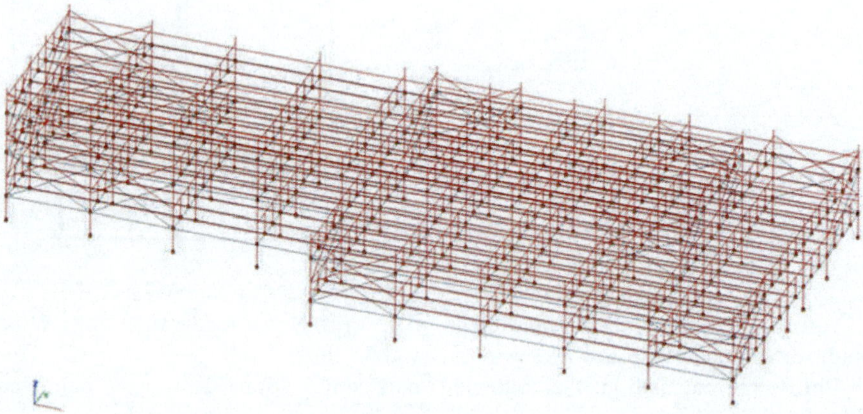

Fig. 5 Self-weight loading scheme. Coating load—0.15 t/m^2

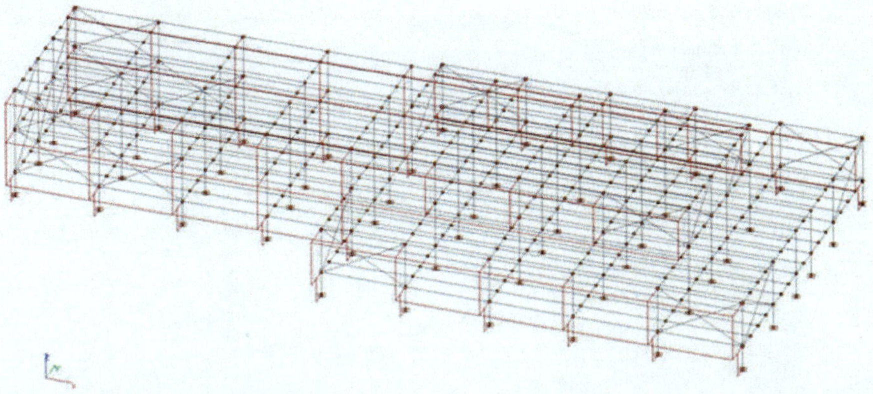

Fig. 6 Wind load scheme option 1

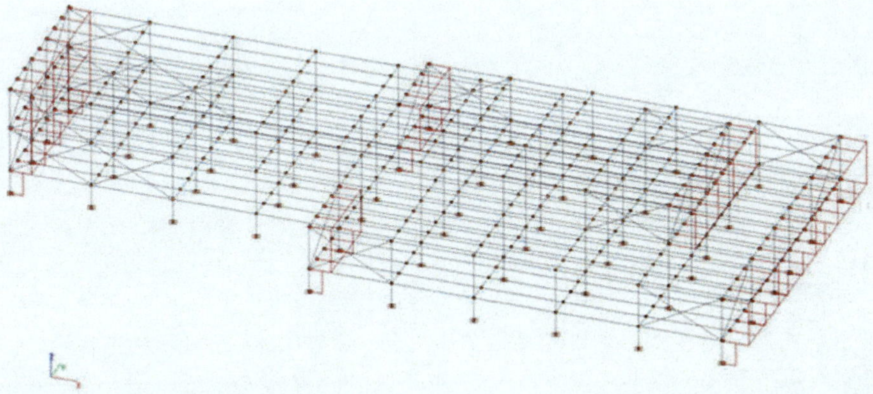

Fig. 7 Wind load scheme option 2

3 Results and Discussion

Change in the value of displacement along the x-axis in the first version of the wind load and an increase in the tension of steel ropes from 1 to 25 t [22] (Figs. 9, 10 and 11).

In Fig. 13 the columns in which the internal forces were checked are marked in red (Fig. 12).

	Load combinations	name
1	(L1)*1+(L4)*1+(L5)*1	constant load
2	(L2)*0.9+(L6)*1+(C1)*1	wind 1 without tension
3	(L2)*0.9+(L6)*1+(L14)*1+(C1)*1	wind 1 - 1 t
4	(L2)*0.9+(L6)*1+(L15)*1+(C1)*1	wind 1 - 3 t
5	(L2)*0.9+(L6)*1+(L16)*1+(C1)*1	wind 1 - 7 t
6	(L2)*0.9+(L6)*1+(L17)*1+(C1)*1	wind 1 - 10 t
7	(L2)*0.9+(L6)*1+(L18)*1+(C1)*1	wind 1 - 17 t
8	(L2)*0.9+(L6)*1+(L19)*1+(C1)*1	wind 1 - 20 t
9	(L2)*0.9+(L6)*1+(L20)*1+(C1)*1	wind 1 - 25 t
10	(L2)*0.9+(L3)*1+(C1)*1	wind 2 without tension
11	(L2)*0.9+(L3)*1+(L7)*1+(C1)*1	wind 2 - 1 t
12	(L2)*0.9+(L3)*1+(L8)*1+(C1)*1	wind 2 - 3 t
13	(L2)*0.9+(L3)*1+(L9)*1+(C1)*1	wind 2 - 7 t
14	(L2)*0.9+(L3)*1+(L10)*1+(C1)*1	wind 2 - 10 t
15	(L2)*0.9+(L3)*1+(L11)*1+(C1)*1	wind 2 - 15 t
16	(L2)*0.9+(L3)*1+(L12)*1+(C1)*1	wind 2 - 20 t
17	(L2)*0.9+(L3)*1+(L13)*1+(C1)*1	wind 2 - 25 t

Fig. 8 Load combinations

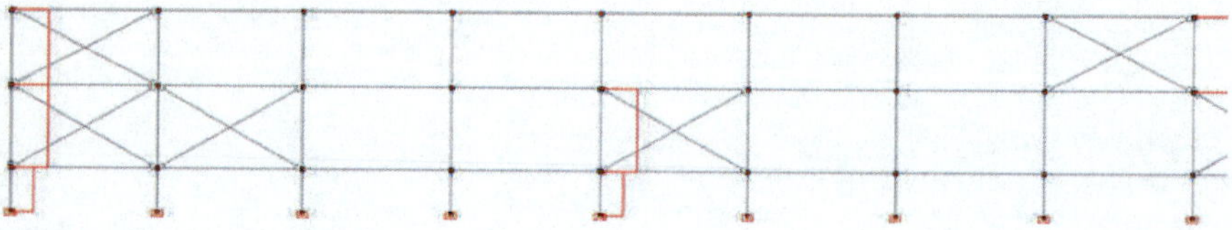

Fig. 9 Wind load direction

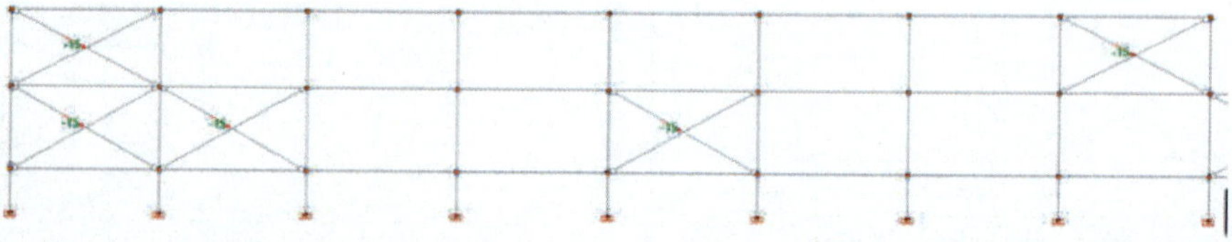

Fig. 10 Rope tension scheme

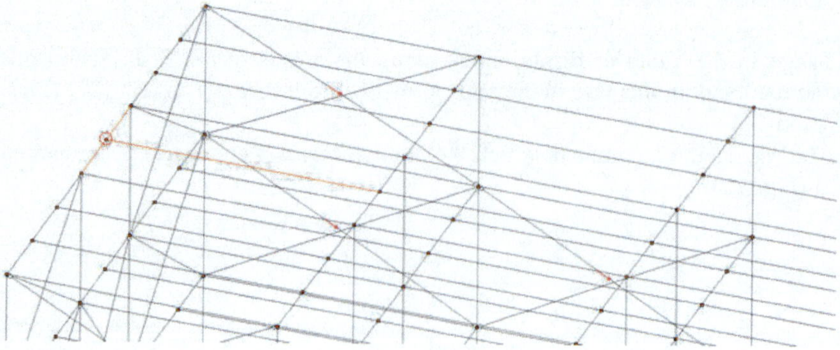

Fig. 11 Node location

Fig. 12 Columns in which the internal forces were compared

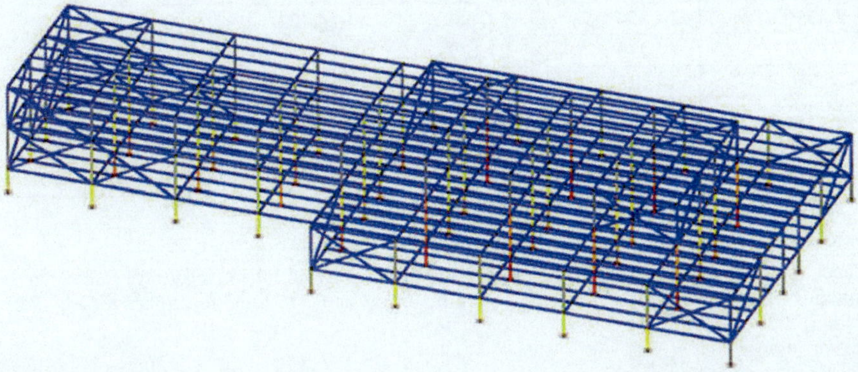

Fig. 13 Color display of efforts N t

4 Conclusions

See Tables 1 and 2.

Table 1 Move the node along the x-axis (Fig. 12)

Distance (mm)	Tensioning the ropes (tonne)
−24.171	0
−24.037	1
−23.727	3
−23.134	7
−22.689	10
−21.948	15
−21.206	20
−20.84	25

Conclusion: due to the tension of the ropes, it is possible to reduce the horizontal movement from 24.171 to 20.84 mm, the difference was 3.31, the movement decreased by 14% [16, 23]

Table 2 Move the node along the x-axis (Fig. 12)

Distance (mm) № 59	Distance (mm) № 60	Distance (mm) № 86	Distance (mm) № 87	Tensioning the ropes (tonne)
−27.346	−14.354	−27.444	−14.402	0
−26.998	−14.279	−27.179	−14.31	1
−26.3	−14.128	−26.397	−14.176	3
−24.905	−13.825	−25.002	−13.873	7
−23.859	−13.599	−23.956	−13.647	10
−22.115	−13.221	−22.212	−13.269	15
−20.371	−12.843	−20.469	−12.891	20
−20.371	−12.843	−20.469	−12.891	25

Conclusion: due to the tension of the ropes, it is possible to reduce the compressive force N in the columns from 27.346 t to 20.371 t mm, the difference was 6.975 t, the force decreased by 26%

References

1. Pham AT, Tan KH (2019) Static and dynamic responses of reinforced concrete structures under sudden column removal scenario subjected to distributed loading. J Struct Eng (United States). https://doi.org/10.1061/(ASCE)ST.1943-541X.0002214
2. Qian L, Li Y, Diao M, Guan H, Lu X (2020) Experimental and computational assessments of progressive collapse resistance of reinforced concrete planar frames subjected to penultimate column removal scenario. J Perform Constr Facil 34:04020019. https://doi.org/10.1061/(ASCE)CF.1943-5509.0001420
3. Alembagheri M, Sharafi P, Hajirezaei R, Tao Z (2020) Anti-collapse resistance mechanisms in corner-supported modular steel buildings. J Constr Steel Res 170. https://doi.org/10.1016/j.jcsr.2020.106083
4. Adam JM, Parisi F, Sagaseta J, Lu X (2018) Research and practice on progressive collapse and robustness of building structures in the 21st century. https://doi.org/10.1016/j.engstruct.2018.06.082
5. Soong TT, Manolis GD (1987) Active structures. J Struct Eng 113:2290–2302. https://doi.org/10.1061/(ASCE)0733-9445(1987)113:11(2290)

6. Gorbacheva GA, Tarasov MV, Smirnov DV, Sanaev VG (2019) Shape memory effect of mycologically destroyed wood. IOP Conf Ser Earth Environ Sci. https://doi.org/10.1088/1755-1315/226/1/012035
7. Feiber SD, de Souza TC, Bressiani L, Balestra CET (2021) Analysis of co2 emissions between construction systems: light steel frame and conventional masonry. Environ Eng Manag J 19:2147–2156
8. Committee 222 (2010) Corrosion of prestressing steels reported by ACI, vol 01, pp 1–42
9. Zhu X, Jing X, Cheng L (2011) A magnetorheological fluid embedded pneumatic vibration isolator allowing independently adjustable stiffness and damping. Smart Mater Struct 20. https://doi.org/10.1088/0964-1726/20/8/085025
10. Lyu CH, Gilbert BP, Guan H, Underhill ID, Gunalan S, Karampour H, Masaeli M (2020) Experimental collapse response of post-and-beam mass timber frames under a quasi-static column removal scenario. Eng Struct 213:110562. https://doi.org/10.1016/j.engstruct.2020.110562
11. Abovskiy NP, Maximova OM (2007) Neuro-prognosis based on step model with teaching for natural tests results of building structures. Opt Mem Neural Netw. https://doi.org/10.3103/s1060992x07010055
12. Azim I, Yang J, Javed MF, Iqbal MF, Mahmood Z, Wang F, Liu Q (2020) Prediction model for compressive arch action capacity of RC frame structures under column removal scenario using gene expression programming. Structures 25:212–228. https://doi.org/10.1016/j.istruc.2020.02.028
13. Venanzi I (2016) A review on adaptive methods for structural control. Open Civ Eng J 10:653–667. https://doi.org/10.2174/1874149501610010653
14. Casciati S, Chassiakos AG, Masri SF (2014) Toward a paradigm for civil structural control. Smart Struct Syst 14:981–1004. https://doi.org/10.12989/sss.2014.14.5.981
15. Pavlenko AD, Rybakov VA, Pikht AV, Mikhailov ES (2016) Non-uniform torsion of thin-walled open-section multi-span beams. Mag Civ Eng 67:55–69. https://doi.org/10.5862/MCE.67.6
16. Reinhorn A, Soong T, Lin R, Riley M, Wang Y, Aizawa S, Higashino M (1992) Active bracing system: a full scale implementation of active control, p 120
17. Dyke SJ, Yi F (2000) On the performance of controlled civil engineering structures. In: Advanced technology in structural engineering. American Society of Civil Engineers, Reston, VA, pp 1–8. https://doi.org/10.1061/40492(2000)7
18. Formisano A, Davino A (2021). Hardness vs strength for structural steels: first results from experimental tests. https://doi.org/10.1007/978-3-030-64908-1_21
19. Mezentseva A, Gel'Manova Z, Konakbayeva A (2020) Improving the reliability of metal structures of buildings and structures during reconstruction. IOP Conf Ser Mater Sci Eng. https://doi.org/10.1088/1757-899X/889/1/012006
20. Sogukpinar H (2020) Effect of hairy surface on heat production and thermal insulation on the building. Environ Prog Sustain Energy 39. https://doi.org/10.1002/ep.13435
21. Krittanawong C, Virk HUH, Kumar A, Aydar M, Wang Z, Stewart MP, Halperin JL (2021) Machine learning and deep learning to predict mortality in patients with spontaneous coronary artery dissection. Sci Rep 11:8992. https://doi.org/10.1038/s41598-021-88172-0
22. Hamilton T, West J, Wouters J (2009) Corrosion of prestressing steels reported by ACI Committee 222. ACI Commitee 222:1–29
23. Ugolev BN (2014) Wood as a natural smart material. Wood Sci Technol 48:553–568. https://doi.org/10.1007/s00226-013-0611-2

Deformation of 2D RC Beam-Column Joint

Sergey Savin, Anastasia Obydina, and Ekaterina Mikhailik

Abstract In the beam-column subassembly and adjacent bar sections of RC frames of frame-braced structural systems, a complex 2D stress–strain state arises. The computational models for such subassemblies presented in the scientific literature are mainly based on a simplified representation of the forces acting in them. These ones allow evaluate the implementation of the characteristic destruction mechanisms and further each of them is analysed separately. Therefore, in this article, a finite difference method is proposed for deformation analysis of a reinforced concrete beam-column monolithic subassembly. A distinctive feature of the proposed solution is the ability to take into account the discrete nature of the reinforcement, as well as the incomplete adhesion of reinforcement to concrete along the contact surface. Also, the article presents an example of calculating the deformed state of a beam-column monolithic subassembly of a RC frame scale model, for which a simulation was previously performed using bar FE. Comparison of the calculation results according to the proposed model and the traditional bar model shows the differences in the deformed state of the 2D RC beam-column joint. This can be explained by taking into account the 2D stress–strain state factors, which are not taken into account in the bar model.

Keywords Reinforced concrete · Beam-column joint · Finite difference · Numerical analysis · Deformation

S. Savin (✉) · A. Obydina · E. Mikhailik
Moscow State University of Civil Engineering, Yaroslavskoe shosse, 26, Moscow 129337, Russia
e-mail: savinsyu@mgsu.ru

S. Savin
Research Institute of Construction Physics of RAACS, Lokomoivny pr, 21, Moscow 127238, Russia

N. Vatin et al. (eds.), *Proceedings of MPCPE 2021*, Lecture Notes in Civil Engineering 182, https://doi.org/10.1007/978-3-030-85236-8_18

1 Introduction

In the reinforced concrete frames of buildings and structures at the beam-column junctions, a complex stress–strain state arises under external forces [1, 2]. In some cases, when the girders adjoin the column only from two opposite faces in the case of a frame-type assembly, such a stress–strain state can be conventionally considered as biaxial, assuming that the stresses are distributed relatively uniformly over the section width (from the plane of action of bending moments).

For a more accurate account of the deformed state of such 2D frame joints, as a rule, approaches are used that are similar to the applied element method [3, 4], which has gained wide recognition in recent years in modeling the deformation and destruction of reinforced concrete bearing systems of buildings and structures under special influences. Better convergence with experimental data when using the applied element method compared to the standard FEM procedure in a bar formulation is achieved by simulating the connection between short sections of structural elements using elastic-yielding springs.

Analysis of the calculated models of the girder-column interface nodes presented in the scientific literature shows that most of them are based on a simplified representation of a complexly stressed element by replacing stresses with generalized forces. Thus, highlighting two characteristic resistance mechanisms, truss and compressed inclined strip, Hwang and Lee [5] assess the possibility of their implementation separately. A similar approach to the analysis of the work of the girder and column interface units, which are somewhat different in their design, can be found in the works of other authors [6–8]. In works [9–11], the behavior of a plane joint element is modeled using elastic ties (springs). The introduction of such elastic-yielding bonds between elements into the computational model allowed Feng and Ning [9] to achieve better quantitative and qualitative convergence with experimental data than when using traditional rod models of the finite element method with rigid nodal joints. However, the use of the described approach is associated with the complexity of determining the compliance parameters of the elements-springs, taking into account the possibility of cracking, leading to the structural orthotropy of the plane-stressed joint [9, 12–14].

Summarizing the results of a brief analysis of the calculated models of the beam-column joints presented in the scientific literature, it can be concluded that the models of such nodes are mainly based on a simplified representation of the forces acting in them and an assessment of the implementation of one of the characteristic fracture mechanisms, the analysis of each of which performed separately. In this regard, the purpose of this work is to develop a generalized design model of a 2D reinforced concrete frame unit, which would take into account the presence and location of longitudinal and transverse reinforcement, as well as various loading options for beams and columns adjacent to the joint.

2 Method

To assess the nature of deformation and destruction of 2D nodes of reinforced concrete frame-tie frames of buildings and structures in this study, we will use a model of a bearing system consisting of universal physically nonlinear rod finite elements of girders and columns and special elements (by the type of super elements [15, 16]), modeling crossbar-column interface nodes. In this case, the stress–strain states of the sections at the ends of the bar elements will be the boundary conditions for calculating the nodal connection. The further solution of the problem of estimating the stress state of a node will be carried out in the "plane stress state" setting, neglecting the stresses from the plane of the load action, assuming that they are small in comparison with the components of the stress state in the plane. As a first approximation, the area bounding the junction of the column and the girder will be considered rectangular.

In contrast to the traditional approach to the calculation using super elements, in this study, the general solution to the problem of the plane stress state of a monolithic girder-column interface is based on the grid method in displacements. The use of the grid method makes it possible to construct an algorithm that is simpler than the FEM for assessing the convergence of the calculation results and adjusting the density of the region partitioning to achieve the required accuracy, as well as taking into account the contact interaction of reinforcement and concrete. The form of writing the problem in displacements, in turn, makes it possible to avoid solving additional problems of evaluating the values of the stress function at the nodes on the contour of the region that bounds the considered rectangular special element of the crossbar and column conjugation.

Following the accepted two-dimensional formulation of the problem of calculating the stress–strain state of the element of the joint of the girder and the column of a reinforced concrete frame, the equilibrium equations of an infinitesimal volume of an orthotropic body can be written in displacements in the following form:

$$\begin{aligned} &\frac{E_x}{1-\nu^2}\frac{\partial^2 u}{\partial x^2}+\frac{E_y\nu_x}{1-\nu^2}\frac{\partial^2 v}{\partial x\partial y}+G_{xy}\left(\frac{\partial^2 u}{\partial y^2}+\frac{\partial^2 v}{\partial x\partial y}\right)=0;\\ &\frac{E_y}{1-\nu^2}\frac{\partial^2 v}{\partial y^2}+\frac{E_x\nu_y}{1-\nu^2}\frac{\partial^2 u}{\partial x\partial y}+G_{xy}\left(\frac{\partial^2 v}{\partial x^2}+\frac{\partial^2 u}{\partial x\partial y}\right)=0, \end{aligned} \tag{1}$$

where u, v are displacements of grid nodes along orthogonal coordinate axes OX and OY, respectively;

E_x, E_y are reduced moduli of material deformations along the orthogonal coordinate axes OX and OY, respectively;

$G_{xy} = \sqrt{E_x E_y}[2(1+\nu)]^{-1}$ is a reduced shear modulus;

$\nu_x = \nu\sqrt{E_x/E_y}; \nu_y = \nu\sqrt{E_y/E_x}$,
ν is Poisson ratio, which takes value ν = 0.2 for concrete.

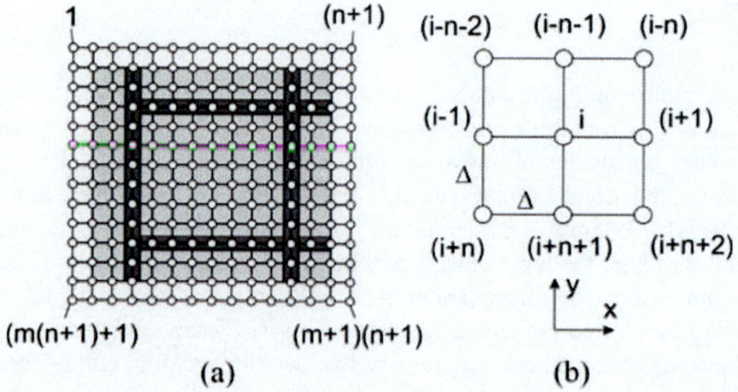

Fig. 1 Scheme for the calculation of a joint element by the method of meshes: general view (**a**); accepted rule of numbering grid nodes (**b**)

In Eq. (1), we neglected the components of the volumetric forces from the self-weight of the material, since they do not make a significant contribution to the general stress state of the element at the loading stages considered in the study.

We divide the area bounding the joint element under consideration by a grid with the same horizontal and vertical spacing so that the centers of gravity of the reinforcing bars coincide with the grid lines (Fig. 1a).

3 Results and Discussion

3.1 Construction of a Computational Model and Finite-Difference Equations of the Grid Method

Let us write down the derivatives in Eq. (1) in finite differences:

$$\begin{aligned}
\frac{\partial^2 u_i}{\partial x^2} &= \frac{u_{(i-1)} - 2u_i + u_{(i+1)}}{\Delta^2};\\
\frac{\partial^2 v_i}{\partial y^2} &= \frac{v_{(i-n-1)} - 2v_i + v_{(i+n+1)}}{\Delta^2};\\
\frac{\partial^2 u_i}{\partial y^2} &= \frac{u_{(i-n-1)} - 2u_i + u_{(i+n+1)}}{\Delta^2};\\
\frac{\partial^2 v_i}{\partial x^2} &= \frac{v_{(i-1)} - 2v_i + v_{(i+1)}}{\Delta^2};\\
\frac{\partial^2 u_i}{\partial x \partial y} &= \frac{u_{(i-n)} - u_{(i-n-2)} + u_{(i+n+2)} - u_{(i+n)}}{4\Delta^2};\\
\frac{\partial^2 v_i}{\partial x \partial y} &= \frac{v_{(i-n-2)} - v_{(i+n)} + v_{(i-n)} - v_{(i+n+2)}}{4\Delta^2},
\end{aligned} \tag{2}$$

where Δ is a dimension of grid bay (Fig. 1b).

For the mixed derivative, we additionally write an expression in one-sided differences, which will allow us to relate the equilibrium equation (2) for the corner points of the contour of the element under consideration to the boundary conditions:

$$\begin{aligned} \frac{\partial^2 u_i}{\partial x \partial y} &= \frac{u_{(i-n)} - u_{(i-n-1)} + u_{(i+n+2)} - u_{(i+n+1)}}{\Delta^2} \\ &= \frac{u_{(i-n-1)} - u_{(i-n-2)} + u_{(i+n+1)} - u_{(i+n)}}{\Delta^2}; \\ \frac{\partial^2 v_i}{\partial x \partial y} &= \frac{v_{(i-n-2)} - v_{(i-1)} + v_{(i-n)} - v_{(i+1)}}{\Delta^2} \\ &= \frac{v_{(i-1)} - v_{(i+n)} + v_{(i+1)} - v_{(i+n+2)}}{\Delta^2}. \end{aligned} \tag{3}$$

Then Eq. (1) taking into account (2) for the i-th point of the considered plane stressed element (except for the corner points) can be rewritten as:

$$\begin{aligned} &4\left(a_{(i-1)}u_{(i-1)} - 2a_i u_i + a_{(i+1)}u_{(i+1)}\right) + \left(b_{(i-n-2)}v_{(i-n-2)} - b_{(i+n)}v_{(i+n)}\right. \\ &\quad \left. + b_{(i-n)}v_{(i-n)} - b_{(i+n+2)}v_{(i+n+2)}\right) \\ &\quad + \left(4d_{(i-n-1)}u_{(i-n-1)} - 8d_i u_i + 4d_{(i+n+1)}u_{(i+n+1)}\right. \\ &\quad \left. + d_{(i-n-2)}v_{(i-n-2)} - d_{(i+n)}v_{(i+n)} + d_{(i-n)}v_{(i-n)} - d_{(i+n+2)}v_{(i+n+2)}\right) = 0; \\ &4\left(c_{(i-n-1)}v_{(i-n-1)} - 2c_i v_i + c_{(i+n+1)}v_{(i+n+1)}\right) \\ &\quad + \left(b_{(i-n)}u_{(i-n)} - b_{(i-n-2)}u_{(i-n-2)} + b_{(i+n+2)}u_{(i+n+2)} - b_{(i+n)}u_{(i+n)}\right) \\ &\quad + \left(4d_{(i-1)}v_{(i-1)} - 8d_i v_i + 4d_{(i+1)}v_{(i+1)} + d_{(i-n)}u_{(i-n)}\right. \\ &\quad \left. - d_{(i-n-2)}u_{(i-n-2)} + d_{(i+n+2)}u_{(i+n+2)} - d_{(i+n)}u_{(i+n)}\right) = 0, \end{aligned} \tag{4}$$

For the corner points of the contour, instead of (2), expressions (4) should be substituted into Eq. (1).

When deriving Eq. (4), the following designations were adopted: ai, bi, ci, di are the coefficients of reduction of the deformability parameters of materials at the nodes intersected by the reinforcing bars, as well as at arbitrary nodes of the element in question in the deformed state and determined from the relations:

$$\begin{aligned} a_i &= \frac{E_{x,i}}{1-v^2} = \frac{E_b + kE_s \frac{A_{sx}}{b_{col}\Delta}}{1-v^2}; \\ b_i &= \frac{E_{y,i}v_{x,i}}{1-v^2} = \frac{E_{x,i}v_{y,i}}{1-v^2} = \frac{E_b + kE_s \frac{A_{sy}}{b_{col}\Delta}}{1-v^2} v_{x,i} = \frac{E_b + kE_s \frac{A_{sx}}{b_{col}\Delta}}{1-v^2} v_{y,i}; \\ c_i &= \frac{E_{y,i}}{1-v^2} = \frac{E_b + kE_s \frac{A_{sy}}{b_{col}\Delta}}{1-v^2}; \\ d_i &= G_{xy,i} = \frac{E_b \sqrt{\left(1 + k\frac{E_s}{E_b}\frac{A_{sy}}{b_{col}\Delta}\right)\left(1 + k\frac{E_s}{E_b}\frac{A_{sx}}{b_{col}\Delta}\right)}}{2(1+v)}, \end{aligned}$$

where E_b, E_s are deformation moduli for concrete and steel rebars respectively;

A_{sx}, A_{sy} are—cross-sectional area of reinforcing bars along the OX and OY axes, respectively, whose axes pass through the i-th mesh node;

b_{col} is a width of column cross section;

Δ is a dimension of grid bay for the beam-column joint under consideration;

k is a coefficient taking into account the transfer of shear forces from concrete to the surface of reinforcing bars and taken in the range from 0 to 1, where 1 corresponds to the full transfer of forces, and 0 corresponds to the complete absence of adhesion between the reinforcement and concrete [17–19].

3.2 Calculation of the Beam-Column Joint Deformed State

Using the relations obtained in the previous section, we will calculate the deformed state of the T-shaped node along the "A" axis of the scale model of the reinforced concrete frame presented in [20]. The frame reinforcement scheme is shown in Fig. 2. The materials of the frames of the first series are following: concrete of compression strength class B25, for which the standard compression resistance is $R_{b,n}$ = 18.5 MPa, and the initial elastic modulus is E_0 = 30,000 MPa. The columns and crossbars are reinforced by spatial reinforcement cages with axial steel wire of class Bp500 ($R_{s,n}$ = 500 MPa) and transverse steel ties of class A300 ($R_{s,n}$ = 300 MPa).

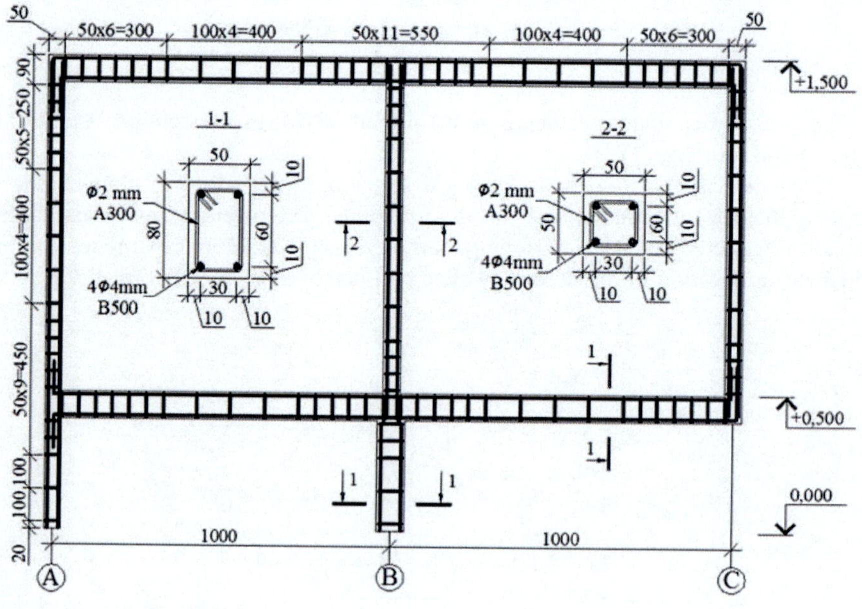

Fig. 2 Reinforcement scheme of 2D scale model of RC moment frame

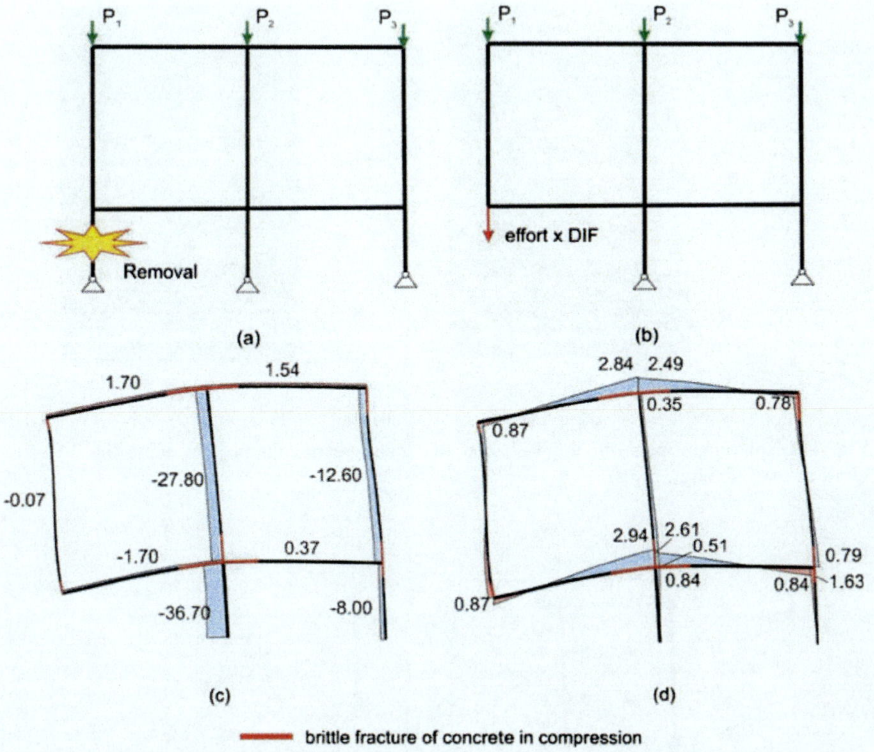

Fig. 3 Calculation scheme of the experimental scale model of the RC frame: primary—"n" (**a**); secondary—"n − 1" (**b**); axial forces, kN (**c**) and bending moments, kNm (**d**) at t = T/4 after sudden corner column removal at the first floor

At the stage of normal operation, axial forces $P_1 = 4$ kN, $P_2 = 20$ kN, $P_3 = 16$ kN have been applied to the upper nodes of the frame as it is shown in Fig. 3. The calculation results for the 2D beam-column joint are presented in the form of displacement plats $U(x)$, m and $W(z)$, m in Figs. 4 and 5.

Therefore, we considered an example of calculating the deformed state of a girder-column node of a scale model of a reinforced concrete frame, for which a simulation was previously, performed using bar FE. Comparing results of calculation allows us to conclude that structural behavior of plane-stressed joint model and bar model differs. This can be explained by taking into account 2D stress–strain state within the 2D model of the finite difference method. That makes such solution more accurate.

In order to simulate the deformed state of a 2D reinforced concrete beam-column joint of the moment frame, the model based on a finite difference method is proposed. A distinctive feature of the proposed solution is the ability to take into account the discrete nature of the reinforcement, as well as incomplete adhesion of the reinforcement to concrete along the contact surface.

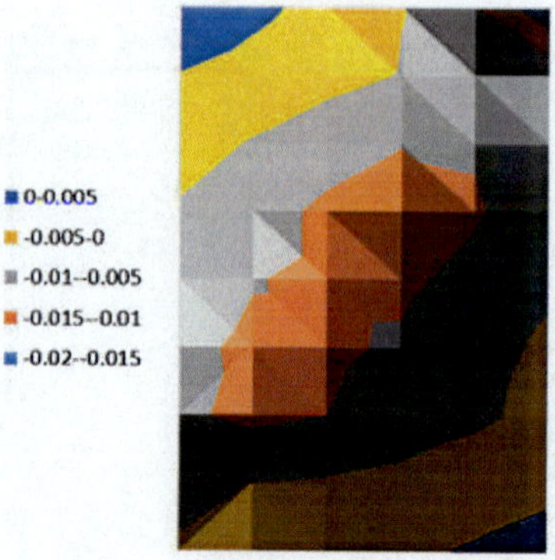

(a)

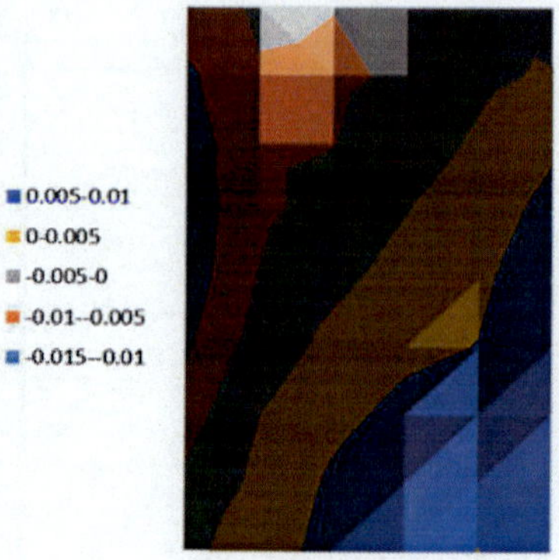

(b)

Fig. 4 Displacement plats for the T-shaped RC frame beam-column joint along the "A" axis: U(x), m (**a**) and W(z), m (**b**)

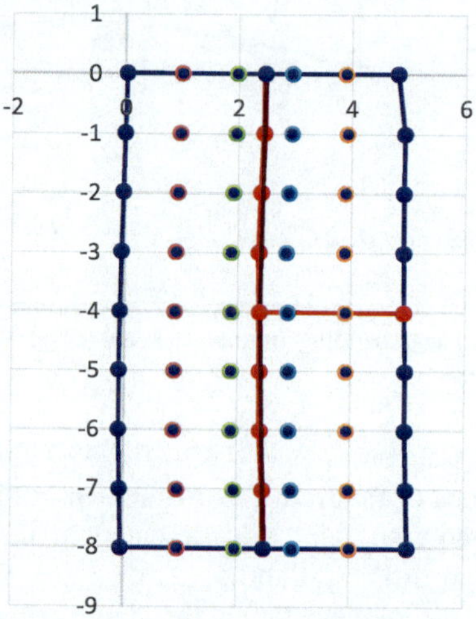

(a)

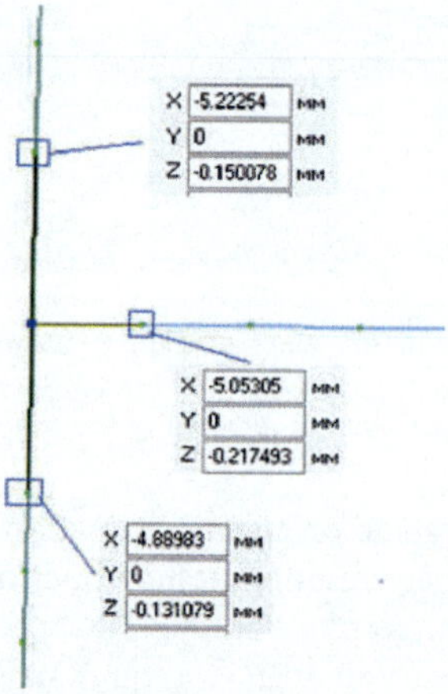

(b)

Fig. 5 Deformed RC beam-column joint under 2D stress–strain state (**a**) and bar model (**b**)

4 Conclusions

Summarizing the results of the study, the following conclusions can be formulated:

1. A computational model of a 2D beam-column joint of a reinforced concrete frame in the form of a finite difference method has been built.

2. Comparison of the calculation results according to the proposed model and the traditional bar model shows the differences in the deformed state of the 2D RC beam-column joint. This can be explained by taking into account the 2D stress–strain state factors, which are not taken into account in the bar model.

References

1. Lukin MV, Popov MV, Lisyatnikov MS (2020) Short-term and long-term deformations of the lightweight concrete. IOP Conf Ser Mater Sci Eng 753:032071. https://doi.org/10.1088/1757-899X/753/3/032071
2. Lisyatnikov MS, Shishov II, Sergeev MS, Hisham E (2020) Precast monolithic coating of an industrial building based on variable-height beam-slabs. IOP Conf Ser Mater Sci Eng 896:012064. https://doi.org/10.1088/1757-899X/896/1/012064
3. Grunwald C, Khalil AA, Schaufelberger B, Ricciardi EM, Pellecchia C, De Iuliis E, Riedel W (2018) Reliability of collapse simulation—comparing finite and applied element method at different levels. Eng Struct 176:265–278. https://doi.org/10.1016/j.engstruct.2018.08.068
4. Tagel-Din H, Meguro K (2000) Nonlinear simulation of RC structures using applied element method. Struct Eng Eng 17:137–148. https://doi.org/10.2208/jscej.2000.654_13
5. Hwang SJ, Lee HJ (2000) Analytical model for predicting shear strengths of interior reinforced concrete beam-column joints for seismic resistance. ACI Struct J 97:35–44. https://doi.org/10.14359/831
6. Tsonos AG (2008) Effectiveness of CFRP-jackets and RC-jackets in post-earthquake and pre-earthquake retrofitting of beam–column subassemblages. Eng Struct 30:777–793. https://doi.org/10.1016/j.engstruct.2007.05.008
7. Hayati N, Hamid A (2015) Seismic performance of interior beam-column joint with fuse-bar designed using ec8 under in-plane lateral cyclic loading. In: International conference on disaster management and civil engineering (ICDMCE'15) Oct 1–3, 2015 Phuket (Thailand). Universal Researchers. https://doi.org/10.17758/UR.U1015302
8. Yu J, Tan KH (2013) Structural behavior of RC beam-column subassemblages under a middle column removal scenario. J Struct Eng 139:233–250. https://doi.org/10.1061/(ASCE)ST.1943-541X.0000658
9. Feng D-C, Xie S-C, Ning C-L, Liang S-X (2019) Investigation of modeling strategies for progressive collapse analysis of RC frame structures. J Perform Constr Facil 33:04019063. https://doi.org/10.1061/(ASCE)CF.1943-5509.0001328
10. Ahmadi R, Rashidian O, Abbasnia R, Mohajeri Nav F, Usefi N (2016) Experimental and numerical evaluation of progressive collapse behavior in scaled RC beam-column subassemblage. Shock Vib 2016:1–17. https://doi.org/10.1155/2016/3748435
11. Feng D-C, Wu G, Lu Y (2018) Numerical Investigation on the progressive collapse behavior of precast reinforced concrete frame subassemblages. J Perform Constr Facil 32:04018027. https://doi.org/10.1061/(asce)cf.1943-5509.0001179
12. Fedorova NV, Ngoc Tuyen V, Yakovenko IA (2020) Strength criterion for a plane stress reinforced concrete element under a special action. Vestn MGSU 1513–1522. https://doi.org/10.22227/1997-0935.2020.11.1513-1522
13. Fedorova NV, Ngoc VT (2019) Deformation and failure of monolithic reinforced concrete frames under special actions. J Phys Conf Ser 1425:012033. https://doi.org/10.1088/1742-6596/1425/1/012033
14. Abdelwahed B (2019) A review on building progressive collapse, survey and discussion. Case Stud Constr Mater 11:e00264. https://doi.org/10.1016/j.cscm.2019.e00264

15. Nagibovich AI (2019) Superelement simulation of dynamic characteristics of large-size combined systems «Foundation—reinforced concrete structures—metal structures». IOP Conf Ser Mater Sci Eng 698:022063. https://doi.org/10.1088/1757-899X/698/2/022063
16. Belostotsky AM, Potapenko AL, Akimov PA (2018) Universal software system "Stadyo" for the numerical solution of linear and nonlinear problems of the field theory, statics, stability and dynamics of spatial combined systems: general parameters and superelemental features. Int J Comput Civ Struct Eng 14:26–41. https://doi.org/10.22337/2587-9618-2018-14-3-26-41
17. Savin SY (2021) Stability of eccentrically compressed reinforced concrete elements under special impacts with account taken of shear deformations. Vestn MGSU 49–58. https://doi.org/10.22227/1997-0935.2021.1.49-58
18. Elsanadedy HM, Al-Salloum YA, Almusallam TH, Ngo T, Abbas H (2019) Assessment of progressive collapse potential of special moment resisting RC frames—experimental and FE study. Eng Fail Anal. https://doi.org/10.1016/j.engfailanal.2019.07.045
19. Almusallam T, Al-Salloum Y, Elsanadedy H, Tuan N, Mendis P, Abbas H (2020) Development limitations of compressive arch and catenary actions in reinforced concrete special moment resisting frames under column-loss scenarios. Struct Infrastruct Eng 16:1616–1634. https://doi.org/10.1080/15732479.2020.1719166
20. Savin SY, Kolchunov VI, Korenkov PA (2020) Experimental research methodology for the deformation of RC frame under instantaneous loss of column. IOP Conf Ser Mater Sci Eng 962:022054. https://doi.org/10.1088/1757-899X/962/2/022054

The Stability of Wood Composite Wall Panels with Elastically Deformable Mechanical Links

B. V. Labudin, E. V. Popov, P. A. Shemelyak, V. V. Sopilov, A. V. Bobyleva, and E. S. Zabbarova

Abstract Composite panels with wooden ribs and covers, made of board materials—are the universal enclosing and supporting structures. When such structures are subjected to vertical loads, the main part of the load is absorbed by the vertical load-bearing ribs. Due to the shear resistance of screw connections, part of the load can be absorbed by the sheathing. With relatively large spacing between the fixing points at floor level, an important factor in determining the required cross-sectional dimensions of ribs is to ensure their stability from the wall plane; and considering the joint operation of ribs and sheathing allows the critical compressive force to be significantly increased. The feature of such structures calculating is the necessity to take into account the ties ductility at the seams of the layers. Because of it, the critical force cannot be calculated by the classical Euler formula as for a composite structure with rigid bonds. The presented mathematical model is based on the solution of determinant equation which is obtained from differential equation system of 3-layer composite element considering both mutual shift of layers and shifts of elastic axis from vertical in a given form. The resulting expression allows to determine the value of critical force taking into account major parameters of the composite structure. Considering the sheathing in the framing operation allows increasing the critical force by 70%, which is mainly influenced by the stiffness of the shear bonds. Increasing the thickness of covers does not have the same tangible effect.

Keywords Wood · Wood composite panels · Compression · Stability · Semirigid connections

---

B. V. Labudin · E. V. Popov (✉) · P. A. Shemelyak · V. V. Sopilov · E. S. Zabbarova
Higher School of Engineering, Northern (Arctic) Federal University, Arkhangelsk 163002, Russia

A. V. Bobyleva
Higher School of Natural Sciences and Technologies, Northern (Arctic) Federal University, Arkhangelsk 163002, Russia

N. Vatin et al. (eds.), *Proceedings of MPCPE 2021*, Lecture Notes in Civil Engineering 182, https://doi.org/10.1007/978-3-030-85236-8_19

1 Introduction

Composite structures, made of various materials, are widely used in the construction of civil building, industrial and agricultural structures [1, 2]. Composite panels with wooden ribs and covers, made of board materials—are the universal enclosing and supporting structures that can be used as covering, floors, wall envelope; and also as the main load-bearing elements in wooden-frame building. Such elements perform the functions of beams, flooring, lining, can provide thermal protection of the building (panels with the insulation inside the structure), and perform the function of horizontal stiffness diaphragms. There are numerous studies devoted to improving the performance of such panels. A new method is proposed in [3] for joining the outer and inner layers of 3-layer panels, which makes it possible to significantly increase the resistance of joints to longitudinal and transverse shear. The bent strength of prefabricated reinforced concrete wall panels with insulation using various connections between the outer bearing reinforced concrete layers is studied in the article [4]. The research for the ultimate compressive load and analysis of the stability of carbon fiber and glass fiber panels with stiffening ribs «Z»-profile and «L»-profile is presented in [5]. The results of studies of shear strength between layers of modular sandwich panels with a sheath made of high density polyethylene depending on the stamping method are presented in article [6]. The development of wood-based materials and structures base has made it not only technically possible but also economically feasible to use ribbed wall panels and volumetric modules in buildings and structures for various purposes [7–9].

The advantages of such panels (Fig. 1) are most evident when it combines load-bearing and enclosing functions, while the main longitudinal ribs take on racks' role, and the covers take on enclosing layers' role.

It is possible to increase the load-carrying capacity of the ribs by effectively engaging the covering panels, for which a rigid adhesive bond at the «rib-sheathing» boundary is traditionally used. At the same time, the use of adhesive bonds significantly complicates the technological process of wall panels' production, which contradicts the main idea of using such structures—simplicity and low production cost. A number of experimental and theoretical studies [10] demonstrated the advisability of taking the sheathing into account when jointing it to the ribs, using semi-rigid mechanical ties. Nails, screws, staples or combined connections based on claw washers are used as mechanical ties. Studies [11] present that taking the participation of the covers due to the screw ties into account gives a significant effect on the stiffness increasing of wall panels during shear when it serve as vertical diaphragms.

However, additional research is needed on the behavior of composite panels, where the ribs are the main load-bearing elements of the frame, and on the influence of the thickness of the covers and the stiffness of the ties on the strength and stability of such panels.

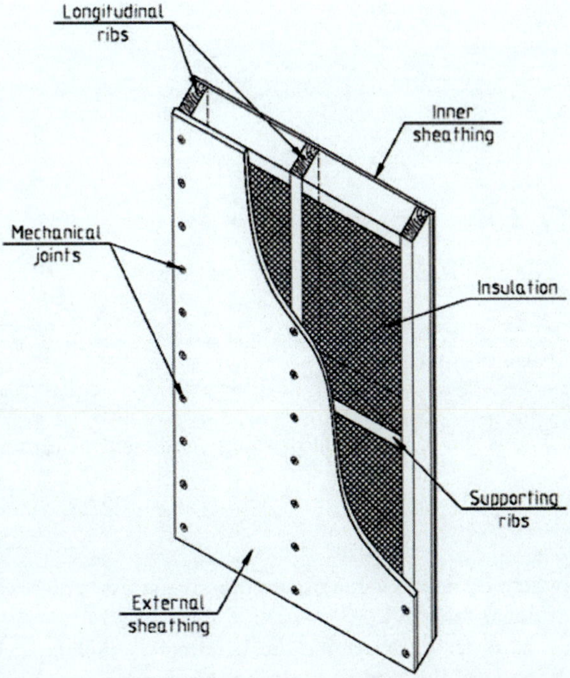

Fig. 1 Ribbed wood composite wall panel

2 Models and Methods

In the engineering design of timber framed ribbed panels it is permissible to consider reduced T- or I-sections where the sheathing acts as the flanges. The normal stresses along the axis of the panel arising in the flanges will be greatest at the ribs and reduce as they move away from the ribs. In the middle of the sheathing the stresses will have a minimum value. Therefore, the reduced width of the sheathing (Fig. 2), rather than the actual width, must be included in the calculation so that the values of the highest longitudinal stresses for the actual and reduced cross-sections are equal [12–14].

According to [15] the possibility of loss of stability in the calculation of centrally-compressed rods is taken into account by the introduction of a bending coefficient, from the condition:

$$\frac{N}{F \cdot \varphi} \leq R_c, \tag{1}$$

where F is the cross-sectional area;

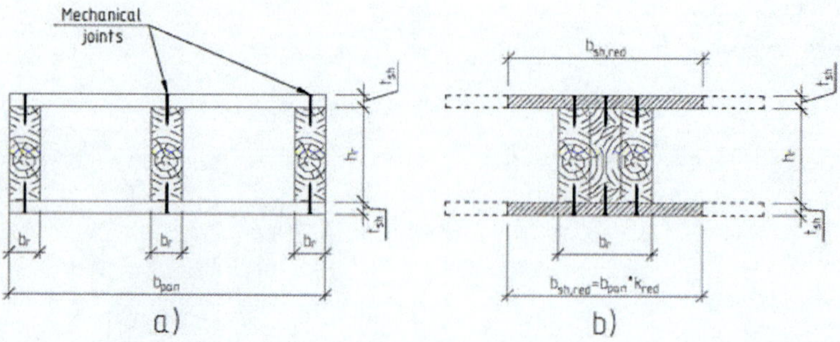

Fig. 2 Schematic of the calculated reduced panel cross-section: **a**—actual cross-section; **b**—design cross-section

φ is the longitudinal bending coefficient determined according to the formula:

$$\varphi = \frac{\sigma_{cr}}{R_c}, \tag{2}$$

where σ_{cr} is the value of compressive stresses corresponding to the loss of stability (critical force N_{cr}).

In order to determine the bending coefficient, it is necessary to determine the fraction of the load that falls on the strut when the critical force N_{cr} is applied. The system of equations for a three-layer composite element [16–21] is presented below:

$$\begin{cases} \frac{T''}{\xi_1} = \Delta_{11}T_1 + \Delta_{12}T_2 + \Delta_{10} \\ \frac{T''}{\xi_2} = \Delta_{21}T_1 + \Delta_{22}T_2 + \Delta_{20} \end{cases}, \tag{3}$$

where T_i is the force at the joints of the composite bar,

Δ_{ij}, Δ_{iy} are coefficients at the unknowns;

Δ_{i0} are free terms;

ξ_i is the stiffness factor of the shear links in the i-joint, determined by Formula (12).

$$\begin{aligned} \Delta_{11} &= \frac{1}{E_1 \cdot A_1} + \frac{1}{E_2 \cdot A_2} + \frac{c_1^2}{\Sigma EI}; \\ \Delta_{22} &= \frac{1}{E_2 \cdot A_2} + \frac{c_1^2}{\Sigma EI}; \\ \Delta_{12} = \Delta_{21} &= \frac{-1}{E_2 \cdot A_2} + \frac{c_1 c_2}{\Sigma EI}; \end{aligned} \tag{4}$$

$$\Delta_{10} = \frac{-N_1^0}{E_1 \cdot A_1} + \frac{N_2^0}{E_2 \cdot A_2} + \frac{M^0 \cdot c_1}{\Sigma EI},$$
$$\Delta_{20} = \frac{-N_2^0}{E_2 \cdot A_2} + \frac{N_3^0}{E_3 \cdot A_3} + \frac{M^0 \cdot c_2}{\Sigma EI}, \tag{5}$$

where E_1, A_1, E_3, A_3—modulus of elasticity and cross-section area of outer and inner sheathing respectively; E_2, A_2—the same, ribs; c_1, c_2—distance from centre of gravity of outer/inner sheathing to rib centre respectively; N_{0i}—longitudinal load of i-th element; M^0—total bending moment from outer load; ΣEI—bending stiffness of a zero shear joint rod.

As we consider the case of central compression and consider the application of external load on the strut with its redistribution onto the sheathing only by the operation of mechanical connections, we assume in further calculations that $M^0 = N_1^0 = N_3^0 = 0$.

The system of Eqs. (3) can be represented by two independent equations for the generalized unknown joint forces $\overline{T}_{12}$:

$$\overline{T}_1'' = \lambda_1^2 \overline{T}_1 + \overline{R}_1; \quad \overline{T}_2'' = \lambda_2^2 \overline{T}_2 + \overline{R}_2 \tag{6}$$

where $\overline{R}_{1,2}$ are the generalized load terms defined by expressions (7);

$\lambda_{1,2}$—the characteristic numbers determined by (8).

$$\overline{R}_1 = \sqrt{\xi_1}\Delta_{10} \cos\varphi + \sqrt{\xi_2}\Delta_{20} \sin\varphi;$$
$$\overline{R}_2 = -\sqrt{\xi_1}\Delta_{10} \sin\varphi + \sqrt{\xi_2}\Delta_{20} \cos\varphi, \tag{7}$$

$$\lambda_{1,2} = \sqrt{\frac{1}{2}\left(\xi_1\Delta_{11} + \xi_2\Delta_{22} \pm \sqrt{(\xi_1\Delta_{11} - \xi_2\Delta_{22})^2 + 4\Delta_{12}^2\xi_1\xi_2}\right)}. \tag{8}$$

The functions of the generalised unknown forces in the joints and $\overline{T}_1\overline{T}_2$ are of the form:

$$\overline{T}_1 = A_1 sh(\lambda_1 x) + B_1 ch(\lambda_1 x) + \int_0^x R_1(t) sh[\lambda_1 \cdot (x - t)]dt;$$
$$\overline{T}_2 = A_2 sh(\lambda_1 x) + B_2 ch(\lambda_1 x) + \int_0^x R_2(t) sh[\lambda_2 \cdot (x - t)]dt. \tag{9}$$

The expressions bellow are used to move from the functions $\overline{T}_1$ and $\overline{T}_2$ to the shear forces $T_1(x)$ and $T_2(x)$:

$$T_1(x) = \sqrt{\xi_1} \cdot \cos\varphi \cdot \overline{T}_1 - \sqrt{\xi_1} \cdot \sin\varphi \cdot \overline{T}_2;$$
$$T_2(x) = \sqrt{\xi_2} \cdot \sin\varphi \cdot \overline{T}_1 - \sqrt{\xi_2} \cdot \cos\varphi \cdot \overline{T}_2. \tag{10}$$

The angle φ is determined by the formula:

$$\varphi = arctg\left(\frac{\lambda_1^2 - \xi \cdot \Delta_{11}}{\xi \cdot \Delta_{12}}\right). \tag{11}$$

The joint stiffness coefficients ξ_i are determined according to the formula:

$$\xi_i = \frac{n \cdot c_{c,i}}{S_{c,i}}, \tag{12}$$

where $c_{c,i}$ is the stiffness factor of the joints in the i-th joint, determined experimentally;

$S_{c,i}$—spacing of connections; n—amount of longitudinal ribs.

Substituting expressions (4) and (12) into (12) and integrating the right part of the equations, the following equalities are obtained:

$$\overline{T}_1(x) = A_1 sh(\lambda_1 x) + B_1 sh(\lambda_1 x) - \left[\frac{\left(\sqrt{\xi_1} \cdot \Delta_{10} \cdot \cos\varphi + \sqrt{\xi_2} \cdot \Delta_{20} \cdot \sin\varphi\right) \cdot (ch(\lambda_1 x) - 1)}{\lambda_1^2}\right], \tag{13}$$

$$\overline{T}_2(x) = A_2 sh(\lambda_2 x) + B_2 sh(\lambda_2 x) - \left[\frac{\left(\sqrt{\xi_1} \cdot \Delta_{10} \cdot \cos\varphi + \sqrt{\xi_2} \cdot \Delta_{20} \cdot \sin\varphi\right) \cdot (ch(\lambda_1 x) - 1)}{\lambda_2^2}\right], \tag{14}$$

The arbitrary constants A_i and B_i are determined from the boundary conditions (shear forces at the supports without shear hindrances is equal to 0):

$$x = 0 \rightarrow \overline{T_1}(0) = \overline{T_2}(0) = 0; x = l \rightarrow \overline{T_1}(l) = \overline{T_2}(l) = 0.$$

Substituting boundary conditions into expressions (13) and (14), the following expression is obtained: $B_1 = B_2 = 0$,

$$A_1 = \frac{\sqrt{\xi_1}\Delta_{10}\cos\varphi \cdot (1 - chl\lambda_1) + \sqrt{\xi_2}\Delta_{20}\sin\varphi \cdot (1 - chl\lambda_1)}{\lambda_1^2 \cdot shl\lambda_1}; \tag{15}$$

$$A_2 = \frac{\sqrt{\xi_2}\Delta_{20}\cos\varphi \cdot (1 - chl\lambda_2) - \sqrt{\xi_1}\Delta_{10}\sin\varphi \cdot (1 - chl\lambda_1)}{\lambda_2^2 \cdot shl\lambda_2}. \tag{16}$$

Substituting values A_1 and A_2 in (13) and (14) and then the values obtained into expressions (10) makes it possible to determine the shear forces T_i in any cross-section along the height of the post. The normal stresses in the ribs are determined by the formula:

$$\sigma_r = \frac{1}{n_r}\left(\frac{-T_1 + T_2}{A_r} \pm \frac{M_r \cdot 0.5 \cdot t_r}{I_r}\right), \tag{17}$$

where M_p is the moment in the rib resulting from the work of the shear bonds, determined by formula (18) (for a symmetrical three-layer panel, M_p = 0); n_t is the number of longitudinal ribs.

$$M_r = -\frac{(T_1 \cdot c_1 + T_2 \cdot c_2) \cdot E_r \cdot I_r}{\Sigma EI} \tag{18}$$

To determine the value of N_{cr}, consider a strut as a composite 3-ply compressed rod, the outer layers of which are sheathing and the inner layers are branches (Fig. 3). The shear bonds in the 2 outside seam are ductility (nails, screws) or rigid (glue) ties. The system of differential equations of a composite compressed rod with account for the longitudinal bend has form:

$$\begin{cases} T_1''/\xi_1 = \Delta_{11}T_1 + \Delta_{12}T_2 + \Delta_{1y}y + \Delta_{10} \\ T_2''/\xi_2 = \Delta_{21}T_1 + \Delta_{22}T_2 + \Delta_{2y}y + \Delta_{20} \\ \Sigma N^0 y'' = \Delta_{y1}T_1 + \Delta_{y2}T_2 + \Delta_{yy}y_2 + \Delta_{y0} \end{cases}, \tag{19}$$

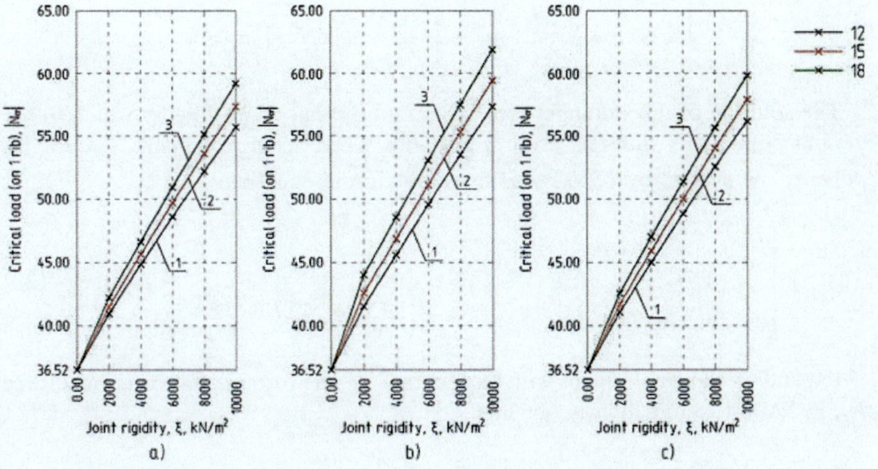

Fig. 3 Dependencies of $|N_{cr}|$ for a single rib on the stiffness of the bracing cladding to ribs: **a**—OSB cladding; **b**—FRC; **c**—particle board; 1—t_{ob} = 12 mm; 2—t_{ob} = 15 mm; 3—t_{ob} = 18 mm

Here:

$$\Delta_{yi} = \Delta_{yi} = \frac{\Sigma N^0}{\Sigma EI} c_i; \quad \Delta_{yy} = \frac{(\Sigma N^0)^2}{\Sigma EI}; \quad \Delta_{y0} = \frac{\Sigma N^0}{\Sigma EI} M^0, \tag{20}$$

where y is the deflection of the element;

ΣN_0—total longitudinal load on the panel.

Since $M^0 = 0$, the coefficients Δ_{i0} and Δ_{y0} are 0, the system is homogeneous. Its solution is $T_i = 0$, y = 0. However, for some total values of the longitudinal force the homogeneous system has solutions other than zero which correspond to the forms of loss of stability of the compressed composite rod. These values of the total longitudinal force ΣN_0 are critical. At articulated end restraint of the rack the boundary conditions are: $x = 0$ $T_i = 0 \rightarrow y = 0$; $x = l \rightarrow T_i = 0$, $y = 0$. The system of differential equations under these boundary conditions has solution [12]:

$$T_i = \alpha_i \sin \chi x; \quad y = \alpha_y \sin \chi x, \tag{21}$$

where $\chi = k\pi/l$, k is a positive integer that takes into account the form of stability loss;

α_i, α_y are constant coefficients.

The system of equations below is obtained by substituting expressions (21) into (19) and reducing by sinχx:

$$\begin{cases} \left(\Delta_{11} + \frac{\chi^2}{\xi_1}\right)\alpha_1 + \Delta_{12}\alpha_2 + \frac{\Sigma N^0}{\Sigma EI} c_1 \alpha_y = 0 \\ \Delta_{21}\alpha_{21} + \left(\Delta_{22} + \frac{\chi^2}{\xi_2}\right)\alpha_2 + \frac{\Sigma N^0}{\Sigma EI} c_2 \alpha_y = 0 \\ \frac{\Sigma N^0}{\Sigma EI} c_1 \alpha_1 + \frac{\Sigma N^0}{\Sigma EI} c_2 \alpha_2 + \left[\frac{(\Sigma N^0)^2}{\Sigma EI} + \Sigma N^0 \chi^2\right]\alpha_y = 0 \end{cases}. \tag{22}$$

The solution of this homogeneous system of ordinary equations in which αi, αy are independent is different from zero only when its determinant is zero. By reducing the system by $(\Sigma N_0)^2/\Sigma EI$, we obtain the equation:

$$\begin{vmatrix} (\Delta_{11} + \chi^2/\xi_1) & \Delta_{12} & c_1 \\ \Delta_{21} & (\Delta_{22} + \chi^2/\xi_2) & c_2 \\ c_1 & c_2 & \left[\Sigma EI + \chi^2(\Sigma EI)^2/\Sigma N^0\right] \end{vmatrix} = 0. \tag{23}$$

Expanding the determinant with respect to ΣN_0, the formula for the critical force ($N_{cr} = \Sigma N_0$) determination is obtained:

$$N_{cr} = -\frac{\chi^2 \Sigma EI^2 \cdot \left(\chi^4 + \chi^2 \xi_1 \Delta_{11} + \chi^2 \xi_2 \Delta_{22} + \xi_1 \xi_2 \Delta_{11} \Delta_{22} - \xi_1 \xi_2 \Delta_{12}^2\right)}{\begin{pmatrix} \chi^2 \left(\chi^2 \Sigma EI - c_1^2 \xi_1 - c_2^2 \xi_2 + \xi_1 \Delta_{11} \Sigma EI + \xi_2 \Delta_{22} \Sigma EI\right) \\ - \xi_1 \xi_2 \cdot \left(c_2^2 \Delta_{11} + c_1^2 \Delta_{22} - 2c_1 c_2 \Delta_{12} - \Delta_{11} \Delta_{22} \Sigma EI + \Delta_{12}^2 \Sigma EI\right) \end{pmatrix}}. \tag{24}$$

The minimum values of the critical force are corresponding to $k = 1$, which occurs while the rod bends by one half-wave of the sine wave form of longitudinal bending.

3 Results and Discussion

There is a timber-framed wall panel, measuring 1.0 × 3.0 m, with three longitudinal ribs (pine wood). The ribs are hinged at the ends. The cross-section of the ribs is 40 × 100 mm. The covers is connected to the ribs by ductility screw ties. The panel ribs are subjected to longitudinal compressive load transmitted from the upstream panels without eccentricity. There are considered panels with double-sided cladding of 12-, 15- and 18-mm thickness made of OSB/3 oriented chipboard, structural plywood and cement-bonded particleboard. Critical force values and bending coefficients are determined depending on bond stiffness, covers thickness and material. The designed width of the covers according to [22] is 0.9 bp = 0.9 m. The results of calculating the critical force $|N_{cr}|$ are shown in the graphs in Fig. 3. The values are given for one rib. The critical force at $\xi = 0$ is determined for the rib without consideration of the sheathing, by the Euler formula.

The presented method makes it possible to determine the value of critical force for the ribs included in the compressed wood-composite wall panels. After determining the values of the critical force $|Ncr|$ according to the Formula (2), it is possible to determine the coefficient of longitudinal bending φ taking into account the participation of the panel covers, and according to the Formula (1) to make a simplified engineering calculation of such structures.

Covers participation together with ribs consideration under compression force can increase the strength and stability of the structure [23, 24]. Depending on the stiffness of the mechanical bonds, the increase in critical force is found to be up to 62% with oriented structural board (OSB) sheathing; up to 69% with plywood (wood-laminated board, glued with the help of urea–formaldehyde composition sheathing) and up to 64% with cement bonded particle board (CBPB). It is evident from the graphs in Fig. 3 that an increase in shear ties stiffness gives a significantly greater increase in $|N_{cr}|$ values than an increase in sheath thickness. Thus, an increase in connection stiffness from 0 to 10,000 kN/m^2 results in increase of $|N_{cr}|$ values by 52–69%, while an increase of cover thickness by 1.5 times (from 12 to 18 mm)— by 6–8%. The use of plywood cladding gives the greatest effect of increasing $|N_{cr}|$, up to 4.5% compared with OSB, up to 2.6% with CBPB.

4 Conclusions

1. The presented mathematical model makes it possible to estimate the stress–strain state of timber-framed wall panels and determine the value of the critical force, taking into account the covers, involved by connection to the ribs using ductility mechanical ties.
2. The two-sided covering of the ribs can significantly increase the critical force value $|N_{cr}|$. The greatest increase in the critical force $|N_{cr}|$ is observed when the mechanical shear ties fixing the sheathing to the ribs are rigid. Increasing the thickness of the sheathing as well as the material used (OSB, plywood, CBPB) has slight effect.
3. In order to increase the strength and stability of timber-wall panels, it is first necessary to reduce the shear resistance of the mechanical ties at the «rib-cover» seams, which can be achieved both by using more shear-resistant ties and by reducing the spacing of the ties.

References

1. Sergeev M, Rimshin V, Lukin M, Zdralovic N (2020) Proceedings conference on multi-span composite beam. IOP Conf Ser Mater Sci Eng (Vladimir). https://doi.org/10.1088/1757-899X/896/1/012058
2. Roshchina S, Lukin M, Lisyatnikov M (2020) Compressed-bent reinforced wooden elements with long-term load. Lect Notes Civil Eng 70:81–91. https://doi.org/10.1007/978-3-030-42351-3_7
3. Naik RK, Panda SK, Racherla VA (2020) New method for joining metal and polymer sheets in sandwich panels for highly improved interface strength. J Compos Struct 251:112661
4. Gombedaa MJ, Naito CJ, Quiel SE (2021) Flexural performance of precast concrete insulated wall panels with various configurations of ductile shear ties. J Build Eng 33:101574
5. Elumalai ES, Krishnaveni GR, Sarath Kumar D, Xavier D, Kavitha G, Seralathan S, Hariram V, Micha Premkumar T (2020) Buckling analysis of stiffened composite curved panels. Mater Today Proc 33(Part 7):3604–3611. https://doi.org/10.1016/j.matpr.2020.05.662
6. Tahmoorian F, Nemati S, Sharafi P, Samali B, Khakpour S (2021) Punching behaviour of foam filled modular sandwich panels with high-density polyethylene skins. J Build Eng 33:101634
7. Kavelin AS (2014) Proceedings conference on Sovremennye problemy promyshlennogo i grazhdanskogo stroitel'stva [Modern problems of industrial and civil construction]. Stroitel'stvo, Rostov, pp 98−100. https://elibrary.ru/download/elibrary_24672247_47513863.pdf (Date of application 29.05.2021)
8. Zhadanov VI, Charikova VV, Xoroshavin EA (2016) Proceedings conference on Konstruktivnye resheniia maloetazhnykh zdanii na osnove drevesiny, primeniaemye v Finliandii [Structural solutions for low-rise buildings based on wood used in Finland]. Universitetskii kompleks kak regional'nyi tsentr obrazovaniia, nauki i kul'tury, Orenburg, pp 520−525
9. Piatikre-stovskii KP, Travush VI (2016) Paneli dlia sten zhilykh domov i pokrytii razlichnykh zdanii iz drevesiny [Panels for walls of residential buildings and coverings of various buildings made of wood]. Zhilishchnoe stroitel'stvo, Moskow, No 4, pp 44–47

10. Popov EV, Filippov VV, Melekhov VI, Labudin BV, Tiurikova TV (2016) Vliianie zhestkosti sviazeĭ sdviga pri raschete rebristykh panelei na dereviannom karkase [Influence of shear bond stiffness in the calculation of ribbed panels on a wooden frame]. Lesnoi zhurnal, Arkhangel'sk, No 4, pp 123–134
11. Baszen M (2017) Semi-rigid behavior of joints in wood light-frame structures. Procedia Eng 172:88–95. https://doi.org/10.1016/j.proeng.2017.02.022
12. Rzhanitsyn AR (1986) Sostavnye sterzhni i plastinki [Composite rods and plates]. Stroiizdat, Moscow, 314 p. https://pnu.edu.ru/media/filer_public/2013/04/10/3-9_pzhanicin_1986.pdf (Date of application 29.05.2021)
13. Naychuk AY (2013) Estimation of load-bearing capacity and stiffness of timber beams with through-thickness cracks. Adv Mater Res. https://doi.org/10.4028/www.scientific.net/AMR.778.361
14. Turkovskij SB, Pogorel'tsev AA (2001) Wooden structures with rigid joints in structures with corrosive medium. Promyshlennoe i Grazhdanskoe Stroit 10–13
15. Roshchina SI, Lukin MV, Lukina AV, Sergeyev MS, Lisyatnikov MS (2015) Experimental research on pressed-bending reinforced timberwork. Int J Appl Eng Res 10:45307–45312
16. Lukin M, Prusov E, Roshchina S, Karelina M, Vatin N (2021) Multi-span composite timber beams with rational steel reinforcements. Buildings. https://doi.org/10.3390/buildings11020046
17. Gribanov AS, Roshchina SI, Naichuk AY, Melekhov VI (2020) Wooden beams with local wood modification. IOP Conf Ser Mater Sci Eng. https://doi.org/10.1088/1757-899X/896/1/012067
18. Azinović B, Serrano E, Kramar M, Pazlar T (2018) Experimental investigation of the axial strength of glued-in rods in cross laminated timber. Mater Struct Constr 51. https://doi.org/10.1617/s11527-018-1268-y
19. Lukina A, Roshchina S, Gribanov A (2021). Method for restoring destructed wooden structures with polymer composites. https://doi.org/10.1007/978-3-030-72404-7_45
20. Aleksiievets VI, Aleksiievets II, Ivaniuk AM, Roshchina SI (2020) Load-carrying capacity of bolted joints of timber structures under static loading. IOP Conf Ser Mater Sci Eng. https://doi.org/10.1088/1757-899X/896/1/012043
21. Turkov AV, Makarov AA (2016) Experimental research of cross beams systems of wooden elements for the static and dynamic loads
22. DIN EN 1995−1−1/A2−2014 Eurocode 5: design of timber structures—Part 1−1: general—common rules and rules for buildings; German version EN 1995−1−1:2004/A2:2014
23. Lukina A, Roshchina S, Gribanov A (2021) Method for restoring destructed wooden structures with polymer composites. Presented at the 2021. https://doi.org/10.1007/978-3-030-72404-7_45
24. Lukin M, Sergeev M, Lisyatnikov M (2021) Non split wooden beam reinforced with composite reinforcement. Presented at the 2021. https://doi.org/10.1007/978-3-030-72404-7_12

Assessment of the Impact of Vibration Indicators on the Foundation of the Existing Building During Pile Driving

Rauan E. Lukpanov, Serik B. Yenkebayev, Duman S. Dusembinov, and Denis V. Tsygulyov

Abstract The paper presents the results of vibromonitoring, the purpose of which was to determine the vibration parameters on the existing residential building from pile driving. Pile driving was assumed in conditions of dense urban development, near the building. The main evaluation parameters of vibration were: vibration displacement (amplitude), vibration velocity, frequency, and vibration acceleration. Vibration parameters were measured at different distances from the excitation source: in the vicinity of the pile, at a distance of 5, 15, 40 m, and near the building. According to the results of the measurements, the relationships of the vibroparameters changes by the driving time were obtained, based on the analysis of which the dependence of the vibroparameters changes depending on the distance from the excitation source was determined. Also, the paper presents statistical data of the obtained results, based on which the analysis of the relationship of the obtained particular values of vibration indicators was carried out. In the conclusion the calculation of the permissible distance of pile driving, made based on the vibration attenuation coefficient, determined by the results of actual measurements, is presented.

Keywords Pile driving · Piles · Vibration monitoring · Vibration acceleration · Displacement amplitude

1 Introduction

Vibromonitoring, designed to assess the maximum permissible pile driving distance, in conditions of dense urban development, has wide application in modern urban planning [1–6]. One of the main techniques for determining the permissible pile driving distance is a theoretical calculation regulated by the All-Union Scientific Research Institute [7–10]. The calculation is based on data on

R. E. Lukpanov · S. B. Yenkebayev · D. S. Dusembinov · D. V. Tsygulyov (✉)
L.N. Gumilyov Eurasian National University, Munaitpassov str, 13, Nur-Sultan 010008, Kazakhstan

N. Vatin et al. (eds.), *Proceedings of MPCPE 2021*, Lecture Notes in Civil Engineering 182, https://doi.org/10.1007/978-3-030-85236-8_20

engineering-geological conditions of construction, structural features of adjacent buildings and structures, and, of course, the conditions of pile driving [11–15]. The data on the engineering-geological conditions are necessary to assess the propagation of vibration indicators through the soil thickness, the design features of the building are necessary to assess the conditions of vibration loadings perception, and the driving data are necessary to assess the initial parameters of the excitation source vibration [16–20].

The object of vibromonitoring is a construction site "Residential complex with detached commercial buildings and parking lot, located in Nur-Sultan".

2 Methods

The vibration monitoring program included the following tasks:

- analysis of initial data: adjacent buildings and structures, pile field, driving conditions, engineering and geological conditions;
- calculation of the permissible distance for pile driving based on theoretical data according to code (VSN 40-97) [21];
- experimental measurements of vibration effects of pile driving on the existing buildings (by several control measurements, subject to the closest pile driving to the inspected object);
- experimental measurements of the vibration effects of pile driving on the soil base (by differently spaced control measurements);
- calculation of the permissible distance for pile driving based on the obtained vibromonitoring data.

Software package "Program for the calculation of foundations and foundations, StroyExpertiza" was used for the automated calculation. The calculation is based on the methodology regulated by VSN 40-97. As the initial data the pile driving parameters, ground conditions, and the characteristic of the object, relative to which the influence of pile driving is estimated, are taken. The automated calculation is used as a preliminary estimation of groundmass vibration propagation (during pile driving), it is necessary to determine the reference points of experimental measuring of vibration influence of pile driving on the ground base.

The measurements were made in two stages:

The first stage—conducting experimental measurements of the vibration effects of pile driving on the existing building;

The second stage—conducting experimental measurements of the vibration effects of pile driving on the subsoil.

Conditions of the first stage:

1. To conduct experimental measurements of the impact on the foundation of the building from pile driving, the worst (unfavorable) condition—the closest possible location of the source of excitation of shock and vibration effects from the building—is taken.

2. To avoid the negative impact of shock and vibration load from pile driving on the existing buildings and structures, the number of measurements should be minimal, but sufficient to obtain reliable results. Thus, to analyze the impact of pile driving on the foundation of the existing building, at least 2 tests are performed (to compare the results). In case of a large gap between the two results, a 3rd control measurement was carried out to eliminate one of the erroneous data.
3. Install the vibration sensors depending on the availability of the structure according to the following mutually exclusive priority: on the foundation structure, on the most massive basement wall structure, on the staircase of the entrance group (facing the source of excitation). Sensors to be installed on the bare structure, according to all the requirements of the manual of the measuring instrument.

Conditions of the second stage:

1. To conduct experimental measurements of shock and vibration effects on the subsoil from pile driving, the worst (unfavorable) condition is taken: the closest possible location of the source of excitation of shock and vibration effects.
2. To avoid the negative impact of shock and vibration load from pile driving on existing buildings and structures, the number of measurements should be minimal, taken equal to one for a single reference point. Detection of random error is performed by comparing the results of measurements on: control points of different distances; equidistant control points of two objects; the results of subsequent measurements at a distance from the existing buildings and structures.
3. For a more detailed analysis of the impact of pile driving (as well as to identify the random error of p. 3) additional tests were conducted at remote sites, in which the vibration impact of the pile driven on the building is minimal. The main condition is the similarity of the geotechnical structure. The additional tests include a greater number of measurements required to compare with the previously obtained results (if necessary, their correction).
4. Vibration sensors are installed on a reinforced concrete slab, the inertia of which allows resisting shock and vibration effects and surface vibrations of the ground. Sensors are installed on the bare structure, according to all the requirements of the manual of the measuring instrument.

Measurements of vibration impact were made by 2 devices Vibrotest-MG4.01 in parallel on 2 control points respectively. Fixing vibration sensors on the control points of the building structures and the groundmass is carried out using anchoring of supporting plates. Figure 1 shows diagrams of the first and second stages with the location of the excitation source (hammering) and control points of measurements. Figure 2a shows the positions of sensors installed on reinforced concrete structures, and Fig. 2b shows the installation of vibration sensors on a reinforced concrete slab of sufficient inertness.

Determination of the minimum permissible pile driving distance to the existing building is performed according to the method of the All-Union Scientific Research

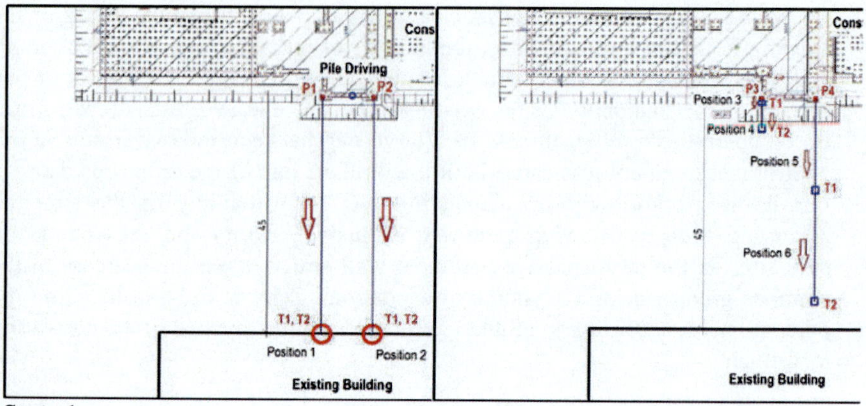

Stage 1 Stage 2

Fig. 1 Layout of vibromonitoring checkpoints

a

b

Fig. 2 Installation of vibration monitoring sensors

Institute [3], based on the obtained vibromonitoring data. The calculation technique is reduced to the determination of two coefficients λ and δ. The coefficient λ characterizes the initial acceleration, depending on the impact energy of the hammer and the ground conditions of the site. The coefficient δ characterizes the effect of attenuation of vibration indicators during the passage of the shock wave through the soil thickness, determined based on the Bouguer-Lambert law.

3 Results

Figure 3 shows the dependences of changes in the measured vibration parameters over time of pile immersion. In Fig. 3a—change in acceleration over time, Fig. 3b—speed over time, Fig. 3c—frequency over time, Fig. 3d—amplitude over time.

Figure 4 shows the dependences of the maximum and average values of the vibration parameters by distance from the excitation source. In Fig. 3a—change of acceleration by distance, Fig. 3b—velocity by distance, Fig. 3c—frequency by distance, Fig. 3d—amplitude by distance. To assess the magnitude of the attenuation of the vibration effect, the changes of private values are shown on the absolute time scale, multiples of 10 s. Table 1 shows the maximum values of vibration indices depending on the distance of pile driving.

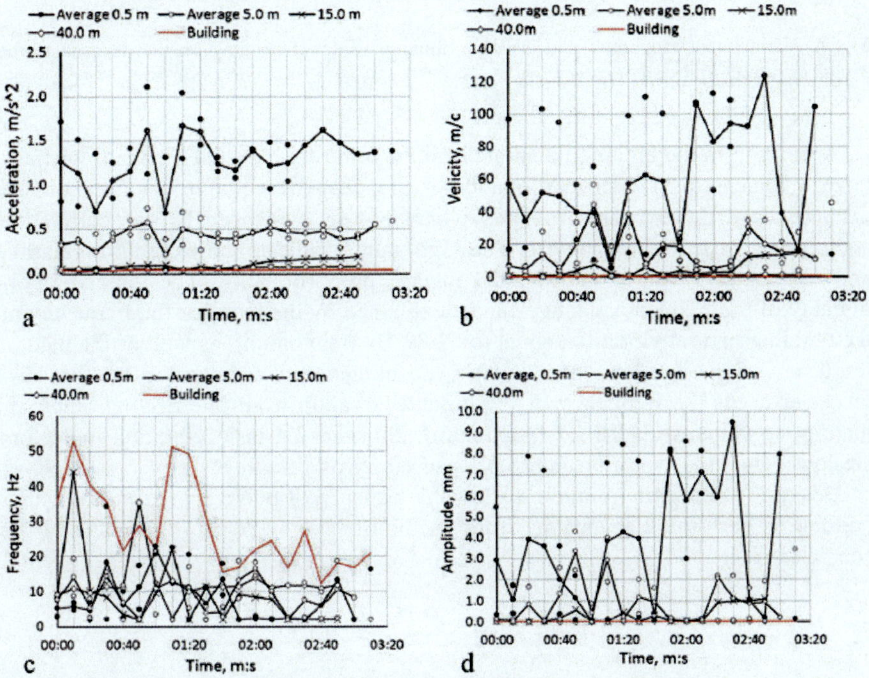

Fig. 3 Variation of vibration indicators by time of measurement (hammering)

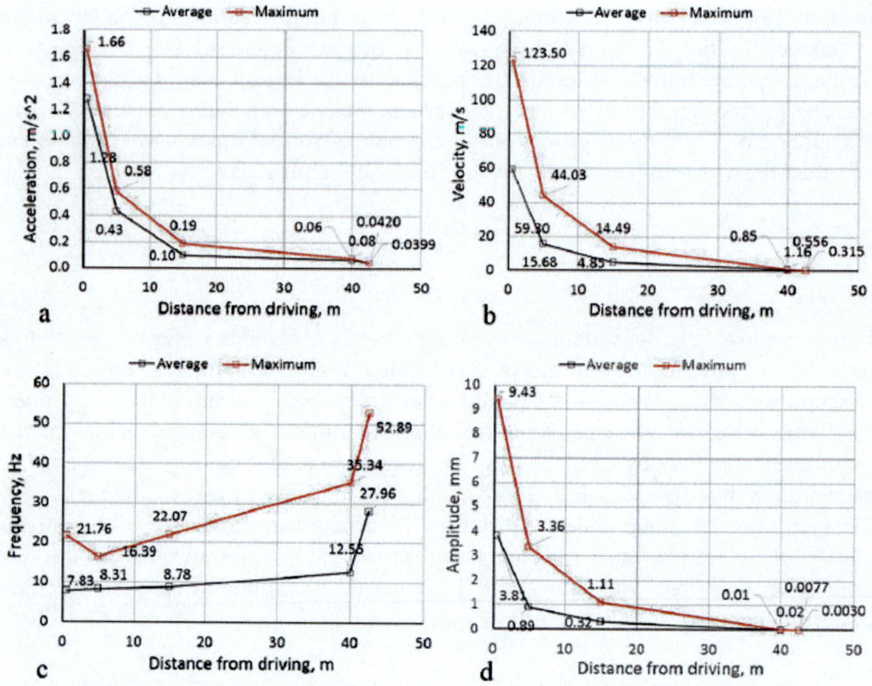

Fig. 4 Variation of maximum and average vibration values depending on the distance to the excitation source (pile driving)

According to the obtained diagrams, all the indicators, except for the frequency, have a logical inverse proportionality of the dependence of the values on the distance from the excitation source, which is also evidenced by the correlation coefficients presented in Table 1. The high values indicate the existing regularity, and the negative values indicate the inverse proportionality. The direct proportionality of the frequency change can be explained by the fact that the frequency of the building natural vibration, equal to 52.89 Hz (maximum), as well as the natural frequency of the soil base of 35.34 Hz (no influence of hammering at 40 m distance), exceeds the frequency of the ground vibration from pile driving, approximating the value of 21.76 Hz (maximum). All other intrinsic vibration indicators are lower than the vibration indicators caused by pile driving.

Determination of the minimum allowable distance of pile driving to the existing building is performed according to the method of the VSN [2]. The value of the coefficient λ is:

$$\lambda = \frac{a_I}{\frac{A_f}{A_s}\alpha_0} = \frac{0.15}{0.7 \cdot 2.1} = 0.10$$

Table 1 Maximum values of vibration indicators

Parameters	Distance (L) from pile driving, m					Correlation coefficient, $\left(\frac{Distance\ (L)}{Parameter\ (a,v,f,A)}\right)$
	0.5	5.0	15.0	40.0	42.5	
	Soil	Soil	Soil	Soil	Soil	
Acceleration (a), m/s^2	1.66	0.58	0.19	0.08	0.042	−0.766/
Speed (v), m/s	123.50	44.03	14.49	1.16	0.556	−0.788
Frequency (f), Hz	21.76	16.39	22.07	35.34	52.89	0.908
Amplitude (A), mm	9.43	3.36	1.11	0.02	0.007	−0.791

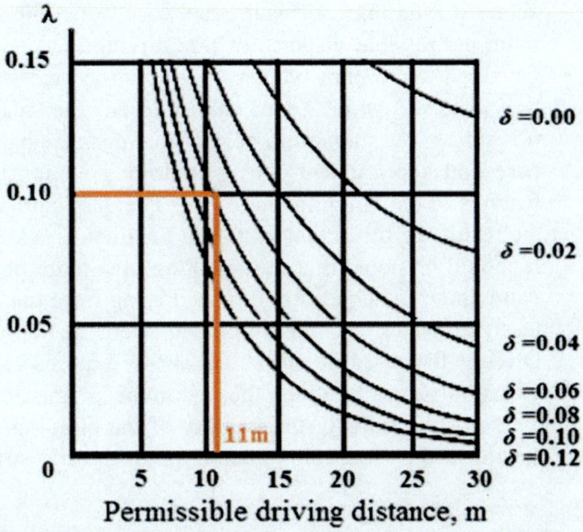

Fig. 5 Nomogram for determining the maximum permissible pile driving distance

where a_I—allowable acceleration of vertical vibrations of the foundation, at which no additional deformation of the foundation occurs, taken according to [3], depending on the type of soil 0.15 m/s^2; α_0—maximum acceleration of soil vibrations at the distance of 0.5 m from the driven pile, taken 2.10 m/s^2 according to the vibromonitoring results; A_f—maximum amplitude of the foundation, equal to 0.055 mm according to vibromonitoring; A_s—maximum amplitude of the soil adjacent to the building, equal to 0.080 mm according to vibromonitoring.

The coefficient of damping of vibrations with distance δ for flowing and soft plastic loams we take 0.07.

The permissible distance is determined from the graph (Fig. 5). If the coefficient $\lambda]] >\ = 0.10$, and coefficient of damping $\delta = 0.08$, then the allowable distance from the piles to the building is 11 m, while the allowable distance according to the results of theoretical calculations based on engineering and geological data, the parameters of pile driving equipment and design features of the building was 14 m. In both cases, the maximum permissible distance does not exceed the design distance of pile driving to the existing building of ~35–40 m.

4 Conclusions

- Measurements of vibration indicators of natural vibrations and forced vibrations from pile driving were made. Measurements were made at different distances, to assess the effect of attenuation of vibration indices.
- According to the results of vibromonitoring values of vibration parameters of adjacent buildings and soil base, based on which the calculation of the maximum permissible distance of pile driving.
- Obtained maximum values of vibration accelerations are from 0.0322 to 0.0554 m/s^2, which does not exceed the maximum allowable value of 0.15 m/s^2. The maximum permissible distance based on theoretical calculations based on geotechnical data, pile-driving equipment parameters, and structural features of the building was 14 m. The maximum permissible distance based on the results of the actual vibration parameters was 11 m.
- It should be noted that the resulting maximum permissible distance of 11 m is valid under the condition of pile driving from the distance (or from distances at least) at which the measurements were made—35–40 m from the building. Driving the piles at closer distances requires repeated measurements of the vibration values to adjust the maximum permissible distance due to changes in the soil conditions in the vicinity of the building, and therefore changes in the spread of the vibration values through the soil strata.

References

1. Lukpanov RE (2015) Comparison of GOST and ASTM as to soil testing by vertically loaded piles. Soil Mech Foundat Eng 52(1):33–37. https://doi.org/10.1007/s11204-015-9303-2. http://link.springer.com/10.1007/s11204-015-9303-2
2. Igoe D, Gavin K (2021) Investigation of cyclic loading of aged piles in sand. J Geotech Geoenvironmental Eng 147(4):04021011. https://doi.org/10.1061/(ASCE)GT.1943-5606.0002451. http://ascelibrary.org/doi/10.1061/%28ASCE%29GT.1943-5606.0002451
3. Lukin MV, Popov MV, Lisyatnikov MS (2020) Short-term and long-term deformations of the lightweight concrete. IOP Conf Ser Mater Sci Eng 753:032071. https://doi.org/10.1088/1757-899X/753/3/032071. https://iopscience.iop.org/article/10.1088/1757-899X/753/3/032071
4. Kovalevsky VV, Glinsky BM, Khairetdinov MS, Fatyanov AG, Karavaev DA, Braginskaya LP, Grigoryuk AP, Tubanov TA (2020) Active vibromonitoring: experimental systems and fieldwork results. Act Geophys Monit (Elsevier) 43–65

5. Colaço A, Alves Costa P, Mont'Alverne Parente C, Silva Cardoso A (2021) Ground-borne noise and vibrations in buildings induced by pile driving: an integrated approach. Appl Acoust 179:108059. https://doi.org/10.1016/j.apacoust.2021.108059. https://linkinghub.elsevier.com/retrieve/pii/S0003682X21001523
6. Massarsch KR, Wersäll C, Fellenius BH (2021) Dynamic ground response during vibratory sheet pile driving. J Geotech Geoenvironmental Eng 147(7) [(ASCE)GT.1943-5606.0002520]. https://doi.org/10.1061/(ASCE)GT.1943-5606.0002520. http://ascelibrary.org/doi/10.1061/%28ASCE%29GT.1943-5606.0002520
7. Magade SB, Ingle RK (2021) Effect of free vibration frequency on modeling of bridge pier. Transp Infrastruct Geotechnol 8(2):296–304. https://doi.org/10.1007/s40515-020-00138-3. https://link.springer.com/10.1007/s40515-020-00138-3
8. Lisyatnikov MS, Shishov II, Sergeev MS, Hisham E (2020) Precast monolithic coating of an industrial building based on variable-height beam-slabs. IOP Conf Ser Mater Sci Eng 896:012064. https://doi.org/10.1088/1757-899X/896/1/012064. https://iopscience.iop.org/article/10.1088/1757-899X/896/1/012064
9. Lodahl MR, Riis AE, Brødbæk KT, Leth CT (2021) Vibrations from driven concrete piles in layered soft soils close to a railway embankment. IOP Conf Ser Earth Environ Sci 710 (1):012057. https://doi.org/10.1088/1755-1315/710/1/012057. https://iopscience.iop.org/article/10.1088/1755-1315/710/1/012057
10. Lin G-L, Lu L-Y, Lei K-T, Liu K-Y, Ko Y-Y, Ju S-H (2021) Experimental study on seismic vibration control of an offshore wind turbine with TMD considering soil liquefaction effect. Mar Struct 77:102961. https://doi.org/10.1016/j.marstruc.2021.102961. https://linkinghub.elsevier.com/retrieve/pii/S0951833921000253
11. Lukpanov RE, Tsigulyov DV, Yenkebayev SB, Askarov DT (2016) Influence of blow energy of the hammer on the bearing capacity of piles during dynamic testing. In: Challenges and innovations in geotechnics—proceedings of the 8th Asian Young geotechnical engineers conference, 8AYGEC 2016, pp 71–74
12. Liming Q, Xuanming D, Changjie Z, Chongrong W, Guangwei C (2021) Numerical and test study on vertical vibration characteristics of pile group in slope soil topography. Earthq Eng Eng Vibr 20(2):377–390. https://doi.org/10.1007/s11803-021-2026-7. https://link.springer.com/10.1007/s11803-021-2026-7
13. Zheng C, Gan S, Luan L, Ding X (2021) Vertical dynamic response of a pile embedded in a poroelastic soil layer overlying rigid base. Acta Geotech 16(3):977–983. https://doi.org/10.1007/s11440-020-01033-4. http://link.springer.com/10.1007/s11440-020-01033-4
14. Iversen KM, Roesen BS (2021) Full-scale static compression pile load test in sand with post installation pre-stressing to engage toe resistance. IOP Conf Ser Earth Environ Sci 710 (1):012007. https://doi.org/10.1088/1755-1315/710/1/012007. https://iopscience.iop.org/article/10.1088/1755-1315/710/1/012007
15. Sayginer O, Chiasera A, Varas S, Lukowiak A, Ferrari M, Bursi O (2020) Design and fabrication of multilayer-driven optomechanical device for force and vibration sensing. Fiber Lasers Glass Photonics Mater Appl II 65. https://doi.org/10.1117/12.2555347. https://www.spiedigitallibrary.org/conference-proceedings-of-spie/11357/2555347/Design-and-fabrication-of-multilayer-driven-optomechanical-device-for-force/10.1117/12.2555347.full
16. Orazova DK, Zhussupbekov AZ, Lukpanov RE, Yenkebayev SB (2016) Evaluation of wind power unit reliability according to the results of field studies on the example of Ereymentau wind power station. In: Challenges and innovations in geotechnics—proceedings of the 8th Asian Young geotechnical engineers conference, 8AYGEC 2016, pp 65–69
17. Lukpanov RE, Awwad T, Orazova DK, Tsigulyov D (2019) Geotechnical research and design of wind power plant 2019:220–227
18. Malekjafarian A, Jalilvand S, Igoe D, Doherty P (2019) On the estimation of foundation damping of mono pile-supported offshore wind turbines. Vibroeng Procedia 23:7–12. https://doi.org/10.21595/vp.2019.20648. https://www.jvejournals.com/article/20648

19. Veshnyakov V (2019) Vibration of adjacent building due to impact pile driving by diesel hammer. https://doi.org/10.5593/sgem2019/1.2/S02.073. https://www.sgem.org/index.php/elibrary?view=publication&task=show&id=5118
20. Khanmohammadi M, Fakharian K (2019) Numerical modelling of pile installation and set-up effects on pile shaft capacity. Int J Geotech Eng 13(5):484–498. https://doi.org/10.1080/19386362.2017.1368185. https://www.tandfonline.com/doi/full/10.1080/19386362.2017.1368185
21. VSN 490–87 Departmental Construction Standards. Design and installation of pile foundations and sheet pile fences in the reconstruction of industrial enterprises and urban development

Complex of Static and Dynamic Tests of Soils with Piles Using Different Methods in Heterogeneous Soil Conditions

Rauan E. Lukpanov, Denis V. Tsygulyov, Serik B. Yenkebayev, and Duman S. Dusembinov

Abstract The paper presents the results of a set of pile soil tests. The tests were carried out with dynamic and static loads using different test methods. The dynamic tests refer to the express methods and the static tests without soil stabilization to the fast methods of bearing capacity assessment. Both tests are relatively fast in execution time, have a measurement error, and therefore need to correct the results. Therefore, the results of the different test methods are evaluated in comparison with the traditional static tests performed up to the complete stabilization of the soil (reference static tests). Tests were carried out during the construction of an oil and gas industrial complex, where the project provided for the use of a large number of piles. Based on the results of statistical processing of partial values of bearing capacity, reliability coefficients of each of the compared tests were obtained. Additional evaluation criteria, in addition to the carrying capacity of the pile, were the cost of the test and the time of its production. The obtained results were projected onto the responsibility classes of buildings and constructions, regulated by the local normative documents. The practical value of the study is the possible adaptation of this method of assessing the bearing capacity in the construction of large industrial and civil facilities, using a large number of piles.

Keywords Pile · Foundation · Static load test · Dynamic load test · Driving

1 Introduction

Today, the use of pile foundations refers to the standard design solutions of buildings and structures for civil and industrial purposes [1–3]. The main standard pile tests for determining the bearing capacity include static and dynamic tests [4–7]. Static tests are the most reliable in comparison with the dynamic ones since the test conditions

R. E. Lukpanov · D. V. Tsygulyov · S. B. Yenkebayev (✉) · D. S. Dusembinov
L.N. Gumilyov Eurasian National University, Munaitpassov str, 13, Nur-Sultan 010008, Kazakhstan

N. Vatin et al. (eds.), *Proceedings of MPCPE 2021*, Lecture Notes in Civil Engineering 182, https://doi.org/10.1007/978-3-030-85236-8_21

have a maximum approximation to the real conditions of loading of soils by piles in the construction of buildings and structures [8–13]. Therefore, the results of static tests are referred to as the reference ones, concerning which the bearing capacity obtained by other methods is compared [14–16]. Dynamic tests refer to the express methods of determining the bearing capacity of piles [17]. Relative to the static tests the dynamic tests give an error in the bearing capacity results, primarily related to the loading conditions of the pile [18–20]. The confidence in the static tests is also confirmed by the reliability coefficients regulated by the standard [21]. Thus, for static tests, the reliability coefficient is 1.2, and for dynamic tests, it is 1.4 [22]. However, the undoubted advantage of dynamic tests is the relative speed of their production, as well as the cost and labor intensity [23]. The latter determines the regulated number of static and dynamic tests: 0.5% of the total number of piles, but not less than 2 static tests for a facility; 1.0% of the total number of piles, but not less than 6 dynamic tests for a facility [24]. The minimum number of dynamic tests is associated with a mandatory statistical analysis, which can be performed to analyze at least 6 private values of the estimated parameters [25].

The tests were carried out at the construction site of an oil and gas complex represented by frame-type industrial buildings made of metal elements (Fig. 1). Static tests were conducted for 33 driven piles C9-30, length 9 m, square cross-section 0.3 × 0.3 m. Dynamic tests were carried out for 128 C9-30 piles, including piles previously subjected to static tests.

The purpose of the tests was to assess the bearing capacity of the piles on the ground, and the purpose of the study was to compare the dynamic test results with the static ones to assess the applicability of the dynamic test method in the engineering and geological conditions of the construction site.

2 Methods

Static and dynamic tests are performed following the requirements of GOST 5686 "Soils. Field pile testing methods". The dynamic tests were carried out using a pile driving unit MSDSh based on SP 49 RH 14, with the weight of the impact part of

Fig. 1 Oil and gas plant

3.5 tons (Fig. 2a). The essence of the static tests is to measure the movement of the pile under step (10 steps) loading to a test load of 150% of the design load. The stabilization criterion is the total displacement of the pile not more than 0.1 mm during the last hour of observation. The maximum dwell time of a step without stabilization is 24 h. The tests, depending on soil conditions, can last from 1 to 3 days. Dynamic tests refer to the express method, the essence of which is to measure the movement of piles when the hammer strikes.

Static tests were performed using an anchor and support stand (Fig. 2b) and a loading platform (Fig. 2c), depending on the conditions of the pile field. During cold periods, the test platform was covered with a protective tent (Fig. 2b) and the test pile was heated for at least 24 h before load application. The need for ground heating is related to the elimination of the effect of pile compression by the ground, which occurs during crystallization of the groundwater in the crialite zone. Compression can lead to distorted test results, and since the effect is short-lived, it needs to be excluded, unlike piles in permafrost conditions, where ground icing is permanent. All measuring instruments were calibrated before the test and during the test were protected from wind and possible precipitation. The piles were tested after resting (3–6 days after driving, depending on soil type).

The design value of the pile on the ground bearing capacity (F_d) according to the results of field tests with dynamic load is determined according to the formula:

$$F_d = \gamma_c \frac{\frac{\mu AM}{2}\left[\sqrt{1 + \frac{4(G(H-h))}{AS_a} \cdot \frac{m_1 + \varepsilon^2(m_2 + m_1)}{m_1 + m_2}} - 1\right]}{\gamma_s}, \tag{1}$$

a b c

Fig. 2 Dynamic and static load test performance

γ_c—working condition factor, taken for an indentation load of 1;

γ_s—safety factor, taken 1.4; η—pile material coefficient, for reinforced concrete piles is taken 150 tf/m^2; A—pile perimeter area, equal to 0.09 m^2; G—a mass of the striking part of the hammer equal to 3.5 tons; H—the actual drop height of the striking part of the hammer, m; h—rebound height of the striking part of the hammer, m; s_a—actual residual refusal, equal to the value of pile immersion from one hammer blow, m; m_1—total mass of the hammer, t; m_2—a mass of the pile and the head, t; ε—impact recovery factor, when driving reinforced concrete piles with percussion hammers using a header with a wooden liner, is taken 0.2.

The load-bearing capacity of the pile according to the results of the static tests is determined based on the maximum permissible displacement of the pile. The maximum permissible displacement depends on the type of structure and the test method in relation to the dwell time of the loading stages. Thus, the most common tests before the stabilization of the soil maximum permissible settlement, with respect to which the bearing capacity is determined is: for buildings with a reinforced concrete frame—20 mm; for buildings with a metal frame of 30 mm. For other types of structures, other criteria of the allowable settlement are regulated.

The comparison is made in the following sequence.

At the first stage of the comparison of private normative and calculated values of the carrying capacity of piles obtained by dynamic and static loading (through the coefficients k_s and k_d). Comparison of normative and calculated values allowed giving an estimate of the true values of the carrying capacity of piles, without taking into account the factors affecting the final result of the test.

At the second stage, a comparison of the obtained average normative and design values of the carrying capacity of piles with the accepted values (using coefficients k'_s and k'_d) was made. The comparison made it possible to assess the conditional and limiting requirements of the normative.

Before the comparison, statistical processing of the test results was performed to exclude the possible test error, as well as to assess the overall ratio of the averaged values of the compared test methods over the entire site. When comparing the test results, the conditional and restrictive requirements regulated by the standard were taken into account: the maximum permissible pile settlement values; the requirement for the number of tested piles, the minimum number of partial values for statistical processing, etc.

3 Results

Comparison of the test results was carried out in the ratio of private values of the carrying capacity of static tests to dynamic tests, expressed by the coefficient k_s and k_d:

$$k_s(k_d) = \frac{F_u^s\left(F_d^s\right)}{F_u^d\left(F_d^d\right)}, \tag{2}$$

where, k_s—coefficient of comparison of particular normative values of load capacity, obtained by dynamic and static tests; k_d—coefficient of comparison of partial design values of carrying capacity obtained by dynamic and static tests; F_u^s—particular normative values of bearing capacity according to the results of static tests, kN; F_u^d—particular normative values of carrying capacity based on the results of dynamic tests, kN; F_d^s—particular design values of the load capacity based on the results of static tests, kN; F_d^d—particular design values of the load capacity based on the results of dynamic tests, kN.

Figure 3 shows diagrams comparing partial values of static and dynamic tests. The coefficient k_s ranges from 0.53 to 1.42. The coefficient k_d ranges from 0.61 to 1.66.

Comparison of the accepted or final test results is carried out in the ratio of averaged and accepted values of carrying capacity of piles, obtained by dynamic and static tests, expressed by the coefficients k_s' and k_d':

$$k_s'(k_d') = \frac{F_u^s\left(F_d^s\right)}{F_u'^d\left(F_d'^d\right)}, \tag{3}$$

where, k_s'—coefficient of comparison of the accepted and averaged standard values of carrying capacity, obtained by dynamic and static tests; k_d'—coefficient of comparison of the accepted and averaged design values of carrying capacity obtained by dynamic and static tests; F_u^s—averaged normative values of bearing capacity based on the results of static tests, kN; $F_u'^d$—accepted normative values of

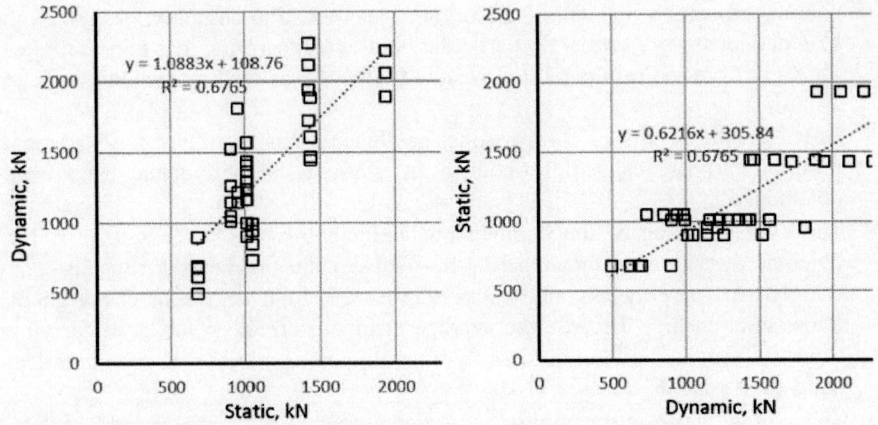

Fig. 3 Comparison of measured static and dynamic load tests data points

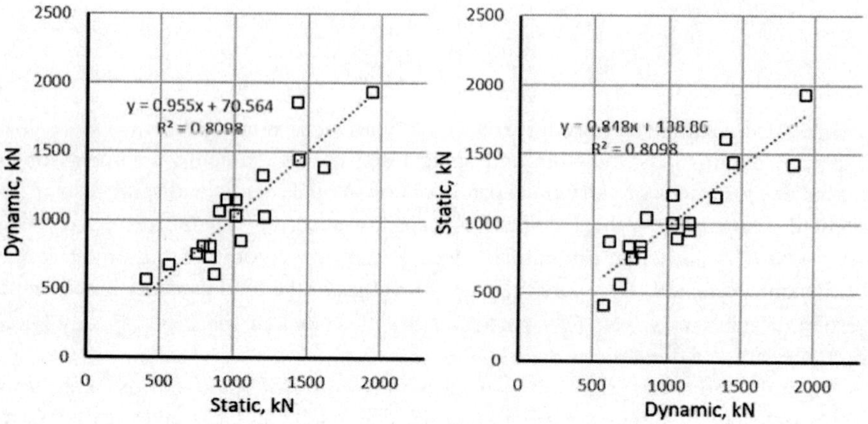

Fig. 4 Comparison of accepted static and dynamic load tests data points

bearing capacity based on the results of dynamic tests, kN; F_d^s—averaged design values of the load capacity based on the results of static tests, kN; $F_d^{\prime d}$—accepted design values of the load capacity based on the results of dynamic tests, kN.

Figure 4 shows diagrams comparing the accepted values of the carrying capacities of static and dynamic tests. The $k_d^{\prime}$ factor ranges from 0.77 to 1.22. The $k_d^{\prime}$ factor ranges from 0.9 to 1.43.

4 Conclusions

- Performed a set of static and dynamic soil tests with piles to determine the bearing capacity at the site of the oil and gas industrial complex.
- Comparisons were made of particular values of carrying capacity of piles obtained from the results of static tests of 33 piles, as well as dynamic tests of 128 piles.
- Based on the results of comparisons, coefficients characterizing the statistical ratio of carrying capacities obtained by static tests to dynamic tests were obtained.
- The average value of the comparison coefficient $\bar{k_s} = 0.88$, less than 1, i.e. dynamic tests showed larger partial normative values of bearing capacity than static ones. Nevertheless, the test results have a high degree of convergence (close relationship), because the average error of normative values of carrying capacity of dynamic tests in these hydrogeological conditions, in the numerical equivalent is 12%.
- The average value of the comparison coefficient $\bar{k_d} = 1.03$, is close to 1, that is, the results have a high degree of convergence of dynamic tests with static ones

(close relationship), close to 100%. The average error of calculated values of dynamic tests carrying capacity in the given hydrogeological conditions, in the numerical equivalent is 3%.

References

1. Zhou J, Zhang X, Zhang L, Dong F, Oh E (2020) Static load tests of driven concrete piles under CFRP confinement. Geomech Geoeng 15(3):159–171. https://doi.org/10.1080/17486025.2019.1635713. https://www.tandfonline.com/doi/full/10.1080/17486025.2019.1635713
2. Du T, Luo W (2020) Experimental study on bearing characteristics of pile foundation in loess area of Longdong. IOP Conf Ser Earth Environ Sci 474:072064. https://doi.org/10.1088/1755-1315/474/7/072064. https://iopscience.iop.org/article/10.1088/1755-1315/474/7/072064
3. Viswanadham BVS, Kumar P (2020) Full-scale load test on bored cast in situ piles—a case study, pp 723–731
4. Lukpanov RE (2015) Comparison of GOST and ASTM as to soil testing by vertically loaded piles. Soil Mech Found Eng 52(1):33–37. https://doi.org/10.1007/s11204-015-9303-2. http://link.springer.com/10.1007/s11204-015-9303-2
5. Lisyatnikov MS, Shishov II, Sergeev MS, Hisham E (2020) Precast monolithic coating of an industrial building based on variable-height beam-slabs. IOP Conf Ser Mater Sci Eng 896:012064. https://doi.org/10.1088/1757-899X/896/1/012064. https://iopscience.iop.org/article/10.1088/1757-899X/896/1/012064
6. Igoe D, Gavin K (2021) Investigation of cyclic loading of aged piles in sand. J Geotech Geoenviron Eng 147(4):04021011. https://doi.org/10.1061/(ASCE)GT.1943-5606.0002451. http://ascelibrary.org/doi/10.1061/%28ASCE%29GT.1943-5606.0002451
7. Feng H, Dai X, Chen S, Chen J (2021) Research on bearing characteristics of open-ended pipe piles under static load. Adv Civil Eng 1–8. https://doi.org/10.1155/2021/5572898. https://www.hindawi.com/journals/ace/2021/5572898/
8. Zhao B, Wang X, Yang M, Liu D, Liu D, Sun S (2020) Experimental study on static load of large-diameter piles in nonuniform gravel soil. Adv Civil Eng 1–15. https://doi.org/10.1155/2020/6291826. https://www.hindawi.com/journals/ace/2020/6291826/
9. Lukin MV, Popov MV, Lisyatnikov MS (2020) Short-term and long-term deformations of the lightweight concrete. IOP Conf Ser Mater Sci Eng 753:032071. https://doi.org/10.1088/1757-899X/753/3/032071. https://iopscience.iop.org/article/10.1088/1757-899X/753/3/032071
10. Iversen KM, Roesen BS (2021) Full-scale static compression pile load test in sand with post installation pre-stressing to engage toe resistance. IOP Conf Ser Earth Environ Sci 710(1):012007. https://doi.org/10.1088/1755-1315/710/1/012007. https://iopscience.iop.org/article/10.1088/1755-1315/710/1/012007
11. Lerch M, Neidhart T, Bubenicek M, Pandrea P (2020) Ableitung der Tragfähigkeit aus Probebelastungen an Betonstopfsäulen. Geotechnik 43(4):298–306. https://doi.org/10.1002/gete.202000030. https://onlinelibrary.wiley.com/doi/10.1002/gete.202000030
12. Kou H, Li W, Chu J, Yang D (2020) Model tests on open-ended concrete pipe piles jacked in sand. Mar Georesour Geotechnol 38(8):939–946. https://doi.org/10.1080/1064119X.2019.1642972. https://www.tandfonline.com/doi/full/10.1080/1064119X.2019.1642972
13. Asghari Pari SA, Habibagahi G, Ghahramani A, Fakharian K (2020) Improve the design process of pile foundations using construction control techniques. Int J Geotech Eng 14(6):636–643. https://doi.org/10.1080/19386362.2019.1655622. https://www.tandfonline.com/doi/full/10.1080/19386362.2019.1655622

14. Khari M, Dehghanbanadaki A, Motamedi S, Jahed Armaghani D (2019) Computational estimation of lateral pile displacement in layered sand using experimental data. Measurement 146:110–118. https://doi.org/10.1016/j.measurement.2019.04.081. https://linkinghub.elsevier.com/retrieve/pii/S0263224119304075
15. Malekjafarian A, Jalilvand S, Igoe D, Doherty P (2019) On the estimation of foundation damping of mono pile-supported offshore wind turbines. Vibroengineering Procedia 23:7–12. https://doi.org/10.21595/vp.2019.20648. https://www.jvejournals.com/article/20648
16. Zhang L, Wang B-H, Wang L-Y, Jing L-P, Fang C, Shan Z-D (2019) Performance improvement in pile anchor system for deep foundation excavation using electroosmotic chemical treatment. Adv Civil Eng 1–10. https://doi.org/10.1155/2019/3168912. https://www.hindawi.com/journals/ace/2019/3168912/
17. Lukpanov RE, Tsigulyov DV, Yenkebayev SB, Askarov DT (2016) Influence of blow energy of the hammer on the bearing capacity of piles during dynamic testing. In: Challenges and innovations in geotechnics—proceedings of the 8th Asian young geotechnical engineers conference, 8AYGEC 2016, pp 71–74
18. Rivera JF, Mejía de Gutiérrez R, Ramirez-Benavides S, Orobio A (2020) Compressed and stabilized soil blocks with fly ash-based alkali-activated cements. Constr Build Mater 264:120285. https://doi.org/10.1016/j.conbuildmat.2020.120285. https://linkinghub.elsevier.com/retrieve/pii/S095006182032290X
19. Xu XB, Zhang TY, Wang JC, Hu MY, Chen KL (2020) Model tests on the bearing capacity of precast open-ended micro pipe piles in soft soil. In: Proceedings of the institution of civil engineers—geotechnical engineering, pp 1–56. https://doi.org/10.1680/jgeen.19.00112. https://www.icevirtuallibrary.com/doi/10.1680/jgeen.19.00112
20. Khanmohammadi M, Fakharian K (2019) Numerical modelling of pile installation and set-up effects on pile shaft capacity. Int J Geotech Eng 13(5):484–498. https://doi.org/10.1080/19386362.2017.1368185. https://www.tandfonline.com/doi/full/10.1080/19386362.2017.1368185
21. Baca M, Brząkała W, Rybak J (2020) Bi-directional static load tests of pile models. Appl Sci 10(16):5492. https://doi.org/10.3390/app10165492. https://www.mdpi.com/2076-3417/10/16/5492
22. Zhusupbekov AZ, Enkebaev SB, Lukpanov RE, Tulebekova AS (2012) Analysis of the settlement of pile foundations under soil conditions of Astana. Soil Mech Found Eng 49(3):99–104. https://doi.org/10.1007/s11204-012-9174-8. http://link.springer.com/10.1007/s11204-012-9174-8
23. Orazova DK, Zhussupbekov AZ, Lukpanov RE, Yenkebayev SB (2016) Evaluation of wind power unit reliability according to the results of field studies on the example of Ereymentau wind power station. In: Challenges and innovations in geotechnics—proceedings of the 8th Asian Young geotechnical engineers conference, 8AYGEC 2016, pp 65–69
24. Zhussupbekov AZ, Tulebekova AS, Lukpanov R, Zhumadilov IT (2016) Comparison analysis of features in Eurocode and Kazakhstan norms requirements. In: Challenges and innovations in geotechnics—proceedings of the 8th Asian Young geotechnical engineers conference, 8AYGEC 2016, pp 251–255
25. Awwad T, Yenkebayev SB, Tsigulyov DV, Lukpanov RE (2019) Analysis of driven pile bearing capacity results by static and dynamic load tests, pp 77–84

Light Steel Framing with Mineral Wool Fire Protection Under Fire Exposure

Marina Gravit and Ivan Dmitriev

Abstract The article calculates the actual limits of fire resistance of light thin-walled structures based on Eurocode 3 and finite element modeling of high-temperature fields in SOFiSTiK PC. Experimental studies of structures with a fire-shield made of mineral wool material MBOR-16F produced by JSC "TIZOL" (Russia) have been carried out. This material is designed specifically for the protection of light-steel structures. The solution of the thermophysical part of the calculation according to Eurocode 3 showed good agreement with the results of experimental data, including samples with effective fire protection, but strongly depend on the calculation step. Accurate results were only achieved with a time step of 1 s. Finite element simulations predicted the correct time to reach the critical temperature of the test sample without any additional assumptions. Further research will be aimed at reducing the consumption of fire-protective material while maintaining the required limits of the fire-resistance of structures.

Keywords Steel construction · Thin walled structures · Cold-formed steel · Structural design · Fire · Fire safety · Fire protection · Fire design

1 Introduction

Bearing and enclosing structures based on the thin-walled galvanized steel profiles are rapidly spreading in construction area. The light gauge steel framing (LGSF or LSF) is widely used in low-rise constructions because of the wide architectural capabilities and excellent technical and economical qualities that allow to operate in dynamic conditions of a changing market with maximum accuracy, flexibility and efficiency.

M. Gravit
Peter the Great St.Petersburg Polytechnic University, Saint Petersburg, Russia

I. Dmitriev (✉)
Technische Universitat Graz, Graz, Austria

N. Vatin et al. (eds.), *Proceedings of MPCPE 2021*, Lecture Notes in Civil Engineering 182, https://doi.org/10.1007/978-3-030-85236-8_22

The classical formulas of structural mechanics and theory of elasticity are commonly used in the calculation of steel sections, however taking into account specificity of thin-walled elements such as reduction of the area for some elements under the load or particular work outside of elastic deformation zone.

The LGSF constructions have great perspectives in the construction area [1, 2], but low level of fire resistance and also insufficient research on both unprotected and different types of fire-protective materials for various structures inhibit their implementation. The fire resistance of thin-walled rods is actively discussed throughout the world but despite the numerous investigations of these structures, this issue has not been fully studied and remains relevant nowadays.

The majority of studies connected with the given designs have theoretical character and some experimental data on strength characteristics (including those connected with issues of local and general stability, resistance to compression, bending, torsion in the whole profile and its elements) only at room temperatures [2–7].

It is necessary to pay more attention to measures for maintaining fire safety and ensuring the standard fire resistance of all structures, especially in case of designing metal frame buildings and structures [8, 9]. This problem is extremely relevant for thin-walled structures due to the high thermal conductivity of steel and the small value of the reduced thickness of the cross-section. Rapid temperature rise in the thin-wall cross section will lead to sharp deterioration of mechanical characteristics.

There is a limited number of works devoted to studying the behavior of thin-walled structures under the influence of high temperatures. There are some studies, which show the influence of the value of the limiting deviation of geometric dimensions [10] or influence on the load-bearing capacity by the value of imperfection of structures [11]. Nowadays, EN1993-1.2 [12] does not include a simplified method for calculating of thin-walled structures, but only offers recommendations for cold-formed sections (class 4) to take the critical heating temperature as 350 °C.

The results of this work, as well as examples of similar studies [13–18], show these values to be unreasonably low. For example, under certain boundary conditions, columns and beams (especially composite ones) made of high-strength steel can retain their bearing capacity at temperatures up to 700 °C.

There is a small number of works concerning the fire resistance of LSF, which includes consideration of the regulatory framework, modelling and experimental studies of authors such as Vatin and Garifullin [19–21], Gravit et al. [22, 23], Naser [24, 25], Chen and Ye [26–30], Dias [31–33]. Only few studies have investigated the behavior of thin-walled steel structures under the fire load, but most of them came to the conclusion that cold-formed steel constructions have higher fire resistance than the Eurocode 3 limit of 350 °C.

The aim of the article is to substantiate the fire resistance limit of composite I-shaped and box-shaped thin-walled steel structures under compression conditions under standard fire load in cases of absence and presence of the effective fire protection.

To achieve this goal, the following tasks are solved:

1. Analytical calculation of the structure on the basis of Eurocode 3;
2. Finite-element modeling of the high-temperature fields for each complex cross-section in SOFiSTiK PC;
3. Fire tests based on theoretical calculation;
4. Deviation analysis of both analytical and finite element calculations with thermocouple results in experimental tests.

Profiles were tested under load. Two cross-sections were considered as the tested samples (Table 1):

1. I-shaped section consists of two C-shaped profiles 380 × 125 × 30 × 3.5 connected through a flange 10 mm by bolt fastening. Static load is equal $E_{fi,d} = 31.0\,\mathrm{t}$
2. Box-shaped section consists of two Σ-shaped profiles 245 × 80 × 20 × 3 connected through a flange 1.5 mm by bolt fastening. Static load is equal $E_{fi,d} = 15.1\,\mathrm{t}$

Profiles were tested both without fire protection and with the use of special flame-protection roll material MBOR-16F produced by TIZOL JSC.

To simplify the calculation scheme of the column with fire protection material produced by TIZOL JSC as well as in accordance with the program of fire tests, the additional fireproof plate "Euro-Lit" and fireproof composition is excluded from the calculation. The critical temperature was calculated using the full analytical calculation method according to the EuroCode [12].

Table 1 The solution of the column fire protection produced by TIZOL JSC and the calculation scheme of the composite section

Picture	Scheme
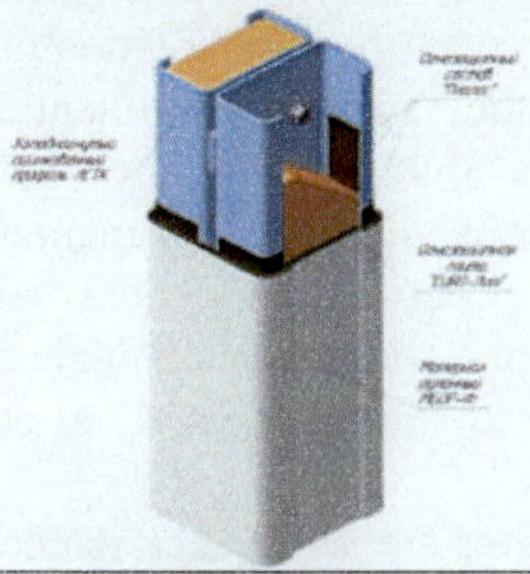	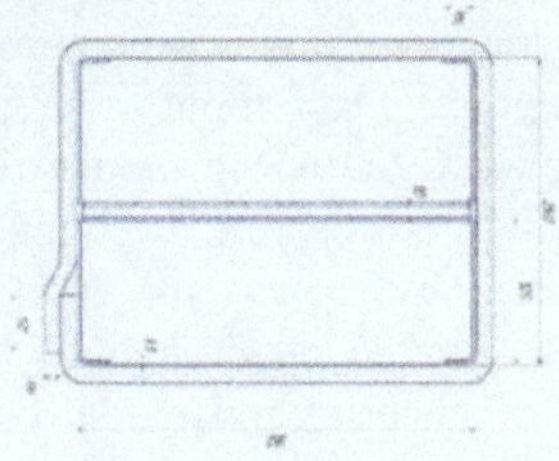
	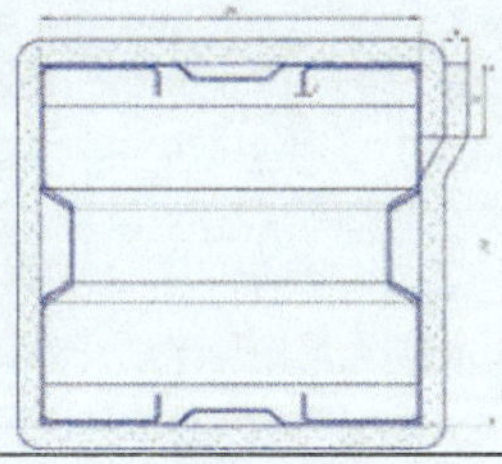

The scientific novelty of this study is the choice of composite cross-section profiles (rather than simple C-shaped profiles) as the object to be tested, which are usually used as supporting structures in buildings and structures. A new effective fire protection roll material MBOR-16F was used as fire protection.

2 Method

2.1 Analytical Calculation

The basic principle of calculation for the steel structures on the fire resistance is to provide necessary and sufficient durability (bearing capacity) throughout the required time period. The most common analytical method is to determine the critical heating temperature of a structure for a given load. The material temperature θd during the fire load have to be less (or equal) to the design critical material temperature.

$$\theta_{\mathrm{d}} < \theta_{cr,d} \tag{1}$$

Based on this thesis, the strength calculation determines the reduced bearing capacity after the necessary time. A comparison on time parameters is used to determine the need of the fire protection for each element.

2.2 Modelling

All calculations of a structure are carried out by a method of finite elements on the basis of spatial model. The Hydra module of the SOFiSTiK software package (ver. 2020) was used for the analysis of temperature distribution over the cross-section of the considered structure. Data input was made using the internal instrumental programming language CADINP in the text editor Teddy. Boundary conditions of the considered models correspond to boundary conditions of the experimental program (temperature is equal to 20 °C, thermal resistance is equal to 9.000 W/K/m^2). The grid is quadrangular. Cell size is not constant with higher density in the area of material and model boundaries. Grid resolution is equal at least 0.01 m.

The properties and geometric dimensions of the structures are set in accordance with the real properties of materials and the samples' overall dimensions. Thermal characteristics of materials specified in the design model. Thermal Conductivity and Heat Capacity of steel are presented in Fig. 1. Numerical dependencies of fire protection are given as Eqs. (2, 3).

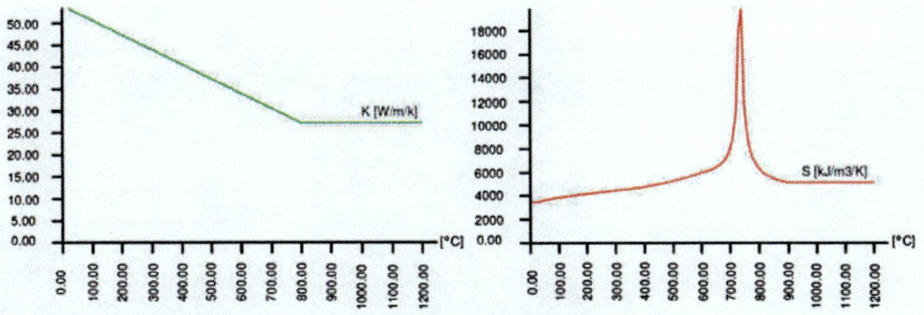

Fig. 1 Thermal conductivity and heat capacity of structural steel

Thermal properties of fire protection:
Specific heat capacity c_p of fire protection in dependence of the temperature θ_p (°C):

$$c_p = 582 + 0.63\left(\theta_p + 273\right) = 754 + 0.63 \cdot \theta_p \, \frac{\mathrm{J}}{\mathrm{kg\,K}} \tag{2}$$

Thermal conductivity λ_p of fire protection in dependence of the temperature θ_p (°C) based on an analysis of the manufacturer's data:

$$\lambda_p = 0.0284 + 0.0002 \cdot \theta_a \, \frac{\mathrm{W}}{\mathrm{m\,K}} \tag{3}$$

Density of fire protection $p_p = 100 \frac{\mathrm{kg}}{\mathrm{m^3}}$.
The system is uniformly heated by the external heat flow, all energy is used to increase the system temperature taking into account its heat capacity. The heating process stops when the temperature at the external boundaries is equalized with the temperature of the external irradiating medium, which corresponds to the standard fire mode and the set temperature in the furnace.

2.3 *Fire Test*

The furnace for fire tests in the Tizol JSC laboratory is a fire chamber lined with fireclay bricks with a loading device for creating compressive forces in the cross-section of the structure and a mechanism for fixation and support. The chrome-aluminum thermocouples were used in the middle section of the structure for the I-beam type and between the profile and the connecting plate for the box type to measure the temperature. The general view of the fire-testing machine in Fig. 2. The thermocouples of I-shaped construction are located in the center between of C-shaped profiles. The thermocouples of box-shaped sections are in between of steel plates.

Fig. 2 Test furnace

The test method consists of determining the time of the limit state of load-bearing capacity (R) due to caving or critical deformation. The fixed time will be considered the actual fire resistance limit of the structure under test. The R limit state according to the Russian government standard GOST 30247.1-94 comes when vertical deformation reaches one hundredth of the length of the structure, which in our case is 30 mm. The standard fire mode according to the Russian government standard GOST 30247.0-94 and international standard ISO 834 was considered:

$$\theta_g = 20 + 345\,\lg(8t + 1) \tag{4}$$

t time (min).
θ_g temperature inside the furnace t (°C).

3 Results

3.1 Analytical Solution

The area A_p of fire protective material is taken as the area of its inner surface. In case of uniform temperature distribution in the cross-section of the steel protected structure, the temperature increase $\Delta\theta_{\alpha,t}$ for the period of time Δt is determined from the following expression (§4.2.5.2, EN 1993-1-2):

$$\Delta\theta_{\alpha,t} = \frac{A_p}{V}\frac{\lambda_p}{d_p} \cdot \frac{(\theta_{g,t} - \theta_{a,t})}{c_a\rho_a\left(1 + \frac{\varphi}{3}\right)} \cdot \Delta t - \left(e^{\frac{\varphi}{10}} - 1\right) \cdot \Delta\theta_{g,t}$$

where

$$\varphi = \frac{c_p\rho_p}{c_a\rho_a} d_p \frac{A_p}{V}$$

A_p	inner surface area of the fire protection material per unit of construction length (m^2/m);
V	volume of the unit of construction length (m^3/m);
c_a	specific heat capacity of steel depending on temperature (J/(kg K));
c_p	specific heat capacity of fire protective material independent of temperature (J/(kg K));
d_p	thickness of the fire protection (m);
Δt	time interval (c), not exceeding 30 s;
$\theta_{a,t}$	steel temperature at time t (°C);
$\theta_{g,t}$	ambient gas temperature at time t (°C);
$\Delta\theta_{g,t}$	ambient gas temperature increase during the time period Δt (K);
λ_p	thermal conductivity of the fire protection system (W/(m K));
ρ_a	steel density (kg/m^3);
ρ_p	fire protective material density (kg/m^3).

The Eurocode prescribes to take the specific heat capacity of the fire protective material independent of the temperature. The thermal conductivity varies depend on the temperature of fire protection. The results depends on the iteration step and the most accurate results were reached only with the time step equals to Δt = 1 s. The solution of the thermophysical problem by the analytical method is given in graphical form on the final graph at the end of the section.

3.2 *Finite Element Solution with the SOFiSTiK PC (ver. 2020)*

The value in the center of the section was taken as the design temperature. The simulation results are graphs with a temperature gradient across the section (Fig. 3). The internal filling of the profile is air.

Similarly, the fire resistance limit was calculated for the box-shaped section. Calculation point is located between steel plates (Fig. 4).

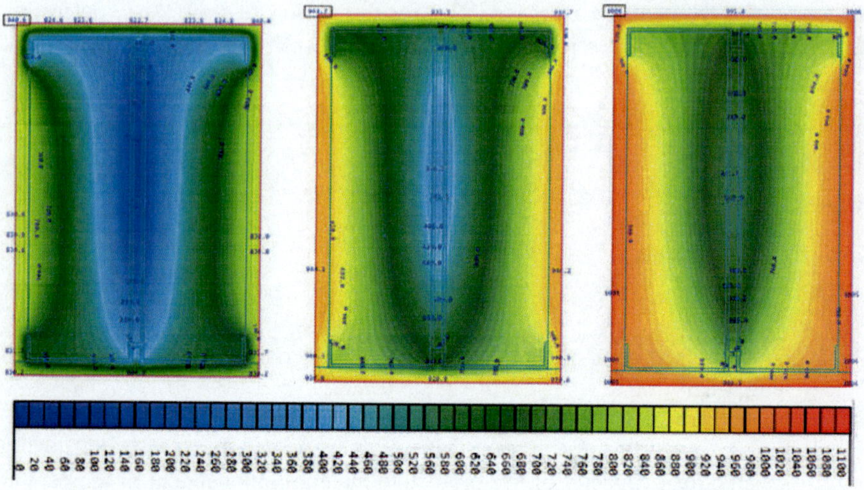

Fig. 3 Temperature gradient at 30, 60 and 90 min

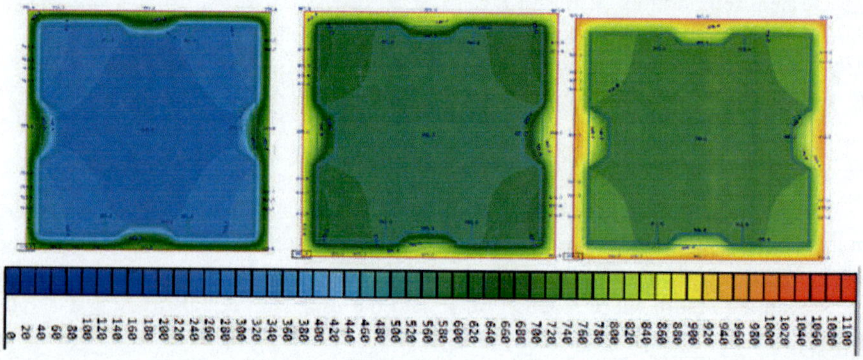

Fig. 4 Temperature gradient at 20, 40 and 60 min

3.3 *Fire Test*

Together with Tizol JSC, fire tests of the profiles with effective fire protection MBOR-16F were conducted. Below you can see photos of the tested samples and a diagram of temperature rise in field tests (Table 2).

Critical deformations of the I-shaped cross section occurred at 74 min 27 s of testing. The average temperature of the cross section according to the results of thermocouples is 567.7 °C and it is accepted as critical. For the box-shaped section, the critical temperature is 624.5 °C and it reached at 57 min 10 s. The accuracy of experimental results is ensured by the sufficient number of thermocouples and the

Table 2 Test samples before and during fire test

No.	Test samples		
1			
2			

experimental error does not exceed 10%. The summary table of the solution of the both static, thermophysical problems and results of fire tests are presented in Fig. 5; Table 3.

The best convergence to the actual results of fire tests was shown by the simulation in the software package. This is due to the implementation of the air gap in the design model inside the fire protected structure and a comprehensive accounting of changing properties of fire protection. Analytical methods do not take these points into account to simplify the calculation scheme.

4 Conclusions

The paper analyzes the applicability of existing numerical methods for calculating of the fire resistance limits of thin-walled constructions, including those with fire protection. This result is familiar with the studies [13–18] and shows that some complex structures (like 2×C columns or beams) made of high-strength steel can retain their bearing capacity at temperatures up to 500–600 °C at least as normal thick-walled construction.

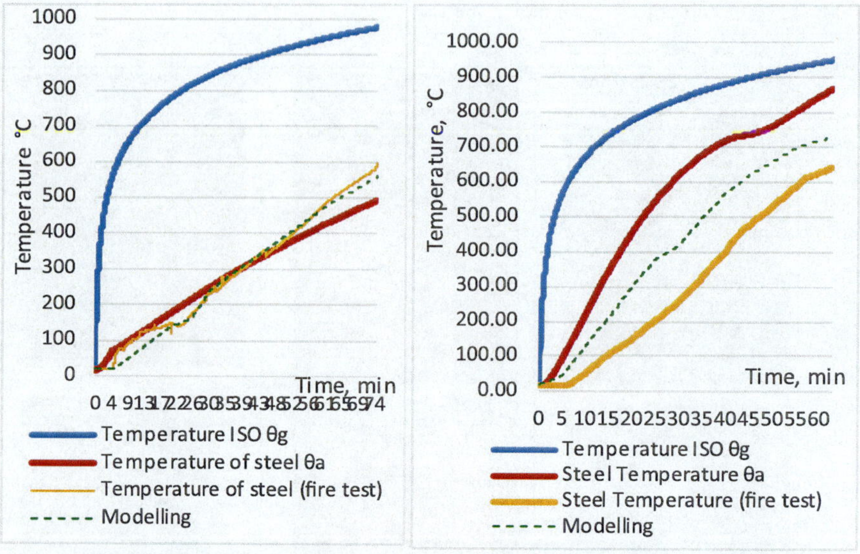

Fig. 5 Fire exposure curves (calculated and actual) for protected samples

Table 3 Data summary on critical temperatures and fire resistance limits for protected samples

Calculation method	Sample 1		Sample 2	
	Calculation	Fire test	Calculation	Fire test
Static calculation. Critical temperatures (°C)				
Critical temperatures (°C)	547.27	567.7	713.08	624.5
Thermophysical calculations. Fire resistance limits (min)				
Analytical solution	86	74:27	49	57:10
Simulation	75		58	

Validation of the static and thermophysical parts of the fire resistance calculation shows that solution of thermophysical problem on the basis of finite element modeling of temperature fields is the most precise method to estimate time of critical deformations in accordance to the criteria of structure heating up to the critical temperature. The most applicable C and Σ cross-sections of building structures with and without fire protection are considered.

The recommendations are given on application of program complexes in solving thermophysical problem of fire resistance calculation.

References

1. Musorina TA, Gamayunova OS, Petrichenko MR, Soloveva E (2020) Adv Intell Syst Comput 1116:429–437
2. Musorina T, Gamayunova O, Petrichenko M (2018) Vestnik MGSU 13:935–943
3. Dinis PB, Camotim D (2010) Thin-Walled Struct 48:771–785
4. Narayanan S, Mahendran M (2003) J Constr Steel Res 59:489–508
5. Pan C-L, Shan M-Y (2011) Thin-Walled Struct 49:363–370
6. Wang H, Zhang Y (2009) J Constr Steel Res 65:1225–1235
7. Zaharia R, Dubina D (2006) J Constr Steel Res 62:240–249
8. Terekh M, Tretyakova D (2020) E3S web of conferences, vol 157
9. Zemitis J, Terekh M (2018) MATEC web of conferences, vol 245
10. Kaitila O (2002) J Constr Steel Res 58:333–351
11. Ranawaka T, Mahendran M (2010) Thin-Walled Struct 48:334–344
12. EN 1993-1-2-2005. Eurocode 3. Design of steel structures Part 1–2. General rules. Structural fire design
13. Kankanamge ND, Mahendran M (2012) Thin-Walled Struct 61:213–228
14. Kankanamge ND (2010) Brisbane. Queensland University of Technology, Australia
15. Laím L, Rodrigues JPC (2011) Proceedings of the 1st Ibero-Latin-American congress on fire safety, vol 1, pp 53–62
16. Lu W, Mäkeläinen P, Outinen J (2008) Proceedings of the fifth international conference on thin-walled structures, pp 713–720
17. Laím L, Rodrigues JPC, Simões DV (2013) Thin-Walled Struct 72:1–13
18. Gravit M, Dmitriev I (2020) Lecture notes in civil engineering, vol 70
19. Vatin N et al (2014) Appl Mech Mater 633–634:1037–1041
20. Garifullin M, Nackenhorst U (2015) Proc Eng 117:1073–1079
21. Garifullin M et al (2016) Proceedings of the international conference on engineering sciences and technologies, ESaT, pp 63–68
22. Gravit M, Dmitriev I, Lazarev Y (2019) Adv Intell Syst Comput 983:929–938
23. Gravit MV, Nedryshkin OV (2019) Proceedings of the 3rd international conference on engineering sciences and technologies, ESaT, pp 375–380
24. Naser MZ, Degtyareva NV (2019) Thin-Walled Struct 136:333–352
25. Naser MZ (2020) V.A Mech Mater 142
26. Zhou H et al (2017) Int J Steel Struct 17:195–204
27. Chen W et al (2020) J Constr Steel Res 164
28. Chen W, Ye J, Zhao Q (2019) Thin-Walled Struct 143
29. Chen W, Ye J, Li X (2019) J Constr Steel Res 158:290–305
30. Chen W, Ye J, Li X (2018) J Constr Steel Res 149:165–179
31. Chen W et al (2013) Thin-Walled Struct 73:145–157
32. Dias Y, Keerthan P, Mahendran M (2019) Fire Saf J 103:1–18
33. Dias Y, Mahendran M, Poologanathan K (2019) Thin-Walled Struct 137:81–93

Effect of Corrosive Damage on the Calculated Scheme of the Compressed Reinforced Concrete Structural Element

M. Berlinova, M. Berlinov, A. Tvorogov, and E. Pechkina

Abstract The article considers the scheme of the effect of corrosive damage with one-sided distribution on one of the faces of the compressed reinforced concrete rod in the absence of a cross load. A practical calculation of the strength of compressed reinforced concrete rods damaged by corrosion is presented, taking into account the effect of corrosive damage on the calculated scheme of the compressed reinforced concrete structural element. The scheme of a compressed concrete rod with one-way uniformly distributed on one of the faces of the chemical corrosion damage parallel to the plane of the system symmetry and in the absence of a cross load is considered. In solving the problem, a general method of calculating the force resistance of reinforced concrete structures damaged by chemical corrosion and the position of existing joint ventures is used.

Keywords Concrete · Strength · Compressed rod · Corrosion of reinforced concrete · Structural element

1 Introduction

In today's environment, in particular, taking into account the existing regulatory framework, in the operation of buildings and structures made of reinforced concrete and when deciding on the appointment for reconstruction or major repairs, the determining element of the approach to assessing the technical condition of the structures is its percentage of wear in the building as a whole. This approach is approximate and can carry additional material costs. Most previous studies of the strength of reinforced concrete constructions do not fully take into account the physical and mechanical characteristics of concrete and reinforcement, which leads to generalized results of the assessment of the carrier ability. For example, in the works of the [1–3] the method of accounting for strength, taking into account the

M. Berlinova (✉) · M. Berlinov · A. Tvorogov · E. Pechkina
Moscow State University of Civil Engineering (National Research University),
Yaroslavskoe shosse, 26, Moscow 129337, Russia

N. Vatin et al. (eds.), *Proceedings of MPCPE 2021*, Lecture Notes in Civil Engineering 182, https://doi.org/10.1007/978-3-030-85236-8_23

kinetics of the concrete strength set. The work considers the possibility of accounting for the depth of corrosive concrete damage in a three-axis tense state. Studies of corrosive damage, their accounting to assess the reliability of the supporting structures of buildings in the event of unilateral damage to compressed concrete columns are considered in this article methodology [4–8].

2 Materials and Methods

At the heart of the study methodology due to unilateral corrosion of compressed reinforced concrete structures, it is necessary to adopt an out-of-centered calculation case and use a common method of calculating the strength and stability of the damaged structural element in accordance with the current provisions of the Russian regulations, which provide an calculation on the strength of rectangular sections of extracenterly compressed elements from the condition:

$$N \cdot e \leq R_b \cdot b \cdot x(h_0 - 0.5x) + R_{sc} \cdot A'_s(h_0 - a'), \tag{1}$$

where N—is a long-range force from an external load; R_b—the initial strength of a compressed zone of concrete; e—distance from the application point of longitudinal force N to the center of gravity of the section stretched or least compressed (with a fully compressed section of the element) fixture equal

$$e = e_0 \cdot \eta + \frac{h_0 - a'}{2} \tag{2}$$

Here η—is a coefficient that takes into account the effect of the longitudinal bending of the element on its carrying ability and is determined according to the [5, p. 8.1.15]; e_0 is determined according to the [5, p. 8.1.7].

The height of the compressed zone x determines:

(a) at $\xi = \frac{x}{h_0} \leq \xi_R$ by formula

$$x = \frac{N + R_s \cdot A_s - R_{sc} \cdot A'_s}{R_b \cdot b}; \tag{3}$$

(b) at $\xi = \frac{x}{h_0} \leq \xi_R$ by formula

$$x = \frac{N + R_s \cdot A_s \cdot \frac{1+\xi_R}{1-\xi_R} - R_{sc} \cdot A'_s}{R_b \cdot b + \frac{2R_s \cdot A_s}{h_0(1-\xi_R)}}; \tag{4}$$

The value of the coefficient η in calculating structures according to the undeformed scheme is determined by the formula

$$\eta = \frac{1}{1 - \frac{N}{N_{cr}}} \tag{5}$$

where

N is a long-lived force from the external load;
N_{cr} conditional critical force defined by the formula

$$N_{cr} = \frac{\pi^2 \cdot D}{l_0^2} \tag{6}$$

Here D is the rigidity of the reinforced concrete element in the maximum strength stage, determined according to the indications of the deformation calculation; l_0—the estimated length of the element, defined as for the elements of the frame structure, considering its deformed condition at the most unfavorable for this element of the load arrangement, taking into account the insupuge of deformations of materials and the presence of cracks.

It is allowed to determine the value of D by formula

$$D = k_b E_b I + k_s E_s I_s \tag{7}$$

where

E_b, E_s modules of elasticity of concrete and fittings respectively;
I, I_s moments of inertia of the area of the section of concrete and the entire longitudinal rebar respectively relative to the axis passing through the center of gravity of the cross-section of the element;

$$k_b = \frac{0.15}{\varphi_l(0.3 + \delta_e)};$$
$$k_s = 0.7$$

φ_l a coefficient that considers the impact of the duration of the load $\varphi_l = 1 + \frac{M_{l1}}{M_1}$, but no more than 2. Here M_1, M_{l1} are the moments regarding the center of the most stretched or least compressed (with a completely compressed stretch) rod, respectively, from the action of the full load and from the action of constant and prolonged loads;
δ_e the relative value of the eccentricity of longitudinal force $\frac{e_0}{h}$, taken at least 0.15 and no more than 1.5.

Along the length of the element, the nature of its deformation and the effect of bends on the value of the bending moment in the calculated segment by calculating the structure as an elastic system.

Calculating the strength of rectangular sections of extra-centered compressed elements with a fixture located at opposite in the plane bending sides of the section, with the eccentricity of longitudinal force and flexibility $e_0 \leq \frac{h}{30}$ is allowed $\frac{l_0}{h} \leq 20$ to produce from the condition

$$N \leq N_{ult}, \tag{8}$$

where

N_{ult} there is a limited value of the longitudinal force that an element defined by the formula can perceive

$$N_{ult} = \varphi \cdot (R_b \cdot A + R_{sc} \cdot A_{s,tot}). \tag{9}$$

Here A—is the area of the concrete section; $A_{s,tot}$—the area of the entire longitudinal fixture in the part of the element; φ—the coefficient taken at a long-term load [5, Table 8.1], depending on the flexibility of the element; in the short-term φ action of the load, the values are determined by linear law, taking $\varphi = 0.9$ at $\frac{l_0}{h} = 10$ and $\varphi = 0.85$ with $\frac{l_0}{h} = 20$.

3 Results and Discussion

As a calculation, a hinge fixed in the support nodes is an extra-centeredly compressed rod, partially damaged by chemical corrosion with a more compressed side (Fig. 1).

Where is: N—external compressive power (—eccentricity of its applications); R_b—the original strength of the compressed zone of concrete; *K*—damage function (K); *h* and *b*—the size of the cross-section of the rod; ℓ_o—distance between support nodes (the length of the rod); a_o, $a_o^{/}$—the thickness of the protective layers of the rebar on the less (or stretched) and more tense edges e_o—the eccentricity of the squeezing force; *x*; h_o—the height of the compressed zone and the working height of the section; *V* and *y*—the axis of counting.

For the practical application of the proposed calculation, considering the reduction of the carrying capacity of the compressed reinforced concrete element, the so-called A.F. Loleit postulate on normal stresses in the cross sections of the rod is used, and the calculation of corrosive damages carried out by the [4] is also introduced to limit the depth of corrosive damages of the height of the compressed section area

$$\delta \leq x, \tag{10}$$

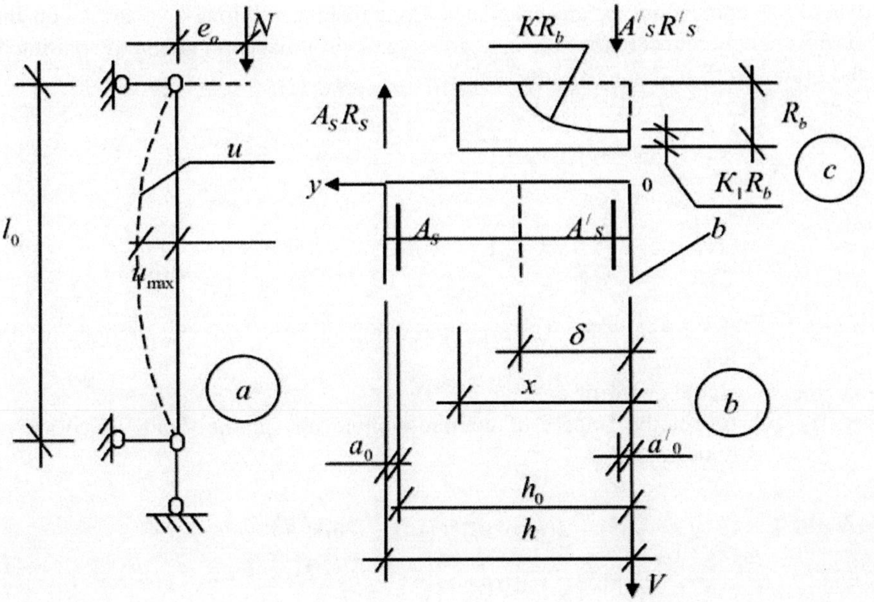

Fig. 1 Calculated schemes: compressed element works **(a)** its cross-section **(b)** and corresponding force resistance **(c)**

determined from the balance of all forces (given the influence of corrosion) on the vertical axis, as well as the refusal to consider the force resistance of the concrete stretched zone. And also: it is recognized to keep the clutch of rebar with concrete; the so-called undeformed calculation scheme (so-called stable strength) is adopted, and the influence of longitudinal bending in the case is taken $\lambda \geq \lambda_{gr}$ into account by increasing the calculated eccentricity

$$e_p = e_o\eta + e_{sl}, \text{ where } 1 \leq \eta = \frac{1}{1 + \frac{P}{P_{kr}}} \leq 2.5, \quad (11)$$

where

e_o there is an eccentricity of the compression application relative to the center of gravity of a less compressed (or stretched) fixture A_s;

e_{sl} accidental eccentricity, rated by [5];

P_{kr} is the corresponding critical force of resilience for Euler.

The function of corrosive concrete damage in the form of:

$$K(y) = \sum_{i=0}^{i=2} a_i y^i, \quad (12)$$

where the coefficients a_i are calculated from fixed conditions—values K_i on the surface of concrete contact with an aggressive environment and geometric pairing at the border with undamaged concrete

$$a_o = K_1; a_1 = \frac{2(1-K_1)}{\delta}; a_2 = -\frac{1-K_1}{\delta^2} \quad (13)$$

$$\text{where is } \delta = \sum_{i=0}^{i=3} q_i \left(\frac{\sigma}{R_b}\right)^i \left[1 - \beta e^{-\alpha(t-t_o)}\right] \text{ at } \beta = 1 - \frac{\delta(t_o, t_o)}{\delta(t_{st}, t_o)} \quad (14)$$

σ and R_b the magnitude of the compressing stresses and the strength of the concrete;

$q_o = \delta_o$ stable depth of damage $\sigma = 0$;

q_1, q_2, q_3 reflect the impact of compression levels on the depth of corrosive damage.

$$q_3 = \frac{(\delta_t - \delta_o) + 2(\delta_o - \delta_l)\frac{\sigma_t}{\sigma_l} - (\delta_o - \delta_l)\left(\frac{\sigma_t}{\sigma_l}\right)^2}{\left(\frac{\sigma_l}{R_b}\right)^2\left(\frac{\sigma_t}{R_b}\right) + 2\left(\frac{\sigma_l}{R_b}\right)\left(\frac{\sigma_t}{R_b}\right)^2 - \left(\frac{\sigma_t}{R_b}\right)^3};$$

$$q_2 = \frac{(\delta_o - \delta_l) - 2\,q^3\left(\frac{\sigma_l}{R_b}\right)^3}{\left(\frac{\sigma_l}{R_b}\right)^2}; q_1 = -2\,q_2\left(\frac{\sigma_l}{R_b}\right) - 3\,q_3\left(\frac{\sigma_l}{R_b}\right)^2 \quad (15)$$

here

σ_l is the meaning of compression stresses, coinciding with the end of linear deformation, the largest compaction of concrete;

σ_t Compression voltage value that matches *max.* the development of cracks, in which the force resistance of concrete is maintained;

δ_l and δ_t corresponding to the lowest and highest values of corrosive concrete corrosive damage in compressed concrete (in the absence of experimental values σ_l and σ_t possible calculation of

$$\sigma_l = 0.35 lgR_b - 0.15\,\text{МПа}; \sigma_t = 0.35\,lgR_b + 0.175\,\text{МПа}[5], \quad (16)$$

but thus, depending on the class of concrete, the interval between σ_l and σ_t that is $\sigma_t - \sigma_l = 0.325$ MPA is fixed; α—empirical at a rate of atnusion; t_0, t, t_{st}—the start time of observation, the current time, the stabilization time.

The corrosive damage to the rebar is considered by the estimated reduction of its area by the reduction of the damage factor

$$\omega_S \leq 1; \tag{17}$$

and now the estimated area lost due to corrosive damage to the force resistance

$$\mathrm{A}_2 = b\int_0^{\delta} K_1(y)\mathrm{d}y = b\sum_{i=0}^{i=2} a_i \int_0^{\delta} y^i \mathrm{d}y = \frac{1}{3}b\delta(2 + K_1), \tag{18}$$

where K_1—is an experimentally mounted numerical value for the outer face of the rod, the corresponding characteristic of the force resistance of the damaged concrete.

Specifically, at K_1

$$\mathrm{A}_2 = \frac{2}{3}b\delta;\ \mathrm{A}_{2,st} = \frac{2}{3}b\delta_{st}. \tag{19}$$

Note that in the absence of the necessary experimental data regarding K_1 for orientation, it is possible to assign K_1 according to the formula [4]:

$$K_1 = 1 - \frac{\delta}{\delta_{st}}, \tag{20}$$

that after substitution (11) in (14) gives

$$\mathrm{a}_0 = 1 - {}^{\delta}/_{\delta_{st}};\ \mathrm{a}_1 = \frac{2}{\delta_{st}};\ \mathrm{a}_2 = -\frac{1}{\delta \times \delta_{st}}; \tag{21}$$

$$\mathrm{A}_2 = \frac{1}{3}b\delta\left(3 - {}^{\delta}/_{\delta_{st}}\right); \tag{22}$$

$$\mathrm{A}_{2,st} = \frac{2}{3}b\delta_{st}; \tag{23}$$

now lost due to corrosive damage, part of the force resistance by compression force will be equal to

$$F_{2,Rb} = R_b b\int_0^{\delta} K_R(y)\mathrm{d}y = R_b b\sum_{i=0}^{i=2} a_{iR} \int_0^{\delta} y^i \mathrm{d}y = \frac{1}{3}b\delta(2 + K_{1,R})R_b \tag{24}$$

or

$$F_{2,R_b} = A_2 R_b. \tag{25}$$

case 1 when $\xi \le \xi_R$ and $N + \omega_S R_S A_S \pm \omega_S^{/} R_S^{/} A_S^{/} = R_b bx - F_{2,R_b}$,

$$x = \frac{N + \omega_S R_S A_S \pm \omega_S^{/} R_S^{/} A_S^{/} + F_{2,R_b}}{bR_b}; \tag{26}$$

case 2 when $\xi > \xi_R$ and $N + \omega_S \sigma_S A_S \pm \omega_S^{/} \sigma_S^{/} A_S^{/} = R_b bx - F_{2,R_b}$,

$$x = \frac{N + \omega_S \sigma_S A_S \pm \omega_S^{/} \sigma_S^{/} A_S^{/} + F_{2,R_b}}{bR_b}; \tag{27}$$

where

σ_S it is calculated according to empirical formulas.

Distance from the center of gravity A_2 to axis V at (13) and (18)

$$r_V = \frac{1}{A_2} \int_0^{\delta} bK(y) y \, dy = \frac{(5 + K_1)}{4(2 + K_1)} \delta, \tag{28}$$

that in a private case $K_1 = 0$ gives both directly and at K_1 by (20) solutions

$$r_V = \frac{5}{8}\delta; \quad r\frac{5}{8_{V,max}}, \tag{29}$$

then lost due to corrosive damage part of the force resistance on the bending moment relative to the center of gravity of the less tense (m/b stretched) longitudinal fixture

$$M^{*} = -F_{2,R_b}(h_0 - r_V), \tag{30}$$

and the overall test of the so-called stable strength (the carrying capacity of the compressed rod, taking into account the longitudinal bend) is carried out by assessing the next inequality

$$P_{kr}^{*}[\eta e_0 + (h_0 - x) + e_{sl}] \le \omega_S^{/} \sigma_S^{/} A_S^{/}\left(h_0 - a_0^{/}\right) + bxR_b\left(h_0 - \frac{x}{2}\right) - F_{2,R_b}(h_0 - r_V), \tag{31}$$

where the critical force of stability P_{kr}^{*} for corrosion-damaged reinforced concrete rod is given a formula.

$$P^*_{kr} = \left(\frac{\Pi}{\mu \times l_0}\right)^2 D^*, \tag{32}$$

where

μ there is a multiplier of the calculation length of the compressed rod, determined by the real conditions of fixing it in knots;

D^* the curved stiffness of the corrosion-damaged rod, calculated relative to the center of gravity of the section [5]. Similarly, it is possible to take

$$D^* = K_b\left(E_b J - A_2 r_W^2\right) + K_S \omega_S E_S J, \tag{33}$$

where

r_W the distance from the center of gravity of area A_2 to the center of gravity of the section;

E_b integral deformation module, reflecting nonlinearity, warp mode in time [4].

$$K_S = 0.7; K_b = \frac{0,15}{\phi_l(0,3 + \delta_l)}; \delta_l = \frac{e_0}{h}; \phi_l = 1 + \frac{\mu_{l,1}}{\mu_1}. \tag{34}$$

Note that the longitudinal bend is not taken into account $\lambda < \lambda_{pr}$.

4 Conclusion

The scheme of a compressed concrete rod with one-way uniformly distributed on one of the faces of the chemical corrosion damage parallel to the plane of the system symmetry and in the absence of a cross load is considered.

Obviously, the consequence of unilateral damage inevitably becomes the adoption of an out-of-centered calculation case, as the damaged concrete is excluded from the work of the structure.

In solving the problem, a general method of calculating the force resistance of reinforced concrete structures damaged by chemical corrosion and the position of existing joint ventures is used [1, 9–21].

References

1. Korol EA, Berlinov MV, Berlinova MN (2016) Kinetics of the strength of concrete in constructions. Procedia Engineering 292–297
2. Berlinov M, Berlinova M (2016) Influence of kinetics concrete hardening on strength of constructions. В сборнике: MATEC web of conferences, p 04014

3. Berlinov MV, Berlinova MN, Tvorogov AV, Pechkina EK (2020) Accounting for corrosive damage to the operated reinforced concrete structures in the conditions of three-axis tense-deformed condition. Constr Archit T 8(3):S40–46
4. Bondarenko VM (2009) Some fundamental questions of the development of reinforced concrete theory. V Intercollegiate Conference, MGAKHIS, Moscow
5. Russian State Standard SP 63_13330_2018
6. Berlinov MV (2018) Strength resistance of reinforced concrete elements of high-rise buildings under dynamic loads. E3S web of conferences, vol 33, p 02049
7. Berlinov MV, Berlinova MN (2019) Durability of reinforced concrete constructions in conditions of prolonged operation. Bull Constr Equipment No. 1:C 60–61
8. Berlinov MV, Berlinova MN, Gregorian AG (2019) Operational durability of reinforced concrete structures. E3S web of conferences, vol 91, p C.02012
9. Tamrazyan AG (2012) Calculation of structural elements with the specified normal distribution and reliability and load bearing capacity, vol 10. Vestnik MGSU, pp S109–115
10. Alexandrovsky SV, Solomonov VV (1972) Dependence deformations the creep of aging concrete from entry-level stresses, inter-industry construction issues. Domestic experience. Reference collection, pp 6–12
11. Bazhenov YM (2002) Concrete technology. ACB, 500 p
12. Berlinova MN, Berlinov MV, Tvorogov AV (2015) Entropy criterion of concrete strength in construction constructions. Sci Rev 22:162–166
13. Zveryaev EM, Berlinov MV, Berlinova MN (2016) The integral method of definition of basic tension condition of anisotropic shell. Int J Appl Eng Res 11(8):5811–5816
14. Tvorogov AV (2015) Diagrams of concrete deformation. In the book: Construction—formation of living environment. Proceedings of the eighteenth international interuniversity scientific-practical conference of students, undergraduates, graduate students and young scientists. Moscow, pp 1005–1007
15. Tvorogova MN (2006) Resistance to the deformation and destruction of reinforced concrete structures, taking into account non-linear and unbalanced properties and modes of heating. Thesis on the Ph.D. degree. Moscow
16. Nazarenko VG, Tvorogova MN, Lukantsov PN (2010) O to the construction of the functions of aging concrete. Concr Reinf Concr 1:6–10
17. Korol EA, Berlinova MN (2018) Development of methods of calculating multi-layered fencing structures with monolithic connection of layers. MISI Publishing House, MOSCOW. (Library of Scientific Developments and Projects of the Moscow State University). Scientific electronic publication
18. Korol EA (2001) Three-layered reinforced concrete structures made of light concrete and features of their calculation. ASV Publishing House, Monograph
19. Berlinov MV (2017) Developing the phenomenological equations triaxial deformation of concrete under dynamic loads In MATEC Web of Conferences. V106, p 4008
20. Korol EA, Berlinov MV, Berlinova MN (2017) The long term stability of multilayer walling structures. In: MATEC web of conferences, vol 106, p 4006
21. Korol E, Kagan P, Barabanova T (2015) Automation of the formation of organizational technological documentation. Appl Mech Mater 738–739:444 p

Complex Application of Rod Composite Reinforcement in Bendable Concrete Elements

S. I. Merkulov and E. K. Akimov

Abstract The complete replacement of steel reinforcement in bent concrete elements with composite reinforcement faces significant technological challenges. In the bent elements, together with the longitudinal rods of composite reinforcement, it is necessary to use transverse steel reinforcement. The anchoring of the transverse rods of composite reinforcement cannot be provided by a reliable attachment to the longitudinal rods, the device of anchors, the manufacture of closed clamps. Modern methods of calculating the strength of inclined sections of bent elements with composite reinforcement are prescribed to perform as for elements with steel transverse reinforcement using the characteristics of composite reinforcement, while not taking into account the features of using composite reinforcement rods for transverse reinforcement of bent elements. The development of calculation methods is hindered by the lack of experimental and theoretical studies of the strength of inclined sections with reinforcement of the support sections with rod composite reinforcement. This paper presents the results of experimental studies of bent concrete elements with longitudinal and transverse composite reinforcement.

Keywords Inclined section · Composite reinforcement · Rebar bends

1 Introduction

The solution to the problem of corrosion safety of reinforced concrete structures operated in aggressive conditions is solved by the use of composite reinforcement. To date, considerable experience has been accumulated in the use of composite reinforcement in a wide range of concrete structures, while composite reinforcement is usually used in combination with steel reinforcement [1–4]. However, the

S. I. Merkulov
Kursk State University, Radishcheva st., 33, Kursk 305000, Russia

E. K. Akimov (✉)
Belgorod State Technological University named after V.G. Shukhov, Kostyukov st., 46, Belgorod 308012, Russia

N. Vatin et al. (eds.), *Proceedings of MPCPE 2021*, Lecture Notes in Civil Engineering 182, https://doi.org/10.1007/978-3-030-85236-8_24

complete replacement of steel rebar with composite rebar faces significant technological challenges. In the bent elements, together with the longitudinal rods of composite reinforcement, it is necessary to use transverse steel reinforcement. A prerequisite for the use of rod composite reinforcement is reliable anchoring, including transverse rods in the event of the formation of inclined cracks. The anchoring of the transverse rods of composite reinforcement cannot be provided with a reliable attachment to the longitudinal rods, the device of anchors, the manufacture of closed clamps. The design of concrete structures with composite reinforcement is carried out in accordance with the code of rules SP 295.1325800.2017, some provisions of which need experimental verification and clarification. Thus, the calculation of the inclined cross-section for the action of transverse forces is prescribed to be performed as a steel transverse reinforcement using the characteristics of composite reinforcement, while not taking into account the noted features of the use of composite reinforcement rods for transverse reinforcement of bent elements.

2 Materials and Methods

2.1 Materials

Currently, experimental and theoretical studies of structures with composite reinforcement are being carried out [5–12]. There are no experimental and theoretical studies of the strength of inclined sections with reinforcement of the support sections with rod composite reinforcement. The authors developed an original method of testing bent beams for crack resistance and strength of inclined sections with composite reinforcement in the form of bends of longitudinal bars of working reinforcement [13–16]. At the Belgorod State Technological University, experimental studies of concrete bending beams with bends of composite reinforcement rods on the support sections were carried out. The main tasks of the tests were to obtain new experimental data on the strength and crack resistance of the bent beams, to identify the features of the joint work of the bends of composite reinforcement rods and concrete in inclined sections, to identify the mechanism of destruction along the inclined section of the bent beams.

The dimensions of the prototypes were 125 × 250 × 1800 mm. Sample materials: concrete of class B25, basalt-plastic reinforcement with sand coating produced by ROCKBAR LLC "Galen" with a diameter of 8 mm. Two rods of longitudinal reinforcement are arranged in a straight line along the entire length of the beam, and the remaining two in the quarters of the span were transferred from the lower zone in the span to the upper face on the support with an exit through the upper part of the beam, the value of the bend of the rods was 20 d. A special feature of the design solution of the experimental beams is the absence of transverse reinforcement along the entire length.

The location of the bends of the composite reinforcement rods was carried out as close as possible to the trajectory of the main tensile stresses. A special feature of the design solution of the experimental beams was the absence of transverse reinforcement along the entire length.

2.2 Methods

Tests of experimental beams are carried out on a test bench, which allows you to set the necessary load application schemes and test mode (Fig. 1). Loading of reinforced concrete elements is carried out by two concentrated forces applied at a distance of 450 mm from the supports. The load is applied in stages of 0.1 of the destructive load determined by preliminary tests, at each stage of loading, an exposure time of 10–15 min was made. To measure the deformations of concrete along inclined sections, strain gauges of the PCB type with a base of 80 mm were applied to the experimental beams, glued to the side surfaces of the beam on both sides of the beams in the support zone along the trajectory of the inclined sections. Also, in parallel with the load cells, hour-type indicators with a division price of 0.001 mm are installed. This arrangement of recording devices will allow you to fully assess the stress–strain state of inclined sections. Also, to measure the deformation of beams along the inclined section, hour-type indicators with a division price of 0.1 mm are installed. To fix the shift of the rebar rods in the concrete, hour-type indicators with a division price of 0.001 mm are installed at the rebar outlets.

Fig. 1 Shell and tube heat exchanger

Fig. 2 Plate heat exchanger

3 Results and Discussion

As a result of tests of bent concrete beams with longitudinal rod composite reinforcement with bends of composite reinforcement on the supporting sections, it was established:

- the breaking load for the prototypes was Nu = 63.0 kN;
- the load of formation of normal cracks was 18.0 kN, inclined cracks-27.0 kN;
- the opening width of the inclined cracks was at a load of 0.4 Nu 0.03 mm; at a load of 0.5 Nu 0.045 mm; at a load of 0.6 Nu 0.07 mm;

The result of experimental studies was an assessment of the adhesion to concrete of the bends of composite reinforcement rods in the zone of action of transverse forces. It was found that there were no deformations of the displacement of the composite reinforcement rods relative to the concrete, and no anchor devices were used on the bends of the composite reinforcement rods. However, a number of studies have shown the need for additional study of the coupling of composite reinforcement with concrete [17–20].

The destruction of the prototypes occurred along an inclined cross-section with the rupture of the bent composite rod (Fig. 2).

4 Conclusion

As a result of experimental studies, the design solution of bent concrete elements with the use of composite rods as longitudinal and transverse working reinforcement was confirmed, which allows to achieve absolute corrosion safety of structures when operating in aggressive environments.

References

1. Akimov EK (2018) Prospects for the use of non-metallic composite reinforcement in bent reinforced concrete elements with bends in support areas science and innovation in construction. Collect Reports Int Sci Pract Conf (to 165th anniversary birth V. G. Shukhov), pp 10–13

2. Rimshin VI, Merkulov SI (2015) Elements of the theory of development of concrete structures with non-metallic composite reinforcement. Ind Civ Constr 5:38–42
3. Merkulov SI, Rimshi VI (2018) Composite materials for concrete and reinforced concrete structures. Curr Issues Archit Constr Collect Mater Int Sci Tech Conf 222–225
4. Subbotin AI, Shutova MN, Shagina A (2019) Analysis of the specifics of using composite reinforcement in the foundations of buildings under construction and reconstruction. Bull Volgogr State Univ Archit Civ Eng Ser Constr Archit 2:37–48
5. Gizdatullin GA, Khusainov RR, Khozin VG, Krasinikova NM (2016) Strength and deformability of concrete structures reinforced with fibre-reinforced polymer bars. Mag Civ Eng 62:32–41. https://doi.org/10.5862/MCE.62.4
6. Krassowska J, Lapko A (2013) The influence of steel and basalt fibers on the shear and flexural capacity of reinforced concrete beams
7. Vetrova OA (2020) Pilot studies of concrete beams deformability reinforced with composite reinforcement. Proc Southwest State Univ 24:103–114. https://doi.org/10.21869/2223-1560-2020-24-1-103-114
8. Smolyago GA, Frolov NV, Poloz MA (2016) Experimental research of RC beams in combination with glass-plastic bars. Bull Belgorod State Technol Univ named after V. G. Shukhov. 2:60–64. https://doi.org/10.12737/23298
9. Stepanova VF, Buchkin AV, Il'in DA (2017) The research of the features of concrete structures with a combined reinforcement (reinforcement of composite polymer and a non-metallic fiber). Acad Archit Constr 1:124–128
10. Merkulov SI, Rimshin VI, Akimov EK (2019) Fire resistance of concrete structures with composite rod reinforcement. Ind Civ Constr 4:50–53. https://doi.org/10.33622/0869-7019.2019.04.50-55
11. Merkulov SI, Akimov EK (2017) Experimental characteristics of basalt-plastic rebar with a diameter of 6 mm. Sci Innov Constr sat Dokl Int Sci Conf Belgorod BSTU Publ house 73–76
12. Rimshin VI, Merkulov SI (2016) About normalization of characteristics of rod non-metallic composite reinforcement. Ind Civ Constr 5:22–26
13. Merkulov SI, Akimov EK (2020) Experimental research methodology of elements with composite fittings with bending at priority sections. Bull Belgorod state Technol Univ named after V. G. Shukhov 9:8–14
14. Merkulov SI, Akimov EK (2020) Strength of inclined sections of beams with composite rod reinforcement. In: Merkulov SI (ed) Secur Constr fund Russ Probl Solut Mater Int Acad Readings, pp 180–185
15. Merkulov SI, Akimov EK (2020) Design parameters of composite rod reinforcement. In: Merkulov SI (ed) Secur Constr fund Russ Probl Solut Mater Int Acad Readings, pp 26–29
16. Merkulov S, Rimshin V, Akimov E, Kurbatov V, Roschina S (2020) Regulatory support for the use of composite rod reinforcement in concrete structures. In: IOP conference series: materials science and engineering. IOP Publishing Ltd. https://doi.org/10.1088/1757-899X/896/1/012022
17. Frolov NV, Poloz MA, Kolesnikova EG (2016) Research of the polymer composite reinforcement test under axial loading. Bull Belgorod State Technol Univ named after V. G. Shukhov 1:74–78. https://doi.org/10.12737/22351
18. Bogdanova ER (2016) Change in properties of adhession between composite polymer reinforcement and concrete under conditions of impact of different environments. Ind Civ Constr 2:39–43
19. Yartsev VP, Nikolyukin AN, Pluzhnikova TM (2018) The impact of freeze-thaw cycles on the bearing capacity and strength deformation of polymerbetonic beams. Bull Tambov State Tech Univ 2:360–366
20. Nikolyukin AN, Yartsev VP, Bondarev BA, Korneeva AO (2019) Modeling the adhesion of reinforcement in concrete based on an artificial neural network. Sci J Constr Archit 2:11–20. https://doi.org/10.25987/VSTU.2019.54.2.001

Using CLT-Panels with Alternatively Angled Layers as Overlappings

Vladimir Melekhov, Alexander Karelskiy, and Dmitriy Lyapin

Abstract CLT-panels are glued wooden structures consisting of several layers of timber arranged orthogonally to each other. The panels are widely used as structures of load-bearing walls and interfloor overlappings. There are many ways to increase the load-bearing capacity of CLT-panels, but they all increase the cost of manufacturing this material. This research proposes a way to increase the bearing capacity of CLT-panels by changing the angle of the wood layers relative to the span. A numerical experiment in the ANSYS program was used as a method for checking the expedience of this modification. The hypothesis was tested for CLT-panels with different number of layers and different fixation conditions. Based on the research results, it was determined that for overlapping slabs supported on four sides, the effect of changing the angle of the layers is negative. For slabs supported in four corners, there is a positive effect. Accordingly, for buildings with beamless overlapping, when the support takes place either directly on the columns or through capitals, the use of modified floors has a certain effect and will lead to savings in material consumption. As a result of the experiment, a new way of using the CLT-panel as an interfloor overlapping is proposed.

Keywords CLT-panels · Timber · Wood structures · Interfloor overlap

1 Introduction

Glued timber structures are becoming increasingly popular in civil and industrial construction. It is connected with a number of advantages: environmental friendliness of the material, economical use of raw materials, better strength indicators in comparison with ordinary wood. CLT-panels (cross-laminated timber panels) are a type of glued timber structures. They are multi-layer panels with a cross-arrangement of layers at an angle of 90° to each other [1–4].

V. Melekhov · A. Karelskiy · D. Lyapin (✉)
Northern (Arctic) Federal University named after M.V. Lomonosov,
Severnaya Dvina Emb. 17, Arkhangelsk 163002, Russia

N. Vatin et al. (eds.), *Proceedings of MPCPE 2021*, Lecture Notes in Civil Engineering 182, https://doi.org/10.1007/978-3-030-85236-8_25

CLT panels are most often used in walls as walls and floors. When used as interfloor overlappings, there are options for supporting the slabs: on both sides; on four sides; in four corners. For example, in a residential complex in Finland four-sided CLT slabs are used [5], while a Vancouver dorm uses cross-laminated panels supported at four corners [6]. The stress–strain state of structures will also differ depending on the boundary conditions [7].

There are many options for increasing the load-bearing capacity of cross-laminated panels by improving their mechanical properties. There is a method for the production of cross-glued panels [8] with the removal of one or more elements of the inner layers. This provides an improved stiffness-to-weight ratio of the panel. The disadvantage of this design is, in addition to reducing the material intensity of the panel, its large deformations when used as an overlapping. In addition, when the individual elements are removed from the inner lamellas, air pockets are formed. This promotes the spread of flame during a fire, which reduces the fire resistance of the entire panel.

In addition, there is the method of manufacturing CLT panels using the sawtooth method of gluing the layers [9]. Each board of the inner layer of the panel contains dowels on both sides. When gluing, the dowels fall into special slots on the boards of the outer panels. After gluing, the resulting panel has better mechanical characteristics due to tighter contact between the wood layers. The disadvantage of this design is the technological complexity of the production of boards for such panels.

There are panels of cross-laminated wood, additionally reinforced with a mesh of rods [10] and two intersecting reinforcing strips [11]. To create such a composite material, grooves are cut in the inner layers of the panel, where a reinforcing mesh (plastic or metal) is installed for the first case and cross reinforcing bars are installed for the second case. The disadvantages of these models are the technological complexity when cutting grooves, as well as additional material consumption for reinforcing elements.

A similar method of improving the mechanical properties of the panel is used in the patent [12], where basalt rods are used to reinforce the panel, which are installed in drilled holes. The disadvantage of this method is the presence of an additional technological process for drilling holes, as well as the consumption of material for basalt rods.

The aim of the work was to develop a way to increase the bearing capacity of overlappings consisting of CLT panels without complicating the production technology and while maintaining all the advantages of this material.

2 Methods

The number of wood layers in a CLT panel is odd. When the overlapping is supported on four sides, the structure will bend in two directions. Layers oriented across the span have a lower modulus of elasticity ("weak") than those located along the span ("strong"). For "weak" layers, the flexural stiffness is significantly less than for "strong" ones and depends on the elastic moduli along and across the fibers. Therefore, for example, for pine, according to [7] the elastic modulus along the fibers (Ex) is equal to 16,600 MPa, and across the fibers (Ey) it is equal to 582 MPa. In each of the two directions of the overlapping, some of the layers will be "strong" and some will be "weak".

To improve the strength and deformation properties of the CLT-panel overlapping, it is proposed to unfold the panel in such a way that the fibers of the layers are located at an angle of 45° to the spans to be covered. Thus, the flexural stiffness for each span will be averaged. New moduli of elasticity (Ex′ and Ey′) for the layers are determined from the system of equations [13–15].

The aim of the work was to develop a way to increase the bearing capacity of overlappings consisting of CLT panels without complicating the production technology and while maintaining all the advantages of this material.

$$\begin{cases} \frac{1}{E_x'} = \frac{\cos^4\varphi}{E_x} + \left(\frac{1}{G} - \frac{2v_x}{E_x}\right)\sin^2\varphi\cos^2\varphi + \frac{\sin^4\varphi}{E_y}, \\ \frac{1}{E_y'} = \frac{\sin^4\varphi}{E_x} + \left(\frac{1}{G} - \frac{2v_x}{E_x}\right)\sin^2\varphi\cos^2\varphi + \frac{\sin^4\varphi}{E_y}, \\ \frac{1}{G'} = \frac{1}{G} + \left(\frac{1+v_x}{E_x} + \frac{1+v_y}{E_y} - \frac{1}{G}\right)\sin^2 2\varphi, \\ v_x' = E_x'\left[\frac{v_x}{E_x} - \frac{1}{4}\left(\frac{1+v_x}{E_x} + \frac{1+v_y}{E_y} - \frac{1}{G}\right)\sin^2 2\varphi\right], \\ v_y' = v_x'\frac{E_y'}{E_x'}. \end{cases} \tag{1}$$

where Ex and Ey are the elastic moduli, respectively, along and across the fibers in the main coordinate system;
Ex′ and Ey′ are the elastic moduli, respectively, along and across the fibers in a new coordinate system rotated by an angle φ relative to the main one;
G is the shear modulus;
vx′ and vy′ are Poisson's ratios in the direction along and across the fibers, respectively, in the main coordinate system.
vx′ and vy′ are Poisson's ratios in the direction along and across the fibers, respectively, in a new coordinate system rotated by an angle φ relative to the main one.

A similar solution is used in the article [16–18]. The authors suggest using CLT-panels with layers rotated to 45° for wall construction.

The research method is carrying out numerical experiments in the ANSYS programming and computing suite [19, 20]. The design model is a 6 × 6 m CLT panel loaded with a distributed load of 3 kPa. For comparison, 3-, 5- and 7-layer panels were considered, the thickness of one layer is 45 mm.

3 Results and Discussion

The first experiment was carried out with CLT overlappings supported on four sides.

Table 1 shows the deflections for a panel with layers arranged orthogonal to the spans (standard layout) and for a panel with layers at an angle of 45° (modified layout). Figure 1 shows a deformation diagram for a standard layer arrangement, and Fig. 2 shows a diagram for a modified arrangement.

Table 1 shows that the modified layer arrangement turned out to be less effective than the standard one. The fewer the number of layers in the panel, the less advantageous the modified version of the arrangement of the layers is.

This is due to the fact that, according to Eq. (1), the modulus of elasticity for "strong" layers decreases significantly (almost 10 times) in transition to the modified coordinate system, and for "weak" layers it increases 2.5 times [21].

The first experiment was carried out with CLT overlappings supported on four sides.

The results of the numerical experiment are presented in Table 2.

As can be seen from Table 2, the modified version of the arrangement of the layers shows the best results for deflections (Figs. 3 and 4).

Table 1 Results of a numerical experiment for a panel supported on four sides

Number of layers of a CLT-panel	Bending (mm)		$\frac{f_2-f_1}{f_1}$ (%)
	Standard layer arrangement (f_1)	Modified layer arrangement (f_2)	
3	18.2	24.0	31.9
5	4.1	4.9	19.3
7	1.6	1.8	15.8

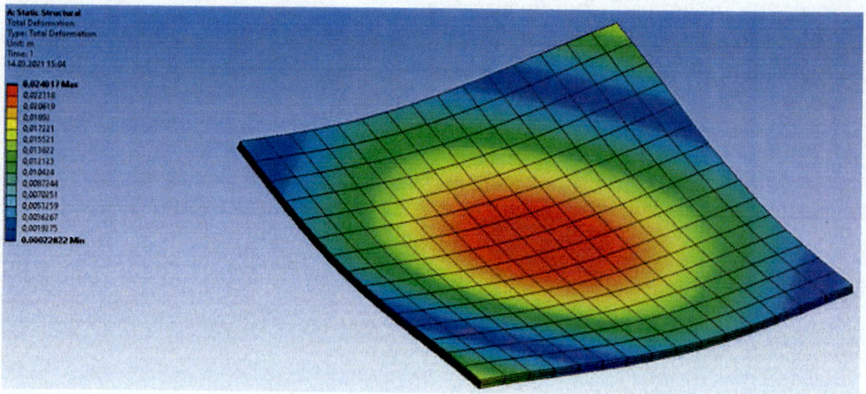

Fig. 1 A deformation diagram for a standard layer arrangement supported on four sides

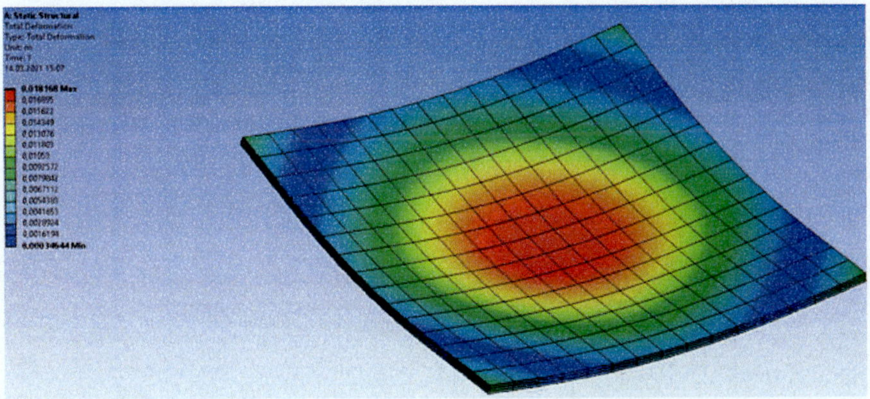

Fig. 2 A deformation diagram for a modified layer arrangement supported on four sides

Table 2 Results of a numerical experiment for a panel supported at four corners

Number of layers of a CLT-panel	Bending (mm)		$\frac{f_2-f_1}{f_1}$ (%)
	Standard layer arrangement (f_1)	Modified layer arrangement (f_2)	
3	53.9	32.3	40.1
5	12.9	8.7	33.6
7	1.5	1.8	20.0

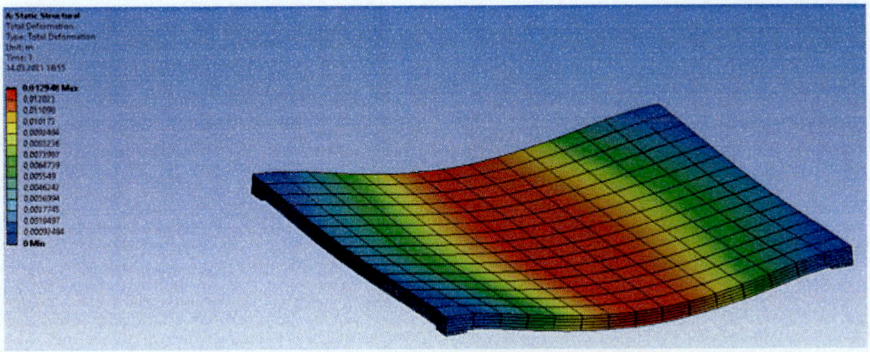

Fig. 3 A deformation diagram for a standard layer arrangement supported on four corners

The data of the numerical experiment confirm the hypothesis that the modified version of the CLT panel is more efficient for use as an overlapping. This is because most of the lamellas are oriented along the grain between the beams [22, 23]. So, the wood grains are directed along the working span. Moreover, thanks to this

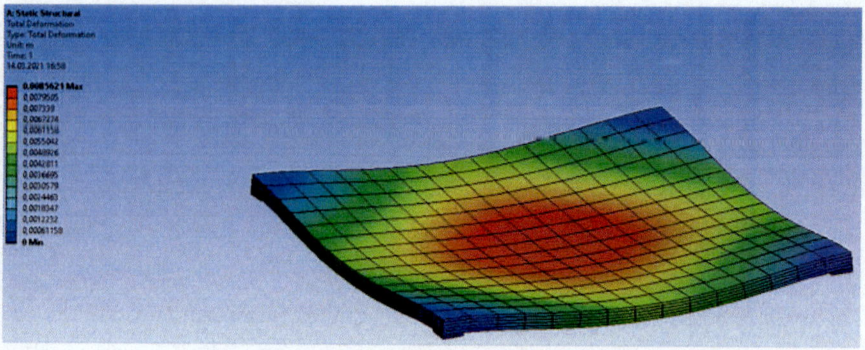

Fig. 4 A deformation diagram for a modified layer arrangement supported on four corners

configuration, better flexural results are achieved due to the higher modulus of elasticity.

4 Conclusion

1. A numerical experiment was carried out to simulate overlappings with different arrangements of CLT-panels.
2. Based on the research results, it was determined that for overlapping slabs supported on four sides, the effect of changing the angle of the layers is negative.
3. For slabs supported in four corners, there is a positive effect. Accordingly, for buildings with beamless overlapping, when the support takes place either directly on the columns or through capitals, the use of modified floors has a certain effect and will lead to savings in material consumption.

References

1. Roshchina S, Lukin M, Lisyatnikov M (2020) Compressed-bent reinforced wooden elements with long-term load. https://doi.org/10.1007/978-3-030-42351-3_7
2. Lukin MV, Roshchina SI, Smirnov EA, Shunqi M (2020) Strengthening of the operated wooden floor beams with external rigid reinforcement. IOP Conf Series Mater Sci Eng. https://doi.org/10.1088/1757-899X/896/1/012065
3. Koshcheev AA, Roshchina SI, Lukin MV, Lisyatnikov MS (2018) Wooden beams with reinforcement along a curvilinear trajectory. Mag Civ Eng 81:193–201. https://doi.org/10.18720/MCE.81.19

4. Steiger R, Serrano E, Stepinac M, Rajčić V, O'Neill C, McPolin D, Widmann R (2015) Strengthening of timber structures with glued-in rods. Constr Build Mater. https://doi.org/10.1016/j.conbuildmat.2015.03.097
5. Roschina SI, Lisyatnikov MS, Koshcheev AA (2019) Technical- and- economic efficiency of reinforced wooden structures. IOP Conf Series Mater Sci Eng. https://doi.org/10.1088/1757-899X/698/2/022005
6. Staub-French S, Pilon A, Poirier E, Fallahi A, Kasbar M, Calderon F, Teshnizi Z, Froese T (2021) Construction process innovation on brock commons Tallwood House. Constr Innov. https://doi.org/10.1108/CI-11-2019-0117
7. Labudin BV, Karelskiy AV, Lyapin DM (2020) Theoretical preconditions for determination of the elastic modulus of clt-panels. https://doi.org/10.4028/www.scientific.net/MSF.992.998
8. Varenik KA, Varenik AS, Sanzharovsky RS, Labudin BV (2019) Wood moisture accounting in creep equations. IOP Conf Series Mater Sci Eng. https://doi.org/10.1088/1757-899X/656/1/012054
9. Buka-Vaivade K, Serdjuks D, Goremikins V, Pakrastins L, Vatin NI (2018) Suspension structure with cross-laminated timber deck panels. Mag Civ Eng. https://doi.org/10.18720/MCE.83.12
10. Labudin B, Popov E, Stolypin D, Sopilov V (2019) The wood composite ribbed panels on mechanical joints. In: E3S web of conferences. https://doi.org/10.1051/e3sconf/20199102021
11. De Luca V, Marano C (2012) Prestressed glulam timbers reinforced with steel bars. Constr Build Mater. https://doi.org/10.1016/j.conbuildmat.2011.11.016
12. Franke S, Franke B, Harte AM (2015) Failure modes and reinforcement techniques for timber beams-state of the art. Constr Build Mater. https://doi.org/10.1016/j.conbuildmat.2015.06.021
13. McConnell E, McPolin D, Taylor S (2014) Post-tensioning of glulam timber with steel tendons. Constr Build Mater. https://doi.org/10.1016/j.conbuildmat.2014.09.079
14. Lisyatnikov MS, Glebova TO, Ageev SP, Ivaniuk AM (2020) Strength of wood reinforced with a polymer composite for crumpling across the fibers. IOP Conf Series Mater Sci Eng. https://doi.org/10.1088/1757-899X/896/1/012062
15. Sun X, He M, Li Z (2020) Novel engineered wood and bamboo composites for structural applications: state-of-art of manufacturing technology and mechanical performance evaluation. https://doi.org/10.1016/j.conbuildmat.2020.118751
16. Ilyichev V, Emelianov S, Kolchunov V, Bakaeva N (2018) Principles of urban areas reconstruction ensuring safety and comfortable living conditions. IOP Conf Series Mater Sci Eng. https://doi.org/10.1088/1757-899X/463/3/032011
17. Gentile C, Svecova D, Rizkalla SH (2002) Timber beams strengthened with GFRP bars: development and applications. https://doi.org/10.1061/(ASCE)1090-0268(2002)6:1(11)
18. Wanninger F, Frangi A (2016) Experimental and analytical analysis of a post-tensioned timber frame under horizontal loads. Eng Struct. https://doi.org/10.1016/j.engstruct.2016.01.029
19. Frese M, Blaß HJ (2011) Statistics of damages to timber structures in Germany. Eng Struct. https://doi.org/10.1016/j.engstruct.2011.02.030
20. Kreher K, Natterer J, Natterer J (2004) Timber-glass-composite girders for a hotel in Switzerland. Struct Eng Int J Int Assoc Bridg Struct Eng. https://doi.org/10.2749/101686604777963964
21. Lukina A, Roshchina S, Gribanov A (2021) Presented at the method for restoring destructed wooden structures with polymer composites. https://doi.org/10.1007/978-3-030-72404-7_45
22. Lukin M, Prusov E, Roshchina S, Karelina M, Vatin N (2021) Multi-span composite timber beams with rational steel reinforcements. Buildings. https://doi.org/10.3390/buildings11020046
23. Lukin M, Sergeev M, Lisyatnikov M (2021) Presented at the non split wooden beam reinforced with composite reinforcement. https://doi.org/10.1007/978-3-030-72404-7_12

Compressed Reinforced Concrete Elements Bearing Capacity of Various Flexibility

A. L. Krishan, V. I. Rimshin, I. L. Shubin, M. A. Astafeva, and A. A. Stupak

Abstract The article considers a calculation method of compress reinforced concrete elements of various flexibility. The calculation method is based on theoretical principles of solid mechanics and on the basic laws of force resistance of compressed reinforced concrete elements. It presents the disadvantages of limited practical application of this calculation method within current design standard. The study describes a new method on the pattern of fasten hinged rod. This rod has a constant cross-section along the length of the element loaded with an eccentrically applied compressive force. The calculation of bending moments and various stiffnesses the compressed rod of normal sections along the length of impact on the curvature rod. This allows an accurate determination of flexible elements deflection, and their bearing capacity. The calculations results clearly show the process of bearing capacity of rod of strength of the normal section loss, and the process loss of stability of the second kind. Comparison of the calculation results with the published experimental data have confirmed better reliability of deformation calculation according to the current design standard.

Keywords Compressed reinforced concrete · Flexibility · Load capacity · Method of deformation calculation

A. L. Krishan (✉) · M. A. Astafeva (✉)
Department of Building Design and Constructions, Nosov Magnitogorsk State Technical University, Uritsky, 11, Magnitogorsk 455000, Russia

V. I. Rimshin · I. L. Shubin
Research Institute of Building Physics of Russian Academy of Architecture and Construction Sciences, 21 Lokomotivny pr., Moscow 127238, Russia

A. A. Stupak
Department of Building Design and Constructions, Nosov Magnitogorsk State Technical University, 38, Lenin Ave., Magnitogorsk 455000, Russia

N. Vatin et al. (eds.), *Proceedings of MPCPE 2021*, Lecture Notes in Civil Engineering 182, https://doi.org/10.1007/978-3-030-85236-8_26

1 Relevance

Specialists in the field of design of reinforced concrete structures are well aware that the calculation of the load capacity of compressed elements according to the non-deformed scheme proposed by the current norms does not always give reliable results [1–14]. For example, when calculating according to SP 63.13330.2018, a practical result can be obtained only for elements that have flexibility in the range of $40 \div 60$. The main reason for this circumstance is the empirical approach to determine the average hardness of the off-center compressed element in determining Euler buckling load. As a result, for elements with flexibility $\lambda < 40$, the calculation gives an underestimated load-bearing capacity. The design load-bearing capacity of compressed structures of $\gamma > 60$ is noticeably overstated, which can contribute to the occurrence of emergency situations.

The purpose of this work is offering a method of deformation calculation of load-bearing capacity of compressed reinforced concrete elements with reference to their flexibility. Previously, a similar technique was developed for calculating tubular concrete columns and reinforced concrete columns with spiral confinement reinforcement [15–18].

2 Statement of the Problem

The basic case is considered here—the force resistance of pin-supported rod at the ends of constant cross section along the length, loaded with compressive force N. Force N is applied to both ends with the same initial eccentricity e_0. The deformation scheme of such rod is shown in Fig. 1, a. The geometric and structural parameters of the compressed element are considered to be known. The section of the reinforced concrete element is considered as to be reduced to concrete, the flexibility of which can be calculated by the formula

$$\lambda_{\text{eff}} = l_0 \cdot \sqrt{\frac{(EA)_{\text{eff}}}{(EI)_{\text{eff}}}}, \tag{1}$$

where l_0 is an estimated length of the rod; $(EI)_{\text{eff}}$, $(EA)_{\text{eff}}$ are effective hardness of the reduced section during bending and compression.

The load-bearing capacity of the given rod may be due to loss of strength of a normal section or loss of stability of a second type. The loss of stability occurs when the deflection of the rod increases while the load decreases. However, it is accepted that the rod in the process of deformation has a uniform curvilinear form of equilibrium, although its parameters change with a change in load.

The calculation is performed with respect to physical and geometric nonlinearities by the method of step-by-step approximations. The strain calculation algorithm can be enlarged as follows. Initially, the normal section of the rod in the middle of its length is considered (section 3 in Fig. 1). Section is preliminary

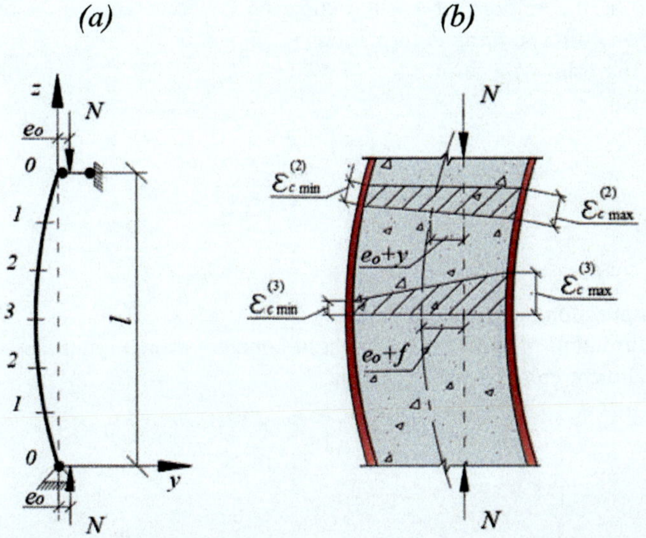

Fig. 1 Deformed state of compressed rod: **a** scheme of loading and deformation of rod divided along the length into 6 sections; **b** strain diagram of relative deformation of concrete in sections 2 and 3

divided into small sections with areas of concrete core A_{cn} and longitudinal reinforcement A_{sk}. The longitudinal deformation of the most compressed fiber of the concrete core $\varepsilon_{c,max}$ is increased step-by-step starting from zero. Strain diagram is taken in accordance with Bernoulli hypothesis. In this case, the relative deformation of the least compressed (stretched) fiber $\varepsilon_{c,min}$, the corresponding deformation $\varepsilon_{c,max}$ should ensure that the equilibrium conditions of the normal section of the calculated element are met:

$$N = \sum_{n} \sigma_{cn} A_{cn} + \sum_{k} \sigma_{sk} A_{sk}; \tag{2}$$

$$N(e_o + f) = \sum_{n} \sigma_{cn} A_{cn} Y_{cn} + \sum_{k} \sigma_{sk} A_{sk} Y_{sk}, \tag{3}$$

where

Y_{cn}, σ_{cn} is the coordinate of the gravity center of n section of the concrete core and the stress of the axial direction at the level of its center of gravity, respectively;

Y_{sk},σ_{sk} is the coordinate of the gravity center of k longitudinal reinforcement and the stress in it;f is a deflection of the element.

Stresses in the middle of each section of concrete core σ_{cn} are calculated depending on corresponding deformations ε_{cn}. For this, the Sargin formula is used, which has the following form:

$$\sigma_c = \frac{k\left(\frac{\varepsilon_c}{\varepsilon_{c1}}\right) - \left(\frac{\varepsilon_c}{\varepsilon_{c1}}\right)^2}{1 + (k-2)\left(\frac{\varepsilon_c}{\varepsilon_{c1}}\right)} \cdot R_c, \qquad (4)$$

where

R_c is compression strength of concrete;
ε_{c1} is deformation of uniaxial compressed concrete at maximum stress;
k is coefficient calculated by formula.

$$k = 1{,}05E_c \frac{\varepsilon_{c1}}{R_c}, \qquad (5)$$

where

E_c is modulus of concrete elasticity.

Stresses in longitudinal reinforcement are found depending on deformations in its gravity center. The relationship between stresses and strains in the reinforcement is recommended as Prandtl diagram.

Curvature of cross-section in middle part of rod $\frac{1}{r_3}$ at $z = l_0/2$ is determined for obtained strain diagram. By analogy with operation [19], curvature of the rod is geometrically connected with inclination of strain diagram in its normal section and is determined by formula

$$\frac{1}{r_3} = \frac{\varepsilon_{c,max}}{x}, \qquad (6)$$

where

x is the height of the compressed section area.

In the first approximation, the deflection value is determined. To do this, you can use the formula obtained under the assumption of bending the rod along the sinusoid.

$$f_1 = \frac{l_0^2}{\pi^2} \frac{\varepsilon_{c,max}}{x}. \qquad (7)$$

Deflection f_1 is overstated due to ignoring the greater hardness of sections located closer to the rod supports. In these sections, the curvature is less.

To clarify deflection, it is proposed to use numerical solution of bending equation of compressed-bent rod [20]. The rod is divided in length into even number of equal sections (see Fig. 1a). Sufficient accuracy for practice is provided by the number of sections $m \geq 6$. In normal sections, curvatures are calculated at the ends of each section $\left(\frac{1}{r_0}\right)$, $\left(\frac{1}{r_1}\right)$, ..., $\left(\frac{1}{r_{m/2}}\right)$. Value of horizontal deviation of rod from y vertical is determined in advance for each of considered sections.

All curves are calculated by Formula (6) after constructing for each section strain diagrams that meets the equilibrium conditions. These conditions are written in the form of Formulae (2) and (3), but with the deflection replaced by the corresponding horizontal y offset. At this stage of calculation, it is necessary to ensure that the longitudinal force N in all normal sections of the rod has the same value. Bending moments, and accordingly curves of sections located closer to the rod supports, decrease.

For the compressed rod under consideration, at the number of partitions $m = 6$ the following formula for deflection calculation is obtained:

$$f = \frac{l_0^2}{266}\left(\frac{1}{r_0} + 6\frac{1}{r_1} + 12\frac{1}{r_2} + 8\frac{1}{r_3}\right). \tag{8}$$

The deflection updating procedure is repeated until the specified calculation accuracy is achieved. With specified deflection, the compressive force N is determined. After that, the longitudinal deformation of the most compressed concrete core fiber $\varepsilon_{c,max}$ is increased again. All calculations listed are repeated until the compressive force N begins to decrease. Maximum value N_u corresponds to the load-bearing capacity of the calculated rod in case of loss of strength or stability of the second type.

3 Comparison of Theory with Experiment

Table 1 shows the results of comparison of experimental data of load-bearing capacity N_u^{exp}.

Experimental data were taken based on the results of published researches of [21] columns of rectangular section with dimensions $b \times h$. The following designations are used here: $A_s = A_s'$ is a cross-sectional area of longitudinal reinforcement located at the least and most compressed faces; $a_s = a_s'$ are distances from the nearest faces to the gravity centers of the corresponding longitudinal reinforcement; $\sigma_y = \sigma_{yc}$ are yield strengths of longitudinal reinforcement during tension and compression. The remaining designations are given in the text. The last column of the table shows the results of comparison of the data of experiments and calculations according to the non-deformed scheme proposed by the current norms of the Russian Federation.

Table 1 also illustrates data that the use of a deformation model in calculations allows you to calculate the load-bearing capacity of compressed elements more

Table 1 Comparison of the experimental data of load-bearing capacity with calculation results to results of calculation N_{u1}^{th}

Sample	l_0, (m)	b × h , (mm)	R_c, (MPa)	Longitudinal reinforcement			e_o, (mm)	N_u^{exp}, (kN)	$\frac{N_u^{exp}}{N_{u1}^{th}}$	$\frac{N_u^{exp}}{N_{u2}^{th}}$
				$d_s = d_{sc}$, (mm)	σ_y, (MPa)	σ_{yc}, (MPa)				
КГ-4-1	4	253 × 119	17,8	9,9	240	240	0	440	0,91	1,06
КГ-4-2	4	253 × 120	18,3	9,8	258	258	0	486	0,83	1,13
КГ-5-1	4,2	240 × 123	15,7	9,4	325	325	0	401	0,92	1,12
КГ-5-2	4,2	240 × 120	13,22	9,6	335	335	0	300	0,82	1,01
КГ-5-5	3,75	240 × 120	18	9,6	330	330	0	440	0,80	1,03
КГ-8-1	4,6	241 × 150	22,0	7,65	354	365	18	320	0,85	1,07
КГ-8-2	4,6	241 × 150	22,0	7,65	372	364	21	290	0,85	1,12
КГ-8-3	4,6	240 × 150	22	7,7	365	365	2,4	290	0,8	1,09
КГ-8-4	4,6	240 × 150	22	18	270	270	25	410	0,84	0,98
КГ-8-5	4,6	242 × 152	22	18	285	285	25	410	0,79	0,94
КГ-8-6	4,6	241 × 150	22	18	290	290	25	430	0,87	1,01
КГ-9-2	3,75	240 × 118	20	9,6	335	335	0	460	0,81	1,03
КГ-9-3	3,75	240 × 120	22,5	9,6	330	330	0	440	0,7	0,89
КГ-10-1	4,65	240 × 150	12,4	19,0	289	271	75	190	0,93	0,97
КГ-11-6	4,65	240 × 150	12,4	18,9	278	278	115	130	1,05	1,04
КГ-12-2	4,6	241 × 150	12,6	11,8	374	374	25	200	0,90	0,92
КГ-12-3	4,6	242 × 150	12,6	11,8	376	376	25	221	0,99	1,01
КГ-14-1	4,55	240 × 150	20,0	12,0	375	375	0	640	0,78	1,08
КГ-14-2	4,55	240 × 150	20,5	12,0	372	372	0	620	0,75	1,03
КГ-15-1	4,68	237 × 150	31	14,1	325	325	22	440	0,90	1,02
КГ-15-3	4,65	240 × 149	31	14,1	327	327	25	386	0,92	0,95
КГ-17-1	4,6	242 × 150	31,5	18,7	270	270	25	444	0,79	0,94
КГ-17-2	4,6	240 × 150	31,5	18,8	245	245	27	434	0,80	0,94
КГ-17-5	4,6	242 × 150	31,5	7,9	350	350	28	280	0,73	0,94

(continued)

Table 1 (continued)

Sample	l_0, (m)	b × h , (mm)	R_c, (MPa)	Longitudinal reinforcement			e_o, (mm)	N_u^{exp}, (kN)	$\frac{N_u^{exp}}{N_{u1}^{th}}$	$\frac{N_u^{exp}}{N_{u2}^{th}}$
				$d_s = d_{sc}$, (mm)	σ_y, (MPa)	σ_{yc}, (MPa)				
КГ-20-1	3,1	243 × 151	40	7,9	465	465	25	650	0,66	1,01
КГ-20-4	3,15	246 × 151	40	7,9	495	495	75	170	1,04	0,92
КГ-20-6	3,1	247 × 151	40	8,0	489	489	0	1303	0,82	1,01
КГ-21-3	2,35	240 × 153	31,5	7,9	387	387	0	1204	0,97	1,02
КГ-21-4	2,35	243 × 152	31,5	7,9	401	401	0	1103	0,88	0,94
КГ-21-6	2,35	242 × 152	31,5	7,8	385	385	25	680	1,16	1,05
КГ-22-1	4,58	240 × 152	34	14	419	419	10	605	0,87	0,90
КГ-22-2	4,58	240 × 153	34	14,1	419	419	11	605	0,86	0,86
КГ-22-3	4,58	244 × 150	34	13,9	435	435	0	899	0,89	0,96
КГ-22-4	4,58	246 × 150	34	14,0	411	411	0	862	0,85	0,91
КГ-22-5	2,35	241 × 151	31	8,0	386	386	25	654	1,07	1,04
КГ-23-1	4,58	243 × 153	36,5	14,0	426	426	22	432	0,78	0,83
КГ-23-2	4,58	243 × 151	36,5	13,9	420	420	25	391	0,75	0,81
КГ-23-3	4,58	250 × 152	36,5	13,9	419	419	23	416	0,74	0,82
КГ-24-1	4,60	242 × 152	26,3	14,2	366	366	0	757	0,78	0,93
КГ-24-2	4,60	251 × 155	26,3	14,1	338	338	0	800	0,76	0,93
КГ-25-1	4,6	244 × 152	35	16	326	326	26	420	0,76	0,95
КГ-27-1	4,58	243 × 152	24	14,0	431	431	0	795	0,84	1,06
КГ-27-2	4,58	246 × 150	24	13,9	428	428	0	824	0,90	1,11
КГ-28-4	4,58	244 × 102	32	9,4	421	421	0	280	0,95	0,78
КГ-28-5	4,58	244 × 100	32	9,4	423	423	16	133	0,65	0,96
КГ-28-6	4,58	243 × 100	32	9,5	420	420	0	300	1,08	0,89
Mean									0,86	0,98
Standard deviate									0,11	0,08

accurately. Using deformation calculation, the following statistical analysis indicators were obtained: maximum and minimum deviations from experiments +13/−22%, average ratio $N_u^{exp}/N_u^{th} = 0.98$, standard deviation –8%. The results of experimental data comparison with calculations according to the methodology of existing norms have significantly worse statistical indicators: maximum and minimum deviations from experiments +16/−35%, average ratio $N_u^{exp}/N_u^{th} = 0.86$, standard deviation –11%.

While calculating compressed rods, it becomes possible to ensure that the load-bearing capacity for the strength of a normal section, with respect to account flexibility, is fully realized. Otherwise, it will not be possible to use the strength properties of concrete in full, and possibly steel as well, which is not economically feasible.

4 Conclusions

Proposed method of deformation calculation makes it possible to increase accuracy of load-bearing capacity determination of compressed reinforced concrete elements significantly. When calculating the deflections of elements, not only bending moments are considered as the influence on their curvature, but also various hardness along the rod length. This corresponds to known features of the force resistance of the compressed rod. Based on the results of calculations, it is possible to analyze the stress–strain state of concrete and reinforcement in full. Using this method, it is not necessary to obtain a formula for calculating the critical force. There is no need to use empirical dependencies, which significantly extends the scope of this technique.

References

1. Bhure N (2018) Steel concrete composite construction—a review. Int J Res Appl Sci Eng Technol 6:564–566. https://doi.org/10.22214/ijraset.2018.11089
2. Du G, Andjelic A, Li Z, Lei Z, Bie X (2018) Residual axial bearing capacity of concrete-filled circular steel tubular columns (CFCSTCs) after transverse impact. Appl Sci 8. https://doi.org/10.3390/app8050793
3. Chuang PH, Kong FK (1997) Large-scale tests on slender, reinforced concrete columns. Struct Eng London 75:410–416
4. de Oliveira WLA, De Nardin S, de Cresce, El Debs ALH, El Debs MK (2009) Influence of concrete strength and length/diameter on the axial capacity of CFT columns. J Constr Steel Res 65:2103–2110. https://doi.org/10.1016/j.jcsr.2009.07.004
5. Han LH (2000) Tests on concrete filled steel tubular columns with high slenderness ratio. Adv Struct Eng 3:337–344. https://doi.org/10.1260/1369433001502265
6. Han LH, An YF (2014) Performance of concrete-encased CFST stub columns under axial compression. J Constr Steel Res 93:62–76. https://doi.org/10.1016/j.jcsr.2013.10.019

7. Lu ZH, Zhao YG (2010) Suggested empirical models for the axial capacity of circular CFT stub columns. J Constr Steel Res 66:850–862. https://doi.org/10.1016/j.jcsr.2009.12.014
8. Uy B, Tao Z, Han LH (2011) Behaviour of short and slender concrete-filled stainless steel tubular columns. J Constr Steel Res 67:360–378. https://doi.org/10.1016/j.jcsr.2010.10.004
9. Uy B (1988) Ductility, strength and stability of concrete-filled fabricated steel box columns for tall buildings|Request PDF. Struct Des Tall Build 7:113–133. https://doi.org/10.1002/(SICI)1099-1794(199806)7:2%3c113::AID-TAL94%3e3.0.CO;2-I
10. Baig MN, Fan J, Nie J (2006) Strength of concrete filled steel tubular columns. Tsinghua Sci Technol 11:657–666. https://doi.org/10.1016/S1007-0214(06)70248-6
11. Claeson C (1998) Structural behavior of reinforced high-strength concrete columns. Goteborg
12. Chang WF, Ferguson PM (1963) Long-hinged reinforced concrete columns. ACI J Proc 60:1–25. https://doi.org/10.14359/7839
13. Du Y, Chen Z, Richard Liew JY, Xiong MX (2017) Rectangular concrete-filled steel tubular beam-columns using high-strength steel: experiments and design. J Constr Steel Res 131:1–18. https://doi.org/10.1016/j.jcsr.2016.12.016
14. Solomin V, Khomiyk V (2013) Ьltress state and strength of reinforced concrete column. Struct Mech Anal Constr 2:11–17
15. Krishan AL, Rimshin VI, Astaf'eva MA, Troshkina EA (2019) Strength and deformability of concrete of compressed and bending reinforced concrete elements with shear reinforcement. Stroit Mater 771:8–11. https://doi.org/10.31659/0585-430x-2019-771-6-8-11
16. Krishan AL, Troshkina EA (2014) Estimation of carrying capacity of eccentrically compressed concrete filled steel tube columns. Adv Environ Biol 8:1974–1977
17. Krishan AL, Rimshin VI, Troshkina EA (2020) Compressed and bending concrete elements with confinement reinforcement meshes. IOP Conf Series Mater Sci Eng. https://doi.org/10.1088/1757-899X/753/2/022052
18. Krishan AL, Rimshin VI, Astafeva MA (2018) Deformability of a volume-compressed concrete. IOP Conf Series Mater Sci Eng. https://doi.org/10.1088/1757-899X/463/2/022063
19. Krishan AL, Rimshin VI, Astafeva MA (2020) Compressed concrete filled steel tube elements: theory and practice. ASV, Moscow
20. Golyshev AB, Bachinsky VYa, Polishchuk VP, KAV, RI (1990) Design of reinforced concrete structures. Reference book. Budivelnik, Kiev
21. Chistyakov EA (1988) Fundamentals of the theory, calculation methods and experimental studies of the bearing capacity of compressed reinforced concrete elements under static loading. Moscow

Spatial Stability Problems of Elastic Beams with Potential Loading: Euler's Problem and Nikolai Paradox

V. V. Lalin, V. S. Nenashev, I. G. Utimisheva, S. F. Diakov, M. V. Sukhoterin, R. B. Orlovich, and E. K. Akimov

Abstract In this paper the variational approach to the stability problems of the elastic bars is reviewed. Commonly problems of that kind are solved by the application of the approximate bar models. In these models some deformations are neglected (i.e. Timoshenko beam or Bernoulli–Euler beam). Another option is to use approximate methods, which neglect the pre-buckling state of the bar (linear formulation). Contrary to common methods mentioned above a geometrically exact beam theory is considered in which the bars made of the elastic material are used. Two spatial stability problems are solved in this paper: the equilibrium stability problem of the cantilever beam axially compressed with the non-follower load and the problem of the lateral-torsional buckling of the cantilever beam under the follower load (torsional moment). Both loads are conservative forces. The precise equations were found to calculate the critical load and the comparison of the obtained critical loads was made for both problems.

Keywords Stability · Lateral buckling · Geometrically exact beam theory · Euler's problem · Nikolai paradox

1 Introduction

In the classical variational approach, there is a method, that makes possible to obtain an exact stability equation like Euler's equations for the second variation of the Lagrange functional of the original static problem [1, 2]. For this purpose, the static problem should be stated as a nonlinear one.

V. V. Lalin · V. S. Nenashev · I. G. Utimisheva (✉) · S. F. Diakov
Peter the Great St. Petersburg Polytechnic University, St. Petersburg, Russia

M. V. Sukhoterin · E. K. Akimov
Admiral Makarov State University of Maritime and Inland Shipping, St. Petersburg, Russia

R. B. Orlovich
West Pomeranian University of Technology, Szczecin, Poland

N. Vatin et al. (eds.), *Proceedings of MPCPE 2021*, Lecture Notes in Civil Engineering 182, https://doi.org/10.1007/978-3-030-85236-8_27

Traditional method to analyze the stability of the elastic structures is to use a variational equation, known as principle of virtual displacements, instead of Lagrange functional and its variations. In that case an approximate method is usually used (i.e. linearized or highly linearized problem [3]) when the solution of the linear static problem [4–10] is used in the stability equation instead of the exact nonlinear solution. As a result, only approximate solutions can be obtained for the linearized stability problems. But without the exact solutions it is hard to evaluate the inaccuracy of the linearized solutions.

As an example, two models of the stability of axially compressed beam regarding the shear stiffness can be mentioned: Engesser's [11] and Harinx's [12]. From the 80s [13, 14] till now [15–18] it is hard to answer which one is more accurate.

In the theory of nonlinear beams there is Nikolai paradox [19] which is known from the 1927. The point of the paradox is that the cantilever beam loaded with the torsional moment can buckle irrespective of the value of the moment. This paradox pushed the development of the stability theory for the structures under the non-potential (nonconservative) load [3, 20, 21]. The Nikolai paradox does not have a common solution so far [21–24].

Nonlinear geometrically exact beam theory with the variational approach and analytical solutions of the nonlinear statical problems are used to analyze the stability of the cantilever beam under the axial compression and under the torque.

2 Methods

In the current paper the geometrically exact nonlinear theory of elastic beams is used [25–31]. According to this theory every point of the beam has six degrees of freedom—three linear and three angular displacements. Angular displacements are independent variables. No conditions are imposed on the displacements, rotation angles and deformations. Special cases of this theory are the well-known Euler–Bernoulli and Timoshenko beam theories.

The material is considered to be elastic according to the Hooke's law.

Geometrically nonlinear static equations according to the geometrically exact theory are:

$$\begin{cases} (\boldsymbol{P}\cdot\boldsymbol{Q})' = 0 \\ (\boldsymbol{P}\cdot\boldsymbol{M})' + \boldsymbol{r}'\times(\boldsymbol{P}\cdot\boldsymbol{Q}) = 0 \end{cases} \quad \begin{cases} \boldsymbol{M} = \boldsymbol{C}\cdot\boldsymbol{Z}^T\cdot\boldsymbol{\varphi}' \\ \boldsymbol{Q} = \boldsymbol{A}\cdot\left(\boldsymbol{P}^T\cdot\boldsymbol{r}' - \boldsymbol{r}_0'\right) \end{cases} \tag{1}$$

where $\boldsymbol{P}$—turn tensor; $\boldsymbol{Q} = N\boldsymbol{e}_1 + Q_y\boldsymbol{e}_2 + Q_z\boldsymbol{e}_3$—force vector; N—axial force; Q_y and Q_z—shear forces respectively along y and z axes; $\boldsymbol{M} = M_t\boldsymbol{e}_1 + M_y\boldsymbol{e}_2 + M_z\boldsymbol{e}_3$—

vector of moments; M_t—torque; M_y and M_z—bending moments respectively around y and z axes; $\boldsymbol{r}$—radius-vector of the beam in actual configuration; $\boldsymbol{Z}$—Zhilin's tensor; $\boldsymbol{A} = EA\boldsymbol{e}_1\boldsymbol{e}_1 + k_yGA\boldsymbol{e}_2\boldsymbol{e}_2 + k_zGA\boldsymbol{e}_3\boldsymbol{e}_3$; $\boldsymbol{C} = GI\boldsymbol{e}_1\boldsymbol{e}_1 + EI_y\boldsymbol{e}_2\boldsymbol{e}_2 + EI_z\boldsymbol{e}_3\boldsymbol{e}_3$—elasticity tensors according to the Hooke's law respectively for forces and moments; $\boldsymbol{\varphi}$—turn vector; Φ—absolute value of a turn vector; $\boldsymbol{r_0}$—radius-vector of the beam in reference configuration; $(\ldots)' = \frac{d}{ds}(\ldots)$, s—arc length of the beam in the reference configuration; vectors $\boldsymbol{e}_1, \boldsymbol{e}_2, \boldsymbol{e}_3$—unit base vectors of the x, y, z system of axes.

Tensors $\boldsymbol{P}$ и $\boldsymbol{Z}$ can be written as follows [26–29]:

$$\boldsymbol{P} = \cos\Phi\boldsymbol{I} + \frac{\sin\Phi}{\Phi}\boldsymbol{I}\times\boldsymbol{\varphi} + \frac{1-\cos\Phi}{\Phi^2}\boldsymbol{\varphi}\boldsymbol{\varphi};$$
$$\boldsymbol{Z} = \frac{\sin\Phi}{\Phi}\boldsymbol{I} + \frac{1-\cos\Phi}{\Phi^2}\boldsymbol{I}\times\boldsymbol{\varphi} + \frac{\Phi-\sin\Phi}{\Phi^3}\boldsymbol{\varphi}\boldsymbol{\varphi}.$$

where $\boldsymbol{I}$—is the unit tensor.

Henceforward the following labels for stiffnesses are used: $k_1 = EA$—axial stiffness;$k_{2y} = k_yGA$—shear stiffness with respect to the y axis;$k_{2z} = k_zGA$—shear stiffness with respect to the z axis;$k_t = GI$—torsional stiffness; $k_{3y} = EI_y$—bending stiffness with respect to the y axis;$k_{3z} = EI_z$—bending stiffness with respect to the z axis.

Essential and natural boundary conditions for the static problem of the cantilever beam under the force and the moment applied at the end-point are:

$$\begin{aligned} &\underline{s=0:} \quad \boldsymbol{r}=0; \boldsymbol{\varphi}=0 \\ &\underline{s=L:} \quad \boldsymbol{P}\cdot\boldsymbol{Q}=\boldsymbol{f}; \boldsymbol{P}\cdot\boldsymbol{M}=\boldsymbol{m} \end{aligned} \tag{2}$$

where $\boldsymbol{f} = \frac{\partial U_1}{\partial \boldsymbol{r}}$; $\boldsymbol{m} = \frac{\partial U_2}{\partial \boldsymbol{\varphi}}\cdot\boldsymbol{Z}^{-1}$—are vectors of the potential loading; $U_1(\boldsymbol{r}), U_2(\boldsymbol{\varphi})$—are the corresponding potential functions, L—length of the beam (Fig. 1).

The exact stability equations for the elastic beams were found with the help of the variational approach:

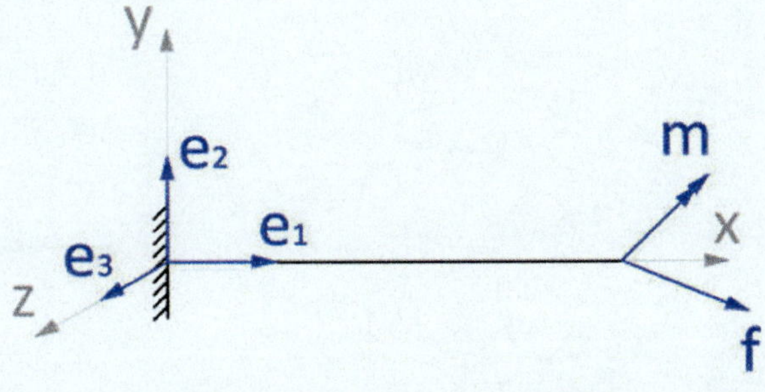

Fig. 1 Cantilever beam under the force and moment applied at the end-point

$$
\begin{aligned}
&\left[\boldsymbol{P}\cdot\boldsymbol{A}\cdot\boldsymbol{P}^T\cdot(\boldsymbol{u}'+\boldsymbol{r}'\times(\boldsymbol{Z}\cdot\boldsymbol{\theta}))-(\boldsymbol{P}\cdot\boldsymbol{Q})\times(\boldsymbol{Z}\cdot\boldsymbol{\theta})\right]'=0\\
&\left[\boldsymbol{P}\cdot\boldsymbol{C}\cdot\boldsymbol{P}^T\cdot(\boldsymbol{Z}\cdot\boldsymbol{\theta})'-(\boldsymbol{P}\cdot\boldsymbol{M})\times(\boldsymbol{Z}\cdot\boldsymbol{\theta})\right]'+\boldsymbol{u}'\times(\boldsymbol{P}\cdot\boldsymbol{Q})+\\
&+\boldsymbol{r}'\times\left[\boldsymbol{P}\cdot\boldsymbol{A}\cdot\boldsymbol{P}^T\cdot(\boldsymbol{u}'+\boldsymbol{r}'\times(\boldsymbol{Z}\cdot\boldsymbol{\theta}))-(\boldsymbol{P}\cdot\boldsymbol{Q})\times(\boldsymbol{Z}\cdot\boldsymbol{\theta})\right]=0,
\end{aligned}
\tag{3}
$$

where u и θ are unknown functions of the stability problem (Fig. 2).

$$
\begin{aligned}
\boldsymbol{u}&=u\boldsymbol{e}_1+v\boldsymbol{e}_2+w\boldsymbol{e}_3\\
\boldsymbol{\theta}&=\theta_1\boldsymbol{e}_1+\theta_2\boldsymbol{e}_2+\theta_3\boldsymbol{e}_3
\end{aligned}
$$

Essential and natural boundary conditions of the stability problem of the cantilever beam under the force and the moment applied at the end-point are:

$$
\underline{s=0:}\quad
\begin{aligned}
&\boldsymbol{u}=0;\boldsymbol{\theta}=0\\
&\left[\boldsymbol{P}\cdot\boldsymbol{A}\cdot\boldsymbol{P}^T\cdot(\boldsymbol{u}'+\boldsymbol{r}'\times(\boldsymbol{Z}\cdot\boldsymbol{\theta}))-(\boldsymbol{P}\cdot\boldsymbol{Q})\times(\boldsymbol{Z}\cdot\boldsymbol{\theta})\right]'=\tfrac{\partial^2U_1}{\partial r^2}\cdot\boldsymbol{u}
\end{aligned}
\tag{4}
$$

$$
\underline{s=L:}\quad\left[\boldsymbol{P}\cdot\boldsymbol{C}\cdot\boldsymbol{P}^T\cdot(\boldsymbol{Z}\cdot\boldsymbol{\theta})'-(\boldsymbol{P}\cdot\boldsymbol{M})\times(\boldsymbol{Z}\cdot\boldsymbol{\theta})\right]\cdot\boldsymbol{Z}=\tfrac{\partial^2U_2}{\partial\varphi^2}\cdot\boldsymbol{\theta}
\tag{5}
$$

Equation (3) can be used for any stability problems of beams as they are presented in general view. The exact solution for the stability problem can be derived if the exact solution of the nonlinear static problem (1), (2) is put in the system (3). The condition of existence of a non-zero solution of the stability problem (3)–(5) gives the exact value for the critical load.

3 Results and Discussion

The exact solution of a spatial stability problem for a cantilever beam under an axial compression.

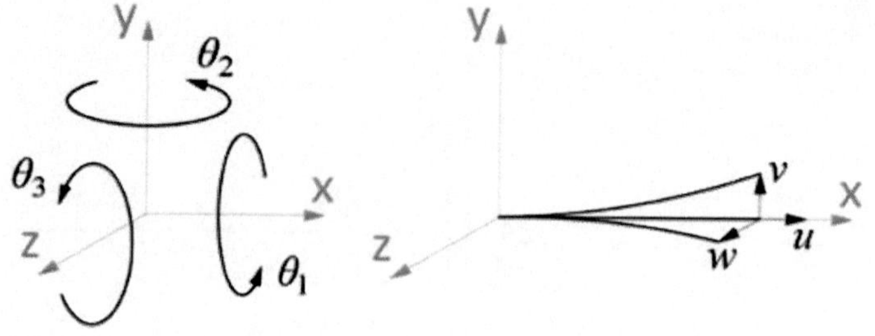

Fig. 2 The unknown functions of the stability problem

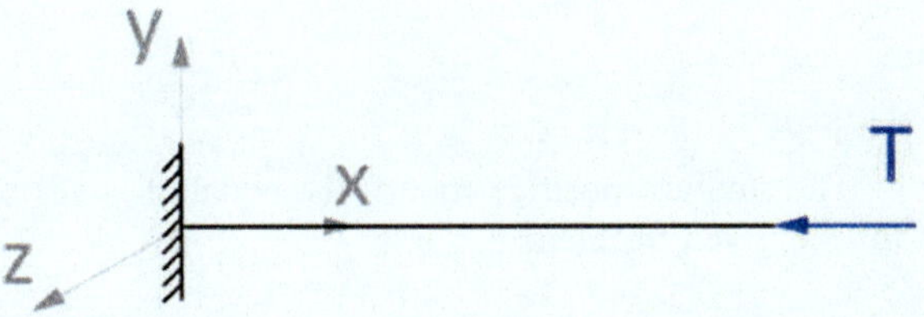

Fig. 3 Cantilever beam under an axial compression

The solution of the spatial stability problem for a cantilever beam under an axial force T was found in this chapter (Fig. 3).

In this problem a non-follow force T is considered. Its potential function is: $U_1(\boldsymbol{r}) = -T\boldsymbol{e}_1 \cdot \boldsymbol{r}$. The other potential functions are: $U_2 = 0, \frac{\partial U_1}{\partial \boldsymbol{r}} = -T\boldsymbol{e}_1, \frac{\partial^2 U_1}{\partial \boldsymbol{r}^2} = 0$.

The solution of a nonlinear static problem (1)–(2) as follows:

$$\begin{aligned} \boldsymbol{r} &= \left(1 - \frac{T}{k_1}\right) s\boldsymbol{e}_1, \quad \boldsymbol{\varphi} = \boldsymbol{0}, \\ \boldsymbol{P} &= \boldsymbol{Z} = \boldsymbol{I}, \quad \boldsymbol{Q} = -T\boldsymbol{e}_1, \quad \boldsymbol{M} = \boldsymbol{0} \end{aligned} \tag{6}$$

If the solution of the nonlinear static problem (6) is put into the equation (3) and regarding the boundary conditions it can be seen that the problem separates into two unrelated equations:

(1) system of equations involving v, θ_3—the unknowns of the stability problem, which are the displacements in plane XOY;
(2) system of equations involving w, θ_2—the unknowns of the stability problem, which are the displacements in plane XOZ.

The solution for both systems of equations can be obtained using the similar method and they both can have a nontrivial solution.

The first system of equations which involves the displacements in plane XOY is:

$$\begin{cases} v' = \mu_3 \theta_3 \\ \theta_3'' + \frac{T}{k_{3z}} v' = 0 \end{cases}$$

This system of equations can be transformed into one differential equation of second order in θ_3:

$$\theta_3'' + \frac{T\mu_3}{k_{3z}} \theta_3 = 0$$

Putting the solution of this equation into the boundary conditions, it can be found out, that the existence of the non-zero solution of the stability problem is the following equation:

$$\cos\sqrt{\frac{T\mu_3}{k_{3z}}}L = 0$$

The smallest positive root of the equation gives us the value of the critical load for the given problem:

$$T_{crz} = \frac{\sqrt{1+\frac{\pi^2 k_{3z}}{L^2}\left(\frac{1}{k_{2y}}-\frac{1}{k_1}\right)}-1}{2\left(\frac{1}{k_{2y}}-\frac{1}{k_1}\right)} \tag{7}$$

The second system of equations can be solved in a similar manner and the second value for the critical load is:

$$T_{cry} = \frac{\sqrt{1+\frac{\pi^2 k_{3y}}{L^2}\left(\frac{1}{k_{2z}}-\frac{1}{k_1}\right)}-1}{2\left(\frac{1}{k_{2z}}-\frac{1}{k_1}\right)} \tag{8}$$

The real value for the critical load is the smallest of two values T_{cry} or T_{crz}.

The exact solution of a spatial stability problem for a cantilever beam under a torque.

The solution of the spatial stability problem for a cantilever beam under a torque M was found in this chapter (Fig. 4).

In this problem an external torque proposed by P.A. Zhilin is considered [29]. The moment is equal to $\boldsymbol{m} = \frac{M}{\Phi}\boldsymbol{\varphi}$, M—absolute value of the vector of the external torque. This moment is a potential follower moment with a potential function $U_2(\boldsymbol{\varphi}) = M\Phi$, the other potential functions are: $U_1 = 0$, $\frac{\partial U_2}{\partial \boldsymbol{\varphi}} = \frac{M}{\Phi}\boldsymbol{\varphi}$, $\frac{\partial^2 U_2}{\partial \boldsymbol{\varphi}^2} = \frac{M}{\Phi}\left(\boldsymbol{I} - \frac{\boldsymbol{\varphi\varphi}}{\Phi^2}\right)$.

The exact solution of a nonlinear static problem as follows:

$$\boldsymbol{r} = s\boldsymbol{e}_1; \boldsymbol{r}' = \boldsymbol{e}_1; \boldsymbol{\varphi} = \frac{Ms}{k_t}\boldsymbol{e}_1; \Phi = \frac{Ms}{k_t}; \boldsymbol{Q} = \boldsymbol{0}; \boldsymbol{M} = M\boldsymbol{e}_1. \tag{9}$$

Tensors $\boldsymbol{P}$ and $\boldsymbol{Z}$ in the current problem are:

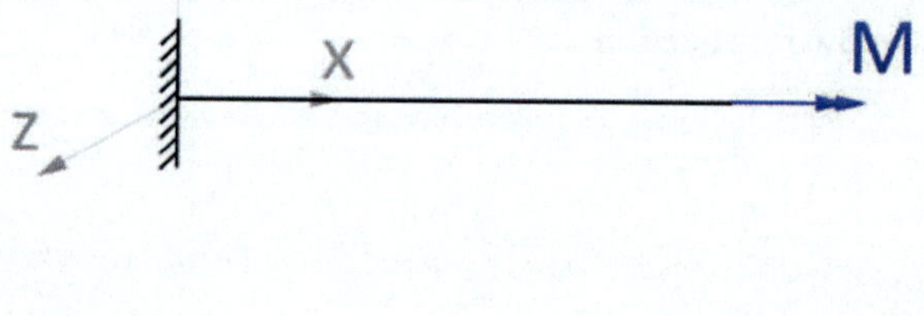

Fig. 4 Cantilever beam under a torque

$$P = \cos\frac{Ms}{k_t}I + \sin\frac{Ms}{k_t}I \times e_1 + \left(1 - \cos\frac{Ms}{k_t}\right)e_1e_1;$$

$$Z = \frac{k_t}{Ms}\left[\sin\frac{Ms}{k_t}I + \left(1 - \cos\frac{Ms}{k_t}\right)I \times e_1 + \left(\frac{Ms}{k_t} - \sin\frac{Ms}{k_t}\right)e_1e_1\right].$$

If the solution of the nonlinear static problem (9) is put into Eq. (3) and regarding the boundary conditions and using the substitution $Z \cdot \theta = \psi$ it can be seen that in order to solve the stability problem it's enough to solve a following system of equations.

$$\begin{cases} \left(k_{3y}\cos^2\Phi + k_{3z}\sin^2\Phi\right)\psi_2' + \\ + \left(k_{3y} - k_{3z}\right)\sin\Phi\cos\Phi\psi_3' + M\psi_3 = B_2 \\ \left(k_{3y} - k_{3z}\right)\sin\Phi\cos\Phi\psi_2' + \begin{pmatrix} k_{3y}\sin^2\Phi + \\ + k_{3z}\cos^2\Phi \end{pmatrix}\psi_3' - M\psi_2 = B_3 \end{cases} \tag{10}$$

where B_2 and B_3—integration constants.

While solving the system of differential equations (10) the problem splits into two separate cases:

(1) $k_t < k_{3z}$; $k_t < k_{3y}$ (bending stiffnesses around z and y axes are greater than torsional stiffness);
(2) $k_{3z} < k_t$ or $k_{3y} < k_t$ (bending stiffnesses around z or y axes are smaller than torsional stiffness).

Solving the system of equations the first case will be considered, when bending stiffnesses around z and y axes are greater than torsional stiffness which is actual to any beam from a gauge chart.

The solution of the system of Eq. (10) is:

$$\psi_2 = C_1\left(\cos\mu s\cos\varphi s - \sqrt{\frac{b}{a}}\sin\mu s\sin\varphi s\right) - $$
$$- C_2\left(\cos\mu s\sin\varphi s + \sqrt{\frac{a}{b}}\sin\mu s\cos\varphi s\right) - \frac{B_3}{M};$$

$$\psi_3 = C_1\left(\cos\mu s\sin\varphi s + \sqrt{\frac{b}{a}}\sin\mu s\cos\varphi s\right) + $$
$$+ C_2\left(\cos\mu s\cos\varphi s - \sqrt{\frac{a}{b}}\sin\mu s\sin\varphi s\right) + \frac{B_2}{M},$$

where $\mu = M\sqrt{ab}$; $\varphi = \frac{M}{k_t}$; $a = \frac{1}{k_{3y}} - \frac{1}{k_t}$; $b = \frac{1}{k_{3z}} - \frac{1}{k_t}$, C_1, C_2—arbitrary constants.

Putting the solution of this equation into the boundary conditions, it can be found out, that the condition of the existence of the non-zero solution is the following equation:

$$\cos \mu L = 1$$

The smallest positive root of the equation gives us the value of the critical torque for the given problem:

$$M_{cr} = \frac{2\pi}{L} \cdot \frac{1}{\sqrt{\left(\frac{1}{EI_z} - \frac{1}{GI}\right)\left(\frac{1}{EI_y} - \frac{1}{GI}\right)}} \tag{11}$$

This formula is the exact value for the critical moment of the cantilever beam under a follow potential torque regarding all the beam stiffnesses involved in the buckling.

4 Discussion

The obtained formulas for the critical forces and moments can be compared with the results of the previous researchers for the approximate beam models. As an example the formulas for the Euler–Bernoulli and Timoshenko beam can be used.

The problem of an axial compression has three well known solutions [6]:

(1) Euler's formula: $T_{crEul} = \frac{\pi^2 EI_z}{4L^2}$. This formula can be obtained from our solution (7) and (8) in case $\frac{1}{EA} \to 0$ and $\frac{1}{k_y GA} \to 0$;
(2) Engesser's formula: $T_{crEng} = \frac{T_{crEul}}{\left(1 + \frac{T_{crEul}}{k_y GA}\right)}$. If we consider that $\frac{1}{k_1} \to 0$, our solution can be written as follows: $T_{crTim} = \frac{k_y GA}{2}\left(\sqrt{1 + \frac{4T_{crEul}}{k_y GA}} - 1\right)$. This formula is an exact solution for the Timoshenko beam [6]. Thus, Engesser's formula can't be obtained from our solution and tends out to be incorrect;
(3) Haringx's formula, which was obtained only for circular springs and was not based on rigorous mathematics. Haringx's formula completely coincide with our solution.

The problem of a torque has several solutions:

(1) Formulas in monograph of Perelmuter and Slivker [3], which are the solutions of the stability problems for the cantilever beam with torque. The problem was solved as a highly linearized problem and with another types of moments. The solutions are $M_{crPS} = \frac{K}{L}\sqrt{EI_y EI_z}$, where K—is a numerical parameter, which differs from the type of the moment. If we consider $\frac{1}{GI} \to 0$, then our Formula (11) can be written as follows: $M_{crL} = \frac{2\pi}{L}\sqrt{EI_y EI_z}$. As it can be seen the general structure of the solutions is similar. To compare the values for the parameter K it is necessary to solve some other problems according to other potential moments, which can be found in the monograph;

(2) Solution of Sergeev-Zhilin. In paper [24] the solution of the stability problem loaded with the same potential torque can be found as in the current work, but for the symmetric cross-section ($EI_y = EI_z$). This solution is the same, as our in the symmetry case condition: $M_{crSym} = \frac{2\pi}{L}\frac{1}{\left|\frac{1}{EI}-\frac{1}{GI}\right|}$.

5 Conclusion

The obtained formulas for the critical forces and moments were obtained from the exact results of the variational calculus and, as a result, are exact for the nonlinear elastic beams.

Formula (7) was obtained for the first time by Kuznetsova (Kushova) and Lalin [32] in a plane statement. The solutions (7) and (8) obtained in this paper prove that for the spatial case the values of the critical load are the same.

From the formula (11) can be found out, that the critical moment does not exist, if one of the bending stiffnesses is smaller than the torsional, which matches, for example, a square cross section with an aspect ratio ≈1,8, but can not match any beam from a gauge chart.

Both formulae (7) and (11) have a limitation: the material of the beam can yielding before buckling. Stability problem, based on rigorous mathematics and considering plasticity for the geometrically nonlinear beam theory is unknown.

References

1. Gelfand IM, Fomin SV (1963) Calculus of variations (trans: Silverman RA; Revis. English Ed.)
2. van Brunt B (2004) The calculus of variations. Springer (2004).
3. Perelmuter AV, Slivker V (2013) Handbook of mechanical stability in engineering (in 3 volumes): vol. 1: general theorems and individual members of mechanical systems. https://doi.org/10.1142/8372
4. Koiter WT (1967) On the stability of elastic equilibrium. National Aeronautics and Space Administration
5. Ziegler H (1968) Principles of structural stability. Blaisdell Publ. Comp. Waltham, Massachusetts, Toronto, London, 192 p
6. Timoshenko SP, Gere JM (2009) Theory of elastic stability. Courier Corporation
7. Wang CM, Wang CY, Reddy JN (2004). Exact solutions for buckling of structural members. https://doi.org/10.1201/9780203483534
8. Trahair NS (2019) Flexural-torsional buckling of structures. CRC press
9. Bažant ZP, Cedolin L (2010) Stability of structures: elastic, inelastic, fracture and damage theories. https://doi.org/10.1142/7828
10. Simitses GJ, Hodges DH (2006) Fundamentals of structural stability. https://doi.org/10.1016/B978-0-7506-7875-9.X5000-2
11. Engesser F (1891) Die knickfestigkeit gerader stäbe. W. Ernst & Sohn

12. Harinx JA (1942) On buckling and the lateral rigidity of helical compression springs. Proc Konjlike Ned Akad Wettenschappen 45:533–539
13. Reissner E (1982) Some remarks on the problem of column buckling. Ingenieur-Archiv 52:115–119
14. Ziegler H (1982) Arguments for and against Engesser's buckling formulas. Ingenieur-Archiv 52:105–113
15. Aristizabal-Ochoa JD (2011) Stability of columns with semi-rigid connections including shear effects using Engesser, Haringx and Euler approaches. Eng Struct 33:868–880
16. Genovese D, Elishakoff I (2019) Shear deformable rod theories and fundamental principles of mechanics. Arch Appl Mech 89:1995–2003. https://doi.org/10.1007/s00419-019-01556-7
17. Blaauwendraad J (2010) Shear in structural stability: on the Engesser–Haringx discord. J Appl Mech 77
18. Li X-F, Lee KY (2018) Effects of engesser's and haringx's hypotheses on buckling of timoshenko and higher-ordershear-deformable columns. J Eng Mech 144. https://doi.org/10.1061/(ASCE)EM.1943-7889.0001363
19. Nikolai EL (1928) On the stability of the rectilinear form of equilibrium of a compressed and twisted rod. Izv Leningr Politekhn Inst 31:3–34
20. Bolotin VV (1963) Nonconservative problems of the theory of elastic stability. Macmillan
21. Radin VP, Samogin YuN, Chirkov VP, SAV (2017) Resheniye nekonservativnykh zadach teorii ustoychivosti [Solution of non-conservative problems of stability theory]. Fizmatlit publ., Moscow
22. Krivtsov AM, Kuzkin VA (2015) O paradokse Nikolai v konservativnikh sistemakh [On the Nicholai's paradox in conservative systems]. In: Zhilin PA (ed) Rigid body dynamics, pp 578–583
23. Mailybaev AA, Seyranian AP (2011) The effect of nonconservative forces on the stability of systems with multiple frequencies and the Nicolai paradox. In: Doklady physics. SP MAIK Nauka/Interperiodica, pp 32–38
24. Sergeev AD (1993) Bifunctsiya ravnovesiya szhatogo sterzhniya, skruchennogo sledyaschim i mertvim momentami [Equilibrium bifurcation of a compressed rod twisted by the following and dead moments]. Wr. SPbSIT, pp 193–195
25. Simo JC (1985) A finite strain beam formulation. The three-dimensional dynamic problem. Part I. Comput Methods Appl Mech Eng 49:55–70
26. Simo JC, Vu-Quoc L (1988) On the dynamics in space of rods undergoing large motions—a geometrically exact approach. Comput Methods Appl Mech Eng 66:125–161
27. Pimenta PM, Yojo T (1993) Geometrically exact analysis of spatial frames. Appl Mech Rev 46:S118–S128. https://doi.org/10.1115/1.3122626
28. Zhilin PA (2007) Applied mechanics-theory of thin elastic rods. Politekhn. Univ. Publ, St. Petersbg
29. Zhilin PA (1996) A new approach to the analysis of free rotations of rigid bodies. ZAMM Zeitschrift fur Angew Math und Mech 76:187–204. https://doi.org/10.1002/zamm.19960760402
30. Eliseev VV (2003) Mechanics of elastic bodies. St. Petersbg. State Polytech. Univ. Publ. House, St. Petersbg
31. Lalin VV, Zdanchuk EV, Kushova DA, Rozin LA (2015) Variational formulations for non-linear problems with independent rotational degrees of freedom. Inz Z 54
32. Kushova DA, Lalin VV (2013) Geometrically nonlinear deformation and stability problems of plane elastic rods with tension-compression, shear and bending stiffness taken into account. Int J Comput Civ Struct Eng 9:178–185

Improvement of Structural and Technological Solutions of Wood-Composite Building Systems Based on the Geodesic Dome

D. A. Zhivotov, Y. I. Tilinin, and V. V. Latuta

Abstract The scientific article presents the materials of theoretical studies of traditional building systems and substantiates the prospects for the practical application of a new construction wood-composite system in the form of a geodesic dome. Proposals for the functional use and organization of the internal space of the geodesic dome are considered. The test of a new nodal joint assembled from fiberglass parts manufactured by the additive method using fiberglass as the starting material is considered. The conducted experiments confirm the prospects of the research topic and give grounds for planning the next stage of research work in the form of field studies.

Keywords Traditional building systems · Wood-composite construction system · Frame · Lattice dome · Functional purpose of the building · Connecting node · Experiment · Wooden elements · Fiberglass parts · Additive technologies

1 Introduction

Building systems are characterized by building materials and technologies used in the construction of buildings [1, 2]. By far the most long-term use in construction is wood, which, at the next stage of technical progress, since the beginning of the eighteenth century, continues to be used in construction as a material for beams of inter-floor floors, rafters and lathing of the attic roof [3, 4] in brick buildings (Fig. 1). The bonding of wood with bio-and water-resistant synthetic adhesives has immeasurably expanded the engineering and architectural capabilities of this natural polymer [5].

D. A. Zhivotov (✉) · Y. I. Tilinin · V. V. Latuta
Saint Petersburg State University of Architecture and Civil Engineering, 4 Vtoraya Krasnoarmeiskaya ul., Saint Petersburg 190005, Russia
e-mail: tilsp@inbox.ru

N. Vatin et al. (eds.), *Proceedings of MPCPE 2021*, Lecture Notes in Civil Engineering 182, https://doi.org/10.1007/978-3-030-85236-8_28

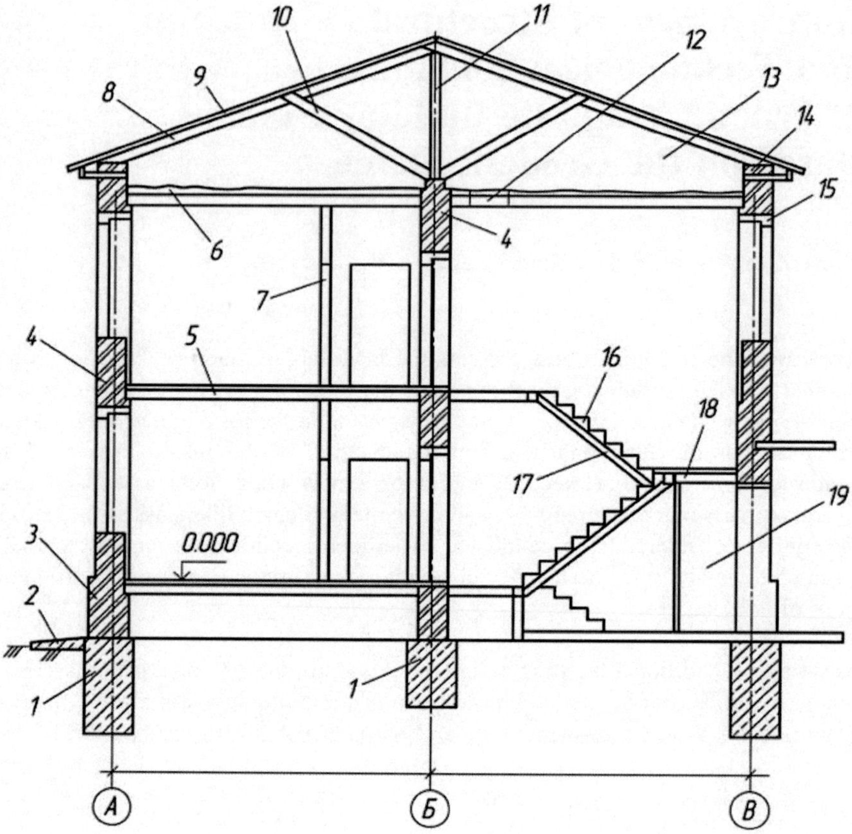

Fig. 1 Structural elements of a brick building: 1—foundation; 2—blind area; 3—basement; 4—brick wall; 5—inter-floor; 6—attic floor; 7—partition; 8—wooden rafters; 9—wooden crate; 10—wooden mower; 11—wooden stand; 12—attic hatch; 13—attic; 14—wooden mauerlat; 15—lintel; 16—stairs; 17—kosour; 18—landing; 19—entrance vestibule

2 Methods

However, in modern construction, brick and wood are replaced by monolithic and precast concrete. Technical progress in the field of construction lifting machines [6], used for the installation of structures and pumps for the supply of concrete and mortar mixtures, has a significant impact on the technologies of construction systems. A significant place in the construction is occupied by light concrete, including gas-silicate concrete blocks, as well as light steel structures and sandwich panels [7].

But despite the variety of building materials and the perfection of machines used for construction, wood is still used in construction not only in the form of

Fig. 2 Building made of structural insulation panels

chipboards on cement and phenol–formaldehyde binder, plywood and glued load-bearing structures, but also as a material of wooden prefabricated prefabricated buildings of the "Module" system type, erected at the end of the twentieth century, or modern buildings made of structural insulation panels [8], consisting of a styrofoam layer on both sides covered with oriented chipboard plates (Fig. 2).

The construction systems discussed above have a number of inconvenient features for construction, for example, in Arctic conditions, this is a small span and functional limitations. An alternative to the considered construction systems, the authors see a construction system made of wooden rod elements in the form of a geodesic dome (Fig. 3), because spatial dome structures have an ideal aerodynamic shape with high resistance to seismic, wind and hurricane impacts [9].

The geodesic dome is a lattice shell in the form of a hemisphere consisting of individual rods that form triangular elements when connected [10]. The advantage of a lattice frame in the form of a geodesic dome is also its lightness and geometric immutability [11]. Scientists continue to study the possibility of using geodesic domes.

In our opinion, such buildings are suitable for polar weather stations or other similar objects. The use of dome frames as the supporting frame of large-span industrial and civil buildings is also considered.

The authors of the article propose to consider the functional purpose of the above examples more broadly, using geodesic domes to cover shift settlements in the Arctic and other Northern regions of the planet.

Climatic conditions in the northern regions significantly complicate the living and working conditions of people. The possibility of creating a special microclimate in the internal space will have a positive impact on the emotional and physical

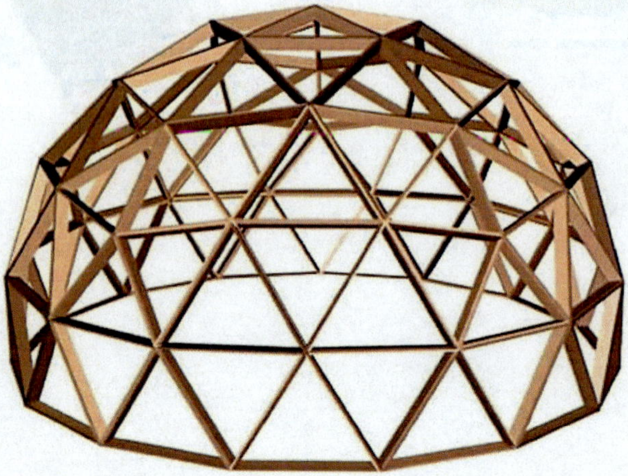

Fig. 3 The frame of the building in the form of a geodesic dome

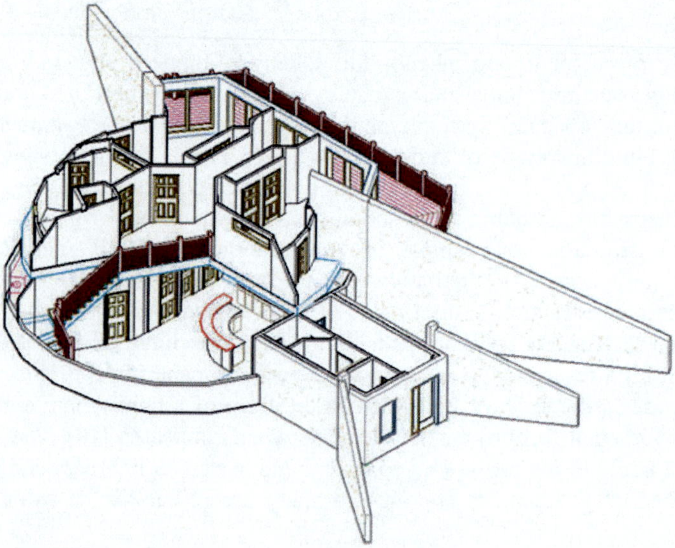

Fig. 4 A variant of the functional filling of the dome space

condition of people. The autonomy of life under such a dome is provided by modern engineering systems and equipment.

The authors of the article, based on the research of domestic and foreign scientists, suggest using the space of the geodesic dome to accommodate residential, administrative, and laboratory research facilities [12] (Fig. 4).

The intrinsic geometry of the geodesic dome is based on using a network of triangles mounted from rods. The rods are located on geodesic lines (the shortest lines connecting two points on a curved-based surface). Such arrangement allows achieving the optimal space filling and the most complete use of the structural strength of the materials [13]. The construction technology consists in the tiered-like erection of rod triangular cells of various types and sizes from bottom to top [14]. The triangular elements converge from the base to the top of the dome, which entails in each tier a change of the rods' size, their inclination angles in relation to the nodal connections, and a different amount of fasteners [15]. Wood is a constantly renewable resource, easy to process and does not leave any inorganic substances after application [16]. Polymer materials are of interest both due to their technical and physic-mechanical properties. The wide range of properties of these materials, changing through using various manufacturing methods, yields a lot of opportunities for scientific and scientific-technical solutions in the future [17]. Scientists continue to study the possibility of using the geodesic dome by changing its shape, configuration, material of manufacture and many other parameters [11]. The honeycomb configuration of the many available [18]. Diamatic dome templates is particularly convenient for converting with mutual element support. This is due to the fact that only three elements of the lattice rods intersect at any vertex, regardless of the number of rod elements used to form the node polygon [19].

The disadvantages of geodesic domes include the fact that the production of modern building materials is aimed primarily at the construction of buildings made of rectangular materials (plywood, glass, rigid insulation mats). The use of the geodesic dome surface is an excellent solution to this issue [19].

3 Results and Discussion

One of the disadvantages of the geodesic dome is the difficulty of connecting six wooden rod elements in one node.

It should be noted that all existing patent solutions use metal for the manufacture of assemblies and connections, in the form of bolts, pins with washers and nuts, which has an extremely negative effect on the state of the dome construction structure as a whole in chemically aggressive environments and low negative temperatures [20].

The authors proposed a new design and technological solution for the nodal connection of the geodesic dome elements. The assembly consists of high-strength composite polymer parts. Figure 5 shows the assembled assembly and fragments of wooden elements, a prototype for laboratory testing.

The production of plastic parts of the assembly is carried out by growing on a 3D printer made of fiberglass.

The main materials for the manufacture of plastic parts of the node connection are:

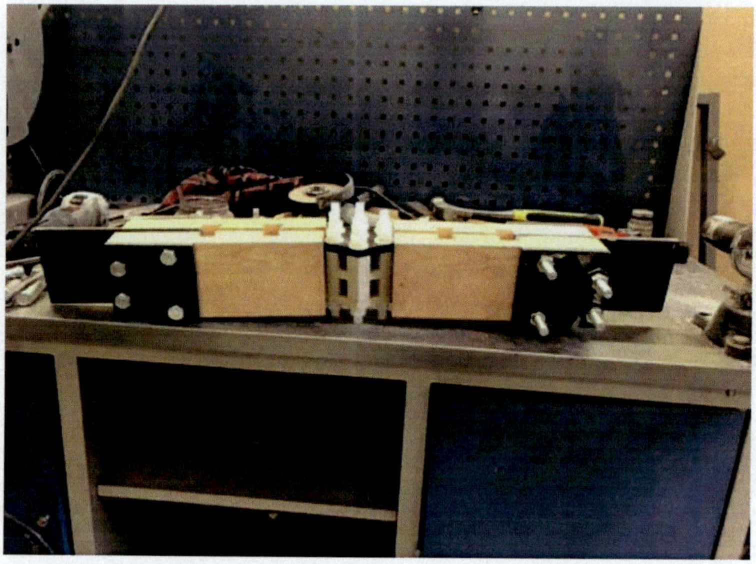

Fig. 5 The connecting assembly with the wooden rods of the dome

- TOTAL GF-30 (N) plastic for the body, covers and connections of the node;
- polyamide d. 12 mm in the form of threaded pins, nuts and washers for the assembly of the node connection.
- For the wooden rod elements, pine wood of normal humidity is used for the rods [6].
- During the preparation of the prototype, steel plates with a thickness of 8 and 10 mm were made, attached to wooden elements on steel bolts and intended for fixing the sample in the Instron 5998 bursting unit, in the mechanical laboratory of SPbGASU (Fig. 6).

The experiment was carried out in compliance with the requirements of the standards on a 20-ton universal electromechanical machine Instron 5998 (Fig. 5) with a maximum breaking load of 200 kN. The equipment of the breaking plant allows you to perform linear movement under the action of a tensile load at a constant speed and perform load measurements.

In the course of the experiment, the mechanical characteristics of the nodal joint made of wood and plastic were studied. The behavior of the material under the influence of load was studied, the stages of deformations and the dependences of displacements were studied.

The results of the experimental data are shown in the form of graphical dependencies of the displacements on the load (Fig. 7). One sample was tested.

The results of the experimental study and the nature of the destruction are combined with each other, indicating the reliability of the data obtained. The graph and Fig. 8 show the changes typical of brittle materials that occur in the body of the node under the influence of load.

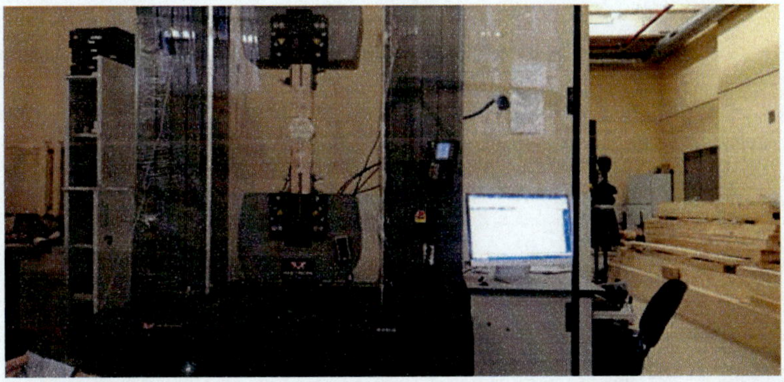

Fig. 6 Testing of the geodesic dome assembly in the Instron 5998 breaking installation, in the mechanical laboratory of SPbGASU

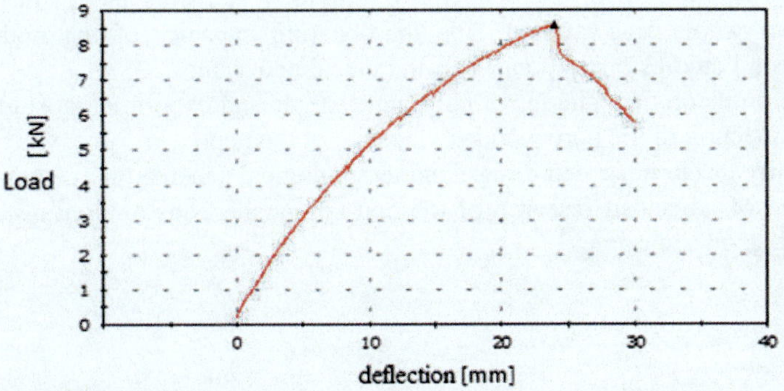

Fig. 7 Dependence of deflection on the applied load

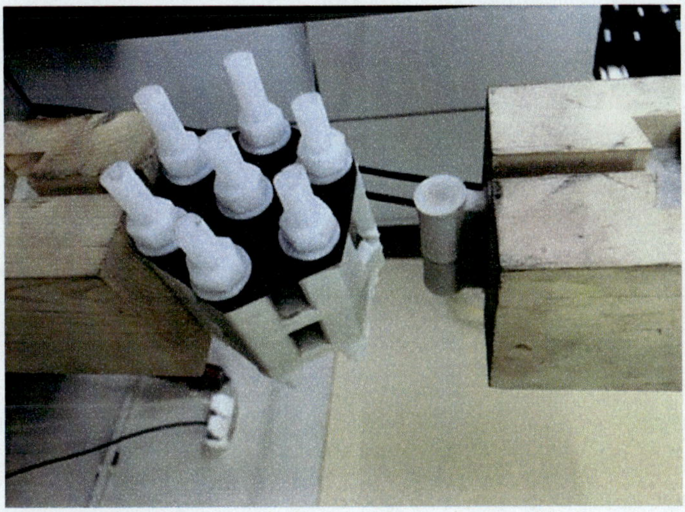

Fig. 8 The destroyed node

The destruction of the body of the nodal joint from GF30 (N) took place gradually, under the influence of the load, the values of deformations gradually increased, passing from the stage of plastic deformations to brittle fracture. Comparison of the results of experimental studies and mathematical results of calculations confirm their identity.

4 Conclusion

The destruction occurred by pulling the blade out of the body of the node when reaching 30% of the expected load value of 8.56 kN, due to the insufficiency of its cross-section. It is worth noting that the design of the body of the nodal joint was calculated according to the characteristics stated by the plastic manufacturer, but the experimental studies of the material properties showed the real values of the limit.

During the experiment, the behavior of individual elements and the design of the node as a whole was studied. The load-bearing capacity of the nodes in the experimental studies corresponded to the calculated values.

Based on theoretical studies of building systems and experimental studies of the nodal connection of the core wooden elements of the geodesic dome, which showed the viability of the proposed design and technological scheme, the authors plan the next stage of advanced research of a wood-composite construction system in the form of a geodesic dome.

References

1. Rybnov E, Egorov A, Hejducki Z, Ghadimian N (2018) Organization and planning of production structures work at large-scale housing construction. Bull Civ Eng 3:98–102. https://doi.org/10.23968/1999-5571-2018-15-3-98-102
2. Tokunova G, Zhludova O, Drozdova I (2017) The influence of the construction sector on the development of Saint-Petersburg agglomeration. Bull Civ Eng 6:339–343. https://doi.org/10.23968/1999-5571-2017-14-6-339-343
3. Golovina SG, SYV (2018) To the issue of studying the joint performance of building materials of enclosing structures in the former profitable houses of the historical center in St. Petersburg. Bull Civ Eng 3:112–117. https://doi.org/10.23968/1999-5571-2018-15-3-112-117
4. Zhivotov D, Pastukh O (2020) Construction of geodesic domes made of wood and composite materials during restoration and conservation of cultural heritage objects. E3S Web Conf 164. https://doi.org/10.1051/e3sconf/202016402020
5. Velichkin V, Zhivotov D, Latuta V, Tilinin Y (2019) Modern technologies for strengthening wooden rafters in apartment buildings of the historical part of St. Petersburg. Bull Civ Eng 5:125–131. https://doi.org/10.23968/1999-5571-2019-16-5-125-131
6. Serov E, Mironova S (2018) Strengthening of glued wooden constructions. Bull Civ Eng 5:68–74. https://doi.org/10.23968/1999-5571-2018-15-5-68-74
7. Yudina A, Tilinin Y (2019) Selection of criteria for comparative evaluation of house building technologies. Archit Eng 4:47–52. https://doi.org/10.23968/2500-0055-2019-4-1-47-52

8. Yudina A, Evtyukov S, Tilinin Y (2019) Development of housing construction technologies in St. Petersburg. Bull Civ Eng 1:110–119. https://doi.org/10.23968/1999-5571-2019-16-1-110-119
9. Yudina A, Sychov S, Gaido A (2020) Construction system for the erection of prefabricated buildings out of factory-made modules. Archit Eng 5:32–37. https://doi.org/10.23968/2500-0055-2020-5-2-32-37
10. Guan Y, Virgin LN, Helm D (2018) Structural behavior of shallow geodesic lattice domes. Int J Solids Struct 155:225–239. https://doi.org/10.1016/j.ijsolstr.2018.07.022
11. Wu Y, Takatsuka M (2006) Spherical self-organizing map using efficient indexed geodesic data structure. Neural Netw 19:900–910. https://doi.org/10.1016/j.neunet.2006.05.021
12. Wei Jp, Tian Lm, Hao Jp (2018) Improving the progressive collapse resistance of long-span single-layer spatial grid structures. Constr Build Mater 171:96–108. https://doi.org/10.1016/j.conbuildmat.2018.03.126
13. Asdrubali F, Ferracuti B, Lombardi L, Guattari C, Evangelisti L, Grazieschi G (2017) A review of structural, thermo-physical, acoustical, and environmental properties of wooden materials for building applications. Build Environ 114:307–332. https://doi.org/10.1016/j.buildenv.2016.12.033
14. Kasyanov NV (2016) The evolution of architectural morphogenesis at the beginning of XXI century in the context of scientific advances. Proc Eng 153:266–270. https://doi.org/10.1016/j.proeng.2016.08.113
15. Lebedev MP, Startsev OV, Kychkin AK (2019) Development of climatic tests of polymer materials for extreme operating conditions. Proc Struct Integr 20:81–86. https://doi.org/10.1016/j.prostr.2019.12.119
16. Kuzmin SA, Egorova AD, Krasilnikov DA, Emelianova ZV (2019) Durability of construction materials modified by polymeric additives. Proc Struct Integr 20:278–283. https://doi.org/10.1016/j.prostr.2019.12.152
17. Halliwell S (2020) Ageing of composites in the construction industry. Ageing Compos 401–420. https://doi.org/10.1533/9781845694937.3.401
18. Rizzuto JP (2018) Experimental investigation of reciprocally supported element (RSE) lattice honeycomb domes structural behaviour. Eng Struct 166:496–510. https://doi.org/10.1016/j.engstruct.2018.03.094
19. Zhivotov D, Latuta V (2020) Using geodesic domes of wood and thermoplastics for rotational camps in the arctic and northern territories. Archit Eng 5:22–28. https://doi.org/10.23968/2500-0055-2020-5-3-22-28
20. Bushin V, Zhivotov D, Podolsky D (2016) Connection point for load-bearing rods for geodesic domes and other spatial structures

Survivability Exposition of a Long-Term Deformable Reinforced Concrete Building Frame Under Accidental Actions

Natalia Androsova and Vitaly Kolchunov

Abstract The survivability study results of the reinforced concrete frame structural multi-storey building system in a non-equilibrium formulation of the problem (initial creep deformations in time and corrosion damage) are presented. A duration preservation analytical assessment of the survivability potential building system in time to the destructive effects of an aggressive environment with the destruction from the structural elements responsible system for the geometric immutability of the structure is given. A technique for determining the stiffness of frame elements taking into account the initial measure of damped simple creep C (t, t_0) is presented. An accidental limit state deformation criterion for the considered non-equilibrium problem is constructed from the condition of maintaining the survivability potential. The calculation of a fragment of a reinforced concrete frame was carried out taking into account the nonequilibrium processes of long-term effects and environmental factors when assessing the exposure of its survivability. The calculation of a reinforced concrete frame fragment, taking into account nonequilibrium processes, showed that the reinforced concrete frame survivability depends on the duration of its operation in the building and the degree of its corrosion damage over time. The period of time during which the considered structural system can turn into a kinematically variable system with a hypothetical removal of one of the load-bearing elements, with a special limiting state, will be about 10 years.

Keywords Progressive collapse · Survivability potential · Reinforced concrete structures · Creep · Accidental actions

N. Androsova (✉) · V. Kolchunov
South West State University, 50 let Oktyabrya street, 94, Kursk 305040, Russia

N. Androsova · V. Kolchunov
Scientific and Research Institute of Construction Physics of RAASN, Lokomotivny proezd, 21, Moscow 127238, Russia

N. Androsova
Orel State University after I.S. Turgenev, Komsomolskaya street, 95, Orel 302026, Russia

N. Vatin et al. (eds.), *Proceedings of MPCPE 2021*, Lecture Notes in Civil Engineering 182, https://doi.org/10.1007/978-3-030-85236-8_29

1 Introduction

Recent studies of the survivability of building structures and the structural safety of bearing systems of buildings are formed on the fundamental provisions of the limit state methods (scientific works by Vedyakov [1], Karpenko et al. [2], Trekin and Kodysh [3], Plevkov et al. [4], Travush et al. [5], Tamrazyan [6], Shapiro [7], Isobe [8], Stylianidis et al. [9], Cha and Ellingwood [10], etc.). When designing, the level of structural safety of load-bearing elements of buildings, despite the inevitable power and environmental wear and operational damage, is usually sufficient for their trouble-free operation during the design period of service. Experimental data show [11] that in the first limiting state there are reserves for design strength up to 25–30%, and in the second for design deformability up to 50%. These reserves can be used in certain design situations, when, in the event of a load-bearing element failure, the mating elements are involved in joint deformation, the forces are redistributed, unloading or reloading the bearing element. Taking into account the nature of the element's operation in an out-of-bounds (special) state will make it possible to use the reserves of the bearing capacity of the structure to the fullest when calculating for progressive collapse. With an accidental action and calculation analysis results from the standpoint of the second limiting state, there is a decrease in the design stiffness of structures, the development of large deformations, which complicate or exclude the further operation of construction objects.

The introduction of an accidental limit state is one of the ways to reduce the cost of ensuring the survivability of buildings in case of emergency impacts. This becomes possible due to a more complete consideration of the features of long-term deformation of structures under the load action and the changes nature in the scheme of building elements operation that is, taking into account physical and structural nonlinearity. In an accidental limiting state, it is possible to admit a greater opening of cracks and the development of deflections, as well as partial destruction of some sections, which contradicts the current criteria for limiting states that ensure the serviceability of structures and, but exclude the onset of geometric variability of the structural system.

Under an accidental limit state [5, 11] the state of structures is taken after exceeding the limit of the bearing capacity in the first and deformability in the second limiting states, in which they do not fully meet the functional requirements, a further insignificant increase in loads and/or impacts leads to their destruction.

Over the past few decades, some results of experimental and theoretical studies have already been accumulated on the resistance study of reinforced concrete structures with a sudden removal of a load-bearing element or connection and the possible patterns establishment of their destruction. The tests were carried out on beam, post-beam, beam-plate structures [12–15]. In these works, the deformation and destruction fundamental features of reinforced concrete structural systems were identified in the sudden removal process of the bearing element (connection). Under the influence of such an emergency load, a dynamic high-speed reloading of the remaining intact structures occurred within a very short period of time.

The increased strength of materials manifested in this case is usually associated with the instantaneous manifestation of viscous resistance forces, which directly perceive external influences and inhibit the development of deformations. However, to date, very few experimental and theoretical studies have been carried out to solve preserving the survivability potential problems of reinforced concrete structural systems in time, taking into account the accumulation of non-equilibrium damage. Certain issues of non-equilibrium processes were considered by Bondarenko [16], Kolchunov et al. [17], Fedorova et al. [12], Tamrazyan et al. [18], Kabantsev et al. [19], Trekin and Kodysh [3], Li et al. [20], Vasanelli et al. [21], etc.

In this regard, the tasks of constructing criteria and taking into account the accidental limiting state of reinforced concrete bending structures, which will make it possible to establish strength and deformation reserves due to the redistribution of forces during prolonged deformation of structures damaged during operation in a supporting spatial system, are relevant. The development of research in this direction will make it possible to more rationally distribute materials and reduce the cost of designing the protection of buildings and structures from progressive collapse without compromising their structural safety.

2 Materials and Methods

2.1 Determination of the Stiffness Elements of the Frame-Bar System Taking into Account the Initial Measure of Damped Simple Creep

Non-equilibrium processes of concrete force resistance, evolving in time, have a force and non-force nature of origin.

Non-equilibrium processes of force origin are characterized by total creep deformations:

$$\varepsilon_{ins} = \varepsilon_{c1} + \varepsilon_{c2} \tag{1}$$

where

ε_{c1} instant (fast flowing) creep deformations;
ε_{c2} long-term (delayed) creep deformations.

It is known that creep deformations (approximately 70%) appear in the first 3–4 months from the moment of loading. In modern conditions, all existing theories of creep are phenomenological in nature, that is, they are based on modeling and describing the most studied experimental phenomena.

The existence of different theories of creep is explained by the fact that the authors, dividing simple creep into linear and nonlinear, use different approaches to describing the components. In all creep equations there is a general quantitative estimate of creep deformations (taking into account long-term loading)—a measure of creep $C(t, t_0)$.

This is a generally accepted parameter of linear creep and is defined as the ratio of creep strains to stress at which this creep occurs:

$$C_{t,t_0}(b) = \frac{\varepsilon_{ins}(t)}{\sigma_b} = \frac{\varepsilon_{c1} + \varepsilon_{c2}}{\sigma_b} \quad (2)$$

The measure of creep deformations $C(t, t_0)$ by the observation t of concrete of a certain composition depends on two factors: the age at the moment of the beginning of loading t_0 and the duration of loading $t - t_0$.

The following prerequisites are accepted:

1. Consideration of the effect of creep in concrete and reinforced concrete structures is based on the phenomenological relationships between stresses and deformations.
2. In the calculations of concrete and reinforced concrete structures, the relationship between stresses and deformations can be taken under the following conditions:
   - under short-term loading—regardless of the concrete class in terms of compressive strength at stresses not exceeding 0.8 of the calculated value of the prismatic strength R_b;
   - under long-term loading—if the stresses in fractions of R_b do not exceed the values in the recommendations.
3. The assumption of V. M. Bondarenko that the concrete creep measures are practically the same regardless of the stress state structure type. The adoption of such an assumption introduces well-known simplifications in solving those problems of the creep theory, which are associated with the study of the structure inhomogeneous stress state or with the simplification in time of the sign of the acting stresses.

Prof. R. S. Sanzharovsky noted that in the creep theory of reinforced concrete structures, three main problems are consistently considered [22]. The first problem is to construct equations for concrete creep under variable (regime) loads. The second is in obtaining a solution to the creep equation. The third problem lies in the calculation of reinforced concrete structures under prolonged loading, including under the accidental action we are considering.

For the practical (quantitative) determination of the creep measure value, one can use the recommendations for taking into account the creep and shrinkage of concrete when calculating concrete and reinforced concrete structures. The concrete creep measure at the time t when it is loaded at the age t_0 is determined by the formula:

$$C(t, t_0) = \left[\frac{1}{E_b(t_0)}\right] - \left[\frac{1}{E_b(t)}\right] + C_{28,besk} \cdot \Omega(t_0) \cdot f(t, t_0), \quad (3)$$

where

$C_{28,besk}$ is the limiting value of the creep measure;
$\Omega(t_0)$ is a function that takes into account the concrete aging effect on the creep measure (t_0 is the loading time of the structure);
$f(t, t_0)$ is a function that takes into account the increase in time of the creep measure (t is the total test time of the sample, the observation time).

The function $\Omega(t_0)$ is determined by the following formula:

$$\Omega(t_0) = c + d \cdot e^{-\gamma \cdot t_0}, \tag{4}$$

The function $f(t, t_0)$ is determined by the formula:

$$f(t, t_0) = 1 - k \cdot e^{-\gamma_1 \cdot (t - t_0)} \tag{5}$$

Coefficients (constant values) are taken: c = 0.5; k = 0.8; γ, γ_1, d are parameters that depend on the surface modulus and are determined according to the tables of the recommendations.

The limiting value of the creep measure $C_{28,besk}$ is calculated by the formula:

$$C_{28,besk} = C^N_{28,besk} \cdot \xi_{2c} \cdot \xi_{3c}, \tag{6}$$

where

ξ_{2c}, ξ_{3c} coefficients depending on the modulus of the open surface of the element M_0 and the relative humidity of the environment;
$C_{28,besk}{}^N$ the limiting value of the concrete creep measure loaded at the age of $t_0 \leq 28$ days.

The reduced cross-section flexural stiffness of the element is calculated by the formula:

$$D = E_{b,red}(t, t_0) \cdot I_{red}. \tag{7}$$

The concrete deformations long-term modulus corresponding to the linear relationship between stresses and deformations at time t at the beginning of loading t_0 is found by the formula:

$$E_b(t, t_0) = \left[\frac{1}{E_b} + C(t, t_0)\right]^{-1}, \tag{8}$$

where

E_b is the modulus of initial elasticity.

2.2 *Deformation Criterion for a Special Limiting State Taking into Account Nonequilibrium Deformation Processes*

An accidental limiting state is a stage in the operation of a structure after reaching the maximum design bearing capacity according to the current criteria for limiting states. For a reinforced concrete bending element, it can occur when the most stressed part of the compressed concrete in the section begins to collapse, and in the tensioned reinforcement the stresses can stabilize and correspond to the yield point, and then decrease. In this case, if the structure deformation is not stopped, the structure will not collapse.

The realization of such a special limiting state is most likely in frame frames with a high degree of static uncertainty in the interaction of all load-bearing elements. Implemented in the transom of the frame, it will lead to an intensive redistribution of efforts in it. This will allow maintaining the frame geometric invariability even with a decrease in the bearing capacity of its individual elements. The increase intensity the in displacements with a decrease in the bearing capacity will depend on the section characteristics and, first of all, on the intensity and scheme of its RC. It follows from what has been said that in order to establish and refine the criteria for a accidental limiting state, both in strength and in deformability, it is necessary to study the stress–strain state of the work of structures in the transcendental stage, when the deformations (deflections) exceeded the normalized ones.

In accordance with SP 385.1325800.2018 "Protection of buildings and structures against progressive collapse. Design rules. Basic statements" (with Amendment No. 1) protection of a building and structure from progressive collapse is ensured if for any elements and their connections the following condition is met:

$$f \leq f_{ult}, \tag{9}$$

where

- f is the movement of the element from the external load;
- f_{ult} the maximum permissible deflections value of the element (for concrete and reinforced concrete structures according to SP 63.13330).

The deformation criterion of a accidental limiting state for the considered non-equilibrium problem from the condition of maintaining the survivability potential:

$$f \leq f_{ult} = \left({}^{1}\!/_{10} \cdots {}^{1}\!/_{30} L\right). \tag{10}$$

Deflection for a bending member:

$$f = S \cdot l_0^2 \cdot \frac{1}{r_{\max}}, \tag{11}$$

The curvature is determined by the formula, taking into account the concrete deformation long modulus:

$$\frac{1}{r_{\max}} = \frac{M_{\max}}{E_{b,red}(t, t_0) \cdot I_{red}}. \tag{12}$$

3 Results

On the example of a long-term deformable fragment of a reinforced concrete frame, a duration quantitative assessment of the survivability potential preservation in time has been made. The calculated fragment of the building's reinforced concrete frame is a two-span two-story reinforced concrete monolithic spatial frame (Fig. 1a). The cross-section of the crossbars has a cross-sectional size of 100 × 120 (b * h) mm and a length of 1000 mm. The section of the racks is 100 × 100 (b * h) mm and the length is 700 mm. For crossbars and posts, concrete of class B20 is adopted. Reinforcement of the racks is symmetrical—with a spatial frame made of Bp500 class reinforcement with a diameter of 4 mm. The crossbars are reinforced with flat frames made of Bp500 class reinforcement with a diameter of 4 mm.

1. *Static calculation of the primary (in the terminology of SP385.1325800.2018 "Protection of buildings and structures against progressive collapse") design scheme of the spatial frame for the design load.* The calculation of the spatial frame according to the primary design scheme (Fig. 2a) under the action of design loads with the use of bar finite elements was carried out using the LIRA-SAPR program. The results of such a calculation (values of forces and displacements) of the frame structure for the design load are shown in Fig. 2b–d.
2. *Calculation of the secondary (in the terminology of SP385.1325800.2018 "Protection of buildings and structures against progressive collapse") design scheme of the spatial frame with a sudden removal of the extreme column of the first floor.* The calculation according to the secondary calculation scheme (Fig. 3a) with switching off the corner column of the first floor was also performed using the LIRA-SAPR program. The calculation results (values of forces and displacements) of the frame structure for the design load are shown in Fig. 3b–d.
3. *Creep measure calculation of the space frame elements.*
   Element open surface modulus: M = 36.67 m^{-1}.
   The value of the function is determined: $\Omega(t_0) = 1.0224$.
   The value of the function f (t, t_0) is determined: f (t, t_0) = 1.0.

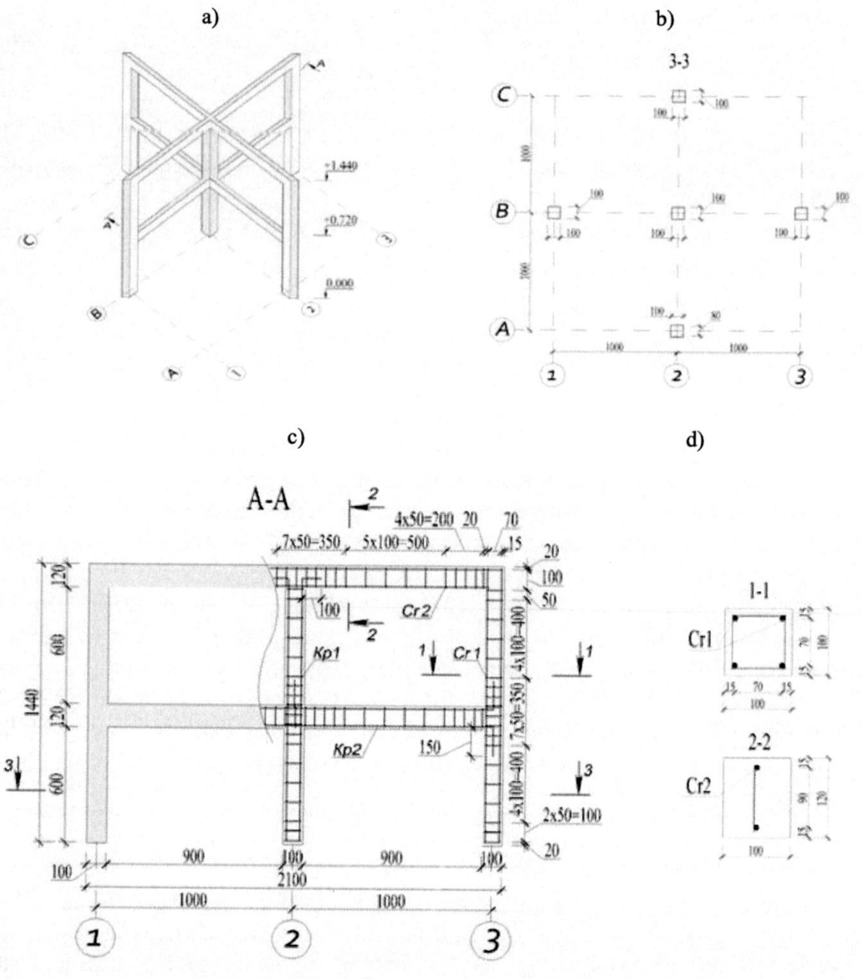

Fig. 1 General view of a spatial two-story reinforced concrete monolithic frame (**a**); the layout of the frame columns (section 3-3) (**b**); formwork drawing and space frame reinforcement along the B (**c**) axis; sections 1-1, 2-2 racks and crossbars of the frame (**d**)

The limiting value of the creep measure is determined: $C_{28,besk} = 1.49 \times 10^{-10} m^2/N$.

The value of the creep measure at $t_0 = 28$ days, t = 3650 days (10 years) will be:

$C(t, t_0) = 1.58 \times 10^{-10} m^2/N$.

4. *Survivability exposition of the spatial reinforced concrete frame.*

Professors Bondarenko V. and Kolchunov V. gave the definition of «survivability exposition»—in the case of a non-equilibrium formulation of the problem

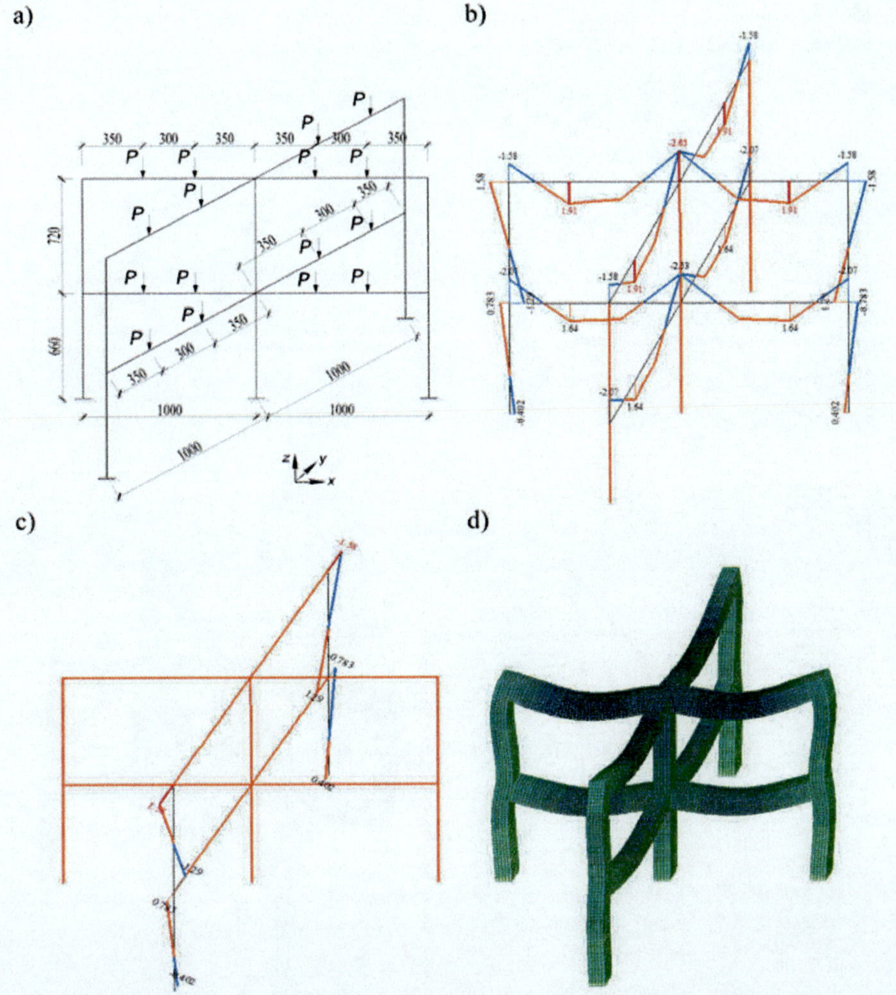

Fig. 2 Primary design scheme of spatial reinforced concrete monolithic frame structure (**a**); moment diagram M_y, kN * m (minimum value −0.736; maximum value 0.529) (**b**); moment diagram M_z, kN * m (minimum value −0.44; maximum value 0.44) (**c**); displacement diagram $f_y + f_z$, mm (maximum value 0.182) (**d**)

(superposition of creep deformations and corrosion damage in time), the survivability potential duration of a building system in time to the destructive effects of an aggressive environment with the disconnection from the structural elements responsible system for the geometric invariability of structures [16]. To assess survivability exposition, a criterial parameter of an accidental limiting state was adopted, in the form of an inequality of curvatures (deflections) in the cross-sections

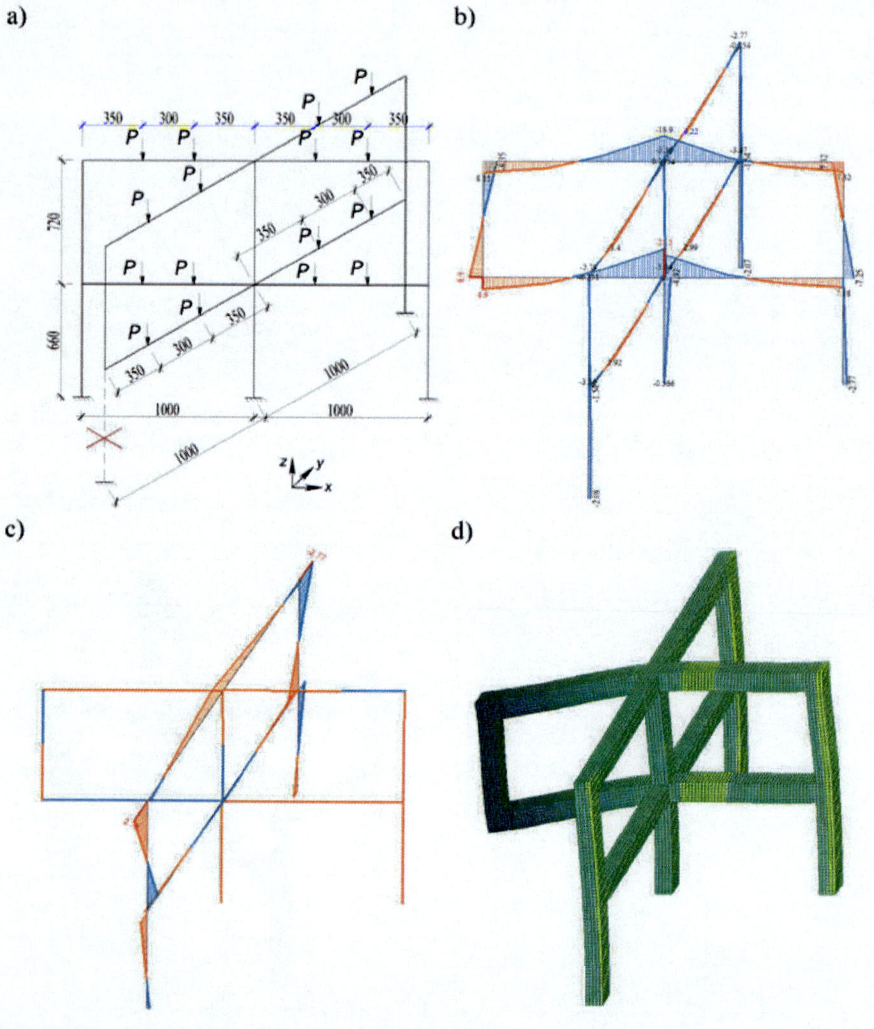

Fig. 3 Secondary design diagram of a spatial reinforced concrete frame (**a**); moment diagram M_y, kN * m (minimum value −4.504; maximum value 2.014) (**b**); moment diagram M_z, kN * m (minimum value −0.527; maximum value 0.527) (**c**); displacement diagram $f_y + f_z$, mm (maximum value 4.27) (**d**)

of crossbars of a long-loaded and corrosively damaged frame after an accidental action in the form of a sudden removal of the extreme rack (formula (10)).

The frame was calculated in two stages. At the first stage, using the primary and secondary design schemes, the calculation for the accidental action under consideration was carried out. As a result of the calculation, it was found that the criterion of a accidental limiting state in the form of maximum deflection is satisfied.

In this case, the girder maximum curvature above the first floor of the spatial frame with the initial concrete deformations modulus will be:

$$\frac{1}{r_{\max}} = \frac{M_{\max}}{E_{b,red} \cdot I_{red}} = \frac{8.9 \times 10^3}{27.5 \times 10^9 \cdot 1.3 \times 10^{-5}} = 0.0249 \, \frac{1}{\text{m}}.$$

Deflection: $f = \frac{1}{4} \cdot 1.0^2 \cdot 0.0249 = 0.00062\,\text{m}$.

The maximum deflection of this girder was 1/160 of its span, which does not exceed the criterion value of 1/30 standardized by SP385.1325800.2018 "Protection of buildings and structures against progressive collapse".

At the second stage, the calculation was carried out for the frame under its long-term loading with the same load and corrosion damage to reinforced concrete, which were determined in time by the method [12]. Based on the calculation results, the curvature of the considered spatial frame was calculated, taking into account nonequilibrium processes (long-term deformation of concrete and corrosion damage in time), which will be:

$$\frac{1}{r_{\max}} = \frac{M_{\max}}{E^*_{b,red} \cdot I_{red}} = \frac{8.9 \times 10^3}{5.14 \times 10^9 \cdot 1.3 \times 10^{-5}} = 0.133 \, \frac{1}{\text{m}}.$$

In this case, the concrete deformations long-term modulus, taking into account the corrosive wear of concrete (at $t_0 = 28$ days; $t = 3650$ days (10 years), was:

$$E^*_b(t, t_0) = \left[\frac{1}{E_b} + C(t, t_0)\right]^{-1} = \left[\frac{1}{27.5 \times 10^9} + 1.58 \times 10^{-10}\right]^{-1} = 5.14 \times 10^9 \, \frac{\text{N}}{\text{m}^2}.$$

Deflection: $f = \frac{1}{4} \cdot 1.0^2 \times 0.133 = 0.0333\,\text{m}$.

The maximum deflection of the girder when calculating according to the secondary design scheme was 1/30 of its span. Consequently, with the accepted class of concrete and frame reinforcement, the level and pattern of the load on the frame and the corrosion damage rate, the survivability exposition of the frame, that is, the period of time during which the structural system can turn into a kinematically variable system with the hypothetical removal of one of the load-bearing elements, with an accidental limiting state, will be about 10 years.

4 Conclusions

1. The proposed method and algorithm for calculating the survivability criterion of a long-term deformable reinforced concrete frame under accidental actions makes it possible to predict the value of this parameter in time for the operated frame-bar frames of buildings and structures.

2. The calculation of a reinforced concrete frame taking into account the non-equilibrium processes of long-term exposure to force and environmental factors when assessing the exposure of its survivability showed that the survivability of a reinforced concrete frame depends on the duration of its operation in the building and the degree of its corrosion damage over time. These factors must be taken into account when calculating the structural system of a building for accidental actions and designing their protection against progressive collapse.

References

1. Vedyakov I (2019) Analysis of regulatory requirements for calculation of building structures for progressive collapse. Bull SRC Stroit 2:15–29
2. Eryshev VA, Karpenko NI, Rimshin VI (2020) The parameters ratio in the strength of bent elements calculations by the deformation model and the ultimate limit state method. IOP Conf Ser Mater Sci Eng 753:022076. https://doi.org/10.1088/1757-899X/753/2/022076
3. Trekin N, Kodysh E (2020) Accidental limit state of reinforced concrete structures and its rating. Ind Civ Constr 4–9. https://doi.org/10.33622/0869-7019.2020.05.04-09
4. Radchenko A, Radchenko P, Batuev S, Plevkov V (2019) Modeling fracture of reinforced concrete structures under impact. Archit Eng 4:22–29. https://doi.org/10.23968/2500-0055-2019-4-3-22-29
5. Travush VI, Fedorova NV (2018) Survivability of structural systems of buildings with special effects. Mag Civ Eng 81:73–80. https://doi.org/10.18720/MCE.81.8
6. Tamrazyan A (2013) Reduce the impact of dynamic strength of concrete under fire conditions on bearing capacity of reinforced concrete columns. Appl Mech Mater 475–476:1563–1566. https://doi.org/10.4028/www.scientific.net/AMM.475-476.1563
7. Shapiro GI, Smirnov AV (2018) Calculation model of typical panel building conjugation with large-span frame constructions. IOP Conf Ser Mater Sci Eng 456. https://doi.org/10.1088/1757-899X/456/1/012090
8. Isobe D (2018) Progressive collapse analysis of structures. Elsevier, pp 1–260. https://doi.org/10.1016/B978-0-12-812975-3.00001-3
9. Stylianidis PM, Nethercot DA, Izzuddin BA, Elghazouli AY (2016) Study of the mechanics of progressive collapse with simplified beam models. Eng Struct 117:287–304. https://doi.org/10.1016/j.engstruct.2016.02.056
10. Cha EJ, Ellingwood BR (2013) Seismic risk mitigation of building structures: the role of risk aversion. Struct Saf 40:11–19. https://doi.org/10.1016/j.strusafe.2012.06.004
11. Kodysh E (2018) Designing the protection of building and structures against progressive collapse in view of the emergence of a special limiting state. Ind Civ Eng 95–101
12. Fedorova NV, Gubanova MS (2018) Crack-resistance and strength of a contact joint of a reinforced concrete composite wall beam with corrosion damages under loading. Russ J Build Constr Archit 6–19
13. Fedorova NV (2016) Analysis of deformation and crack formation of multistory monolithic reinforced concrete frame-bar structural systems under limit and beyond-limit conditions. Ind Civ Eng 8–13
14. Adam JM, Parisi F, Sagaseta J, Lu X (2018) Research and practice on progressive collapse and robustness of building structures in the 21st century. Eng Struct 173:122–149. https://doi.org/10.1016/j.engstruct.2018.06.082

15. Fedorova N, Kolchunov V, Tuyen VN, Dinh Quoc P, Medyankin M (2020) The dynamic effect in a structural adjustment of reinforced concrete structural system. IOP Conf Ser Mater Sci Eng 869. https://doi.org/10.1088/1757-899X/869/5/052078
16. Bondarenko V (2013) The concept and directions of development of the theory of structural safety of building and structural under the influence of force and environmental factors. Ind Civ Eng 28–31
17. Kolchunov VI, Savin SY (2018) Survivability criteria for reinforced concrete frame at loss of stability. Mag Civ Eng 80:73–80. https://doi.org/10.18720/MCE.80.7
18. Tamrazyan AG, Mineev MS, Zhukova LI (2019) Influence of chloride corrosion on probabilistic assessment of bearing capacity of beamless slabs overlap. IOP Conf Ser Mater Sci Eng 661. https://doi.org/10.1088/1757-899X/661/1/012052
19. Kabantsev O, Mitrovic B (2017) Modelling post-critical deformation processes of flat reinforced concrete elements under biaxial stresses. MATEC Web Conf 117:1–7. https://doi.org/10.1051/matecconf/201711700071
20. Li J, Yao Y (2001) A study on creep and drying shrinkage of high performance concrete. Cem Concr Res 31:1203–1206. https://doi.org/10.1016/S0008-8846(01)00539-7
21. Vasanelli E, Micelli F, Aiello MA, Plizzari G (2013) Long term behavior of FRC flexural beams under sustained load. Eng Struct 56:1858–1867. https://doi.org/10.1016/j.engstruct.2013.07.035
22. Sanjarovskiy R, Ter-Emmanuilyan T, Manchenko M (2015) Creep of concrete and its instant nonlinear deformation in the calculation of structures. In: CONCREEP 10, pp 238–247. American Society of Civil Engineers, Reston, VA. https://doi.org/10.1061/9780784479346.028

Force Resistance of a Non-linearly Deformable Reinforced Concrete Beam with Corrosion Damage Under Dynamic Load

M. Berlinov and M. Berlinova

Abstract A nonlinear method of dynamic calculation of a reinforced concrete beam is developed, based on a discrete model with corrosion damage and rheology of deformation. Corrosion damage is taken into account by reducing the coefficients to the calculated mechanical characteristics of concrete and reinforcement. Presented the results of the calculation of a reinforced concrete beam with and without taking into account corrosion. Noted that the corrosion damage and the non-linearity and non-equilibrium deformation of reinforced concrete structures under dynamic influences allows us to more fully accounting the real work of materials. Without taking into account corrosion Analysis of the results of the calculation of a reinforced concrete beam in the resonant region of vibrations at the first frequency of natural vibrations shows that the amplitude-frequency characteristic becomes nonlinear and the frequency of resonant vibrations becomes dependent on the amplitude of vibrations. The comparison of the calculation results with the experimental data showed a satisfactory convergence.

Keywords Reinforced concrete · Corrosion damage · Phenomenological equations · Nonlinear deformation · Stress · Dynamic loads · Creep · Vibro-creep processes

1 Introduction

In the conditions of long-term operation of reinforced concrete structures in an aggressive environment under various loads [1, 2], it is necessary to assess their stress–strain state as a result of corrosion damage, taking into account the reduction in the cross-sectional area of concrete and reinforcement.

Corrosion damage to reinforced concrete elements [3, 4] can affect the strength of the material, change the calculation schemes, redistribute forces in the sections of

M. Berlinov (✉) · M. Berlinova
Moscow State University of Civil Engineering (National Research University), Yaroslavskoe shosse, 26, Moscow 129337, Russia

N. Vatin et al. (eds.), *Proceedings of MPCPE 2021*, Lecture Notes in Civil Engineering 182, https://doi.org/10.1007/978-3-030-85236-8_30

the structure and disrupt the joint work of concrete with reinforcement, as well as lead to other consequences that reduce the design life of the structures and other operational characteristics [5, 6]. The most unfavorable result of the development of the corrosion process of reinforced concrete structures is a decrease in their load-bearing capacity and suitability for normal operation, which leads to non-compliance with safety requirements and limit conditions under operating loads during the entire period of operation [7–20].

2 Materials and Methods

Based on the generally accepted practice [9], the force in the cross section of the element before the formation of cracks is the sum of the forces perceived by the reinforcement and concrete, then it is obvious that the following formula can be obtained taking into account the corrosion damage in concrete and reinforcement:

$$\sigma_x(t)A_x = \sigma_{b,x}(t)\,K^*_{b,x}(t)A_{b,x} + \sigma_{s,x}(t)\,w_{s,x}(t)A_{s,x}, \tag{1}$$

here: $\sigma_x(t)$—the average stress in the cross section of reinforced concrete element; $\sigma_{s,x}(t)$ and $\sigma_{b,x}(t)$—the average stresses in the reinforcement and concrete $A_{b,x}$ and $A_{s,x}$—correspond to the values of the area of the reinforcement and concrete; $K_{b,x}(t)$ —a coefficient that takes into account the degree of corrosion damage to concrete, which changes over time of observation; $w_{s,x}(t)$—a similar coefficient for corrosion damage to reinforcement.

The coefficient that takes into account the degree of corrosion damage to concrete is defined as:

$$K^*(z) = \left\{1 - \left[\frac{p}{\delta(t,t_0)}\right]^2\right\} + \frac{2p}{\delta(t,t_0)^2}z - \frac{1}{\delta(t,t_0)^2}z^2 \tag{2}$$

where $\delta(t, t_0)$—the value of the damage; z—the vertical coordinate within the height of the cross-section; $x = \delta(t,t_0) + p$—the height of the compressed concrete zone; p—the undamaged part of the cross-section.

The area of the reinforcement that has been subjected to corrosion damage should be considered as the reduction in accordance with the following expression:

$$A_s = \omega_s A_{so} \tag{3}$$

A_{so}—the area of intact reinforcement, ω_s—the coefficient that consider the reduction in the area of the reinforcement as a result of corrosion, determined by the formula:

$$w_{s,} = \left(1 - \frac{2f}{D} + \frac{16f^2}{\pi D^2}\right) \tag{4}$$

$$\text{where is:} f = \frac{k}{\sqrt{a}} t^n \tag{5}$$

k and n are empirical coefficients, a is the value of the protective layer of concrete, mm, t is the time.

In the case of a crack, to determine the stresses and deformations, you can use the well-known assumption of V. I. Murashev:

$$\sigma_{s,x} = \psi_{s,x}\mu_{s,x}w_s(t)\sigma_{s,x}; \quad \varepsilon_{s,x} = \frac{\sigma_{s,x}}{E_s} = \frac{\psi_{s,x}\mu_{s,x}w_s(t)\sigma_{s,x}(t)}{E_s} \tag{6}$$

Ψ_s—is a linearized coefficient that takes into account the unevenness of the compressed concrete zone; μ_s—is the reinforcement coefficient.

The current state of dynamic methods for calculating reinforced concrete structures of operated buildings caused by seismic or equipment impacts does not fully correspond to the actual work of materials. This is due to the fact that the nonlinearity and rheology of reinforced concrete deformation are considered only indirectly by modern calculation methods.

In this article, an attempt is made to construct a calculation method based on modern phenomenological equations that characterize the work of reinforced concrete under external loading with corrosion damage, and taking into account the intensification of rheological processes (vibration creep) under dynamic influences. Let us illustrate the calculation method by the example of a reinforced concrete beam with a harmonic source of external disturbances, bearing in mind that the proposed approach can be extended to other structures.

Considering the above, the hypothesis of hysteresis energy losses during uniaxial deformation can be extended to the case of a volume stress-deformed state, moreover, the following reasoning can be used to substantiate this assumption. Each component of the stresses within the elementary volume of the body during the dynamic operation of the material will form a closed hysteresis loop in the corresponding coordinate plane.

As initial assumptions, we will use the phenomenological equations of the mechanical state of materials, while taking into account the energy dissipation during vibrations. We will take the computational model in the form of a discrete scheme (Fig. 1a), which allows us to use the Dalembert principle. The resolving equations of the dynamic process with a harmonic source of external influences will take the following form:

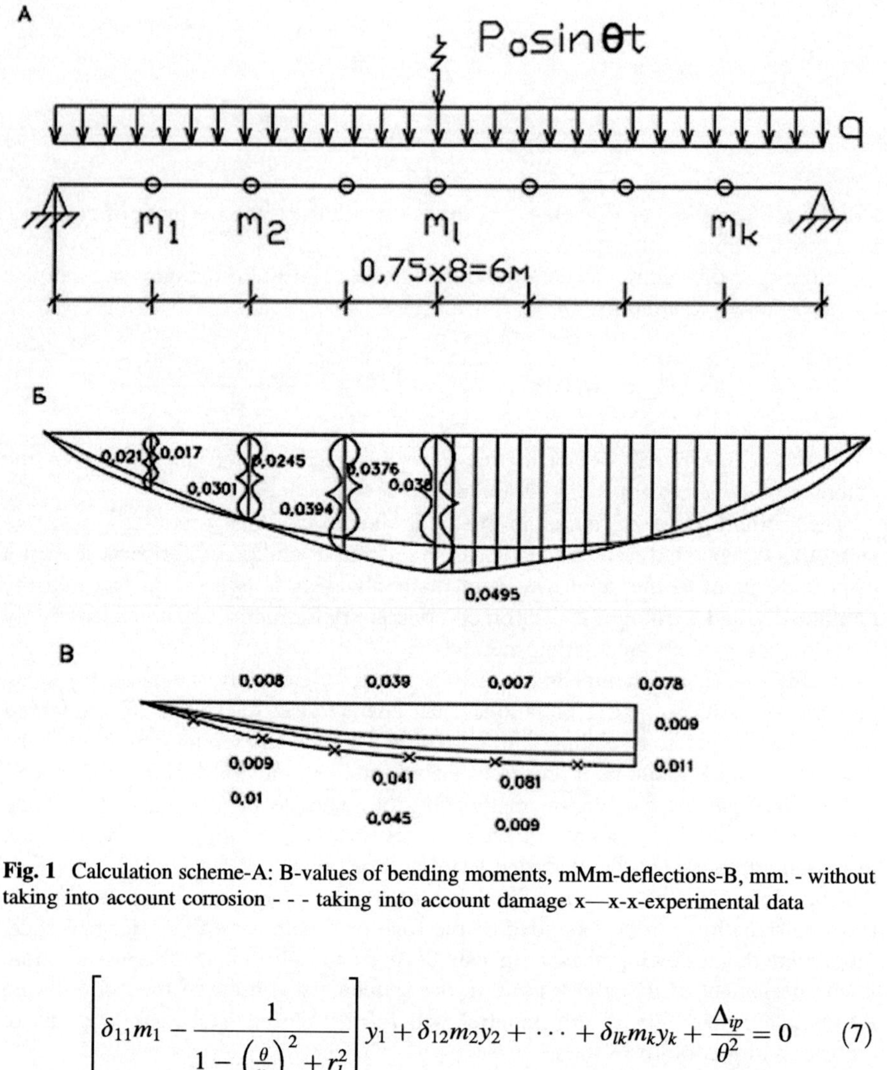

Fig. 1 Calculation scheme-A: B-values of bending moments, mMm-deflections-B, mm. - without taking into account corrosion - - - taking into account damage x—x-x-experimental data

$$\left[\delta_{11}m_1 - \frac{1}{1-\left(\frac{\theta}{\omega_i}\right)^2 + r_l^2}\right] y_1 + \delta_{12}m_2y_2 + \cdots + \delta_{lk}m_ky_k + \frac{\Delta_{ip}}{\theta^2} = 0 \qquad (7)$$

here: m_k—is the mass of the beam section concentrated at the point k; θ is the frequency of the driving force; ω_i is the frequency of natural oscillations of the corresponding current; r_l^2—is the coefficient of inelastic resistance reflecting mechanical energy losses during vibrations, determined from experimental data; y_k is the amplitude value of the oscillations of the k-th mass; Δ_{kp} is the static displacement caused by the amplitude value of the disturbing force.

The unit displacements, δ_{ik} included in formulas (7), under conditions of corrosion damage and nonlinear and rheological deformation will depend on the level of the stress state. We will evaluate the specified factor using the method of integral estimates, then:

$$\delta_{lk} = \frac{\overline{M}_l \overline{M}_k}{\hat{B}(A, t)} \tag{8}$$

here: $\overline{M}_l$ and $\overline{M}_k$—are, respectively, plots of moments from the unit forces applied at points l and k; $\hat{B}(A, t)$ is an integral operator that estimates the rigidity of the beam section under consideration.

Given the dependencies (1), (2), and (6), we assume the relationship between stresses and deformations in the following form:

$$\sigma(t) = \varepsilon(t)E(t)K^*(z)\varphi(\varepsilon) - \int_{t_0}^{t} \varepsilon(\tau)\varphi(E)\frac{R(t,\tau)K^*(z)}{K(\theta,\sigma)}d\tau, \tag{9}$$

where E(t) is the modulus of elastic instantaneous deformations; $\varphi(\varepsilon)$ is a single function of nonlinearity, determined on the basis of experimental data; R(t,τ) is the resolvent of the creep measure; K(θ,σ) is the coefficient of vibration creep.

The value of the rigidity B (A, t) can be obtained using the well-known reasoning in the following form:

$$\hat{B}(A,t) = \left[\mathrm{E}(t)K^*(z)\varphi(\varepsilon) - \int_{t_0}^{t} \varphi(\varepsilon)\frac{R(t,\tau)K^*(z)}{K(\theta,\sigma)}d\tau\right]\left[\frac{bx^3}{12} + bx\left(g_3\frac{x}{2}\right)^2\right] + \mathrm{E}_s^{'}\omega s,\mathrm{x(t)}\mathrm{A}_s^{'}(g_0 - a')^2 + \frac{E_s\omega s,\mathrm{x(t)}A_s}{\psi}(h_0 - g_0)^2 \tag{10}$$

here: b is the width of the cross-section of the beam; x is the height of the compressed concrete zone; h_0 and g_0 are, respectively, the distance from the center of gravity of the stretched reinforcement and the center of gravity of the reduced section to the compressed face; E_s, $E_s^{'}$ is the elastic modulus of the compressed and stretched reinforcement; A_s, $A_s^{'}$ is the areas of the reinforcement.

The value of the vibration creep coefficient is determined by a well-known method, depending on the level of active stresses and the frequency of vibrations.

$$\mathrm{K}(\sigma_{max}, \sigma_{min}, \theta, t_0) = \frac{\boldsymbol{\Phi}_{\mathbf{no}}(\boldsymbol{\theta}_{\mathbf{min}})}{\boldsymbol{\Phi}_{\mathbf{no}}(\boldsymbol{\theta})} \tag{11}$$

θ_min = (1.5–3) 1/s—the minimum circular oscillation frequency at which the phenomenon of vibration creep is observed.

This approach turns the system of equations (7) into a nonlinear one, the solution of which is carried out using the method of integral estimates, which provides for the fixation of rheological and corrosion processes based on the method of consecutive approximations. This technique allows you to replace the solution of a nonlinear system of equations with multiple solutions of systems of linear equations with constantly updated values of the rigidity characteristics, and the use of the step-by-step method makes it possible to view and change the mode of action of the external load and the values of corrosion damage over time long-term operation.

3 Results and Discussion

As an illustration, we present the results of the calculation of a reinforced concrete beam with a span of 6 m. made of concrete of class B 12.5. Compressed and stretched zone reinforced of class A-400. Dimensions of the cross-section of the beam b × h = 60 × 25 cm. The amount of **reinforcement** compressed and stretched zone $A_s' = 1.57\ cm^2$, (2Ø 10); As = 15.2 cm^2, (4Ø 22). The resolution of the creep measure was determined from the integral equation:

$$L(t,\tau) - R(t,\tau) = \int_0^t L(\xi,\tau)R(t,\xi)d\xi, \tag{12}$$

$$\text{where:} \quad L(t,\tau) = -E(\tau)\frac{\delta}{\delta\tau}\left[\frac{1}{E(\tau)} + \hat{C}(t,\tau)\right] \tag{13}$$

$$\text{here:} \quad E(\tau) = E(\infty)\left(1 - \beta e^{-dt}\right); \tag{14}$$

$$C(t,\tau) = \frac{1}{E(t_0)} - \frac{1}{E(t)} + C_0\left[1 - e^{-r(t-t_0)}\right] + A\left[e^{-r_2 t_0} - e^{r_2 t}\right]. \tag{15}$$

t_0 and t—loading and observation time.

Values of other parameters included in Eqs. (7), (10), (14) accepted by the following E(∞) = 2.7 × 10^6 MPa; C_0 = 1.02 × 10^{-4} MPa; A = 0.82 × 10^{-4} MPa; α = 0.1 day; $\gamma_1 = \gamma_2 = 0.014$ day, γl = 0.06. The deformation nonlinearity function was assumed in the following form:

$$\varphi(t) = \{1 + \eta[\varepsilon(t)]^m\}. \tag{16}$$

In formula (10), the values of n = 0.172; m = 0.22. The law of change of the external load **P = P_oSinθt**, Po = 0.002 Mn, θ = 365 rpm. The value of the static evenly distributed load q = 0.003 Mn/m. For a mildly aggressive environment, the corrosion rate of reinforcement was assumed to be 0.01 mm/year, and of concrete 0.5 mm/year.

The results of the calculation according to the proposed method are shown in Fig. 1b, B. Comparing the plot of the bending moments calculated according to the proposed method with the values obtained without taking into account corrosion damage (Fig. 1b), the following conclusions can be drawn: taking into account corrosion and non-linearity of deformation leads to a decrease in the maximum bending moment up to 21% during the operating time t = 50 years.

The values of dynamic deflections calculated taking into account corrosion and vibration creep give a better match with the results of experimental studies (Fig. 1c). The difference was about 18%, Note that the calculation of a similar design for seismic effects will differ from the above device only in the form of a system of resolving equations (7), which (in this case, with synchronous displacements of the supports) will take the form:

$$\sum_{i=1}^{n} \delta_{lk} m_l \frac{d^2}{dt^2}(y_l + \Delta) = \frac{1}{1 - \left(\frac{\theta}{\omega}\right) + \gamma_e^2} y_1, \tag{17}$$

where Δ is the value of the dynamic displacement of the supports, determined by experimental accelerograms for a given seismic zone of construction.

Analysis of the results of the calculation of a reinforced concrete beam in the resonant region of vibrations at the first frequency of natural vibrations shows that the amplitude-frequency characteristic becomes nonlinear, i.e., the frequency of resonant vibrations becomes dependent on the amplitude of vibrations (Fig. 2), which is explained by the nonlinearity of the deformation of the structural material. Omitting the intermediate and final calculations here, we note that some simplifications were allowed during the calculations, and an approximate solution was obtained as a result of the decomposition, when the system (7) was resolved into trigonometric series using the harmonic balance method.

Fig. 2 Amplitude-frequency response

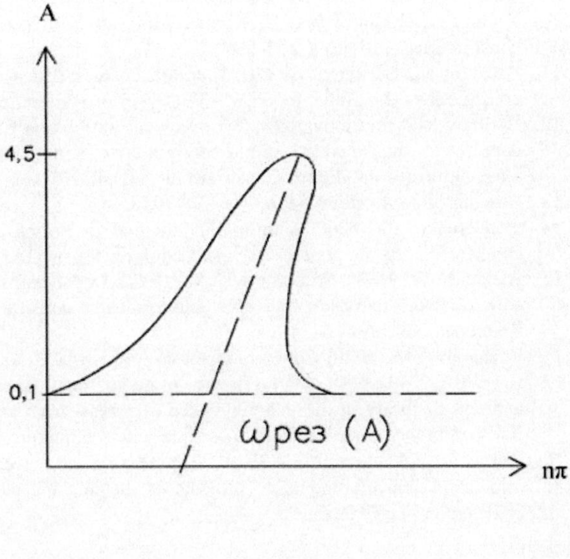

4 Conclusion

It should be noted that taking into account the corrosion damage and the non-linearity and non-equilibrium deformation of reinforced concrete structures under dynamic influences allows us to more fully take into account the real work of materials. This, in turn, makes it possible to find additional reserves of load-bearing capacity, thereby resulting in material savings.

References

1. Berlinov MV (2018) Strength resistance of reinforced concrete elements of high-rise buildings under dynamic loads. In: E3S web of conferences, 2018, vol 33, p 02049
2. Berlinov MV, Berlinova MN (2019) Durability of reinforced concrete constructions in conditions of prolonged operation. Bull Constr Equip 1(1013): 60–61
3. Berlinov MV, Berlinova MN, Gregorian AG (2019) Operational durability of reinforced concrete structures. In: E3S web of conferences, 2019, vol 91, p 02012
4. Bondarenko VM, Rimshin VI (2015) The theory of dissipative power resistance of reinforced concrete. Monograph, p 287
5. Zverjaev EM, Berlinov MV, Berlinova MN (2016) The integral method of definition of basic tension condition of anisotropic shell. Int J App Eng Res 11(8):5811
6. Korol EA, Berlinov MV, Berlinova MN (2016) Kinetics of the strength of concrete in constructions. Procedia Eng 165:292
7. Korol EA, Kharkin YuA (2011) Book of reports of XX Russian-Polish-Slovak workshop "theoretical fundamentals of construction", 401
8. Bondarenko VM, Kolchunov VI (2004) Calculation models of the power resistance of reinforced concrete monograph, p 112
9. Tamrazyan AG (2012) Calculation of structural elements with the specified normal distribution and reliability and load bearing capacity. Vestnik MGSU 10:109–115
10. Korol EA, Berlinov MV, Berlinova MN (2016) Kinetics of the strength of concrete in constructions. Proc Eng 292–297
11. Berlinov M, Berlinova M (2016) Influence of kinetics concrete hardening on strength of constructions. In: Всборнике: MATEC web of conferences, 2016, p 04014
12. Berlinov MV, Berlinova MN, Tvorogov AV, Pechkina EK (2020) Accounting for corrosive damage to the operated reinforced concrete structures in the conditions of three-axis tense-deformed condition. Constr Archit 8(3):40–46
13. Russian State Standard SP 63_13330_2018
14. Bondarenko VM (2009) Some fundamental questions of the development of reinforced concrete theory. In: V Intercollegiate conference—MGAKHIS, Moscow, 2009
15. Alexandrovsky SV, Solomonov VV (1972) Dependence deformations the creep of aging concrete from entry-level stresses, inter-industry construction issues. Domestic experience. Reference collection 6–12
16. Bazhenov YM (2020) Concrete technology. M.: ACB 500p
17. Korol EA, Vu Dyn T (2019) Influence of geometrical parameters of the cross section, strength and deformability of the materials used on stress-strain state of three-layered reinforced. In: IOP conference series: materials science and engineering, 2019, vol 661, p 012121
18. Korol OA, Dudina AG (2019) Engineering and technical support of territories for implementation of renovation projects of the housing stock. In: E3S web of conferences, 2019, vol 97, p 06027

19. Korol EA, Kustikova UO, Vu Dyn T, Nguen Hu K (2019) Finite element analysis of three-layer concrete beam with composite reinforcement. In: E3S web of conferences, 2019, vol 97, p 02023
20. Korol EA, Maylan AR, Petrosyan RS, Antoniadi DD (2020) Optimal order of construction of facilities in complex development with minimal additional costs. In: IOP conference series: materials science and engineering, 2020, vol 896, p 012052

Compressed Knots on Glued-In Rods of Wooden Structures

Aleksandr Pogoreltsev

Abstract Large-span glulam structures with glued-in rods rigid bonds are being widely used in Russia. When designing the joints perceiving compressive forces are commonly used the glued-in rods combined with a steel plate. The compression is transmitted to wood both via rods and the embedded part. Russian standards prescribe not to take into account the work of the embedded part because the joint work is poorly studied. However full-scale structural tests have shown that the joint bearing load capacity that preserve compressive forces significantly exceeds the calculated values determined without taking into account the join work of glued-in rods and steel plates. The manuscript represents the results of numerical and experimental studies of the joint operation of glued-in rods and steel embedded parts in compressed bonds of timber structures. Compressed joints of wooden structures at the built in Russia objects are considered.

Keywords Bearing load capacity · Compressive forces · Glued-in rod · Join work · Glulam structures · Large-span · Steel embedded parts

The widespread use of glued wooden structures in the construction of large-span buildings and structures has raised the issues of the development of node joints, taking significant forces. In Russia under the leadership of S. B. Turkovsky there was developed a system of knotted joints on glued rods—"TSNIISK system".

The system is based on the rods, primarily of periodic profile reinforcing steel, glued on epoxy compounds in the holes in the wood. With a competent design, the load-bearing capacity of such assemblies is determined by the strength of the rods.

Historically, the first studies carried out at the beginning of the XX century concerned the longitudinal reinforcement of bending structures, in which the joint work of the reinforcement with wood was provided using various mechanical means or glue (Granholm H. et al.) [1]. With the advent of epoxy adhesives, intensive research began on reinforced wooden structures, including prestressed ones, both in our country (V. F. Bondin, I. M. Linkov, V. Yu. Shchuko, etc.) and

A. Pogoreltsev (✉)
JSC «Research Center «Construction» (Kucherenko TSNIISK), Moscow, Russia

N. Vatin et al. (eds.), *Proceedings of MPCPE 2021*, Lecture Notes in Civil Engineering 182, https://doi.org/10.1007/978-3-030-85236-8_31

abroad (Bohannan B., Bond D., Lantos, G., Peterson, J. et al.) [2–6]. The studies concerned the effect of various reinforcement parameters on the bearing capacity.

In what follows, the use of glued-in rods for local timber reinforcement and in the nodes of wooden structures was also considered [7]. Abroad, glue-down rods with a metric thread were initially investigated, the diameter of the holes for which is less than the outer diameter of the thread [8]. Then, studies of glued single and group threaded rods with holes larger than the outer diameter of the rods began and continue today [9–13]. A number of tests included the study of rods glued perpendicular to the grain of the wood [14].

In contrast to foreign researchers who considered mainly threaded rods, the USSR conducted studies of joints with glued rods of periodic profile reinforcement used for reinforced concrete structures. In 1970 Pospelov N.D. rigid joints of beams with reinforcing rods glued along the fibers were proposed.

In 1981, in the Soviet Union in the new edition of the standard "Construction norms and regulations" SNIP II-25-80 "Wooden structures" appeared section "Connections on the glued steel bars, working on pulling or pushing through", which dealt with reinforcement rods glued along the fiber of the wood.

At the Volokolamsk plant of glued wooden structures in 1975, under the scientific supervision of S. B. Turkovsky. began the massive use of the connection on rods glued along the grain of the wood. But the disadvantages of such compounds discovered at the same time led to the need for their improvement. At TsNIISK, new joints were proposed, in which the glued-in rods were located not along the grain of the wood, but intersected the fibers at an angle.

The result of the research was a new system of nodal joints—the "TsNIISK system", which used reinforcing bars glued at an angle to the wood fibers [15–17]. Gluing the rods perpendicular to the wood fibers is a particular case of gluing at an angle.

In Russia, large-span structures made of glued wood with "CNIISK system" nodes on glued-in rods are widely used [18–22]. When designing the nodes that absorb compressive forces, glued rods combined with a steel embedded part are often used. Compressive forces are transmitted to the wooden element both through the glued-in rods and through the embedded detail. Examples are support assemblies for beams, arches, and frames, attachment assemblies for sprengel (Fig. 1) and compressed truss struts, cornice assemblies for frames (Fig. 2), etc.

Russian design standards "Code of rules SP64.1330.2011. Wooden structures" (hereinafter SP 64) and "Code of rules SP 382.1325800.2017 Glued wooden structures on glued rods. Calculation methods" (hereinafter SP 382) prescribe not to take into account the work of wood for crushing under the embedded part, which is due to the small study of joint work. However, full-scale tests of structural fragments have shown that the bearing capacity of nodes that perceive compressive forces significantly exceeds the calculated values determined without taking into account the joint work of glued-in rods and steel plates.

In 2007, control tests were carried out to confirm the load-bearing capacity of the tie-rod fastening in the support node of the roof truss at the Gorki-2 sports complex in the Moscow region (Fig. 3). The unit was calculated without taking into account

Fig. 1 Nodes of structures on glued-in rods that perceive compressive forces—the Aquatics Palace in Kazan with a crossbar in the form of a truss girder

Fig. 2 Nodes of structures on glued-in rods that perceive compressive forces of the frame with a span of 36 m of training ice hall of the sports complex "Yantar" in Moscow

the work of the wood for crushing under the embedded plate. The test fragments had a cross-section of 2 × (120 × 800) mm, a length of 1 m and simulated the fastening of the tightening to the wood with 4 Ø25 rods made of A400 reinforcement 500 mm long, glued at an angle of 25° to the direction of the fibers and united by the embedded part.

Fig. 3 Control tests of the fastening of the tightening in the support node of the truss covering the sports complexes in the village. Gorki-2, Moscow region: **a** general view of trusses in the process of installation; **b** support node; **c** fragment testing

Two fragments were tested:

Fragment No. 1 repeated the technical solution adopted in the project in terms of embedded parts, to which reinforcing bars were welded (Fig. 3b). The compressive forces were transmitted to the wood by the glued-in rods and the stop of the embedded part.

Fragment No. 2 was made with a 10 mm gap between the embedded part and the wood. Compressive forces were transmitted to the wood only by the glued-in rods.

The compressive force transmitted from the embedment to the wood in the support assembly was created in a Man-500 press with a capacity of 5000 kN (Fig. 3c). The design load on a fragment is equal to 500 kN, control—750 kN, which corresponds to the design load at the mode of loading A—linearly increasing load at standard tests. Loading was performed in steps of 100 kN or 1/5 of design load.

When fragment No. 1 was tested up to a load of 800 kN or 1.6 of the design load, uneven distribution of forces between the glued-in rods was observed, which leveled out later on. Plastic deformations expressed in bending of the plates of the embedded part were recorded at a load of 1500 kN. Destruction was caused by shearing of wood along the fibers at a load of 1530 kN, which is more than 2 times higher than the reference load. At the same time, the carrying capacity of the glued-in rods was not exhausted.

Fragment No. 2 fractured at a load of 1100 kN from punching the glued-in rods, which was almost 1.5 times higher than the control load.

The analysis of the test results confirmed the correctness of the design of the unit, the glued-in rods provide the required bearing capacity, and the work of timber for compression under the embedded part increases the bearing capacity of the unit 1.4 times.

The goal is to develop a methodology for taking into account the joint work of glued-in rods for punching and wood under the embedded part for compression in nodes that perceive compressive forces. Below are the results of preliminary numerical studies by the finite element method.

For a preliminary assessment of the effect on the bearing capacity of the joint work of glued rods and wood under the plate, the stress–strain state of model specimens was investigated, in which compressive forces are perceived only by the glued-in rods and jointly by the glued-in rods and the wood under the plate.

Model specimens with dimensions of 140 × 400 × 500 mm are equipped with two glued-in rods with a diameter of 16 mm with a gluing depth of 15d or 240 mm and an embedded part with dimensions "d × specimen width × 7d" or 16 × 140 × 112 mm (Fig. 4). The angle of inclination of the glued-in rods to the direction of the fibers was taken to be from 0° to 90° with a step of 15°.

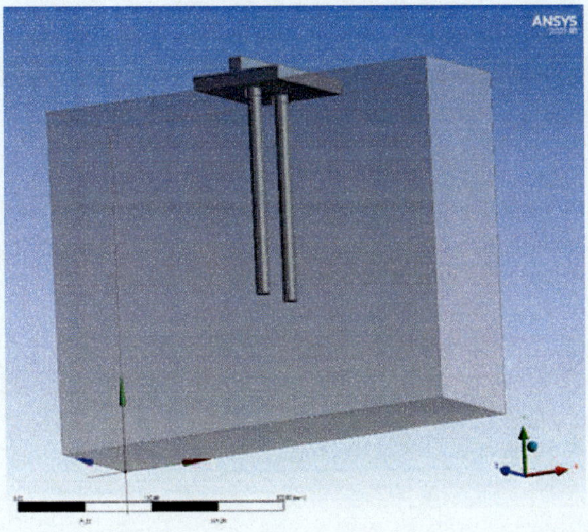

Fig. 4 Sample-model for the study of the stress–strain state of a node that perceives a compressive force

The dimensions of the samples were assigned based on the possibility in the future to make samples at different angles of sticking of the rods for experimental verification of the numerical analysis data from one workpiece with a cross section of 140 × 500 mm.

For each angle of inclination of the fibers to the direction of force and the glued in rods, 2 options were simulated:

- with the transfer of force only through two glued-in rods;
- with the transfer of force through the glued-in rods and at the steel plate—wood contact.

The load was taken to be equal to the design load determined in accordance with the design standards. In accordance with SP 64 and SP 382, for rods glued at an angle to the wood fibers, the bearing capacity of one glued in rod is determined by the formula (1):

$$T = R^A \pi d_1 l_r k_c k_\sigma k_d m_{pr} \Pi m_i \leq F_a R_a \tag{1}$$

where R^A = 6.8 MPa—design strength of wood in pull-out or push-in of a rod bonded-in at an angle to the grain or loading mode A (Table 4 SP 64);

d_1—hole diameter, m;

l_r—design length of the rod, m,

$$l_r = l - l_o \leq 30d; \tag{2}$$

l—anchorage length of rod, m;

l_o = $3d$—depth of a possible reduction in the strength of the adhesive layer due to welding; for rods without welding l_o = 0;

d—diameter of bonded-in rod, m;

k_c—factor for uneven distribution of shear stress depending on the anchorage length of the rod, which shall be determined by the formula

$$k_c = a_c - b_c(l_r/d); \tag{3}$$

a_c = 1.2;

b_c = 0.02;

k_σ—factor depending on the sign of normal stresses along the grain in the installation area of the rods;

k_d—factor depending on the rod diameter

$$k_d = a_d - b_d d; \tag{4}$$

a_d = 1.12;

b_d = 10;

m_{pr} = 0.66—load-duration factor, corresponding to the mode of duration B under the combined action of constant and snow load;

Πm_i = 1—product of end-use modification factors;

F_a—rod section area, m^2;

R_a—design strength of the material of the rod, MPa.

For punching bars $k_\sigma = 1$.

The design bearing capacity of a glued-in rod with a diameter of 16 mm and a gluing depth of 240 mm is 58.4 kN, while the maximum compressive stresses in the rods are 292 MPa. The design bearing capacity of the unit without taking into account the compressive work of the wood under the plate is 116.8 kN.

Design codes SP 64 and SP 382 indicate that rods glued at an angle to the fiber of less than 20° are considered as glued along the fibers, at an angle of 20° or more - as glued at an angle to the fibers. The rods glued across the fibers are a special case of the rods glued at an angle to the fibers. The design resistance of the seams of the glued-in rods is taken to be equal to the design resistance of wood when chipping along the fibers at an angle of inclination of not more than 20°, and the design resistance when punching and pulling out the rods at an angle of more than 20°.

Studies have shown that when forces are transmitted only through rods glued at an angle to the wood fibers—45°–90°, the distribution of normal compressive stresses along the length of the rods does not depend on the angle of inclination of the fibers and differs by no more than 5% (Fig. 5).

The distribution of normal stresses along the length of the rods glued along the grain of the wood—0°, is characterized by a more intense decrease in stresses at the initial section of the rod. This is due to the fact that with such rods, the wood has a higher shear stiffness, since along the entire perimeter of the glued-in rod, the shear

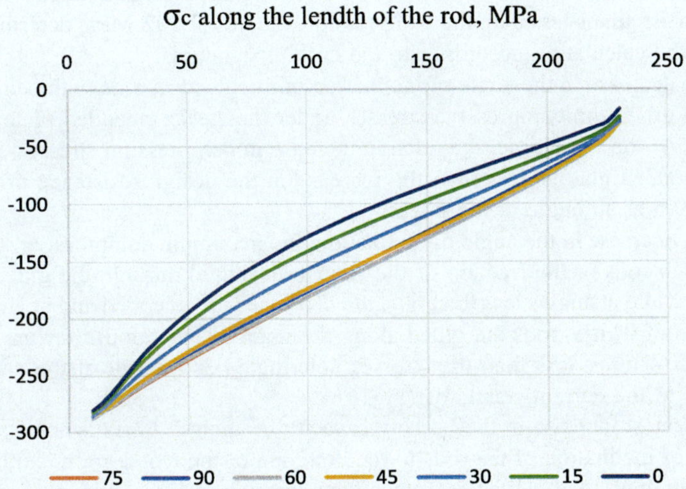

Fig. 5 Distribution of compressive stresses along the length of the glued-in rod without taking into account the work of compression of the wood under the steel embedded part at the angles of inclination of the rods to the direction of the fibers from 90° to 0°

modulus of the wood in the plane with axes along—across the fibers, which in the calculations is taken equal to $G_{0.90}$ = 500 MPa. With obliquely glued rods for a part of the perimeter of the rods, the shear modulus of wood in a plane with axes along—across the fibers, for the other part, the shear modulus of wood in a plane with axes across—across the fibers, which in calculations is assumed to be $G_{90.90}$ = 50 MPa.

For gluing angles of 15° and 30°, the distribution of normal stresses is intermediate.

The coefficient k_c, which takes into account the uneven distribution of shear stresses along the length of the glued-in rod, in the norms of SP 64 and SP 382 does not depend on the angle of inclination of the rod to the wood fibers. Analysis of the results of numerical studies of compressed nodes, in which the compression of wood under the plate is not taken into account, showed that for rods glued at an angle of 40°–90° to the wood fibers, the coefficient k_c can be assumed to be the same. At angles of inclination less than 40°, the coefficient k_c should be determined depending on the value of the angle of inclination.

When the glued-in rods and the steel plate work together at a given loading level of 116.8 kN, the average compressive stresses under the plate fluctuate from 2 to 8 MPa (Fig. 6a). Depending on the level of compression stresses of the wood under the plate, the forces transmitted to the glued in rods are reduced (Fig. 6b).

The minimum compressive stresses under the steel embedded plate (about 2 MPa) arise when the angle of inclination of the glued-in rods to the wood fibers is 75°, when the compressive strength of the wood is close to a minimum, and is 1.5 times less than the design resistance to compression across the fibers in the supporting parts of the structures R_{cm90} = 3 MPa. In this case, the maximum normal compressive stresses in the glued-in rods are equal to 185 MPa, which is 2 times less than the design resistance of the A400 reinforcement, equal to 375 MPa, and more than 1.5 times less than the compression stresses of 292 MPa, determined as a result of the calculation according to the current standards.

With a decrease in the angle of inclination of the glued-in rods to the direction of the wood grain, the compressive stresses under the steel embedded plate increase (Fig. 7a). At the same time, the relative increase in compression stresses under the steel embedded plate is less than the increase in the design resistance of wood to compression at an angle to the fibers.

With a decrease in the angle of inclination, the maximum compressive stresses in the glued-in rods or the fraction of the force transmitted through the glued-in rods decrease, while at angles less than 60°, the decrease is accelerated and at an angle of inclination of 0° (the rods are glued along the grain of the wood) they are equal to 72 MPa or 4 times less than the stresses determined as a result of the calculation according to the current standards (Fig. 7b).

The greatest increase in the bearing capacity of about 2 times is achieved when the angle of inclination of the rods to the direction of the wood grain is 60°. In this case, about half of the load is transmitted through the glued-in rods, and half through the wood, which works for crushing under the steel embedded plate.

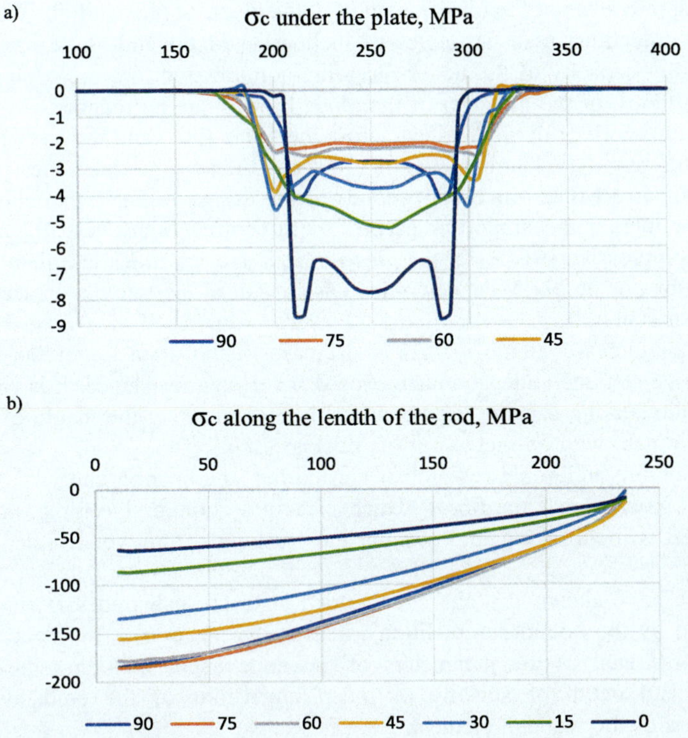

Fig. 6 Distribution of normal compressive stresses during joint operation of glued rods and wood under the steel embedded part at angles of inclination of rods to wood fibers from 90° to 0°: **a** along the length of the glued-in rod; **b** under the steel embedded part

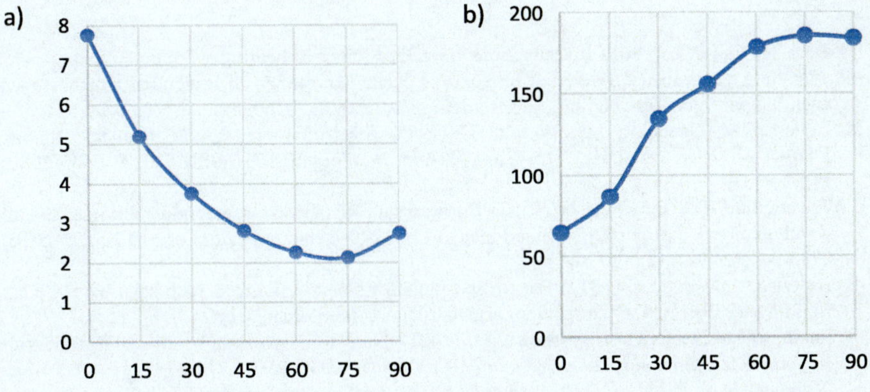

Fig. 7 Dependence of compressive stresses on the angle of inclination of the glued-in rods to the direction of the wood grain: **a** σ_c in wood under the steel embedded plate; **b** $\sigma_{c,max}$ in the glued-in rod

Taking into account the joint operation of the glued-in rods with the wood under the steel embedded plate at angles of inclination of the rods less than 75°, the load-bearing capacity of the nodes receiving compressive forces can be increased by 1.5–2 times, or the diameter of the glued-in rods can be reduced.

The joint work of the rods glued across the fibers (90°) and the steel embedded plate is different. At a loading level of 116.8 kN, the average shear stress under the plate is about 3 MPa, which corresponds to the design resistance to compression across the fibers in the supporting parts of the structures. In this case, the maximum normal compressive stresses in the glued-in rods are 1.5 times less than the compressive stresses of 292 MPa, determined as a result of the calculation according to the current standards.

With an increase in the load or a decrease in the diameter of the rods, the compressive stresses under the plate exceed the design resistance. It is possible to increase the bearing capacity of the assembly by increasing the length of the steel embedded plate with an increase in its thickness.

Thus, before additional research is carried out, the bearing capacity of a node with rods glued across the fibers, which perceives compressive forces, should be determined without taking into account the crushing of the wood under the steel plate.

Studies have shown that the stress–strain state of such nodes is significantly influenced by the boundary conditions, that is, the position of the node in a real structure, as well as the parameters of the node itself, including the number, diameter and depth of sticking of rods, dimensions of the embedded plate, dimensions of the wooden element.

Continuation of numerical studies and experimental confirmation of the results obtained are required.

References

1. Dorn H, Egner K (1961) Investigations on finger joints in bearing wooden structural parts after long years in use (Untersuchungen von Keilzinkenstößen in tragenden Holzbauteilen nach langjähriger Verwendung). https://doi.org/10.1007/BF02609522
2. Bulleit WM, Sandberg LB, Woods GJ (1989) Steel-reinforced glude laminated timber. J Struct Eng (United States) 115:433–444. https://doi.org/10.1061/(ASCE)0733-9445(1989)115:2(433)
3. Triantafillou TC, Deskovic N (1992) Prestressed FRP sheets as external reinforcement of wood members. J Struct Eng (United States) 118:1270–1284. https://doi.org/10.1061/(ASCE)0733-9445(1992)118:5(1270)
4. De Luca V, Marano C (2012) Prestressed glulam timbers reinforced with steel bars. Constr Build Mater 30:206–217. https://doi.org/10.1016/j.conbuildmat.2011.11.016
5. Lukin M, Sergeev M, Lisyatnikov M (2021) Non split wooden beam reinforced with composite reinforcement. https://doi.org/10.1007/978-3-030-72404-7_12
6. .Koscheev AA, Lukin MV, Roshchina SI (2019) Investigation of glued in bars strength and deformability indexes within framework of research of wood reinforcement with cable reinforcement. In: IOP conference series: materials science and engineering. https://doi.org/10.1088/1757-899X/687/3/033023

7. Steiger R, Serrano E, Stepinac M, Rajčić V, O'Neill C, McPolin D, Widmann R (2015) Strengthening of timber structures with glued-in rods. Constr Build Mater 97:90–105. https://doi.org/10.1016/j.conbuildmat.2015.03.097
8. Lukina A, Roshchina S, Gribanov A (2021) Method for restoring destructed wooden structures with polymer composites. https://doi.org/10.1007/978-3-030-72404-7_45
9. Göldi M, Sell J, Strässler H (1979) The shearing strength of glue-joints of preimpregnated wood—contribution to the development of weatherproof gluelams (Scherfestigkeit der Klebverbindung von vorimprägniertem Holz—Beitrag zur Entwicklung wetterbeständigen Brettschichtholzes). https://doi.org/10.1007/BF02607422
10. Azinović B, Serrano E, Kramar M, Pazlar T (2018) Experimental investigation of the axial strength of glued-in rods in cross laminated timber. Mater Struct Constr 51. https://doi.org/10.1617/s11527-018-1268-y
11. Tlustochowicz G, Serrano E, Steiger R (2011) State-of-the-art review on timber connections with glued-in steel rods. Mater Struct Constr 44:997–1020. https://doi.org/10.1617/s11527-010-9682-9
12. Rossignon A, Espion B (2008) Experimental assessment of the pull-out strength of single rods bonded in glulam parallel to the grain. https://doi.org/10.1007/s00107-008-0263-3
13. Parida G, Johnsson H, Fragiacomo M (2013) Provisions for ductile behavior of timber-to-steel connections with multiple glued-in rods. J Struct Eng (United States) 139:1468–1477. https://doi.org/10.1061/(ASCE)ST.1943-541X.0000735
14. Steiger R, Gehri E, Widmann R (2007) Pull-out strength of axially loaded steel rods bonded in glulam parallel to the grain. Mater Struct Constr 40:69–78. https://doi.org/10.1617/s11527-006-9111-2
15. Turkovskij SB, Pogorel'tsev AA (2001) Wooden structures with rigid joints in structures with corrosive medium. Promyshlennoe i Grazhdanskoe Stroit 10–13
16. Lukin M, Prusov E, Roshchina S, Karelina M, Vatin N (2021) Multi-span composite timber beams with rational steel reinforcements. Buildings. https://doi.org/10.3390/buildings11020046
17. Lisyatnikov MS, Glebova TO, Ageev SP, Ivaniuk AM (2020) Strength of wood reinforced with a polymer composite for crumpling across the fibers. In: IOP conference series: materials science and engineering. https://doi.org/10.1088/1757-899X/896/1/012062
18. Turkovskij SB, Pogorel'tsev AA, Eknados'yan IL (2003) Selection of design scheme of lens-shaped trusses from adhesive wood. Stroit Mater 18–20
19. Preobrazhenskaya IP, Pogorel'tsev AA, Turkovskij SB (2003) Development of design and construction of potassium chloride storehouse with framework from prefabricated wood frames with size of 63 m. Stroit Mater 14–16
20. Gorpinchenko VM, Pogorel'tsev AA, Ecknadosyan IL (2005) Large-scale tests provided for a block, consisting of two wooden lens-type roof trusses of the sports complex "Strogino." Promyshlennoe i Grazhdanskoe Stroit 38
21. Roschina SI, Lisyatnikov MS, Koshcheev AA (2019) Technical- and- economic efficiency of reinforced wooden structures. In: IOP conference series: materials science and engineering. https://doi.org/10.1088/1757-899X/698/2/022005
22. Roshchina S, Lukin M, Lisyatnikov M (2020) Compressed-bent reinforced wooden elements with long-term load. Lect Notes Civ Eng. https://doi.org/10.1007/978-3-030-42351-3_7

Numerical Simulation of Steel Columns of Industrial Buildings with Local Mechanical Damage During Explotation

Nina Buzalo, Ivan Gontarenko, Boris Chernykhovskiy, and Anastasia Alekseeva

Abstract This paper presents the results of a numerical study of the stress–strain state of a model of an eccentric compressed I-section column with damage in the ANSYS software, taking into account the nonlinear behavior of the material and contact interactions. A CAD-model of the investigated sample of the structure, as well as a finite-element design model, including finite elements of a volumetric type, has been completed. The planning of a numerical experiment was carried out, the parameters of damage affecting the results of the study were determined. The number of experiments covering the range of variation of the identified influencing parameters of damage was determined. On the basis of the developed computer model, a series of numerical experiments was carried out to assess the effect of the magnitude of local mechanical damage on the force resistance of eccentric compressed elements, which will allow experts to give recommendations on its reliability and the possibility of further explotation when conducting a technical examination of the explotated steel structures of industrial buildings.

Keywords Nonlinear system · Innovation development · Mechanical properties · Local damage · Stress concentrators · Full factorial experiment · Eccentric compression · Stress–strain state · Computer simulation · Nonlinear behavior

1 Introduction

For an objective assessment of the influence of damage parameters on the bearing capacity of the bearing elements of metal structures and when performing verification calculations, it is necessary to have as much information as possible. Many elements of explotated metal structures have zones in which local stress concentrations are observed. This usually occurs near cuts, cracks, in welded seams, in places of mechanical damage, like in Fig. 1. The appearance of stress concentration

N. Buzalo · I. Gontarenko (✉) · B. Chernykhovskiy · A. Alekseeva
Platov South-Russian State Polytechnic University (NPI), 132, Prosveshcheniya, Rostov Region, Novocherkassk 346428, Russian Federation

N. Vatin et al. (eds.), *Proceedings of MPCPE 2021*, Lecture Notes in Civil Engineering 182, https://doi.org/10.1007/978-3-030-85236-8_32

Fig. 1 Local mechanical damage to the column at the crane trestle section of the fire cutting section of a metalworking enterprise

in some cases, for example, with alternating loads, can lead to a decrease in the strength of the element and its brittle failure.

In places of stress concentration, the Bernoulli hypothesis (flat cross-section hypothesis) is unfair, and the formulas of the beam theory are inapplicable [Neuber H. Stress Concentration. OGIZ, 1947]. Stresses near stress concentrators must be determined by methods of elasticity theory, computer simulation or experimentally.

An example is the articles [1–6]. The authors of this article performed computer modeling and structural analysis of an eccentrically compressed I-beam bisteel rack.

The objective of the research is to study the stress distribution near the local damage of the eccentrically compressed I-section rack on physical and numerical models.

This paper presents the results of a numerical study of the stress–strain state of a model of an eccentrically compressed damaged I-section rack in the ANSYS software, taking into account the nonlinear properties of the material and contact interactions [7, 8].

On the basis of the developed computer model, a series of numerical experiments was performed to assess the effect of the magnitude of local mechanical damage on the force resistance of eccentrically compressed elements. This will allow the engineers of the construction expertise to give recommendations on their reliability and the possibility of further operation during the technical examination of the operated metal structures of industrial buildings. Information on the effect of stress concentration on the bearing capacity of an element can be obtained by conducting experimental and numerical studies of structures and their models, which are associated with significant financial and time costs. Experiment planning methods are used to reduce the cost of such research.

2 Method

In the presented work, a model of a rack with local mechanical damage to the wall (shown in a Fig. 2) is investigated. The number of factors studied is 3, the levels and intervals of variation are presented in Table 1. The factors determining the process are indicated in the following order: $\tilde{x}_1$—the length of the column wall cutout, $\tilde{x}_2$—the column wall cutout width, $\tilde{x}_3$—the force in the column section from the external load. Required number of experiments $2^3 = 8$.

As the main technique for planning a numerical experiment, a full factorial experiment of the type 2^k was chosen, which allows, with a known number of factors, to find the number of experiments necessary to implement all possible levels of factors [9]. For this, a formula $N = 2^k$ is used, where N—is the number of experiments, k—is the number of factors, 2—is the number of levels.

For a linear model, it is enough to indicate only two levels: the upper and lower boundaries of the variation interval. For simplicity, coded values of the levels are assumed: the upper level corresponds to +1, the lower –1. The main level, which is necessary to determine the variation interval, is taken equals zero. Thus, the factorial space of the design is established [10–12].

To study the influence of various factors on the distribution of stresses in the sections of damaged struts, a factorial design is used, built in such a way that only one variable varies in each experiment, and the values of the remaining independent variables remain constant. By varying each independent variable in turn, we establish the required dependencies. Factors are encoded according to the formula:

$$x_j = \frac{\tilde{x}_j - \tilde{x}_{j0}}{I_j}, \quad (1)$$

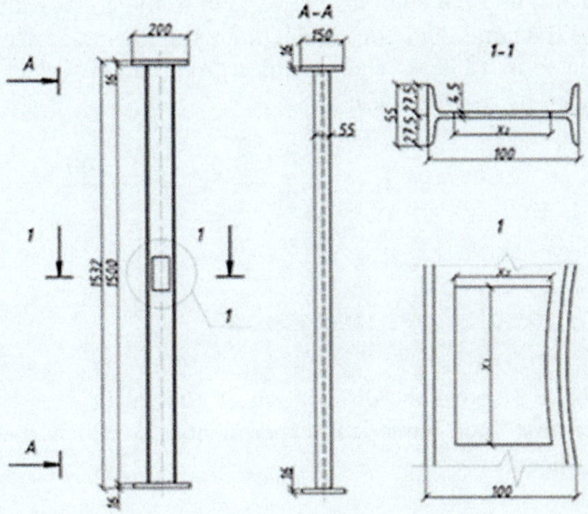

Fig. 2 Geometric dimensions of the rack model

Table 1 Variation levels and intervals

Levels	Factor		
	$\tilde{x}_1$, mm	$\tilde{x}_2$, mm	$\tilde{x}_3$, mm
Main level	120	40	675
Upper level	140	50	775
Lower level	100	30	575
Variation interval	20	10	100

Table 2 Experiment planning matrix 2^3

№ test	x_1	x_2	x_3
1	+	+	+
2	−	+	+
3	+	−	+
4	−	−	+
5	+	+	−
6	−	+	−
7	+	−	−
8	−	−	−

where

x_j is coded value of the factor,
$\tilde{x}_j$ natural value of the factor,
$\tilde{x}_{j0}$ natural value of the main level,
I_j variation interval,
j number of the factor.

The experimental conditions are presented in tabular form (Table 2) in the form of an experiment planning matrix, where the rows correspond to different experiments, and the columns correspond to the coded values of the factors.

The conditions for conducting experiments in natural values of factors are presented in Table 3. The transition from natural values of factors to coded values is given by the formulas:

$$x_1 = \frac{\tilde{x}_1 - 120}{20},\ x_2 = \frac{\tilde{x}_2 - 60}{10},\ x_3 = \frac{\tilde{x}_3 - 775}{100}$$

3 Results and Discussion

Figure 3 shows the distribution of equivalent stresses according to von Mises for the first four numerical experiments, in which the highest stress values were obtained.

Table 3 Natural values of the factors

№ test	x_1	x_2	x_3
1	140	50	775
2	140	30	775
3	100	50	775
4	100	30	775
5	140	50	575
6	140	30	575
7	100	50	575
8	100	30	575

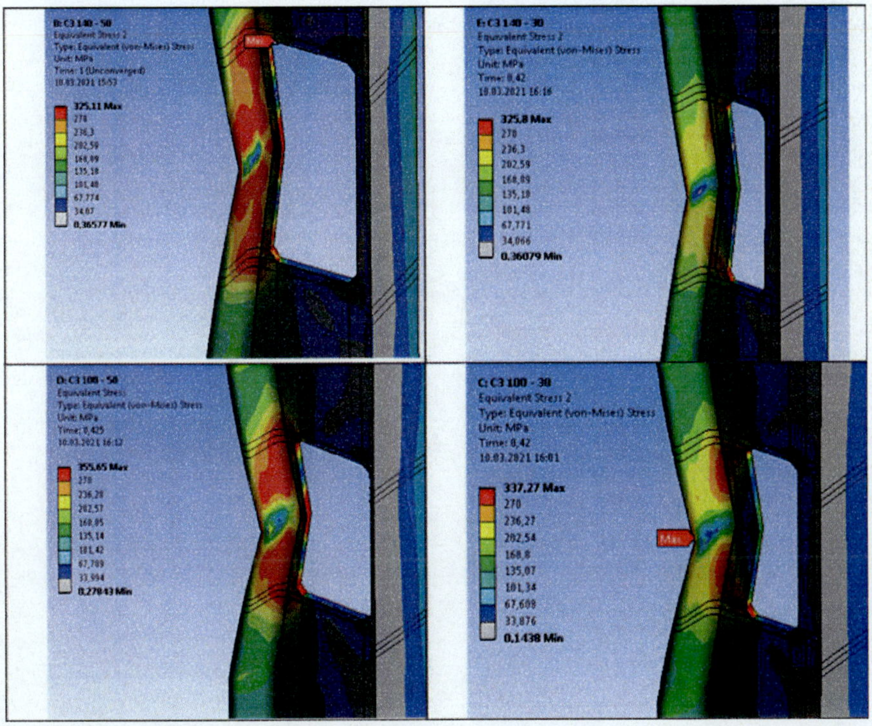

Fig. 3 Distribution of equivalent stresses in racks with localized damage

To determine the displacements of the rack, the point of the damaged section was selected, shown in Fig. 4.

Figure 5 shows the dependences of the displacements of the selected point on the applied forces for the first four numerical experiments, in which the highest stress values were obtained [13].

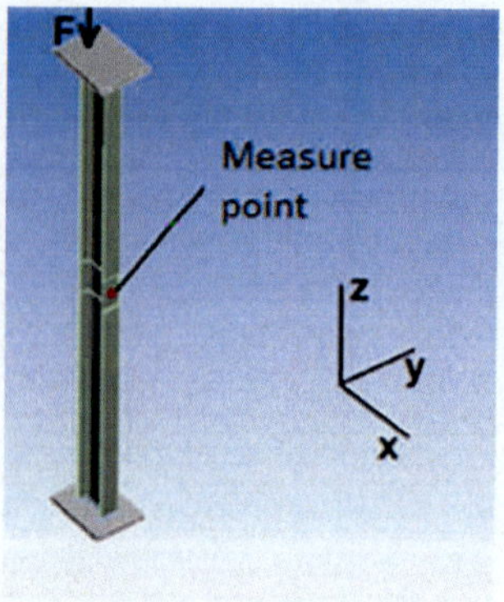

Fig. 4 Point of the section at which the displacements are measured

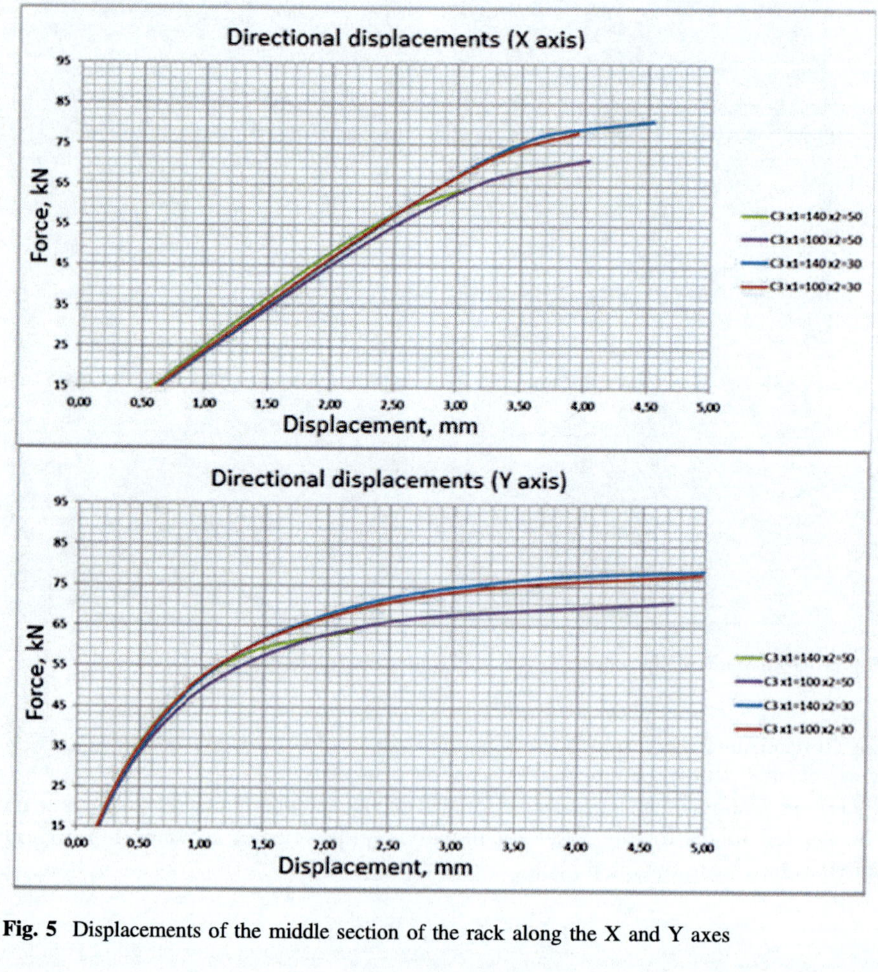

Fig. 5 Displacements of the middle section of the rack along the X and Y axes

4 Conclusions

The study performed allows us to determine the influence of the parameters of the damaged rack on the stress concentration for any point of the damaged section and to evaluate the rack displacements under load.

In the presence of a stress concentration, the deformation of the damaged sample increases in comparison with the deformation of an undamaged element [14, 15], which is associated with the appearance of plastic deformations at the stress concentrator. The part of the damaged section, where the stress concentration is observed, works outside the elastic model. This suggests that, in reality, the load-bearing section is even smaller than the actual section of the damaged rack. The described phenomenon is especially dangerous for steel structures exposed to alternating loads, or operated in cold areas. Generally, the fatigue strength of any steel element is lower than the yield strength. If an alternating load causes plastic deformations in the damaged structure [16–21], then low-cycle fatigue failure of the element becomes probable. High-cycle fatigue failure is also likely if the amplitude of the acting stresses in the damaged element is close to or higher than the fatigue limit for the element in question. The combination of the above factors with low temperatures also negatively affects the durability of metal structures, since at low temperatures the steel becomes more brittle.

Acknowledgements The reported study was funded by RFBR, project number 20-38-90056\20 and project number 20-38-90046\20.

References

1. Evtushenko SI, Petrov IA, Shutova MN, Chernykhovsky BA (2021) Bearing capacity of eccentrically compressed bisteel columns. Mag Civ Eng 102. https://doi.org/10.34910/MCE.102.1
2. Buzalo N, Gontarenko I, Chernikhovski B (2020) Force resistance of steel columns of industrial buildings with corrosion damage. In: IOP conference series: materials science and engineering, vol 896, p 012044. https://doi.org/10.1088/1757-899X/896/1/012044
3. Evtushenko SI, Petrov IA, Alexeev SA (2019) Optimization task when calculating the bi-steel thin-walled rod. https://doi.org/10.1063/1.5138445
4. Umnyakova NP (2018) Numerical modeling of heat transfer through the air interlayer considering the surface radiation. In: IOP conference series: materials science and engineering, vol 456, p 012096. https://doi.org/10.1088/1757-899X/456/1/012096
5. Ilyashenko AV (2008) About eccentric compression of thin-walled bars composed of initially curved finned elements. Int J Comput Civ Struct Eng 4:71–74
6. Przybylski J, Kuliński K (2019) Shape enhancement of an eccentrically loaded column using piezoelectric actuator. Eng Struct 189:644–654. https://doi.org/10.1016/j.engstruct.2018.07.014
7. Buzalo N, Gontarenko I, Chernykhovsky B (2020) Reducing the force resistance of steel columns in industrial buildings with corrosion damage during operation. Constr Archit 8:9–13. https://doi.org/10.29039/2308-0191-2020-8-4-9-13

8. Vorontsov GV, Petrov IA, Alexeev SA (2008) Stiffness matrices of spatially loaded non-linearly deformable rods. Part 2. News High Educ Inst. North Caucasus Reg Ser Eng 2:15–17
9. Chepurnenko A, Neumerzhitskaya N, Turko M (2018) Finite element modeling of the creep of shells of revolution under axisymmetric loading. https://doi.org/10.1007/978-3-319-70987-1_86
10. Trubina D, Abdulaev D, Pichugin E, Garifullin M (2014) The loss of local stability of thin-walled steel profiles. Appl Mech Mater 633–634, 1052–1057. https://doi.org/10.4028/www.scientific.net/AMM.633-634.1052
11. Zarifyan AZ, Petrov IA (1999) Determination of the optimum sizes of cross section compressed H-shaped bisteel columns. Eazy Build Constr 8:1–14
12. Petrov IA (2000) Optimization of sections of eccentrically compressed bistal columns. SRSTU (NPI), Novocherkassk
13. Ragheb WF (2015) Local buckling of welded steel I-beams considering flange–web interaction. Thin-Walled Struct 97:241–249. https://doi.org/10.1016/j.tws.2015.09.026
14. Lalin VV, Rozin LA, Kushova DA (2013) Variational functionals for two-dimensional equilibrium and stability problems of Cosserat-Timoshenko elastic rods. Mag Civ Eng 36:87–96. https://doi.org/10.5862/MCE.36.11
15. Lalin VV, Zdanchuk EV, Kushova DA, Rozin LA (2015) Variational formulations for non-linear problems with independent rotational degrees of freedom. Mag Civ Eng 56:54–65. https://doi.org/10.5862/MCE.56.7
16. Evtushenko SI, Fedorchuk VE, Alexeeva AS, Chernyckhovsky BA (2018) Strain analysis of supporting profile vulcanizing press. Mater Sci Forum 931:200–206. https://doi.org/10.4028/www.scientific.net/MSF.931.200
17. Evtushenko SI, Petrov IA, Shutova MN, Alekseeva AS (2017) The comparative analysis of different computations methods of strength of materials by the example of calculations of the axle beam. In: IOP conference series: materials science and engineering, vol 177, p 012023. https://doi.org/10.1088/1757-899X/177/1/012023
18. Buzalo NA, Alexeeva AS, Chernykhovsky BA (2019) Modeling of damages of steel structures of industrial buildings. J Phys Conf Ser 1425:012057. https://doi.org/10.1088/1742-6596/1425/1/012057
19. Chernykhovsky B, Evtushenko S, Alekseeva A (2021) Finite element analysis of a frame connection of a girder with a rack under beyond design basis impact. In: IOP conference series: materials science and engineering, vol 1030, p 012045. https://doi.org/10.1088/1757-899X/1030/1/012045
20. Buzalo NA, Alexeeva AS, Chernykhovsky BA (2019) Modelling framework joints in ANSYS software complex. In: IOP conference series: materials science and engineering, vol 687, p 033036. https://doi.org/10.1088/1757-899X/687/3/033036
21. Buzalo N, Tsaritova N, Chernykhovsky B (2019) Swivel joints of transformable net shells structures. Mater Sci Forum 945:276–280. https://doi.org/10.4028/www.scientific.net/MSF.945.276

Numerical Studies of the Optimal Parameters of the Flange of a Two-Layered Reinforced Concrete Element of a T-section with High Strength Concrete in a Compressed Zone

Yuri Rogatnev, Oleg Sokolov, Jeremy Minani, Dmitry Panfilov, and Yuri Ivanov

Abstract A numerical study of the stress–strain state of two-layered flexible reinforced concrete T-section elements with high-strength concrete –B 90 in the compressed zone and concrete B 35 in the tensional zone has been carried out. 10 models with longitudinal reinforcement made of steel rebar of A-500C, AT-800 and 5 classes with glass-composite reinforcement have been developed. The bearing capacity of the elements has been investigated. The ability of the flanges to redistribute stresses along their width has been determined. The value of the bearing capacity and the height of the compressed zone was determined according to SP 63.13330 and SP 295.1325800. The cases of the distribution of deformations beyond the boundary of the contact of two layers are considered. The influence of the transition of compressive stresses across the boundary of two layers on the bearing capacity and on the redistribution of compressive stresses along the width of the flange is determined. The influence of the modulus of elasticity and the cross-sectional area of the working reinforcement on the bearing capacity of the elements and on their ability to redistribute stresses was also established. The relationships between the deflection values and the external load are determined for all considered elements. The nature of stress redistributions along the flange width at different load levels has also been established. An analysis was carried out in order to select the optimal class of working reinforcement for the elements under study.

Keywords High strength concrete · T-beam · Effective width flange · High strength reinforcement · Composite reinforcement

Y. Rogatnev · O. Sokolov (✉) · J. Minani · D. Panfilov · Y. Ivanov
Voronezh State Technical University, 84, 20 letiya Oktyabrya St., Voronezh, Russian Federation
e-mail: osokolov@vgasu.vrn.ru

N. Vatin et al. (eds.), *Proceedings of MPCPE 2021*, Lecture Notes in Civil Engineering 182, https://doi.org/10.1007/978-3-030-85236-8_33

1 Introduction

In the practice of designing bending reinforced concrete elements, the T-section is found as separate elements, as part of a monolithic beam floor and as a design section of precast floor slabs. The wide range of elements is explained by the rationality of using the cross section. A single-span T-beam is considered, the flange of which is made of high-strength concrete of class B90, and the web is made of concrete of normal strength B35. Previous studies of one-layered high strength concrete bending elements made it possible to establish their effectiveness [1–3].

The bending T-flange distributes compressive stress across its entire width. However, their distribution in the direction from the middle of the web to the edge of the flange has a damping character. At large overhangs of the flange, the difference between the two values becomes significant, and the bearing capacity of the element does not correspond to the assumed values. Experimental studies of elements of different design, method of support and nature of loading made it possible to establish the limiting values of the width of the flange [4–6]. These studies are the basis for limiting the values of the T-flange width used in the calculation in domestic and foreign regulatory documents. These values are structured in the form of dependencies of the flange overhang or its width on other geometric and design parameters of the element. Thus, in SP 63.13330.2018, the width of the flange overhang for an individual element should not exceed 1/6 of the element span and no more than 6 h_f flanges for $h_f \geq 0.1$ h, 3 h_f flanges for $0.05h \leq h_f \leq 0.1h$, where h_f is the thickness shelves, h—cross-sectional height. In EN 1992–1-1, the effective flange width for a single element

$$b_{eff} = 0.2 \cdot l_0 \cdot b_w \tag{1}$$

Effective compression flange widths of isolated beams shall not exceed 4 b_w for Japan concrete codes—"Japan Society of Civil Engineers-Standard Specification for Concrete Structures". As for Japan codes the max value of effective flange width in ACI 318–19 also shall not exceed 4 times web width. But National Standard of the People's Republic of China—GB 50,010 have the same values as for SP 63.13330. In addition, another way to assess the ability to redistribute stresses is to look for the value of the useful shelf width. The effective flange width is an abstract value at which the values of the compressive stresses above the edge would be equal to the values at the edge of the flange [7]. It is noted that the effect on the redistribution of stresses along the width of the flange is mainly exerted by its rigidity. Research of prestressed T-beams shows efficient distribution of stresses in such elements, even at values of the flange width, which were much higher than the maximum permissible [8].

In this regard, the determination of the limiting value of the flange width for a two-layered T-element with high-strength concrete in a compressed zone is an urgent task. Preliminary numerical studies have shown the effectiveness of their use with high-strength reinforcement AT-800 without prestressing [9]. Using of composite reinforcement also shows effectiveness in bending elements [10].

Thus, the task of the current work is to determine the limiting value of the width of the flange of the elements under study and to assess the ability to redistribute compressive stresses along its width for various types of reinforcement.

2 Methods

There are a large number of software packages based on the finite element method. Some of them make it possible to determine with sufficient accuracy the stress–strain state of the structure under study [11, 12]. The current problem was solved in the Ansys 2020 R2 software package. This software package allows you to take into account the nonlinear work of materials of reinforced concrete structures [13].

The investigated beams were tested with a concentrated load in one third of the span over the web. The span was 2050 mm. In this case, the supports on one side were hinged, and on the other, hinged-fixed (Fig. 1).

The physical and mechanical properties of concrete of normal strength class B35, reinforcement AT-800, A-500C and glass-composite reinforcement were obtained as a result of tests and taken from the works devoted to the determination of the stress–strain state of double-layered structures [9, 14]. In view of the fact that the redistribution of stresses along the width of the flange is influenced by its crack resistance, it became necessary to investigate high-strength concrete jointly for axial tension and compression (Fig. 2). The test results of high-strength concrete are presented in Table 1.

The course of the numerical experiment will be similar in structure to that of [3]. The author set himself the task of determining the influence of a compressed flange on the bearing capacity of a T-beam. To solve this problem, the elements were tested with a sequential increase in the width of the flange, while with a constant flange thickness and span. In his work, the author, within the framework of one series, considered elements with an overhang from $2.5h_f$ to $6.8h_f$. In the current work, the flange overhang will be considered in the range from 6 to $10h_f$, since the concrete used by the authors was of low grade. In parallel with the study of the effect of the compressed flange on the bearing capacity of the element, the useful width of the element flange is considered. The coefficient of completeness of the stress diagram is used for analysis of distribution compressive stresses over a compressed flange [15, 16].

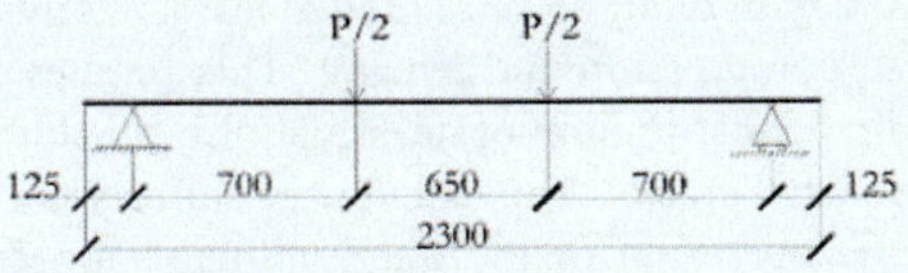

Fig. 1 Design diagram of a T-beam

Fig. 2 Compression and tensile testing of high-strength concrete specimens

Table 1 Physical and mechanical characteristics of high-strength concrete

Rb (MPa)	εb (%)	Rbt (MPa)	εbt	Eb (MPa · 10^{-3})
65.02	0.156	5.36	0.0128	41.68

$$\omega_\sigma = \frac{\int_{-\frac{bf}{2}}^{\frac{bf}{2}} \sigma_x d_x}{b_f \sigma_{\max}} \tag{2}$$

where

σ_x the ordinates of the stress plot in the flange.
$\sigma_{\max}$ maximum flange stresses above the rib.

In this case, the useful width of the flange was determined by the formula

$$b_{eff} = b_f \cdot \omega_\sigma \tag{3}$$

The analysis of compressive stresses, in the current work, was carried out on a cross-section in the middle of the beam span, at the ultimate load. In this case, the step of taking readings was equal to 1/50 b_f (Fig. 3). In Ansys 2020 R2, the path tool was used to determine the stresses along the cross-section at the center of the beam (Fig. 3).

Since the current work considers elements with flanges whose overhangs exceed the maximum permissible values, it is obvious that the flanges work simultaneously in compression and tension. This implies the formation of additional irregularities in the distribution of deformations, in addition to those that are formed as a result of normal cracks. This unevenness affects not only the operation of the flange, but also the operation of the longitudinal reinforcement of the flange. When stresses are distributed between the rods of the longitudinal reinforcement of the flange,

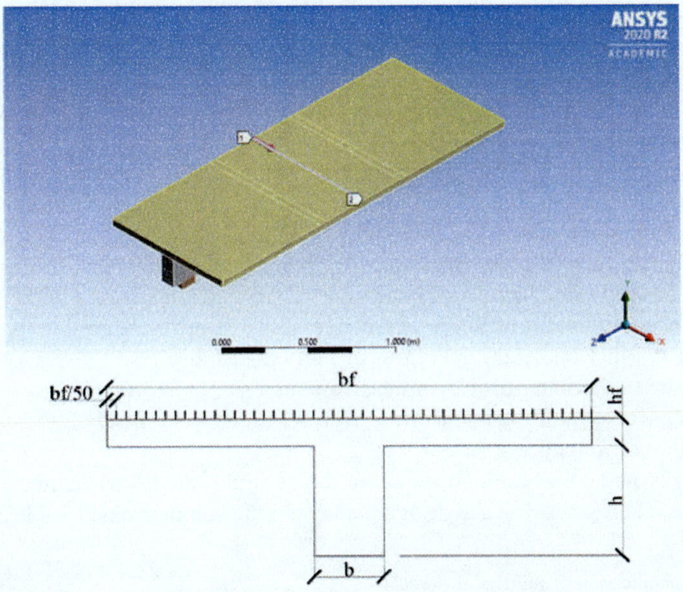

Fig. 3 Determination of the completeness factor of the stress diagram

sufficiently distant rods work ineffectively and do not affect the bearing capacity of the beam. For isolated, single-span elements, it is recommended to install longitudinal reinforcement within the web. The device of longitudinal reinforcement in the overhangs of the flange is possible at a distance equal to h_f in each direction from the web with a shortage of bearing capacity of 5–6%. In the current work, the longitudinal reinforcement, which affects the bearing capacity, is installed along the axis of the beam web. The rest of the reinforcement is located outside the border of effective work and does not affect the bearing capacity.

An important design parameter when determining the maximum permissible values of the width of the flange is the percentage of working reinforcement. First, its value should exceed the maximum permissible, since the effect of a compressed flange on the bearing capacity can be estimated at the value of the bending moment

$$M = M_{ult}$$

$$M_{ult} = R_b \cdot b_f \cdot x \cdot \left(h_0 - \frac{x}{2}\right) + R_{sc} \cdot A_s' \cdot (h_0 - a') \quad (4)$$

In this case, it is necessary to strictly observe the selected percentage of reinforcement for elements with the same type of longitudinal reinforcement and, when determining it, take into account the compressed overhangs of the flanges. This approach makes it possible for the beams to set the height of the compressed zone constant, and the only variable parameter that affects the bearing capacity from

Eq. (4) is the flange width. The values of the limiting percent of reinforcement for different types of reinforcement of double-layered elements with high-strength concrete in the compressed zone are determined in [9].

To achieve the required limiting state in the elements under study, it is required to ensure the strength of the remaining stressed zones. One of these is the contact zone between the flange and the web. To ensure strength, in previous studies and with flange overhang values ≥4.75 h_f, the flange had transverse reinforcement [3]. This reinforcement had a positive effect on the redistribution of compressive stresses. In the current work, the transverse rebar in the flange was arranged with a step of 100 mm, the diameter of the rods is 10 mm, the reinforcement class is A-500C. Longitudinal rods in the overhangs of the flange were arranged in order to ensure its stability and joint operation of individual transverse rods, while slightly affecting the redistribution of compressive stresses.

The transverse reinforcement of the web was selected from the need to prevent fracture along the inclined section.

Based on the above assumptions about the design of the tested beams, 15 models were made. Their geometric and design parameters are summarized in Table 2 (Fig. 4).

Table 2 Specification of numerical models

Name of element	Class of reinforcement	Percentage of reinforcement	b_f (mm)	h_f (mm)	b (mm)	h (mm)
DTB-A1-1	A 500C	3.8	700	50	100	200
DTB-A1-2			800			
DTB-A1-3			900			
DTB-A1-4			1000			
DTB-A1-5			1100			
DTB-A2-1	AT 800	2.2	700			
DTB-A2-2			800			
DTB-A2-3			900			
DTB-A2-4			1000			
DTB-A2-5			1100			
DTB-A3-1	AGC	2.2	700			
DTB-A3-2			800			
DTB-A3-3			900			
DTB-A3-4			1000			
DTB-A3-5			1100			

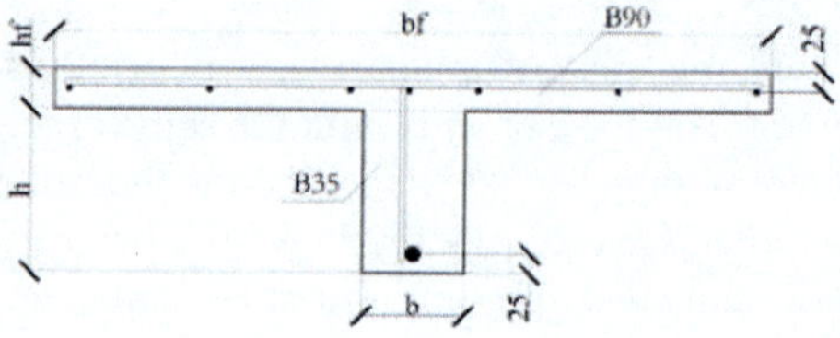

Fig. 4 Cross section of test items

3 Results and Discussion

The element is loaded with a certain step until it breaks. Evidence of reaching the limit state is the message of the software package about exceeding the maximum permissible values of the increment of plastic deformations, and the relationship between the displacement values and the degree of load indicates the nature of destruction necessary for the study (Fig. 5). The relationship between stresses and deformations for the working reinforcement was determined by a two- or three-line diagram, if the rebar reached the limit values of deformations, these diagrams would have repeated the relationship between displacements and the load. For all three types of reinforcement, the increase in bearing capacity stops at flange overhang values of $8h_f$ (Table 3). At the same time, in the DTB-A3 elements, a higher level of increase in the bearing capacity is observed with an increase in the width of the flange. So elements DTB-A1 and DTB-A2 showed the level of increase in bearing capacity of 3.5–3.8%, and DTB-A3—7.6%. However, the value of the bearing capacity of the elements DTB-A1 and DTB-A2 is higher (Fig. 6). It should be added that the bearing capacity of DTB-A2-5 not only did not increase compared to DTB-A2-3, but also turned out to be less, therefore it will not be taken into account in the analysis.

Despite the same approach to the design of elements, the boundary of the height of the compressed zone turned out to be at different levels. So, for DTB-A1 elements, the border of the compressed zone passed in the middle of the rib at a level of 125 mm, for DTB-A3 elements, the border passed through the place of contact of the flange and the web at a level of 50 mm, and for DTB-A2 elements at 75 mm. Only two items from the entire experiment showed significantly different values. So

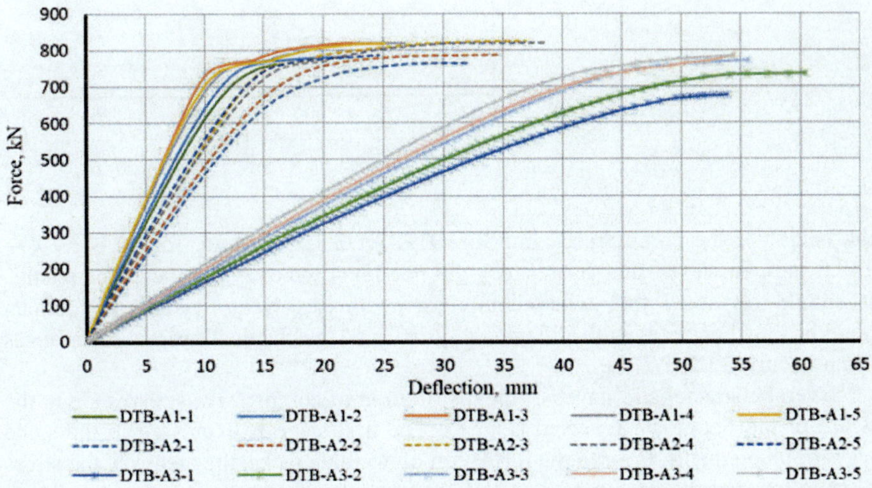

Fig. 5 Relationship between deflection values and external force

Table 3 Numerical results

Name of element	P_{ult} experiment (kN)	f (mm)	ω_σ	b_{eff} (mm)	x (mm)	P_{ult} SP (kN)	x SP (mm)
DTB-A1-1	763.25	16.11	0.9	630	125	453.9	28.77
DTB-A1-2	783.46	22.16	0.81	648	135.4	491.67	27.34
DTB-A1-3	819.207	23.55	0.743	668	93.8	534.05	26.29
DTB-A1-4	818,576	26.77	0.71	710	140.6	572.57	25.35
DTB-A1-5	818.077	25.11	0.648	712.8	125	612.15	24.63
DTB-A2-1	764.22	32.04	0.879	615.3	72.9	420.64	26.46
DTB-A2-2	787.42	34.8	0.903	722.4	78.1	455.32	25.57
DTB-A2-3	823.6	36.9	0.83	747	72.9	491.07	24
DTB-A2-4	821.1	38.7	0.742	742	78,1	527.93	23.21
DTB-A2-5	786.13	24.7	0.68	748	78.1	565.95	22.63
DTB-A3-1	676.65	53.9	0.872	610.4	52.1	457.22	29
DTB-A3-2	735.2	60.4	0.865	692	52.1	534.73	29.87
DTB-A3-3	784.11	55.67	0.869	782.1	46.9	615.37	30.71
DTB-A3-4	784.13	54.4	0.859	859	46.9	699.03	31.54
DTB-A3-5	784.56	54.4	0.85	935	62.5	785.58	32.35

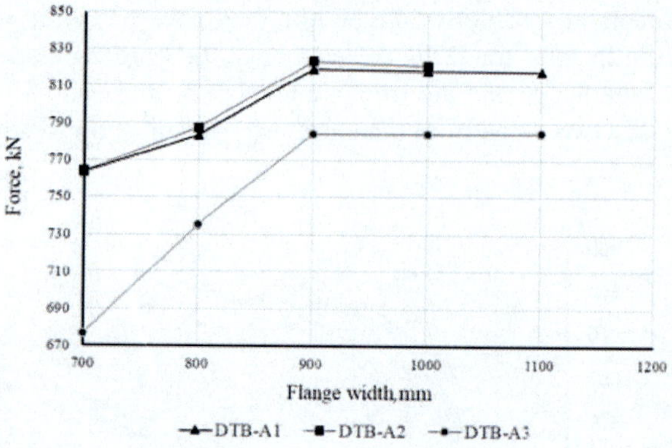

Fig. 6 Relationship between the flange width and the external load

the height of the compressed zone for DTB-A1-3 is 93.8 mm, for DTB-A1-4—140.6 mm. However, this discrepancy did not affect the overall logic of the results. It should be noted that when compressive stresses went beyond the contact boundary of concretes with different physical and mechanical properties, a stress jump occurred (Fig. 7).

Several characteristic stages of the distribution of compressive stresses along the width of the flange are revealed (Fig. 8). The first stage is expressed in the same way for all elements. Prior to the formation of normal cracks, the nature of the stress distribution and the degree of non-uniformity of deformations do not change with

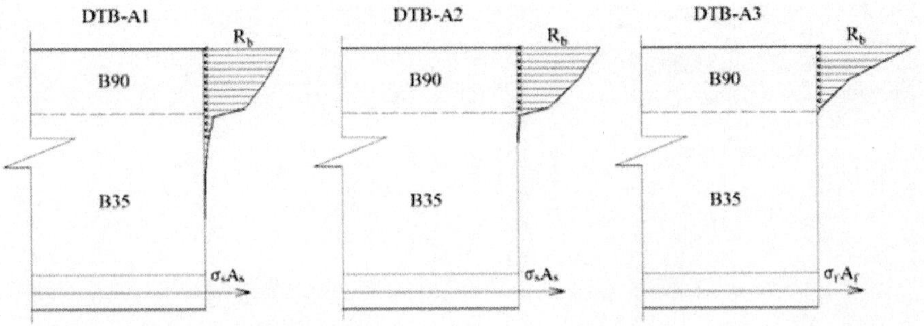

Fig. 7 Stress–strain state of elements

increasing load. The flange works in accordance with the theory of elasticity [17]. After the formation of a normal crack, the degree of unevenness increases, and local deformation jumps are formed. The previously determined height of the compressed zone in the middle of the cross-section, located in the middle of the span of the beams, could give grounds to assume that the entire flange is compressed along its height. However, tensile stresses propagate along the bottom fibers of the flange, and the neutral axis is nonlinear. In this regard, during the second stage, cracks develop across the flange. In turn, cracks in the flange are the cause of the formation of new jumps in deformations. Therefore, in some elements, it would be possible to distinguish a third stage—after the propagation of the crack along the entire overhang of the flange.

The character of stress redistribution can be qualitatively assessed by the coefficient of completeness of the compressive stress diagram (Table 3). The value of the coefficient decreased with an increase in the flange overhang to one degree or another, regardless of the type of longitudinal reinforcement. However, the decrease in the values of the coefficient occurs more intensively in elements with steel reinforcement. In elements with glass-composite reinforcement, a similar decrease in the coefficient is found only in the DTB-A3-5 element. This explains the higher level of increment in the bearing capacity of the elements. The reason for these phenomena is the level of stress redistribution deep into the web. In cases of passing the contact boundary of two concretes with different physical and mechanical properties, a stress jump occurs, which forms the second component of the compressive stress diagram. The general view of the diagram is visually similar to a funnel, the neck of which limits the redistribution of compressive stresses into the depth of the element. A decrease in the ability of an element to distribute stress along its height affects the redistribution of stresses along the width of the compressed flange. When analyzing the values of the completeness coefficients of the stress diagram, it is necessary to pay attention to the crack resistance of the flange. An increase in crack resistance can reduce the level of uneven stress distribution. In order to save the resources of the computational model, more than three rods were not arranged in the overhangs of the flanges. In this regard, it can be assumed that

Fig. 8 Characteristic stages of the distribution of compressive stresses along the width of the flange

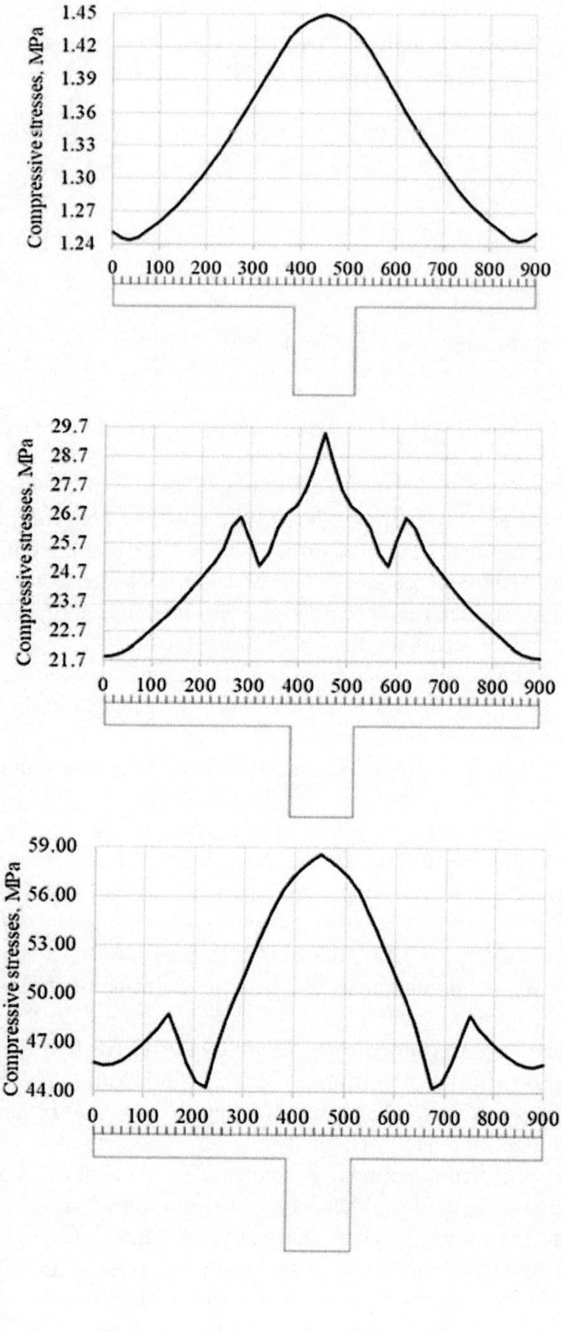

the coefficient of completeness of the stress diagram for elements with a flange width of 900 mm and more, when additional reinforcement is introduced into the flange, will increase by no more than 5%.

Several factors have an effect on the bearing capacity of T-elements with high-strength concrete in the compressed zone. So from Fig. 6 you can see the effect of the modulus of elasticity and the percentage of working reinforcement on the element.

The outline of the diagram of the compressive stresses of the DTB-A3 elements had a truncated appearance, which explains the lower bearing capacity. However, the DTB-A2 elements, in which the percentage of working reinforcement was lower in comparison with the DTB-A1, with a similar modulus of elasticity, turned out to be the most durable. The distribution of deformations across the boundary of two layers was accompanied by a stress jump. This is how the second component of the general compressive stress diagram is formed. The height of its action—x 'affects the bearing capacity of the element. This height is determined

$$x' = x - h_{hsc} \tag{5}$$

where

h_{hsc} high-strength concrete layer thickness.

It is the value of the x′ value that determines the losses from the area of the compressive stress diagram. Let us clearly consider the stress–strain state of two-layered elements DTB-A1 and DTB-A2 in comparison with single-layered elements made of high-strength concrete (Fig. 9).

In Fig. 9 shaded area—losses from the area of compressive stresses during the distribution of deformations beyond the boundary of two layers [18–20]. Thus, a larger percentage of the working reinforcement of the DTB-A1 elements negatively affected the bearing capacity due to an increase in the height of the compressed zone, and therefore x′.

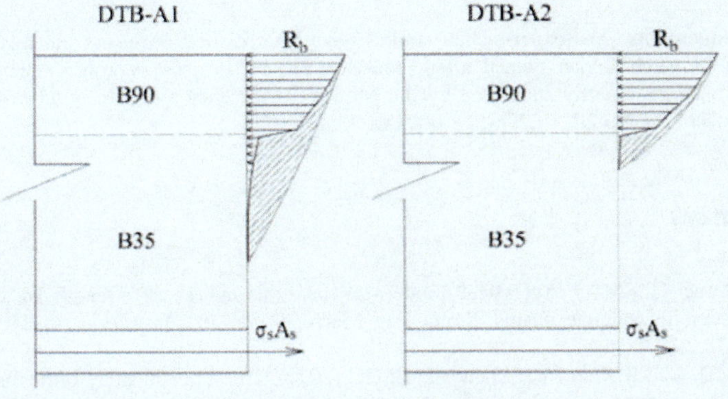

Fig. 9 Comparison of the area of compressive stress diagrams

The distribution of stresses along the height of the section responds to the distribution of stresses along the width of the flange. This can be concluded when comparing the values of the quantities x' and ω_σ.

4 Conclusion

Within the framework of a numerical experiment, the estimated maximum flange overhang in terms of bearing capacity was determined—$8h_f$. This confirms the assumption that the shelf efficiency increases with an increase in the elastic modulus of the concrete. It should be noted that the appearance of cracks reduces the flange stiffness and its ability to redistribute stresses.

Determination of the ability to redistribute stresses allows us to conclude that it is necessary to enter reduction factors when calculating the bearing capacity, the values of which must be verified experimentally.

When studying the operation of a compressed flange of two-layered T-section elements, it is necessary to take into account the ability of the element to distribute compressive stresses by height of the section. Numerical studies have established a shortage of bearing capacity and weak ability to redistribute stresses along the width of the flange with a significant height of the compressed zone extending beyond the boundary of two layers.

An increase in the class of working reinforcement makes it possible to mitigate the consequences of a stress jump, and therefore to reduce the relative losses of the bearing capacity and increase the ability of the flange to redistribute stresses. The use of glass-composite working reinforcement makes it possible to achieve high rates in the redistribution of stresses along the width of the flange.

The most effective version of the execution of two-layered bending reinforced concrete T-section elements with high-strength concrete in the compressed zone is presented together with the working reinforcement of the AT-800 class. This approach allows you to achieve more efficient flange operation combined with a high load-bearing capacity.

Acknowledgements The experimental studies have been carried out using the facilities of the Collective Research Center named after Professor Yu. M. Borisov, Voronezh State Technical University, which is partly supported by the Ministry of Science and Education of the Russian Federation, Project No 2021-2296-512-0001-060.

References

1. Campione G (2021) Analytical model for flexural and shear strength of normal and high-strength concrete beams. Struct Eng Mech 78:199–207. https://doi.org/10.12989/sem.2021.78.2.19
2. Sun MD, Gao R, Gao MC, Chen YT, Li CG (2017) Experimental study of flexural behavior of reactive powder concrete beams reinforced with high-strength steel bars. Bridg Constr 47:25–30. https://doi.org/10.3969/j.issn.1003-4722.2017.02.005

3. Dok G, Caglar N, Ilki A, Yilmaz C (2021) Residual load bearing capacity and failure mechanism of impacted high-strength reinforced concrete shear beams. Eng Fail Anal 121:105185. https://doi.org/10.1016/j.engfailanal.2020.105185
4. Graf O (1910) Versuche Mit Eisenbetonbalken. Armierter Beton 3:451–465
5. CEB-Arbeitskreis 9 "T-Balken" (1958) Studie über die Berechnung von T-Querschnitten (Plattenbalkenquerschnitten) unter Berücksichtigung der derzeitigen Normen. CEB-Bull d 'Inf 8:3–32
6. CEB-Arbeitskreis 9 "T-BalkenBericht der Kommision auf der 5 (1959) Anlage 1: Berücksichtigung der Biegesteifigkeit der Platte. Anlage 2: Plattenbalken-Modelle aus bewehrtem Gips (Versuchsprogramm). CEB-Bull d 'Inf 18:1–41
7. Chiewanichakorn M, Aref AJ, Chen SS, Ahn I-S (2004) Effective flange width definition for steel-concrete composite bridge girder. J Struct Eng 130:2016–2031. https://doi.org/10.1061/(asce)0733-9445(2004)130:12(2016)
8. Xiao X, Huang EH, Ni YS (2015) Theoretical analysis and experiment of the effect of flange width on beam-plates system of prestressed concrete. Jilin Daxue Xuebao (Gongxueban)/J Jilin Univ (Eng Technol Ed) 45:1784–1790. https://doi.org/10.13229/j.cnki.jdxbgxb201506008
9. Rogatnev YuF, Ivanov YuV, Sokolov OO (2020) Numeric modelling of stress-strain state of t-section element with high-strength concrete in compressed zone exposed to momentary load. Izv Vuzov Stroitelstvo 9:36–47
10. Rahal KN, Rumaih HA (2011) Tests on reinforced concrete beams strengthened in shear using near surface mounted CFRP and steel bars. Eng Struct 33:53–62. https://doi.org/10.1016/j.engstruct.2010.09.017
11. Zhao ZZ, Kwan AKH, He XG (2004) Nonlinear finite element analysis of deep reinforced concrete coupling beams. Eng Struct 26:13–25. https://doi.org/10.1016/j.engstruct.2003.08.014
12. Mostofinejad D, Talaeitaba SB (2011) Nonlinear modeling of RC beams subjected to torsion using the smeared crack model. Proc Eng 1447–1454. https://doi.org/10.1016/j.proeng.2011.07.182
13. Willam KJ, Warnke ED (1975) Constitutive model for the triaxial behaviour of concrete. Proc Intl Assoc Bridg Structl Engrs 19:1–30
14. Khorokhordin AM, Usachev AM, Korotkih DN (2018) Comparative assessment of mechanical properties of polymer composite reinforcement. Stroit Mater 761:71–75 (2018). https://doi.org/10.31659/0585-430x-2018-761-7-71-75
15. Zou B, Chen A, Davalos JF, Salim HA (2011) Evaluation of effective flange width by shear lag model for orthotropic FRP bridge decks. Compos Struct 93:474–482. https://doi.org/10.1016/j.compstruct.2010.08.033
16. Sayhood EK, Mohammed NS, Hanon AS (2020) Effect of web openings on effective flange width of reinforced concrete T-section. In: IOP conference series: materials science and engineering. https://doi.org/10.1088/1757-899X/737/1/012002
17. Timoshenko SP, Goodier JN (1970) Theory of elasticity, 3rd edn. McGraw-Hill
18. Lisyatnikov MS, Shishov II, Sergeev MS, Hisham E (2020) Precast monolithic coating of an industrial building based on variable-height beam-slabs. In: IOP conference series: materials science and engineering, p 012064. IOP Publishing Ltd. https://doi.org/10.1088/1757-899X/896/1/012064
19. Polikutin A, Potapov Y, Levchenko A, Perekal'skiy O (2019) The stress-strain state of normal sections rubcon bending elements with mixed reinforcement. In: Advances in intelligent systems and computing. Springer, Berlin, pp 586–599. https://doi.org/10.1007/978-3-030-19868-8_56
20. Polikutin AE, Levchenko AV, Oforkaja TO (2018) Comparison of the strength of normal sections of rubber concrete rectangular and T-beams. In: IOP conference series: materials science and engineering. Institute of Physics Publishing. https://doi.org/10.1088/1757-899X/463/2/022092

Study of the Mass Effect of a Complex Node of UnderGround Pipelines of Orthogonal Configuration Based on Real Earthquake Records

Diyorbek Bekmirzaev, Ibrakhim Mirzaev, Ruslan Kishanov, Nodirakhon Mansurova, and Shoista Sabirova

Abstract This paper considers the class of problems of dynamics of spatial systems of underground pipelines under the action of three-component seismic waves propagating in the ground on the basis of instrumental records of real earthquakes. For complex systems of underground pipelines as a method for solving the problems of nonstationary dynamics on spatial coordinates of pipelines the discretization by finite element method is used, and on time variable the implicit finite-difference scheme of Newmark type. The OX-axis is assigned as the direction of propagation of the three-component seismic waves. In order to account for the angle of incidence of seismic waves, the structure of the underground pipeline is rotated by a given angle relative to the direction of the OX axis. The conducted theoretical and computational-experimental studies solve the problem of assessment of stress–strain state of underground pipelines of complex orthogonal configuration under seismic effect arbitrarily directed to the main axes of the pipeline. In the calculations, a digitized seismogram record of the Chilean earthquake of January 24, 2016 was taken as a seismic waves. Multi-factor calculations were performed for different soil types and seismic waves incidence angles. The effect of a complex pipeline assembly on the dynamic process in the pipeline was investigated. Influence of boundary conditions and compound node on pipeline waves formation has been determined. Dangerous points of occurrence of maximum stresses in underground pipeline systems under the action of seismic waves propagating in space at an arbitrary angle of attack have been determined.

Keywords Underground pipeline · Seismic wave · Interaction in the pipe-soil system · Orthogonal configuration · Numerical method

---

D. Bekmirzaev (✉) · I. Mirzaev · R. Kishanov
Academy of Sciences of the Republic of Uzbekistan Institute of Mechanics and Seismic Stability of Structures Named After M.T. Urazbayev, Tashkent, Uzbekistan

N. Mansurova
Namangan Engineering Construction Institute, Namangan, Uzbekistan

S. Sabirova
Tashkent State Transport University, Tashkent, Uzbekistan

N. Vatin et al. (eds.), *Proceedings of MPCPE 2021*, Lecture Notes in Civil Engineering 182, https://doi.org/10.1007/978-3-030-85236-8_34

1 Introduction

Analysis of the consequences of strong earthquakes shows that the seismic resistance of underground structures depends on the direction of the seismic wave. Since during earthquakes underground structures can be exposed to waves from arbitrary directions, a determination of the stress–strain state of underground pipelines in the presence of longitudinal, torsional, and transverse oscillations is relevant and serves to determine the possible seismic hazard [1–6].

In the seismodynamics of underground structures, the nature of the pipeline system interaction with the environment and the structural features of the object are important [3, 4]. Numerous data on the consequences of recent strong earthquakes related to the study of life support systems such as underground gas, water, and oil pipelines and tunnels for various purposes were analyzed in order to supplement the developed theory with new data and to assess the operation effectiveness.

At present, the case-history of strong earthquakes that have occurred in the world shows that a preliminary hazard assessment of earthquakes and timely consideration of appropriate measures (problems of assessing seismic risk and their mitigation) are of great importance for the reduction of catastrophic aftermath. Thus, it is necessary to ensure the strength and stability of underground structures subjected to seismic loads. In this sense, the implementation of targeted scientific research to improve numerical methods for solving problems, to calculate the earthquake resistance of underground pipeline systems under random seismic effects is a relevant issue [6–20]. In works of S. Sarioletlagh, M. Nekooei, A. Aziminejad, D. Ha, T. H. Abdoun, M. J. O'Rourke, M. D. Symans, M. Saberi, H. Arabzadeh, A. Keshavarz, P. Vazouras, S. A. Karamanos, et al. the use of different mechanical mathematical models was analyzed and a number of urgent problems of underground and ground structures were solved [5–12, 14–20].

2 Methods

This paper is devoted to the study of the underground pipelines dynamics based on the seismodynamic theory of underground structures, using a mathematical model of the theory of rods, considered in [3, 4] for the case of rod point displacements under combined action of longitudinal and transverse forces.

Figure 1 shows a system of pipelines and wells of orthogonal configuration. The well is modeled as a solid body rigidly connected to the pipe-line; the well interacts with soil. Let us study the effect of the well mass on the stress state of an underground pipeline, the ends of which are fixed to the ground, under the action of a three-dimensional wave recorded during an earthquake.

Consider a section of underground pipeline 172 m long, to which 4 nodes are connected at the marks of 41, 84, 89 and 132 m (see Fig. 1).

Fig. 1 Section of a complex system of underground pipelines of orthogonal configuration

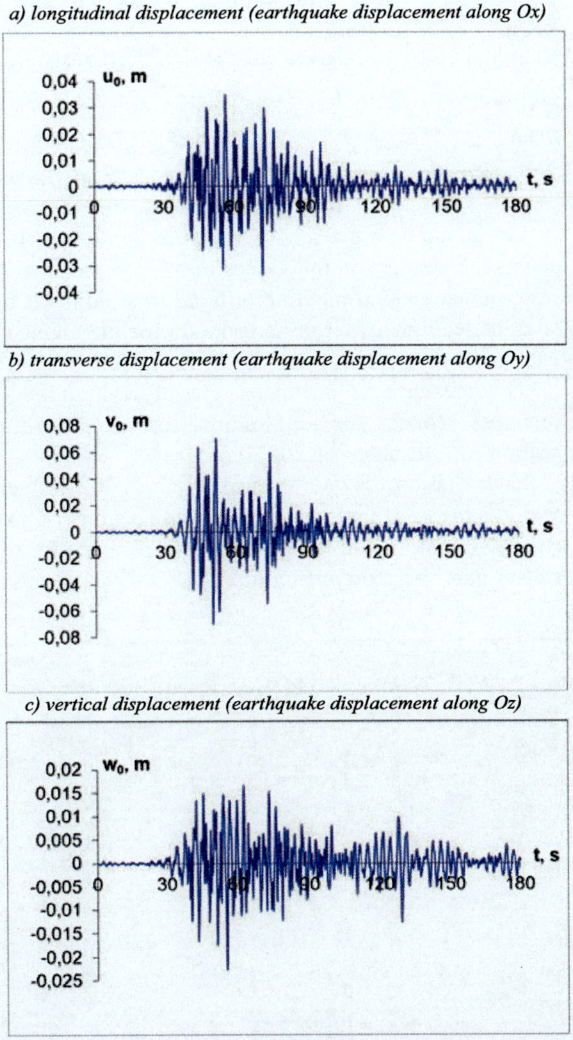

Fig. 2 Digitized records of seismograms of January 24, 2016 Chile earthquake

Let the external influence (seismogram) have the form shown in Fig. 2. The head of the laboratory at the University of Berkeley Sh. Takhirov kindly provided the digitized records of the earthquake seismograms to us.

3 Results and Discussion

As an example, let us consider the problem with the following initial data: the mechanical and geometrical parameters of the underground pipeline and soil are taken as: $E = 2{\cdot}10^5$ MPa; $\rho = 7.8{\cdot}10^3$ kg/m^3; $D_H = 0.5$ m; $D_B = 0.49$ m; $\mu_{soil} = 0.2$; $\mu_{pipe} = 0.3$; $l = 172$ m; in a straight section $k_x = 1.5{\cdot}10^4$ kN/m^3; $k_{y,z} = 3.9{\cdot}10^4$ kN/m^3; in a complex section $k_x = 0.5{\cdot}10^4$ kN/m^3; $k_{y,z} = 1.3{\cdot}10^4$ kN/m^3; $C_p = 1000$ m/s.

Initial data for a well are: $E = 2.5{\cdot}10^4$ MPa; $\rho = 2.5{\cdot}10^3$ kg/m^3; $D_H^{uz} = 4$ m; $D_B^{uz} = 3.5$ m; $I_x^{uz} = I_y^{uz} = m_1/12{\cdot}H_{uz}^2 + 1/2{\cdot}m_2{\cdot} R_H^{uz2}$ m^2; $I_z^{uz} = m_1/2{\cdot}(R_H^{uz2} + R_B^{uz2}) + m_2{\cdot}R_H^{uz2}$ m^2; $H_{uz} = 1.5$ m; $m_1 = \pi{\cdot}H_{uz}/4{\cdot}(D_H^{uz2} - D_B^{uz2}){\cdot}\rho$; $V_1 = \pi{\cdot}H_{uz}/4{\cdot}(D_H^{uz2} - D_B^{uz2}) + 2{\cdot}\pi{\cdot}\ R_H^{uz2}{\cdot}h_{uz}$; $m_2 = \pi{\cdot}R_H^{uz2}{\cdot}h_{uz}{\cdot}\rho$; $m_{uz} = V_1{\cdot}\rho$; $k_x^{uz} = 0.5{\cdot}10^4$ kN/m^3; $k_{y,z}^{uz} = 1.3{\cdot}10^4$ kN/m^3, the mass of the node (a well) is $m_{uz} = 2 \times 10^3$ kg.

Let us analyze the results obtained, presented in the form of graphs. Figure 3 shows the changes in the values of the compressive (tensile) stresses along the axis of the underground pipeline at fixed time points. Compressive (tensile) stresses in straight sections of an underground pipeline change insignificantly from the presence of a nodal mass (see Fig. 3, in sections x = 41 m, $x = 132$ m).

The presence of compensators of a complex configuration in underground pipelines reduces the longitudinal forces of the underground pipeline near this section to a distance of 20–30 m.

Under compressive (tensile) stresses, the influence of the nodes ($x = 41$ m, $x = 132$ m) is insignificant in straight sections of an underground pipeline. In this case, the seismic wavelength makes it possible for the nodal mass interacting with soil to gain the required inertia.

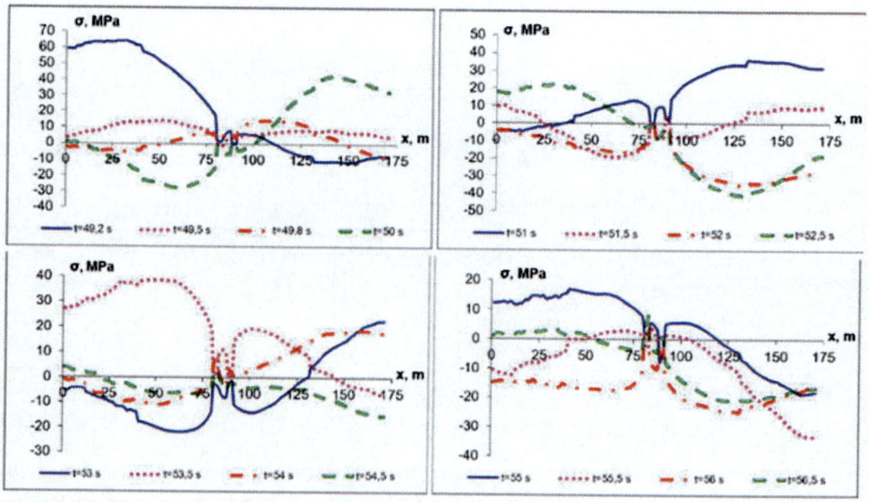

Fig. 3 Changes in compressive (tensile) stresses along the underground pipeline axis at fixed time points

Figures 4 and 5 show changes in the values of total stresses (σ_y^+, σ_y^-) along the axis (bending relative to the *Oy* axis) of the underground pipeline at fixed time points.

Here the nodes in straight sections of the pipeline play an important role since the total stresses increase near the nodes of the underground pipeline (see Figs. 4 and 5, sections $x = 41$ m and $x = 132$ m).

This is due to bending deformations, which have large values near the nodal mass since the nodal mass has a large value and a large contact surface with soil and, therefore, the nature of this point motion changes the stresses.

In sections of an underground pipeline ($x = 81$ m and $x = 92$ m) of a complex configuration, the values of the total stresses decrease.

Figures 6 and 7 show changes in the values of the total stresses (σ_z^+, σ_z^-) along the axis (bending relative to the *Oz* axis) of the underground pipeline at fixed time points.

In straight sections near the nodes of the underground pipeline, the total stresses increase (Figs. 6 and 7, in sections $x = 41$ m and $x = 132$ m). In sections of an underground pipeline ($x = 81$ m and $x = 92$ m) of a complex configuration, the values of total stresses decrease due to the possibility of special deformation in this section caused by the combined action of different waves.

As seen from the figures, the presence of large total stresses near the nodal mass is related to large deformations near the nodal mass since it has a sufficiently large value and a large contact surface with soil and, therefore, the nature of the motion of this point changes the stresses (Figs. 4 and 7).

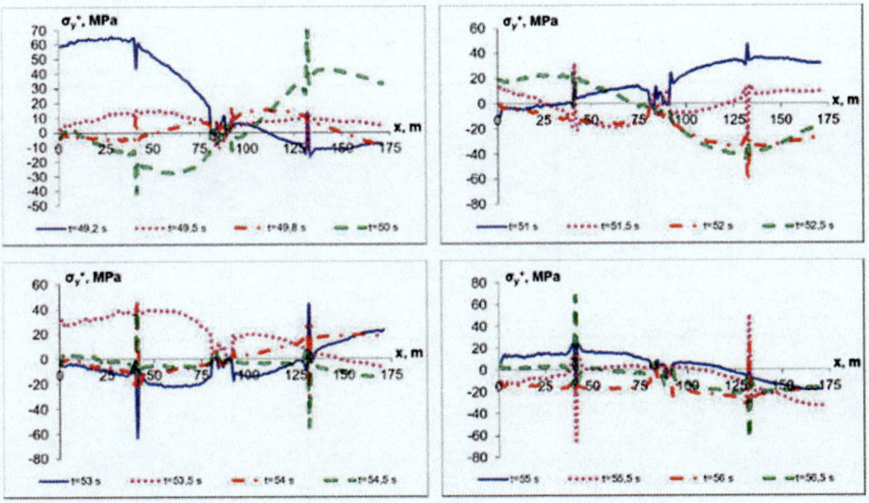

Fig. 4 Changes in total stresses (σ_y^+) along the axis (bending relative to the *Oy* axis) of an underground pipeline at fixed time points

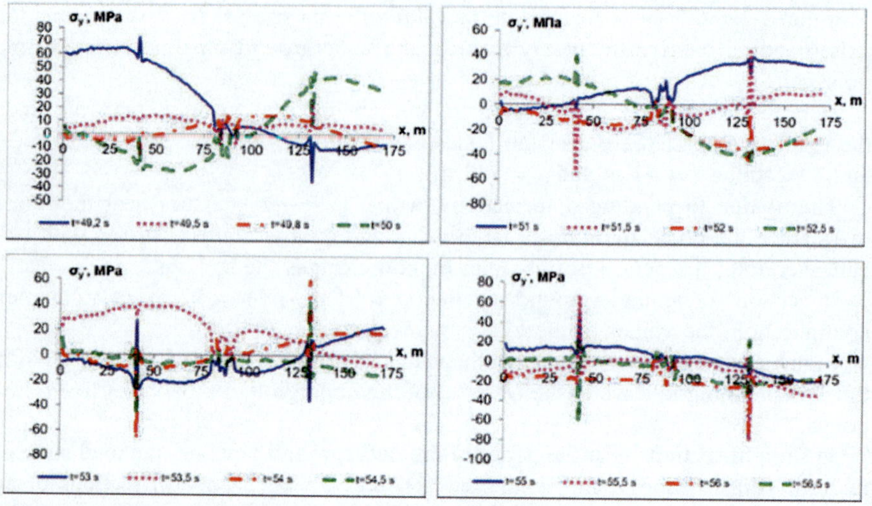

Fig. 5 Changes in total stresses (σ_y^-) along the axis (bending relative to the *Oy* axis) of an underground pipeline at fixed time points

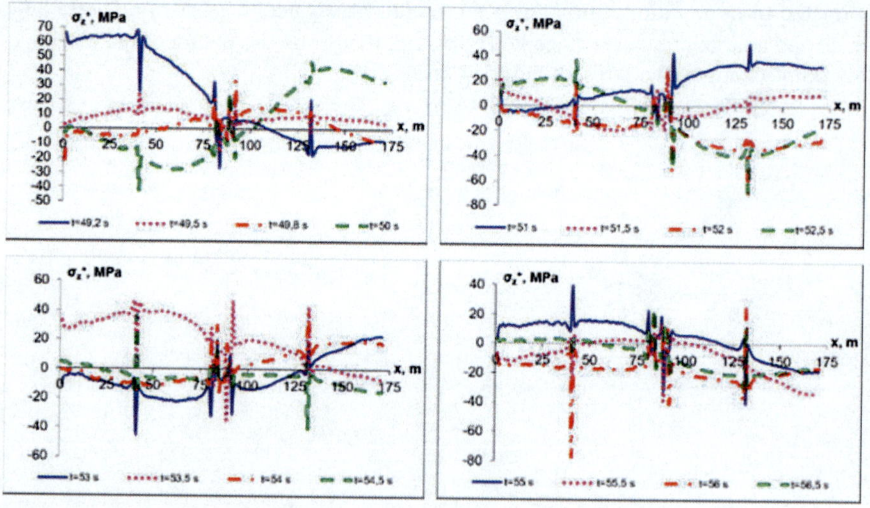

Fig. 6 Changes in total stresses (σ_z^+) along the axis (bending relative to the *Oz* axis) of an underground pipeline at fixed time points

The presence of large nodal masses leads to an increase in the total stresses at the joints to this mass even under long seismic waves. The wells built in the pipelines have a small weight and, under the action of long seismic waves, do not lead to a significant increase in the total stresses at the well joint (see Figs. 4 and 7).

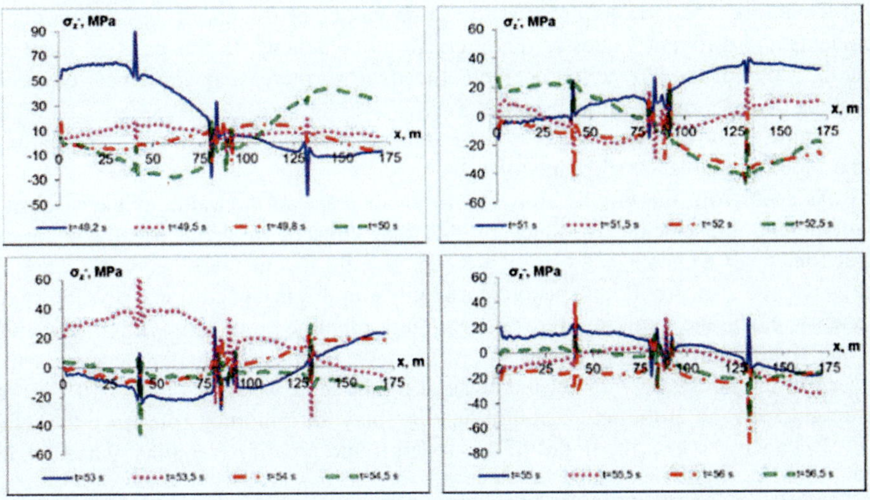

Fig. 7 Changes in total stresses (σ_z^-) along the axis (bending relative to the Oz axis) of the underground pipeline at fixed time points

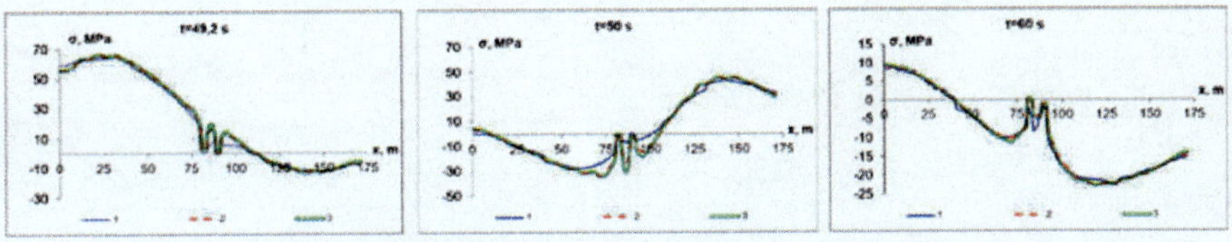

Fig. 8 Changes in compressive (tensile) stresses along the axis of the underground pipeline at fixed time points: 1 – $\mu = 0$, mass of the node $m_{uz} = 2 \times 10^3$ kg; 2 – $\mu = 1300$ kNs/m^3, mass of the node $m_{uz} = 1.1 \times 10^2$ kg; 3 – $\mu = 2600$ kNs/m^3, mass of the node $m_{uz} = 1.1 \times 10^2$ kg

Underground structures during an earthquake interact with the surrounding soil; the interaction forces have viscoelastic properties.

From this standpoint, we will analyze the stress–strain state of an underground pipeline during viscoelastic interactions with the surrounding soil based on real earthquake records.

Figure 8 shows the changes in the values of the compressive (tensile) stresses along the axis of the underground pipeline at fixed time points.

As seen from the graphs, the effect of the node on the compressive (tensile) stresses, i.e., on the stresses caused by the axial force of the underground pipeline only, is insignificant (see Fig. 8).

The magnitude of the interaction viscosity changes the picture: the greater the value of the interaction viscosity coefficient, the greater the energy exchange between the pipeline and soil.

Therefore, the stresses differ at the same values of the elastic interaction coefficients and different values of the interaction coefficients. In this case, an increase in the value of the interaction coefficient leads to an increase in stress near the peak values of deformation in a seismic wave.

Figure 9 shows the changes in the values of the total stresses (σ_y^+, σ_y^-) along the pipeline axis at fixed points of time.

As seen from the graphs above, with an increase in the value of the viscosity coefficient, the calculated values of the total stresses in geometrically complex sections ($x = 81$ m, $x = 84$ m, $x = 89$ m, $x = 92$ m) increase as well. This is because the viscosity (like a piston in a viscous liquid) is an additional possibility of energy exchange between the underground pipeline and soil with a constant coefficient of elastic interaction.

Here the nodes ($1 - \mu = 0$, the mass of the node (a well) $m_{uz} = 2 \times 10^3$ kg) in straight sections of the underground pipeline play an important role since the total stresses increase near the nodes of the underground pipeline (see Figs. 9 and 10, in

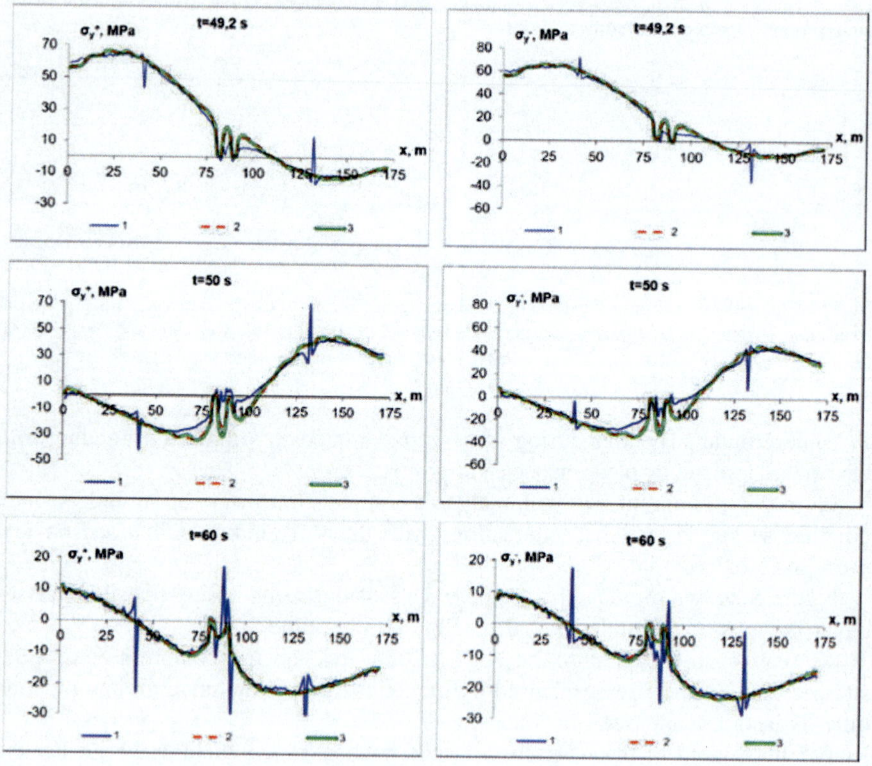

Fig. 9 Changes in total stresses (σ_y^+, σ_y^-) along the axis of the pipeline at fixed time points: $1 - \mu = 0$, mass of the node $m_{uz} = 2 \times 10^3$ kg; $2 - \mu = 1300$ kNs/m^3, mass of the node $m_{uz} = 1.1 \times 10^2$ kg; $3 - \mu = 2600$ kNs/m^3, mass of the node $m_{uz} = 1.1 \times 10^2$ kg

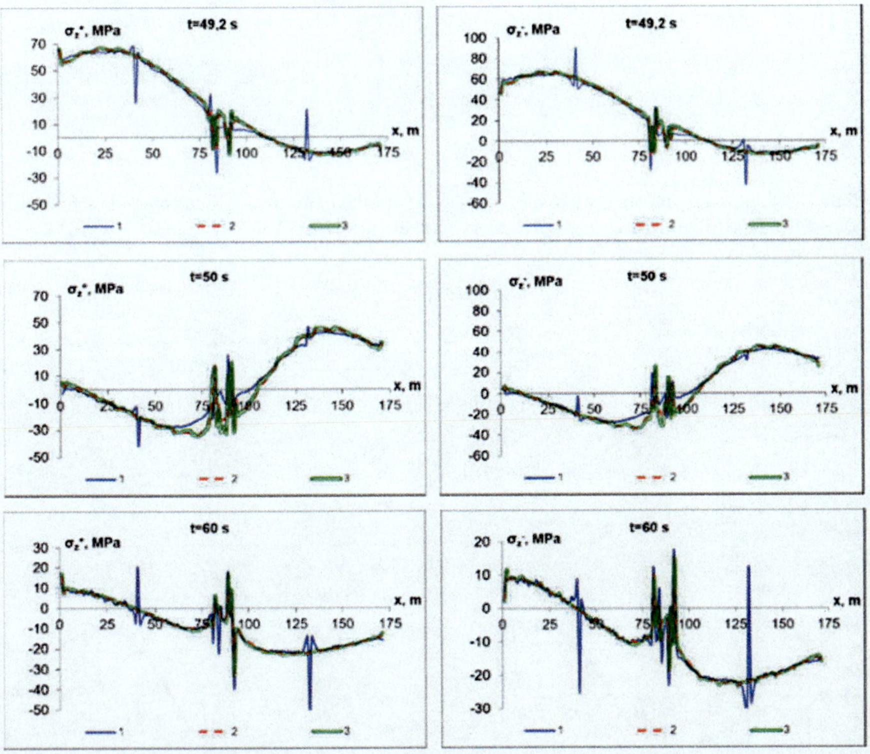

Fig. 10 Changes in total stresses (σ_z^+, σ_z^-) along the axis of the pipeline at fixed time points $1 - \mu = 0$, mass of the node $m_{uz} = 2 \times 10^3$ kg; $2 - \mu = 1300$ kNs/m^3, mass of the node $m_{uz} = 1.1 \times 10^2$ kg; $3 - \mu = 2600$ kNs/m^3, mass of the node $m_{uz} = 1.1 \times 10^2$ kg

sections $x = 41$ m and $x = 132$ m). From the graphs, it is seen that the values of total stresses in the sections of the underground pipeline (see Fig. 10, $1 - \mu = 0$, the mass of the node (a well) $m_{uz} = 2 \cdot 10^3$ kg; $x = 81$ m and $x = 92$ m) of a complex configuration are reduced due to the possibility of special deformation in this section caused by the combined action of different waves and the potential bending of transversely located underground pipelines.

From the graphs in Fig. 10 it is seen, that at the peak of the accelerograms, the total stresses have greater values in relation to other time points (see Fig. 4, in sections $x = 41$ m, $x = 81$ m, $x = 84$ m, $x = 89$ m, $x = 92$ m, $x = 132$ m).

Figure 11 shows the changes in the values of the compressive (tensile) stresses along the axis of the underground pipeline (the values of the node masses do not change) at fixed time points.

Figures 12 and 13 show changes in the values of total stresses (σ_y^+, σ_y^-, σ_y^+,-σ_y^-) along the pipeline axis at fixed time points.

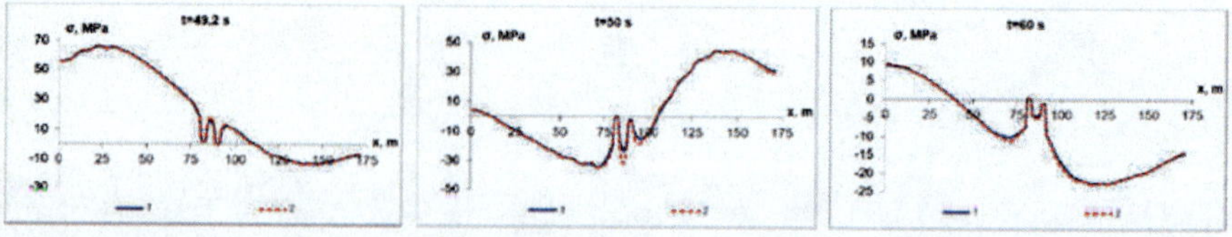

Fig. 11 Changes in compressive (tensile) stress along the axis of the underground pipeline at fixed time points: 1 – μ = 1300 kNs/m^3, m_{uz} = 1.1 × 10^2 kg; 2 – μ = 2600 kNs/m^3, m_{uz} = 1.1 × 10^2 kg

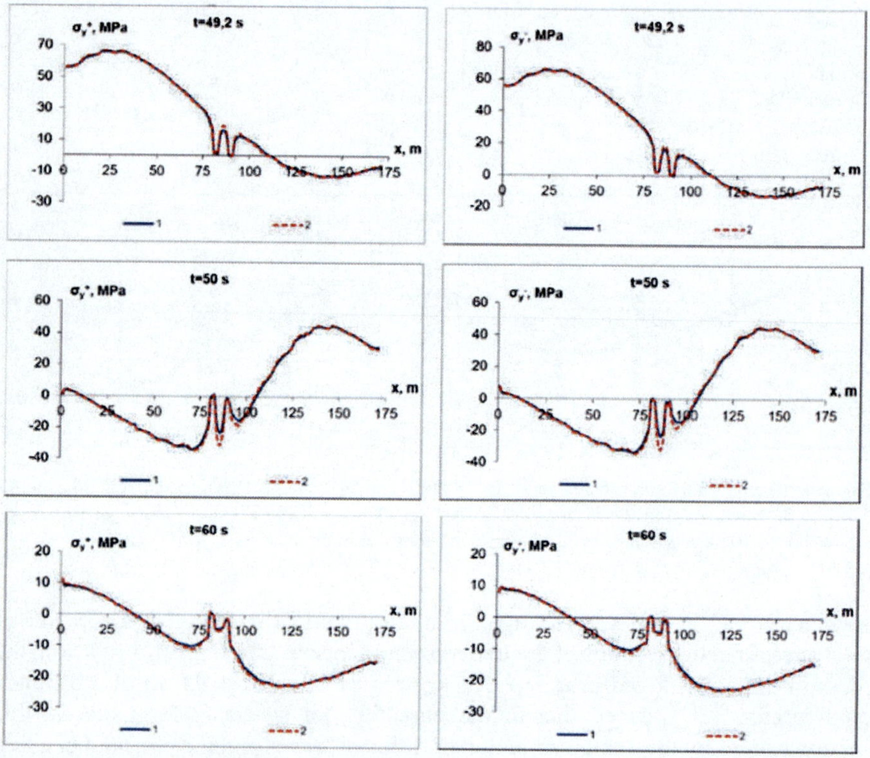

Fig. 12 Changes in total stresses (σ_y^+, σ_y^-) along the pipeline axis at fixed time points: 1 – μ = 1300 kNs/m^3, m_{uz} = 1.1 × 10^2 kg; 2 – μ = 2600 kNs/m^3, m_{uz} = 1.1 × 10^2 kg

Analysis of the numerical results shows that viscoelastic interaction has a significant effect on the stress–strain state of underground pipelines. In the cases considered above, the difference between the results of elastic and viscoelastic solutions is 15–20% (see Figs. 12 and 13).

When calculating the seismic resistance of underground pipelines, it is necessary to take into account the exchange (transfer and entrainment) of energy due to the difference in the velocities of soil particles and the pipeline.

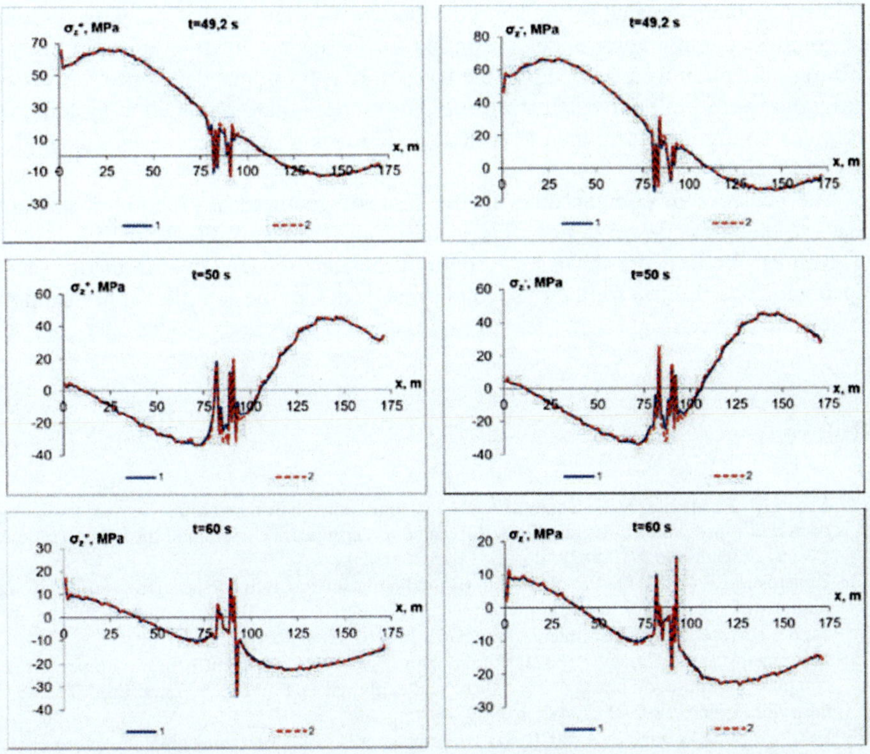

Fig. 13 Changes in total stresses (σ_z^+, σ_z^-) along the axis of the pipeline at fixed time points: 1 – μ = 1300 kNs/m^3, m_{uz} = 1.1 × 10^2 kg; 2 – μ = 2600 kNs/m^3, m_{uz} = 1.1 × 10^2 kg

The soil viscosity increases the exchange of energy between the underground pipeline and soil. It is advisable to determine the coefficients of elastic and viscous interaction in dynamic problems of the pipeline-soil interaction conducting dynamic experiments.

When designing underground pipelines in seismic areas, it is necessary to provide conditions for damping vibrations. A significant reduction in seismic loads on underground pipelines can be reached by using compensators of various designs.

4 Conclusions

Complex configurations of pipeline systems are investigated in orthogonally connected nodes. The theoretical and computational-experimental studies conducted, solve the problems of assessing the stress–strain state of U-shaped pipelines of complex orthogonal configuration based on real earthquake records.

New problems related to the study of the stress–strain state of pipeline systems of complex configuration under the influence of seismic loads propagating in an arbitrary direction were solved. On the basis of the algorithms developed, universal programs were built for calculating the stress–strain state of pipeline systems of complex configuration on the effect of seismic loads propagating at various angles to the pipeline axis.

The presence of compensators in the form of geometrically complex sections leads to the stress decrease near this section to a distance of 20–30 m.

Results obtained present a new contribution into seismodynamic theory; they open wide possibilities for their use in optimal design of the complex of life support in seismic zones.

References

1. Saberi M, Behnamfar F, Vafaeian M (2015) A continuum shell-beam finite element modeling of buried pipes with 90-degree elbow subjected to earthquake excitations. Int J Eng Trans A Basics. https://doi.org/10.5829/idosi.ije.2015.28.03c.02
2. Khachiyan EE (2015) On the possibility of predicting seismogram and accelerogram of strong motions of the soil for an earthquake model considered as an instantaneous rupture of the earth's surface. Seism Instruments. https://doi.org/10.3103/s0747923915020036
3. Bekmirzaev DA, Mirzaev I (2021) Earthquake resistance assessment of buried pipelines of complex configuration based on records of real earthquakes. Soil Mech Found Eng 57:26–31. https://doi.org/10.1007/s11204-021-09697-0
4. Bekmirzaev DA, Mirzaev I (2020) Dynamic processes in underground pipelines of complex orthogonal configuration at different incidence angles of seismic effect. Int J Sci Technol Res
5. Zhuang H, Hu Z, Wang X, Chen G (2015) Seismic responses of a large underground structure in liquefied soils by FEM numerical modelling. Bull Earthq Eng. https://doi.org/10.1007/s10518-015-9790-6
6. O'Rourke M, Vargas-Londono T (2016) Analytical model for segmented pipe response to tensile ground strain. Earthq Spectra. https://doi.org/10.1193/050415EQS064M
7. Nishonov N et al (2020) Behaviour and calculation of polymer pipelines under real earthquake records. In: IOP conference series: materials science and engineering. https://doi.org/10.1088/1757-899X/869/5/052076
8. Stewart H, O'Rourke M, Ha Abdoun T, D Van Laak P, O'Rourke T (2006) Split-containers for centrifuge modeling of permanent ground deformation effects on buried pipeline systems. In: Physical modelling in geotechnics, 6th ICPMG'06—proceedings of the 6th international conference on physical modelling in geotechnics. https://doi.org/10.1201/noe0415415866.ch104.
9. Fard SS, Nekooei M, Oskouei AV, Aziminejad A (2019) Experimental and numerical modeling of horizontally-bent buried pipelines crossing fault slip. Lat Am J Solids Struct. https://doi.org/10.1590/1679-78255463
10. Saberi M, Arabzadeh H, Keshavarz A (2011) Numerical analysis of buried pipelines with right angle elbow under wave propagation. Proc Eng. https://doi.org/10.1016/j.proeng.2011.07.412
11. Kosimov E et al (2021) Comparison of the impacts of harmonic and seismic waves on an underground pipeline during the Gazli earthquake. In: IOP conference series: materials science and engineering. https://doi.org/10.1088/1757-899x/1030/1/012082

12. Takada S, Hassani N, Fukuda K (2002) Damage directivity in buried pipelines of kobe city during the 1995 earthquake. J Earthq Eng. https://doi.org/10.1080/13632460209350407
13. Bekmirzaev D, Mirzaev I, Mansurova N, Kosimov E, Juraev DP (2020) Numerical methods in the study of seismic dynamics of underground pipelines. In: IOP conference series: materials science and engineering, vol 869. https://doi.org/10.1088/1757-899X/869/5/052035
14. O'Rourke TD, Jung JK, Argyrou C (2016) Underground pipeline response to earthquake-induced ground deformation. Soil Dyn Earthq Eng. https://doi.org/10.1016/j.soildyn.2016.09.008
15. Sultanov K, Khusanov B, Rikhsieva B. (2021) Elastic wave propagation in a cylinder with external active friction. J Phys Conf Ser 1901. https://doi.org/10.1088/1742-6596/1901/1/012125012125
16. Vazouras P, Karamanos SA, Dakoulas P (2010) Finite element analysis of buried steel pipelines under strike-slip fault displacement. Soil Dyn Earthq Eng. https://doi.org/10.1016/j.soildyn.2010.06.011
17. Vazouras P, Karamanos SA, Kakoulas P (2012) Mechanical behavior of buried steel pipes crossing active strike-slip faults. Soil Dyn Earthq Eng. https://doi.org/10.1016/j.soildyn.2012.05.012
18. Banushi G, Nunziante S, Klaus T (2018) Innovative analysis of a buried operating pipeline subjected to strike-slip fault movement. Soil Dyn Earthq Eng. https://doi.org/10.1016/j.soildyn.2018.01.015
19. Ghaznavi OA, Vatani Oskouei A (2017) The effect of P-wave propagation on the seismic behavior of steel pipelines. Periodica Polytech Civ Eng. https://doi.org/10.3311/PPci.9866
20. Trifonov OV, Cherniy VP (2012) Elastoplastic stress-strain analysis of buried steel pipelines subjected to fault displacements with account for service loads. Soil Dyn Earthq Eng. https://doi.org/10.1016/j.soildyn.2011.10.001

Dowel Connections with Local Wood Modification

Artem Strekalkin, Artem Koshcheev, Svetlana Roshchina, and Anatoly Naichuk

Abstract The article is devoted to the study of the behavior of the elements of dowel joints and modified wood of the node. Various methods of joining wooden structures are considered. The basis is a comparison of the joint of SHERPA-type wooden structures with the designed new type of connectors made of a cheaper metal alloy and local modification of wood. The influence of wood modification on the strength and deformability of the knot has been investigated. Various methods of wood modification are considered, conclusions are drawn about the possibility of using various modifiers within the framework of this work. Strength characteristics of the materials used in this compound are given. The values of the aluminum alloys of the initial compound and the compound to be used in the current study are compared. A comparative analysis of the deformability indicators of different variants of the assembly is carried out. The results of calculation by the finite element method in the ANSYS Software package are compared. Conclusions are drawn about a comparative increase in strength and a decrease in deformability. The direction of the further course of research is set.

Keywords ANSYS · Construction · Dowel · Finite element · Joint · Simulation · Structure · Timber · Wood · Modification · Buildings

1 Introduction

Structures made of solid and glued wood have been widely used in many industries and economy for more than one century. Moreover, quite often this material is used in buildings with increased chemical aggression. Due to its operational properties, the resistance of wood to this type of aggression is much higher than that of stone, reinforced stone, steel and reinforced concrete structures [1].

A. Strekalkin (✉) · A. Koshcheev · S. Roshchina
Vladimir State University Named After Alexander and Nikolay Stoletovs, Vladimir, Russia

A. Naichuk
Brest State Technical University, Brest, Belarus

N. Vatin et al. (eds.), *Proceedings of MPCPE 2021*, Lecture Notes in Civil Engineering 182, https://doi.org/10.1007/978-3-030-85236-8_35

When designing structures, as well as nodes and joints of elements made of wood, it is necessary to take into account its rational use. This is achieved through the use of new materials, improvement of existing design solutions, various methods of strengthening and modifying wood. The development of wooden structures tends to create systems, types, elements and types of connections that would allow saving wood while increasing the load-bearing capacity and taking into account its physical and mechanical characteristics in the best way [2–12].

In the past few decades, much attention has been paid to the construction of industrial and civil buildings from solid and glued timber. An important factor in the reliable functioning of building elements as a single system is the correct and rational solutions to their joints. In wooden structures, the pairing of elements is divided into two main groups:

1. glueless connections on mechanical bonds;
2. adhesive connections.

Connections of the first group in wooden structures have been used for more than one hundred years. In such joints, the transfer of forces between the wood and the connecting elements is exclusively in the contact area. The breadth of application of such links is due to their universality. In addition, due to the development of automation and industrialization, their manufacture and reliability have also increased significantly. This type of ties most often works for bending and pulling. Figure 1 shows the main types of dowels:

The versatility and ease of installation of joints of elements of wooden structures is an important issue in the design of a building. Currently, there are various options for mating individual structures together, but one of the most promising are SHERPA connections, shown in Fig. 2.

This type of joint allows you to join the main and secondary beams, beams and columns with each other, while the installation of the connection parts takes place on the ground, and during installation, it remains only to "put" one part on the second.

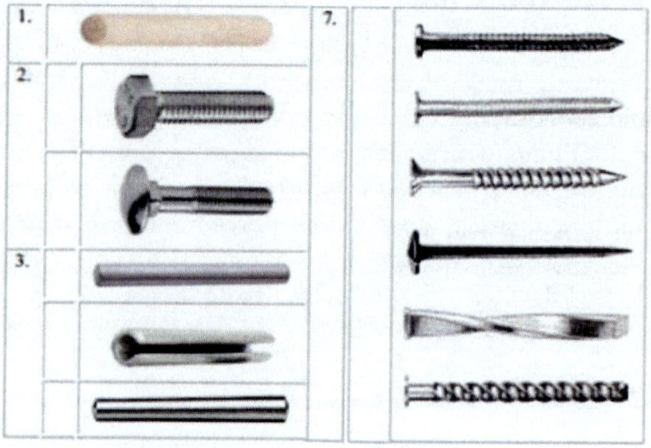

Fig. 1 The main types of dowels used in the joints of wood structures (*1* oak dowel, *2* bolts, *3* pins, *4* screws, *5* self-tapping screws, *6* ordinary screws, *7* nails)

Fig. 2 SHERPA connection

In this case, the fastening of the connectors to the wooden elements is carried out using special SHERPA screws, which are screwed along the body of the structure at a certain angle. In this connection, as in most others, the loss of bearing capacity occurs due to the formation of plastic hinges in the steel dowels, which, in turn, develop due to the crushing of the wood in the dowel socket.

One of the ways to increase the strength and reduce the deformability of wood is modification with various compositions. At present, 4 main methods of modification can be distinguished: thermochemical (radiation-chemical), thermomechanical, chemical and chemical–mechanical [4, 13–16].

2 Methods

Within the framework of research carried out on the basis of VlSU laboratories, the possibility of modifying the joints of elements of wooden structures, in particular SHERPA joints, in order to increase the bearing capacity and reduce deformability is being studied.

Based on the modification technology and the availability of components, several compositions are considered for modification:

1. Compound based on ED-20 resin;
2. Compound based on glue based on urea–formaldehyde resin grade KF-Zh;
3. Compound based on dimethylacrylic polyester [17, 18, 21–26].

Table 1 shows the characteristics of these and other possible modifiers.

As a result of previous studies, it was found that the best modifying composition for wood in this study will be dimethacrylic polyester, since it has a low viscosity and high physical and mechanical characteristics of the samples under study [19, 20, 27–30].

The joint of the main and secondary solid wood floor beams is taken as a modeled joint.

The section of the main beam (h × b) is 200 × 150 mm, the secondary one is 200 × 100 mm.

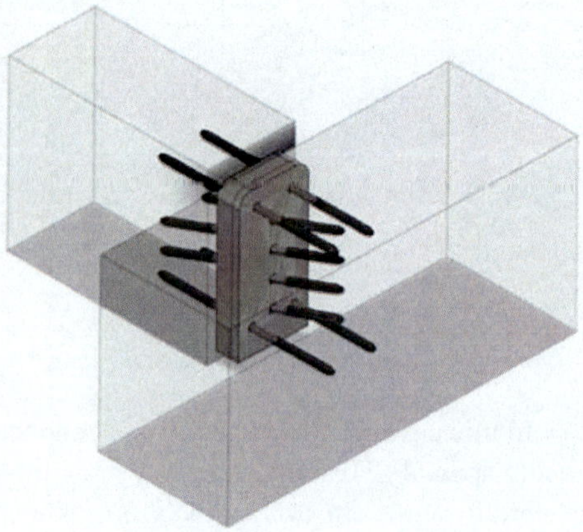

Fig. 3 Simulated SHERPA connection

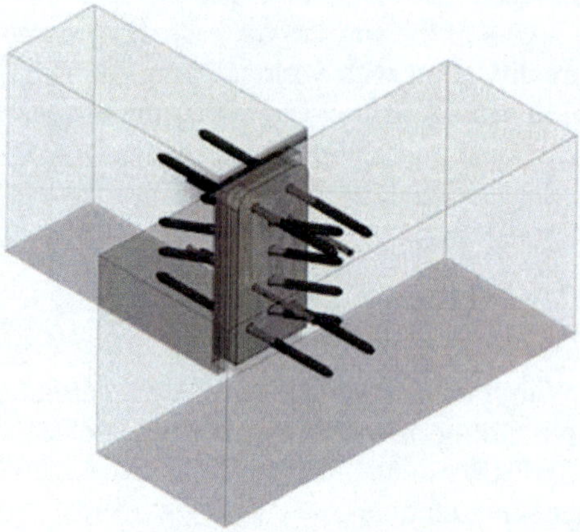

Fig. 4 Simulated stud connection with wood modification fittings

The length of the screws is 80 mm, the diameter is 8 mm, their location takes into account the requirements of the relevant regulatory documents.

The load was set stepwise on the area of the upper side of the secondary beam with a step of 5 Kn.

At the first stage, for this, numerical studies are carried out and their results are compared for several types of compounds, the general view of the models of which is shown in Figs. 3 and 4. The geometry was set using the AUTODESK FUSION 360 software package.

4. The connectors are made of aluminum EN AW-6082 according to EN 755-2. Below is the chemical composition of the alloy (Table 2)

The closest alloy in composition in Russian regulatory documents is AK 6 (1360).

In this case, the following strength characteristics are set: yield point or proportionality limit with a residual deformation tolerance of 0.2%: 245 MPa;

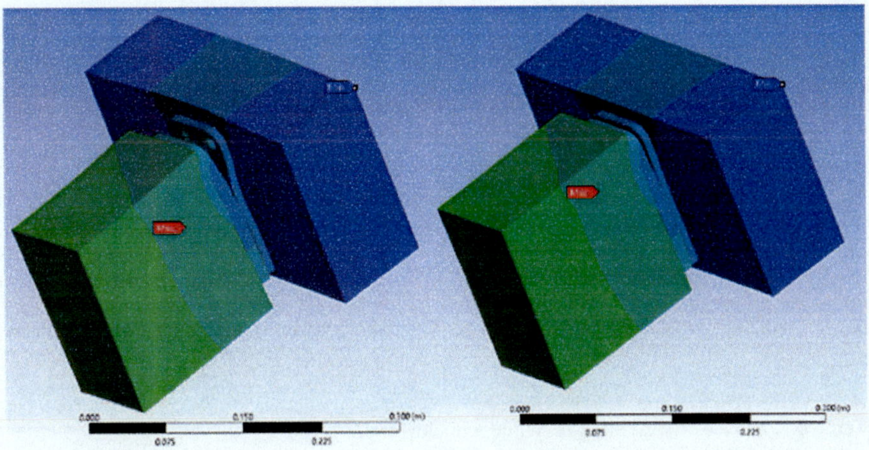

Fig. 5 Deformed models of the investigated connections. *1* SHERPA joint, *2* stud joint with modified wood

short-term strength limit: 365 MPa; elongation after rupture: 7%. The calculation was carried out in the ANSYS software package.

3 Results and Discussion

Based on the results of the simulation, the following conclusions can be drawn:

1. Failure of the joint without wood modification is of a plastic nature and begins with deformation of the wood of the nail socket, and, as a consequence, the formation of a plastic hinge in the nails;
2. The tensile strength of aluminium parts is not achieved in both cases;
3. When modifying wood, an increase in the strength of the entire joint by 20–23% is achieved, the deformability is reduced by 8–10% Below are the calculation models of the compounds under study, as well as the isofields of their deformations (Fig. 5).

4 Conclusions

Based on the results of the studies carried out, it can be concluded that local modification of the wood of the joint of the nail joint is a justified way to increase the strength of reducing the deformability of the nail joint. For further research, a

Table 1 Characteristics of pot life and viscosity of some modifiers

Composition name	Brookfield dynamic viscosity, Pa · s	Pot life at 20 °C, min
ED-20	13–20	60–120
XZ 92,598.00	3–9	25
Etal-370	0,64–1	60–120
KF-Zh	0,35–0,7	600
PN-1	0,25–0,3	15
FR-12	0,09–015	120–240
Dimethylacrylic polyester	0,07–0015	–
Phenolic alcohols	0001067	–
Methyl methacrylate	0,0006	–

Table 2 Chemical composition of aluminum alloy 6082

Chemical composition, %									
Silicon	Iron	Copper	Manganese	Magnesium	Chromium	Zinc	Titanium	Others	
07–1,3	0,5	0,1	0,4–1	0,6–1,2	0,25	0,2	0,1	Each 0,05	Sum 0,15

number of factors have been established, a change in which can affect the strength characteristics of the system under study. Here are a few of them:

1. It is possible to change and determine the optimal angle of inclination of the screws, as well as their location together with the fittings for modifying wood.
2. In both cases, the tensile strength of aluminum connectors is not reached, which allows us to that the cost of the alloy may decrease due to a decrease in the proportion of elements such as silicon, chromium, zinc and titanium.
3. The variant with the installation of the joint in the groove, made in one of the joined wooden elements, indicates the possibility of using this joint in buildings with increased chemical aggressiveness with reliable insulation of the elements. The author of the article believes that further research will be aimed at solving technological problems associated with modifying the assembly, as well as determining the rational parameters and materials of the joint elements.

Acknowledgements The reported study was funded by RFBR, project number 20-38-90113.

References

1. Labudin BV, Popov EV, Stolypin D, Sopilov V (2019) The wood composite ribbed panels on mechanical joints. In: E3S web of conferences, p 02021. EDP Sciences. https://doi.org/10.1051/e3sconf/20199102021
2. Buka-Vaivade K, Serdjuks D, Goremikins V, Pakrastins L, Vatin NI (2018) Suspension structure with cross-laminated timber deck panels. Mag Civ Eng 83:126–135. https://doi.org/10.18720/MCE.83.12

3. Yao Z, Wu D, Chen C, Zhang M (2013) Creep behavior of polyurethane nanocomposites with carbon nanotubes. Compos Part A Appl Sci Manuf 50:65–72. https://doi.org/10.1016/j.compositesa.2013.03.015
4. Yahyaei-Moayyed M, Taheri F (2011) Experimental and computational investigations into creep response of AFRP reinforced timber beams. Compos Struct 93:616–628. https://doi.org/10.1016/j.compstruct.2010.08.017
5. Fossetti M, Minafò G, Papia M (2015) Flexural behaviour of glulam timber beams reinforced with FRP cords. Constr Build Mater 95:54–64. https://doi.org/10.1016/j.conbuildmat.2015.07.116
6. Ruslantsev AN, Portnova YM, Tairova LP, Dumansky AM (2016) Analysis of mechanical properties anisotropy of nanomodified carbon fibre-reinforced woven composites. In: IOP conference series: materials science and engineering. Institute of Physics Publishing. https://doi.org/10.1088/1757-899X/153/1/012003
7. Telichenko VI, Rimshin VI, Karelskii AV, Kurbatov BVLAVL (2017) Strengthening technology of timber trusses by patch plates with toothed-plate connectors. J Ind Pollut Control 33:1034–104
8. Popov E, Ruslanova A, Sopilov V, Zdralovic N, Mamedov S, Labudin B (2020) Contact interaction of a claw washer with wood at limiting shear. Lesn Zhurnal (Forest J) 178–189. https://doi.org/10.37482/0536-1036-2020-4-178-189
9. Khelifa M, Celzard A (2014) Numerical analysis of flexural strengthening of timber beams reinforced with CFRP strips. Compos Struct 111:393–400. https://doi.org/10.1016/j.compstruct.2014.01.011
10. D'Ambrisi A, Focacci F, Luciano R (2014) Experimental investigation on flexural behavior of timber beams repaired with CFRP plates. Compos Struct 108:720–728. https://doi.org/10.1016/j.compstruct.2013.10.005
11. Vodiannikov MA, Kashevarova GG (2017) Analysis of wood structure connections using cylindrical steel and carbon fiber dowel pins. In: IOP conference series: materials science and engineering. Institute of Physics Publishing. https://doi.org/10.1088/1757-899X/205/1/012031
12. Kasal B, Pospisil S, Jirovsky I, Heiduschke A, Drdacky M, Haller P (2004) Seismic performance of laminated timber frames with fiber-reinforced joints. Earthq Eng Struct Dyn 33:633–646. https://doi.org/10.1002/eqe.368
13. Vatin NI, Usanova KY (2019) BIM end-to-end training: from school to graduate school. In: Advances and trends in engineering sciences and technologies iii—Proceedings of the 3rd international conference on engineering sciences and technologies, ESaT 2018, pp 651–656. CRC Press, Balkema. https://doi.org/10.1201/9780429021596-102
14. Sergeev M, Rimshin V, Lukin M, Zdralovic N (2020) Multi-span composite beam. In: IOP conference series: materials science and engineering. IOP Publishing Ltd. https://doi.org/10.1088/1757-899X/896/1/012058
15. Gribanov AS, Strekalkin AA, Kudryatseva AA, Zdralovic N (2020) CFRP composites for strengthening wooden structures. In: IOP conference series: materials science and engineering. IOP Publishing Ltd. https://doi.org/10.1088/1757-899X/896/1/012114
16. Roshchina S, Lukin M, Lisyatnikov M (2020) Compressed-bent reinforced wooden elements with long-term load. In: Lecture notes in civil engineering. Springer, Berlin, pp 81–91. https://doi.org/10.1007/978-3-030-42351-3_7
17. Basterra LA, Balmori JA, Morillas L, Acuña L, Casado M (2017) Internal reinforcement of laminated duo beams of low-grade timber with GFRP sheets. Constr Build Mater 154:914–920. https://doi.org/10.1016/j.conbuildmat.2017.08.007
18. Kim YJ, Harries KA (2010) Modeling of timber beams strengthened with various CFRP composites. Eng Struct 32:3225–3234. https://doi.org/10.1016/j.engstruct.2010.06.011
19. Matveev RP, Labudin BV, Morozov VS, Orlov AO (2017) Numerical analysis of strength and rigidity of the biomechanical system "Bone-Apparatus." Hum Ecol (Russian Fed) 0:58–64. https://doi.org/10.33396/1728-0869-2017-4-58-64

20. Karelskiy AV, Zhuravleva TP, Labudin BV (2015) Load-to-failure bending test of wood composite beams connected by gang nail. Mag Civ Eng 54:77–85. https://doi.org/10.5862/MCE.54.9
21. Labudin BV, Popov EV, Nikitina TA (2019) Notes for calculated resistance to tension for laminated wood. In: IOP conference series: materials science and engineering. IOP Publishing Ltd. https://doi.org/10.1088/1757-899X/687/3/033028
22. Labudin BV, Popov EV, Sopilov VV (2019) Stability of compressed sheathings of wood composite plate-ribbed structures. In: IOP conference series: materials science and engineering, p 033041. IOP Publishing Ltd. https://doi.org/10.1088/1757-899X/687/3/033041
23. Romanovich A, Kleshcunov Y, Vlasov A (2019) On potentiality and practicability of installing flooring suspended in geodesic domes by means of cable system. In: IOP conference series: materials science and engineering. IOP Publishing Ltd (2019). https://doi.org/10.1088/1757-899X/687/3/033025.
24. Modin A, Lukin M, Vlasov A, Hisham E (2020) Energy-efficient indicators of panel housing mass construction in the climatic conditions of central Russia. In: IOP conference series: materials science and engineering. IOP Publishing Ltd. https://doi.org/10.1088/1757-899X/896/1/012063
25. Lisyatnikov MS, Shishov II, Sergeev MS, Hisham E (2020) Precast monolithic coating of an industrial building based on variable-height beam-slabs. In: IOP conference series: materials science and engineering. IOP Publishing Ltd. https://doi.org/10.1088/1757-899X/896/1/012064
26. Lisyatnikov MS, Glebova TO, Ageev SP, Ivaniuk AM (2020) Strength of wood reinforced with a polymer composite for crumpling across the fibers. In: IOP conference series: materials science and engineering. IOP Publishing Ltd. https://doi.org/10.1088/1757-899X/896/1/012062
27. Gribanov A, Glebova T, Roschina S (2020) Restoration of destructive wood in supporting zones of wooden beams. In: Lecture notes in civil engineering. Springer, Berlin, pp 157–166. https://doi.org/10.1007/978-3-030-42351-3_14
28. Koshcheev AA, Roshchina SI, Naichuk AY, Vatin NI (2020) The effect of eccentricity on the strength characteristics of glued rods made of steel cable reinforcement in solid wood. In: IOP conference series: materials science and engineering, p 012059. IOP Publishing Ltd. https://doi.org/10.1088/1757-899X/896/1/012059
29. Gribanov AS, Roshchina SI, Naichuk AY, Melekhov VI (2020) Wooden beams with local wood modification. In: IOP conference series: materials science and engineering. IOP Publishing Ltd. https://doi.org/10.1088/1757-899X/896/1/012067
30. Lukin M, Prusov E, Roshchina S, Karelina M, Vatin N (2021) Multi-Span composite timber beams with rational steel reinforcements. Buildings 11:1–12. https://doi.org/10.3390/buildings11020046

Influence of Floor Slabs to the Progressive Collapse-Resistant Ability of Reinforced Concrete Frame Structures

Sergey Osykov and Aleksandr Trofimov

Abstract When conducting nonlinear static progressive collapse analysis, dynamic effects caused by sudden column loss are considered by using dynamic increase factors (DIF). The values of the DIF depend both on the characteristics of individual structural elements and on the type of the structure itself. This paper presents the design approach for determining the DIF to consider the contribution of floor slabs to the progressive collapse-resistant ability of reinforced concrete frame structures. A simplified approach is based on the energy conservation principle. In this method, the work done by gravity loads is equal to the strain energy of a structure. Proposed method is validated through a several numerical examples consisting of nonlinear dynamic and nonlinear static analyses of reinforced concrete two-way slab structure. Different column loss scenarios on the first floor are considered to evaluate the DIF values. The comparison of the DIF values obtained according to the proposed method with the values obtained according to American standards is conducted. The results show that the presented approach allows to determine the DIF with sufficient accuracy, and confirm the significant role of floor slabs in the progressive collapse resistance of the whole structure.

Keywords Progressive collapse · Dynamic increase factor · Energy conservation principle · Nonlinear static analysis · Column loss · Two-way slab

1 Introduction

Progressive collapse belongs to a phenomenon in which initial local damage of a structural element causes dynamic effects. In static analyses such effects are considered using dynamic increase factors (DIF). The values of the DIF depend both on the characteristics of individual structural elements and on the type of the structure itself [1–3].

S. Osykov (✉) · A. Trofimov
Saint Petersburg State University of Architecture and Civil Engineering, 4 Vtoraya Krasnoarmeiskaya St, 190005 Saint Petersburg, Russia

N. Vatin et al. (eds.), *Proceedings of MPCPE 2021*, Lecture Notes in Civil Engineering 182, https://doi.org/10.1007/978-3-030-85236-8_36

Wide research has been conducted into investigation of DIF values. In [4–9] energy conservation principle has been used to assess the collapse resistance of building frames. The energy-based method was also used to consider the rise-time effect on the DIF [10] and to develop analytical expression for DIF for reinforced concrete (RC) frames under catenary mechanism [11]. In [12] DIF with a new max (M_u/M_p) parameter for nonlinear static (NS) analysis was presented. Differences in results from progressive collapse estimation of the frame using the pulldown and pushdown analysis were compared in [13]. Effects of post-elastic stiffness and damping ratio on DIF have been investigated in [14–17]. Target displacement method for prediction the DIF value was presented in [18].

In these studies, numerical models of frame structures didn't include floor slabs as finite elements—floor slabs were replaced with self-weight loads applied to the beams. In [19], progressive collapse-resistant capacity of the steel frame with composite floor slabs is investigated. Two types of numerical models were considered: the first with composite floor slabs, and the second without. For the first type the DIF value was 1.42 and for the second type was 1.12.

As seen, neglecting the contribution of slabs in structural response is possible in the case of precast concrete floor slabs, but in the case of pure frame system with composite floor slabs it can lead to an underestimation of the progressive collapse resistance of the whole structure. Therefore, steel moment-frames undergoing large deformations (i.e., under catenary mechanism) have a beneficial tension effect of membrane action, which could enhance structural integrity. It is assumed that, in RC two-way slab structures under the beam mechanism (i.e., for small deformations) floor slabs increase the range of elastic response, which leads to a rise the DIF value.

The aim of this paper is to develop an analytical expression for determining the DIF, considering the contributions of floor slabs in structural integrity of RC two-way slab structures for NS analysis and to validate the proposed method.

2 Methods

One of the analytical expressions of the DIF can be approach based on the energy conservation principle [7, 11]. In this method, the work done by instantly imposed gravity loads W should be equal to the strain energy of a structure U, i.e., W = U. The work done by gravity loads is equal to

$$W = P_d \cdot \Delta_u, \quad (1)$$

where P_d and Δ_u are, respectively, nonlinear dynamic loading and ultimate vertical displacement.

Strain energy of a structure may be expressed as

$$U = \int_{0}^{\Delta_u} P_s(\Delta) d\Delta, \tag{2}$$

where P_s is the equivalent nonlinear static loading estimated at the displacement Δ_s.

In energy-based method, the equivalent static loading P_s causes the displacement $\Delta_s = \Delta_u$ to be the same as the instantaneous applied dynamic loading P_d in the interval $[0; \Delta_u]$. In this case, the forced-based DIF can be calculated as

$$\text{DIF} = \frac{P_s}{P_d}. \tag{3}$$

In diagram form, the equality W = U can be illustrated in Fig. 1 as the equality of the areas ABCD and AEFD.

In Fig. 1, P_y is the yielding loading corresponding to the start of the yielding phase in the longitudinal reinforcement, Δ_y is the yielding displacement. The value of $0.5P_y$ is taken equal to half of P_y, based on the assumption that with elastic response of the structure without energy dissipation the DIF = 2 [2].

It should be noted that energy-based assessment of DIF is a simplified approach. Since empirical resistance curves differ from idealized bilinear curves (see Fig. 1), there may be some errors in determining the value of the DIF.

In [7], analytical expression of DIF for linear static (LS) analysis of RC frames is proposed. Based on [7] and the energy equilibrium principle the expression of DIF for NS pushdown analysis can be obtained as

$$P_d \cdot \Delta_u = 0.5 \cdot P_y \cdot \Delta_y + 0.5 \cdot (P_y + P_s) \cdot (\Delta_u - \Delta_y), \tag{4}$$

$$P_s = \frac{2P_d \cdot \Delta_u - P_y \cdot \Delta_u}{\Delta_u - \Delta_y}. \tag{5}$$

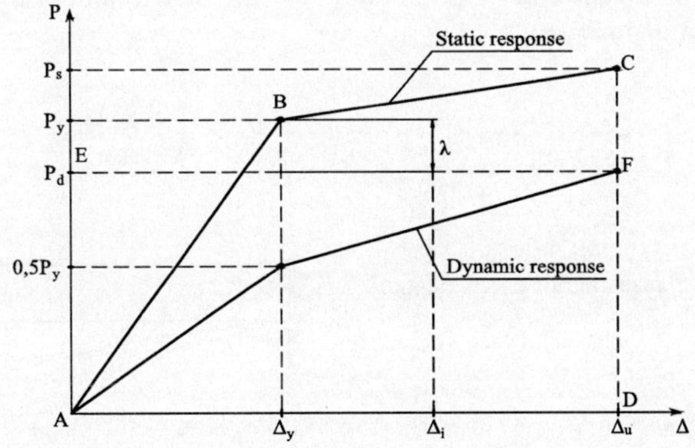

Fig. 1 A schematic explanation of the DIF

The yielding loading P_y is defines as

$$P_y = \lambda \cdot P_d, \tag{6}$$

where λ is a ratio of yielding loading to dynamic loading.

Substituting Eq. 6 into Eq. 5 and considering Eq. 3 the DIF can be calculated as

$$\mathrm{DIF} = \frac{\Delta_u \cdot (2 - \lambda)}{\Delta_u - \Delta_y}. \tag{7}$$

According to the [1], the DIF value is dependent on the ductility factor μ, which can be expressed as a ratio of chord rotation angles (see Fig. 2).

In Fig. 2, θ_y is the yield rotation angle, θ_u is ultimate plastic rotation angle, L is beam length. Since the values of ultimate plastic rotation angles θ_u for two-way slabs given in [20] are 0.02 ÷ 0.05, a simplification is acceptable, in which $\mathrm{tg}\theta = \Delta/L \approx \theta$. Hence, θ_y and θ_u are defined as

$$\theta_y = \frac{\Delta_y}{L}, \tag{8}$$

$$\theta_u = \frac{\Delta_u}{L}. \tag{9}$$

Ductility factor μ may be expressed as

$$\mu = \frac{\Delta_u}{\Delta_y} = \frac{\theta_u}{\theta_y}. \tag{10}$$

Substituting Eq. 10 into Eq. 7 the DIF is defined as

$$\mathrm{DIF} = \frac{\mu \cdot (2 - \lambda)}{\mu - 1}. \tag{11}$$

According to [20], the value of yield rotation angle θ_y for RC beam can be obtained as

$$\theta_y = \left(\frac{M_y}{E_b I_g}\right) l_p, \tag{12}$$

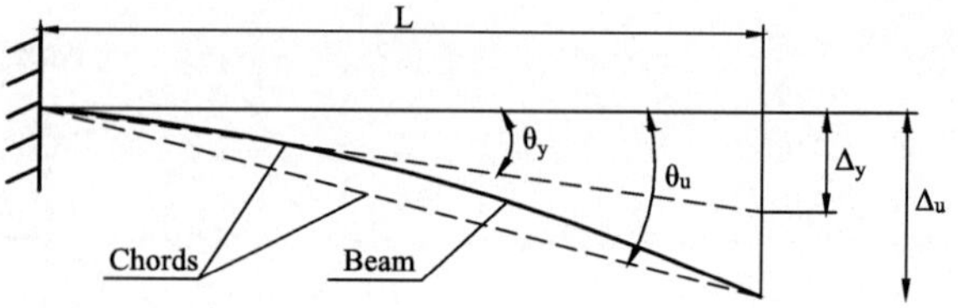

Fig. 2 A schematic explanation of chord rotation

where M_y is yield moment capacity of the T-beam, E_b is initial concrete modulus, l_p is plastic hinge length in the beam, I_g is moment of inertia of T-beam.

Yield moment capacity may be defined as

$$M_y = \alpha_m \cdot R_{bd} \cdot b_w \cdot d^2, \tag{13}$$

where R_{bd} is dynamic compressive strength of concrete, b_w is web width, d is the effective depth of T-beam, $\alpha_m = \xi_T \cdot (1-0.5 \cdot \xi_T)$.

Ratio of neutral axis depth to effective depth of T-beam can be expressed as.

$$\xi_T = \frac{R^w_{sd} \cdot A^w_s + R^f_{sd} \cdot A^f_s}{R_{bd} \cdot b_w \cdot d}, \tag{14}$$

where R^w_{sd} and R^f_{sd} are dynamic strength of longitudinal tension reinforcement in the web and flange respectively, A^w_s and A^f_s are areas of longitudinal tension reinforcement in the web and flange respectively.

A width of effective flange on each side of the web should be equal to the smallest of:

half the distance between adjacent webs;

one-sixth of the smallest beam length among all adjacent to the removed column.

The effective stiffness E_bI_g according to the laboratory test data could be approximated as $0.2E_bI_g$. The plastic hinge length can be calculated as

$$l_p = C \cdot \left(1 + \frac{0.5 \cdot N}{N_u}\right) \cdot (0.9 + 6.1 \cdot \xi) \cdot \left(\frac{l_0}{d}\right)^{1/4} \cdot d, \tag{15}$$

where C is the factor for support cross-section is equal to 0.7, N is the axial force from N_S or N_D preliminary analysis, N_u is the ultimate axial force, l0 is length of the area adjacent to the plastic hinge which is equal to one quarter of beam length.

Ratio of neutral axis depth to effective depth of rectangular cross-section (without flanges) is defined as

$$\xi = \frac{\left(\frac{N}{b_w \cdot d} + R^w_{sd} \cdot \rho' - R^w_{scd} \cdot \rho\right)}{R_{bd}}, \tag{16}$$

where R^w_{scd} is dynamic strength of compression reinforcement in the web, ρ' and ρ are ratios of A_s' to bwd and A_s to bwd respectively, A_s' and A_s are areas of longitudinal tension and compression reinforcement respectively.

The ultimate axial force N_u for rectangular cross-section is defined as

$$N_u = \varphi \cdot \left(R_{bd} \cdot A + R^w_{sd} \cdot A_{s,tot}\right), \tag{17}$$

where A is cross-sectional area of a concrete element, $A_{s,tot}$ is the sum of areas of longitudinal tension and compression reinforcement, φ is the factor equal to 0.9.

Since the value of λ is unknown, taking into account the results of [19] and considering that $1 < DIF \leq 2$, for practical purposes $\lambda = 1.05 \div 1.1$.

The finite element software SAP2000 is used to validate the proposed approach. The numerical example is based on 4 × 3 bay of 4-storey RC two-way slab structure. The height of the first storey is 4 m and the next storeys are 3.5 m. Material properties are taken, multiplied by the strength increment factors. The compressive strength of concrete is 21.3 MPa, yield strength of rebar is 500 MPa. Concrete elastic modulus is 30×10^3 MPa. Detail of beam and column cross-section is shown in Fig. 3.

Column removal scenarios include a corner column, internal column and middle external column on the first storey, as shown in Fig. 4. Combination of gravity loads includes the structural weight, dead loads and live loads. The values of dead and live loads are 5 kN/m^2 each.

Material nonlinearity is considered by using plastic moment hinges for beams and fiber hinges for columns. Moment hinges are assigned to beam ends and at the midspan of the beams. Fiber hinges are assigned to columns ends. The mechanical properties of moment hinges, taken according [20], is shown in Fig. 5. The plastic hinge length at columns is equal to half the column cross-section. To simulate RC floor slabs, nonlinear layered shell element is utilized. Stress–strain relationship for concrete and steel reinforcement is shown in Fig. 6. The model of the structure is supposed to be fixed on the ground level.

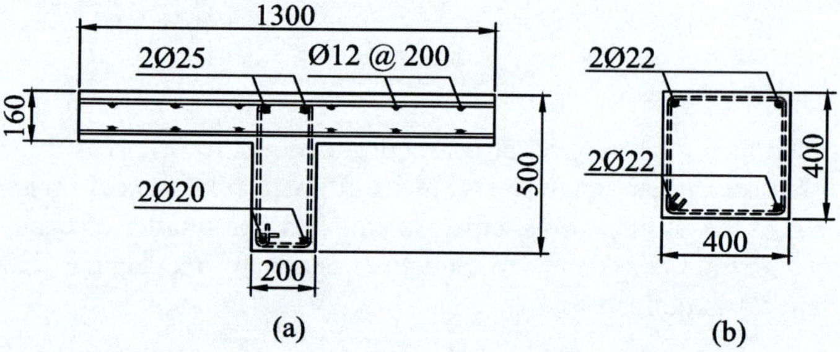

Fig. 3 Detail of two-way slab (**a**) and column (**b**) cross-section

Fig. 4 Plan dimensions and column removal locations

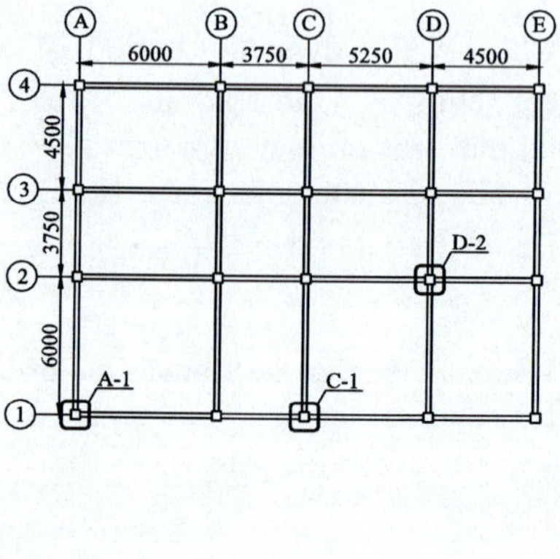

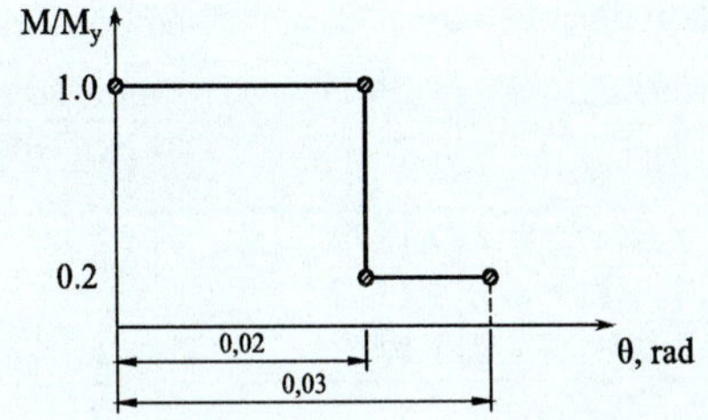

Fig. 5 The plastic moment hinge model

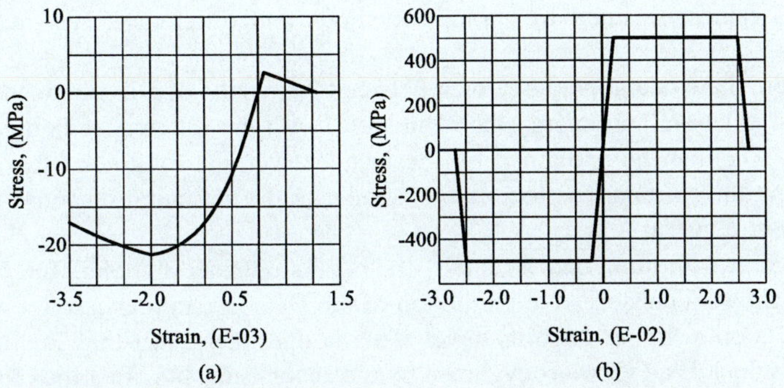

Fig. 6 Stress–strain relationship for concrete **(a)** and steel reinforcement **(b)**

The experimental procedure to determine the DIFexp is consisted of the following steps:

for each column removal case nonlinear dynamic (ND) analysis is performed according to the guideline [1] (with P-delta option chosen) and the values of plastic rotation angles of the beams adjacent to the upper node of the removed column are recorded;

NS pushdown analysis is performed, in which an increased gravity loads is applied according to the guideline [1] until the plastic rotation angle matched to the value obtained in Step 1;

the DIFexp values are calculated according to the Eq. 3s.

In addition to the numerical verification, the DIFUFC values are determined using the equation for RC frames from [1] with the ductility factor μ, obtained by the proposed approach

$$DIF_{UFC} = 1.04 + \frac{0.45}{\frac{\theta_u}{\theta_y} + 0.48}. \tag{18}$$

Table 1 Dynamic increase factors for various column-removed cases

Case	DIFexp	DIF	DIFUFC	Differences (%)	
				$\frac{DIF_{exp}-DIF}{DIF_{exp}}$	$\frac{DIF_{exp}-DIF_{UFC}}{DIF_{exp}}$
D-2	1.508	1.447	1.184	+4.2	+21.5
C-1	1.338	1.381	1.174	−3.2	+12.2
A-1	1.421	1.401	1.177	+1.4	+17.1

3 Results and Discussion

The DIF values obtained by various methods and differences are presented in Table 1.

As seen, there is a good agreement between the results of DIF obtained by using numerical example and energy-based method, with the differences being no more than 5%. This can be explained by the fact that vertical motion of the structure, occurred at the column loss location, is similar to the structural response mode for the element removal.

Compared with numerical example, the results obtained by using the procedure from [1] shows a conservative prediction of the DIF, i.e., in this case the structural response is on a higher ductility level. This is due to the fact that Eq. 18, which represents the DIFUFC, was developed by the analysis results of various RC frames without floor slabs. The obtained differences between DIFexp and DIFUFC confirm the assumption about the significant contribution of floor slabs in the progressive collapse resistance.

4 Conclusion

According to analytical analysis and numerical verification the following conclusions can be made:

1. The floor slabs increase structural integrity of RC frames wherein the progressive collapse-resistant ability would be underestimated if the contribution of slabs is omitted.
2. Compared with ND analysis the proposed method allows to determine the DIF for NS analysis with adequate precision. Procedure from existing design codes gives a conservative prediction of the DIF.
3. A simplified energy-based method can be used to determine the dynamic effects in frame structures instead of using computationally costly and time-consuming ND analysis.

Further work will be aimed at establishing the limits of applicability of the proposed method and determining the factors that influence on DIF.

References

1. UFC4-023-03 (2016) Unified facilities criteria (UFC): design of building to resist progressive collapse. Dep Def
2. McKay A, Marchand K, Diaz M (2012) Alternate path method in progressive collapse analysis: variation of dynamic and nonlinear load increase factors. Pract Period Struct Des Constr 17:152–160. https://doi.org/10.1061/(ASCE)SC.1943-5576.0000126
3. (Max) Liu M, McKay A, Marchand K, Diaz M (2016) Discussion of "alternate path method in progressive collapse analysis: variation of dynamic and nonlinear load increase factors". Pract Period Struct Des Constr 21. https://doi.org/10.1061/(asce)sc.1943-5576.0000166
4. Tsai M-H, Lin B-H (2008) Investigation of progressive collapse resistance and inelastic response for an earthquake-resistant RC building subjected to column failure. Eng Struct 30:3619–3628. https://doi.org/10.1016/j.engstruct.2008.05.031
5. Tsai MH, Lin BH (2009) Dynamic amplification factor for progressive collapse resistance analysis of an RC building. Struct Des Tall Spec Build 18. https://doi.org/10.1002/tal.453
6. Naji A, Irani F (2012) Progressive collapse analysis of steel frames: simplified procedure and explicit expression for dynamic increase factor. Int J Steel Struct 12. https://doi.org/10.1007/s13296-012-4008-0
7. Li Y, Lu X, Guan H, Ye L (2014) An energy-based assessment on dynamic amplification factor for linear static analysis in progressive collapse design of ductile rc frame structures. Adv Struct Eng 17:1217–1225. https://doi.org/10.1260/1369-4332.17.8.1217
8. Tsai MH (2011) Analytical load and dynamic increase factors for progressive collapse analysis of building frames. In: AEI 2011 Building integration solutions—Proceeding of AEI 2011 conference, vol 41168, pp 172–179. https://doi.org/10.1061/41168(399)21
9. Liu M, Pirmoz A (2016) Energy-based pulldown analysis for assessing the progressive collapse potential of steel frame buildings. Eng Struct 123. https://doi.org/10.1016/j.engstruct.2016.05.020
10. Tsai M-H (2016) Application of the work-energy principle to assess the risetime effect on the dynamic response amplification under column loss, 1338–1346. https://doi.org/10.14264/uql.2016.685
11. Li Y, Lu X, Guan H, Ye L (2014) Progressive collapse resistance demand of reinforced concrete frames under catenary mechanism. ACI Struct J 111. https://doi.org/10.14359/51687029
12. Liu M (2013) A new dynamic increase factor for nonlinear static alternate path analysis of building frames against progressive collapse. Eng Struct 48:666–673. https://doi.org/10.1016/j.engstruct.2012.12.011
13. Liu M (2015) Pulldown analysis for progressive collapse assessment. J Perform Constr Facil 29. https://doi.org/10.1061/(asce)cf.1943-5509.0000459
14. Mashhadi J, Saffari H (2017) Effects of postelastic stiffness ratio on dynamic increase factor in progressive collapse. J Perform Constr Facil 31. https://doi.org/10.1061/(asce)cf.1943-5509.0001109
15. Mashhadi J, Saffari H (2017) Modification of dynamic increase factor to assess progressive collapse potential of structures. J Constr Steel Res 138. https://doi.org/10.1016/j.jcsr.2017.06.038
16. Mashhadi J, Saffari H (2016) Effects of damping ratio on dynamic increase factor in progressive collapse. Steel Compos Struct 22. https://doi.org/10.12989/scs.2016.22.3.677
17. Amiri S, Saffari H, Mashhadi J (2018) Assessment of dynamic increase factor for progressive collapse analysis of RC structures. Eng Fail Anal 84. https://doi.org/10.1016/j.engfailanal.2017.11.011
18. Mahmoudi M, Teimoori T, Kozani H (2015) Presenting displacement-based nonlinear static analysis method to calculate structural response against progressive collapse. Int J Civ Eng 13:478–485. https://doi.org/10.22068/IJCE.13.4.478

19. Shi F, Wang L, Dong S (2017) Progressive collapse assessment of the steel moment-frame with composite floor slabs based on membrane action and energy equilibrium. Open Constr Build Technol J 11. https://doi.org/10.2174/1874836801711010200
20. Seismic evaluation and retrofit of existing buildings (2014) American society of civil engineers (ASCE). https://doi.org/10.1061/9780784412855

Spatial Vibrations of High-Rise Buildings Using a Plate Model

M. Usarov, G. Ayubov, D. Usarov, and G. Mamatisaev

Abstract The paper is devoted to the numerical solution of the problem of transverse oscillations of a multi-storey building within the framework of a continuous plate model under seismic effects. Cantilevers anisotropic plate is proposed as a building dynamic model, the theory of which is developed in the framework of a three-dimensional dynamic theory of elasticity and takes into account not only structural forces and moments but also the bimoments. The proposed plate model of a building allows us to take into account and study all types of different spatial oscillations of the building structure under the impacts different in direction. Formulas are given for the reduced density, elastic moduli, and shear of the plate model of the building. The base acceleration, given in time by a harmonic law, is taken as a seismic impact. The problem is solved by the finite difference method. Examples are considered and numerical results are obtained. Waveform, displacements, and accelerations distribution diagram of multi-storey high-rise buildings under transverse oscillations are plotted. Numerical results were obtained and the points of occurrence of the maximum displacement values were established.

Keywords Building · Plate model · Seismic load · Equations of motion · Boundary conditions · Numerical solution

M. Usarov · G. Ayubov · G. Mamatisaev (✉)
Institute of Mechanics and Seismic Stability of Structures of the Academy of Sciences of the Republic of Uzbekistan, Tashkent, Uzbekistan
e-mail: umakhamatali@mail.ru

D. Usarov
Fergana Polytechnic Institute, Fergana, Uzbekistan

N. Vatin et al. (eds.), *Proceedings of MPCPE 2021*, Lecture Notes in Civil Engineering 182, https://doi.org/10.1007/978-3-030-85236-8_37

1 Introduction

In the modern stage of earthquake-resistant construction, various underground and surface shell and plate structures interacting with the ground are widely used. Buildings and structures with various design solutions constitute a complex spatial mechanical system "structure-soil".

At the present stage of the development of science, the issues of seismic resistance of hydraulic structures interacting with the soil base are urgent and complex problems of the mechanics of a deformable solid. Design models of underground structures should reflect the real conditions of their interacting with the ground. In the works of the authors of works [1, 2], a dynamic theory of seismic resistance of underground structures interacting with soil is proposed.

In [3], the solution to the problem of optimizing projects of industrial buildings designed in seismically hazardous zones was considered. Economic efficiency was taken as an optimality criterion, depending on certain variable parameters adopted at the design stage.

In [4], the problem of assessing the impact of the destruction of wall panels in assessing the strength of large-panel structures was solved. Forces exceeding permissible values in structural elements are calculated.

The studies in [5–8] are devoted to the improvement of the box-type model of the building structure, taking into account the contact conditions between the elements of the panels and beams. Equations of motion of box-type elements and graphs of displacements of plates and beams are constructed. The article deals with the problem of forced vibrations of a building of spatial box type, which consists of rectangular panels and interacting beams under a dynamic action given by the foundation displacement according to the sinusoidal law. The method of finite differences was used to solve the problem.

The studies in [9, 10] develop the methods for dynamic spatial calculation of the structure based on the finite difference method in the framework of the theory of bimoments, taking into account the spatial stress–strain state.

References [11, 12] are devoted to the numerical solution of the problem of transverse and longitudinal vibrations of buildings and structures on the basis of a plate model developed within the framework of the bimoment theory of plates. The problems of transverse and longitudinal vibrations of buildings and structures are solved using the developed model, within the framework of the bimoment theory of plates [13, 14].

The data on the mechanism of tsunami wave formation and destruction were analyzed in [15], the recommendations for tsunami-resistant construction were generalized. A solution was proposed to mitigate damage from strong earthquakes and high tsunami waves.

Reference [16] discusses the main issues of determining the reinforcement parameters of reinforced concrete structures during their inspection. The basic ways to solve these problems are analyzed. The most reliable and accurate methods for determining the parameters of reinforcement are shown.

In [17–19], the dynamic characteristics and vibrations of various axisymmetric and plane structures are considered, taking into account various geometries, spatial factors and inelastic properties of materials. The solution to the problem is performed by the finite element method and by the expansion of the solution in terms of natural vibration modes. Various mechanical effects associated with the geometry of the structure and inelastic properties of the material are revealed.

The study in [20] is devoted to the method of static accounting for higher vibration modes in the problems of dynamics of building structures under external harmonic load. With a computational software package, the displacements of nodes and internal forces in the elements of the structures under consideration were determined.

2 Methods

In this work, we propose a cantilever anisotropic plate as dynamic models of a building, the theory of which was developed within the framework of the three-dimensional dynamic theory of elasticity and takes into account not only the forces and moments of the structure, but also bimoments. The proposed lamellar model of a building allows one to take into account and study all types of various spatial vibrations of the building structure under influences of various directions. The formulas for the reduced density, elasticity and shear moduli of the plate model of the building are given.

To describe the movement of the building plate, we introduce a Cartesian coordinate system with variables x_1, x_2 and z. The origin is located in the lower left corner of the median surface of the continual plate. Let us direct the axes OX_1 and OX_2 along the length and height, and the OZ axis along the thickness (width of the building) of the plate model.

Suppose that the seismic movement of the ground occurs in the direction of the OZ axis (the width of the building). Based on this, as an external influence on the lower clamped edge, we set the acceleration of the base $\ddot{u}_0(t)$ in the form:

$$\ddot{u}_0(t) = a_0 \cos(p_0 t), \tag{1}$$

where $a_0 = k_c g$ and $p_0 = 2\pi\omega_0$ are respectively the maximum acceleration and frequency of the subgrade.

At the base of the building, the boundary conditions for flexural-shear vibrations are as follows:

$$\tilde{\psi}_1 = 0,\ \ \tilde{\psi}_2 = 0,\ \ \tilde{\beta}_1 = 0,\ \ \tilde{\beta}_2 = 0,\ \ \tilde{u}_1 = 0,\ \ \tilde{u}_2 = 0,\ \ \tilde{r} = 0,\ \ \tilde{\gamma} = 0,\ \ \tilde{W} = 0. \tag{2}$$

On the free side faces of the building, we have the conditions for the equality of forces, moments and bimoments and force factors to zero:

$$\mathit{aligned}M_{11} = 0,\ M_{12} = 0,\ P_{11} = 0,\ P_{12} = 0,\ Q_{13} = 0,\ \tilde{p}_{13} = 0,\ \tilde{\sigma}_{11} = 0,\ \tilde{\sigma}_{12} = 0,\ \sigma^{*}_{11} = 0.\mathit{aligned} \tag{3}$$

On the free upper edge of the building, we have the following conditions:

$$\mathit{aligned}M_{12} = 0,\ M_{22} = 0,\ P_{12} = 0,\ P_{22} = 0,\ Q_{23} = 0,\ \tilde{p}_{23} = 0,\ \tilde{\sigma}_{11} = 0,\ \tilde{\sigma}_{12} = 0,\ \sigma^{*}_{22} = 0.\mathit{aligned} \tag{4}$$

The problem is solved by the finite difference method. The finite-difference equations of motion of transverse vibrations of buildings are as follows:

$$\begin{gathered}
\frac{(M_{11})^k_{i+\frac{1}{2},j}-(M_{11})^k_{i-\frac{1}{2},j}}{\Delta x_1}+\frac{(M_{12})^k_{i,j+\frac{1}{2}}-(M_{12})^k_{i,j-\frac{1}{2}}}{\Delta x_2}-(Q_{13})_{i,j}=\rho\frac{H^2}{2}\frac{\left(\tilde{\psi}_1\right)^{k+1}_{i,j}-2\left(\tilde{\psi}_1\right)^{k}_{i,j}+\left(\tilde{\psi}_1\right)^{k-1}_{i,j}}{\Delta t^2},\\
\frac{(M_{12})^k_{i+\frac{1}{2},j}-(M_{12})^k_{i-\frac{1}{2},j}}{\Delta x_1}+\frac{(M_{22})^k_{i,j+\frac{1}{2}}-(M_{22})^k_{i,j-\frac{1}{2}}}{\Delta x_2}-(Q_{23})_{i,j}=\rho\frac{H^2}{2}\frac{\left(\tilde{\psi}_2\right)^{k+1}_{i,j}-2\left(\tilde{\psi}_2\right)^{k}_{i,j}+\left(\tilde{\psi}_2\right)^{k-1}_{i,j}}{\Delta t^2},
\end{gathered} \tag{5}$$

$$\begin{aligned}
&\frac{(Q_{13})^k_{i+\frac{1}{2},j}-(Q_{13})^k_{i-\frac{1}{2},j}}{\Delta x_1}+\frac{(Q_{23})^k_{i,j+\frac{1}{2}}-(Q_{23})^k_{i,j-\frac{1}{2}}}{\Delta x_2}\\
&=\rho\frac{(\tilde{r})^{k+1}_{i,j}-2(\tilde{r})^{k}_{i,j}+(\tilde{r})^{k-1}_{i,j}}{\Delta t^2}+\rho\ddot{u}_0(t),
\end{aligned} \tag{6}$$

$$\begin{gathered}
\frac{(P_{11})^k_{i+\frac{1}{2},j}-(P_{11})^k_{i-\frac{1}{2},j}}{\Delta x_1}+\frac{(P_{12})^k_{i,j+\frac{1}{2}}-(P_{12})^k_{i,j-\frac{1}{2}}}{\Delta x_2}-3(\tilde{p}_{13})_{i,j}=\rho\frac{H^2}{2}\frac{\left(\tilde{\beta}_1\right)^{k+1}_{i,j}-2\left(\tilde{\beta}_1\right)^{k}_{i,j}+\left(\tilde{\beta}_1\right)^{k-1}_{i,j}}{\Delta t^2},\\
\frac{(P_{12})^k_{i+\frac{1}{2},j}-(P_{12})^k_{i-\frac{1}{2},j}}{\Delta x_1}+\frac{(P_{22})^k_{i,j+\frac{1}{2}}-(P_{22})^k_{i,j-\frac{1}{2}}}{\Delta x_2}-3(\tilde{p}_{23})_{i,j}=\rho\frac{H^2}{2}\frac{\left(\tilde{\beta}_2\right)^{k+1}_{i,j}-2\left(\tilde{\beta}_2\right)^{k}_{i,j}+\left(\tilde{\beta}_2\right)^{k-1}_{i,j}}{\Delta t^2},
\end{gathered} \tag{7}$$

$$\begin{aligned}
&\frac{(\tilde{p}_{13})^k_{i+\frac{1}{2},j}-(\tilde{p}_{13})^k_{i-\frac{1}{2},j}}{\Delta x_1}+\frac{(\tilde{p}_{23})^k_{i,j+\frac{1}{2}}-(\tilde{p}_{23})^k_{i,j-\frac{1}{2}}}{\Delta x_2}-\frac{4(\tilde{p}_{33})_{i,j}}{H}\\
&=\rho\frac{(\tilde{\gamma})^{k+1}_{i,j}-2(\tilde{\gamma})^{k}_{i,j}+(\tilde{\gamma})^{k-1}_{i,j}}{\Delta t^2}+\frac{1}{3}\rho\ddot{u}_0(t),
\end{aligned} \tag{8}$$

where M_{ij}, Q_{i3} $(i,j=1,2)$ are bending moments and shearing forces; P_{ij}, $(i,j=1,2)$ are longitudinal bimoments, p_{i3} $(i,j=1,2)$; p_{33}—intensities of transverse bimoments.

The system of equations of motion with respect to three generalized functions are $\tilde{u}_1$, $\tilde{u}_2$, $\tilde{W}$ approximated in the form

$$\begin{aligned}(\tilde{u}_1)_{i,j} &= \frac{1}{2}\left(21\left(\tilde{\beta}_1\right)_{i,j} - 7\left(\tilde{\psi}_1\right)_{i,j}\right) - \frac{1}{30}H\left(\frac{\partial \tilde{W}}{\partial x_1}\right)_{i,j},\\ (\tilde{u}_2)_{i,j} &= \frac{1}{2}\left(21\left(\tilde{\beta}_2\right)_{i,j} - 7\left(\tilde{\psi}_2\right)_{i,j}\right) - \frac{1}{30}H\left(\frac{\partial \tilde{W}}{\partial x_2}\right)_{i,j},\end{aligned} \tag{9}$$

$$\left(\tilde{W}\right)_{i,j} = \frac{21}{4}(\tilde{\gamma})_{i,j} - \frac{3}{4}(\tilde{r})_{i,j} - \frac{H}{20}\left(\frac{E_{31}}{E_{33}}\left(\frac{21}{2}\frac{\partial \tilde{\beta}_1}{\partial x_1} - \frac{7}{2}\frac{\partial \tilde{\psi}_1}{\partial x_1}\right)_{i,j} + \frac{E_{32}}{E_{33}}\left(\frac{21}{2}\frac{\partial \tilde{\beta}_2}{\partial x_2} - \frac{7}{2}\frac{\partial \tilde{\psi}_2}{\partial x_2}\right)_{i,j}\right). \tag{10}$$

3 The Results and Discussion

We conventionally assume that the mechanical and geometric characteristics of the materials of the room panels are the same: the modulus of elasticity $E = 20{,}000$ MPa, density $\rho = 2700$ kg/m^3, and Poisson's ratio $v = 0.3$.

We present the results of calculations of forced vibrations of a building within the framework of a thick plate model for the following dimensions of slabs and buildings: the length and width of buildings $a = 30$ m, $H = 11$ m, the height of one floor of the building $b_1 = 3$ m, the thickness of the bearing external and internal walls $h_1 = 0.25$ m and $h_2 = 0.2$ m, overlap thickness $h_{per} = 0.2$ m.

Based on the above initial data on the expressions for determining the density and elastic modulus of the plate model of the building, given in [7], we find the following characteristics of the building materials:

$$E_1^{bd} = 2600\ \text{P},\quad E_2^{bd} = E_3^{bd} = 2000\ \text{P}, G_{12}^{bd} = 480\ \text{P},\quad G_{13}^{bd} = 520\ \text{P},\quad G_{23}^{bd} = 200\ \text{P},$$

$$\rho_{bd} = 451\ \ \text{kg/m}^3,\ \ v_{21} = 0.3,\ \ v_{31} = v_{23} = 0.4\ .$$

In the calculations, the seismicity factor and the base acceleration frequency are given: $k_c = 0.1$ and $\omega_0 = 9.5$ Hz. Dimensionless variables $x = x_1/a$, $y = x_2/b$, $\tau = ct/H$, where $c = \sqrt{E/\rho}$. The calculation step for dimensionless coordinates is taken as $\Delta x = 1/60$, $\Delta y = 1/30$. The stability of the calculation in dimensionless time is ensured according to an explicit scheme with a step of $\Delta = 0.01$.

Table 1 shows the maximum values of the generalized displacement and acceleration $\tilde{r}$ and $\ddot{\tilde{r}}$ obtained during forced lateral vibrations of multi-storey buildings at the extreme points in the upper levels of the floors.

In Fig. 1 shows the vibration waveform of a sixteen-story building, and Fig. 2 shows a diagram of the distribution of acceleration in a sixteen-story building. In Figs. 3 and 4 show the graphs of oscillations of displacements and accelerations of the upper floors over the dimensionless time of a sixteen-story building. Figures 3 and 4, it can be seen that the maximum displacement of the upper floor is equal to

Table 1 Maximum values of displacements and accelerations of multi-storey buildings with lateral vibrations

Number of floors	$\tilde{r}$, 10^{-3} m	$\ddot{\tilde{r}}$, m/s^2
16	0.8984	−2.5240
20	1.5506	−2.5451
22	−2.2687	−5.1227
24	−1.6073	2.9966
28	−3.4151	−4.0074
32	2.8451	3.0118

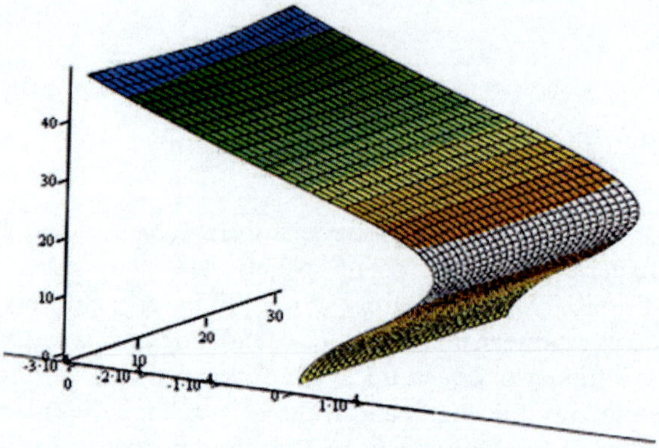

Fig. 1 Waveform of a sixteen-story building

$r = 0.8984 \times 10^{-3}$ m, and the accelerations of the upper floors of the sixteen-story building are equal to $\ddot{r} = -2,5240$ m/s^2.

In Figs. 5 and 6 shows, respectively, the waveform and the diagram of the distribution of acceleration in a twenty-storey building. Figures 7 and 8 show graphs of the generalized displacement r and acceleration in time of a twenty-story building, obtained with lateral vibrations located at the level of the building floors. As you can see, the maximum displacement of the top floor of the building is $r = 1.5506 \times 10^{-3}$ m, and the acceleration is $\ddot{r} = -2,5451$ m/s^2.

In Figs. 9 and 10 shows, respectively, the waveform of a twenty-two-story building and a diagram of the distribution of acceleration in a twenty-two-story building. In Figs. 11 and 12 shows the graphs of oscillations of displacement and acceleration of the upper floors for the dimensionless time of a twenty-two-story building. As you can see, the maximum displacement of the upper floor of the building is, $r = -2.2687 \times 10^{-3}$ m, and the acceleration is $\ddot{r} = -5,1227$ m/s^2.

In Fig. 13 shows the waveform of a twenty-four-story building, and Fig. 14 shows a diagram of the distribution of acceleration in a twenty-four-story building. In Figs. 15 and 16 show the graphs of oscillations of displacements and accelerations of the upper floors for the dimensionless time of a twenty-four-story building. Figures 15 and 16, it can be seen that the maximum displacement of the upper floor is equal to $r = -1.6073 \times 10^{-3}$ m, and the accelerations of the upper floors of a twenty-four-story building are equal to $\ddot{r} = 2,9966$ m/s^2.

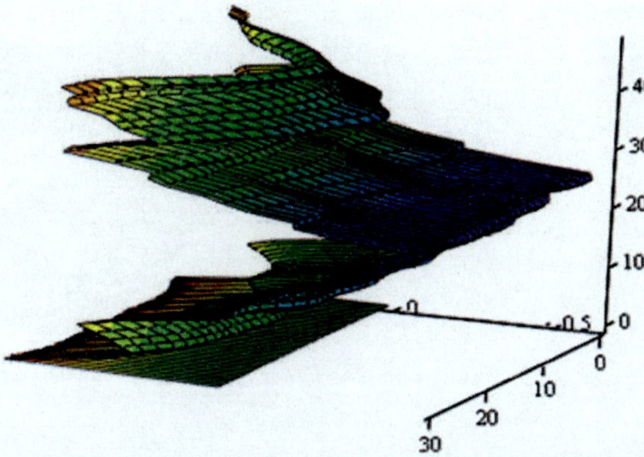

Fig. 2 Plot of acceleration distribution in a sixteen-story building

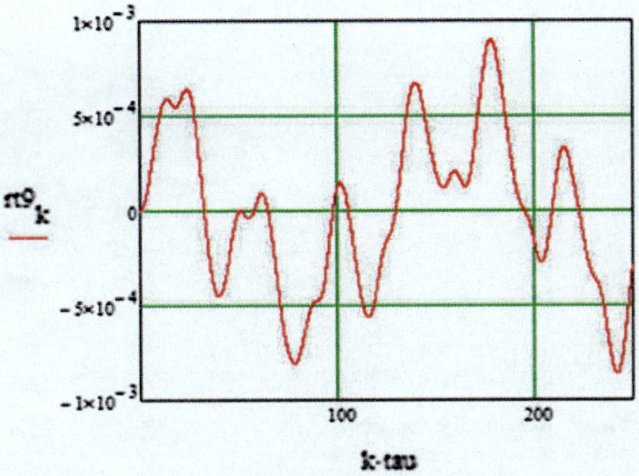

Fig. 3 Moving the top floor of a sixteen-story building, m

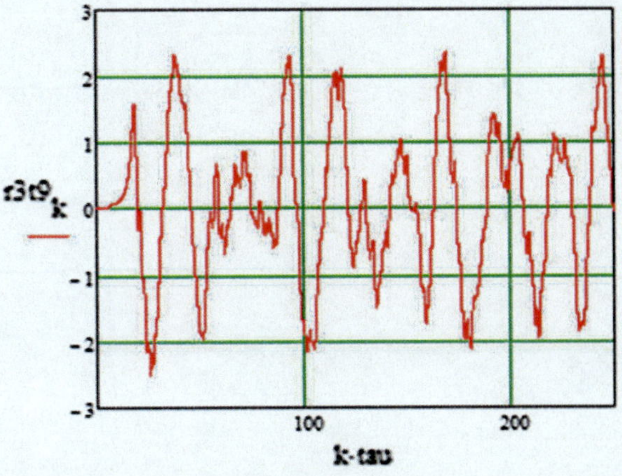

Fig. 4 Acceleration $\ddot{r}$ of the top floor of a sixteen-story building, m/s^2

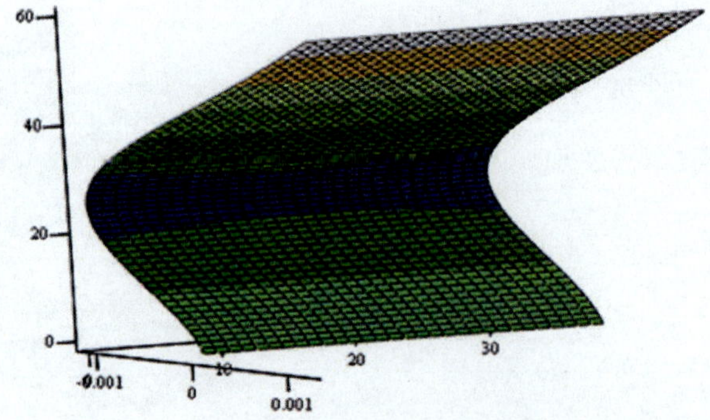

Fig. 5 The waveform of a twenty-story building

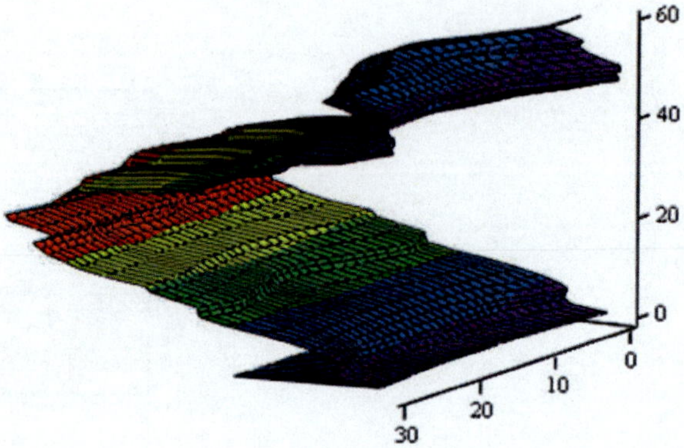

Fig. 6 Diagram of the distribution of acceleration in a twenty-story building

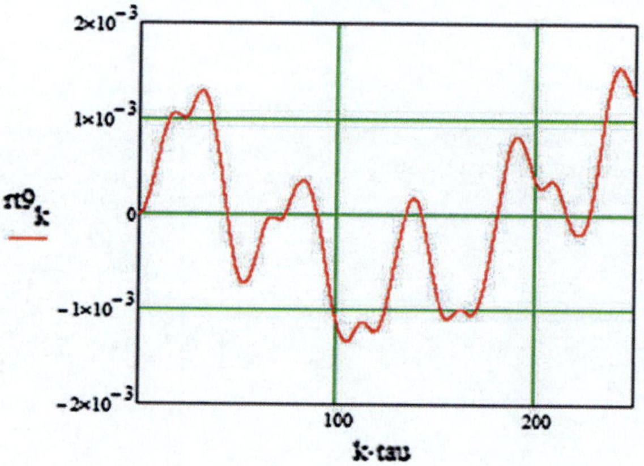

Fig. 7 Moving the top floor of a twenty-story building, m

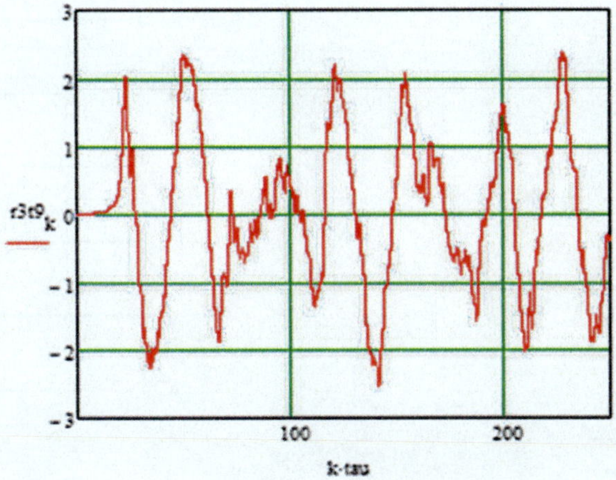

Fig. 8 Acceleration $\ddot{r}$ of the top floor of a twenty-story building, m/s^2

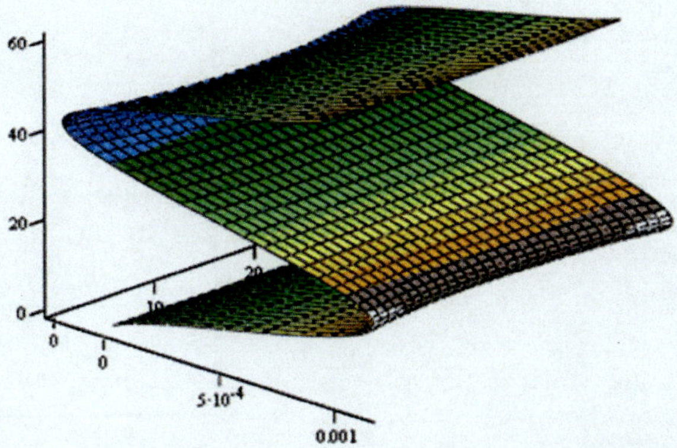

Fig. 9 The waveform of a twenty-two-storey building

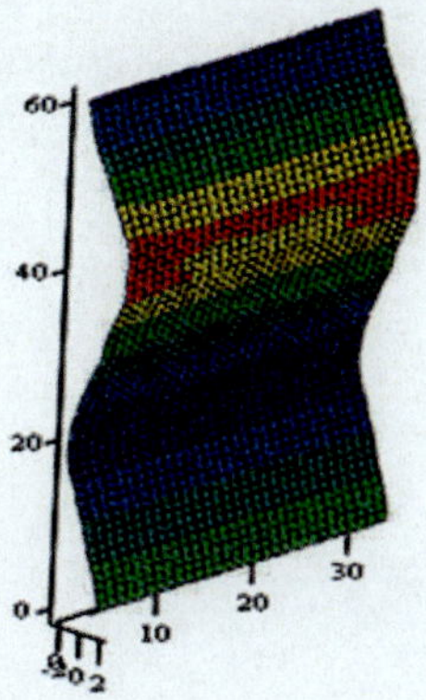

Fig. 10 Plot of acceleration distribution in a twenty-two-storey building

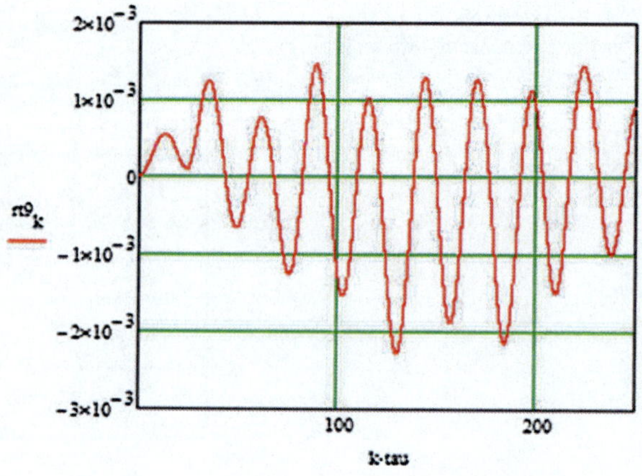

Fig. 11 Moving the top floor of a twenty-two-story building, m

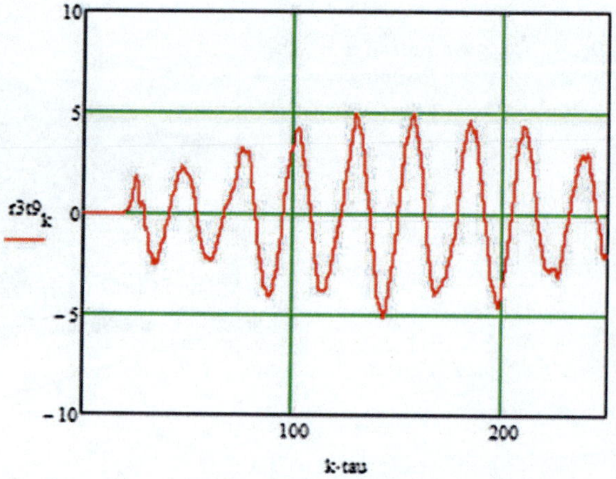

Fig. 12 Acceleration $\ddot{r}$ of the upper floor of a twenty-two-storey building, m/s^2

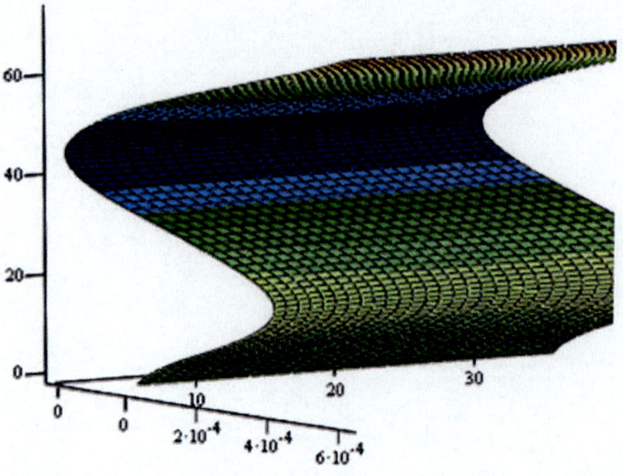

Fig. 13 The waveform of a twenty-four-storey building

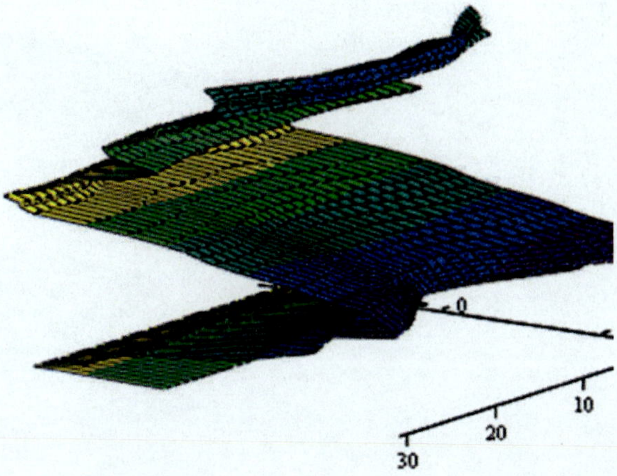

Fig. 14 Plot of acceleration distribution in a twenty-four-storey building

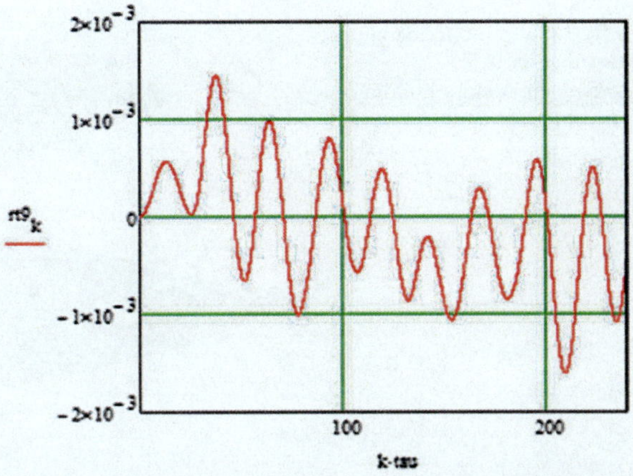

Fig. 15 Moving the top floor of a twenty-four-story building, m

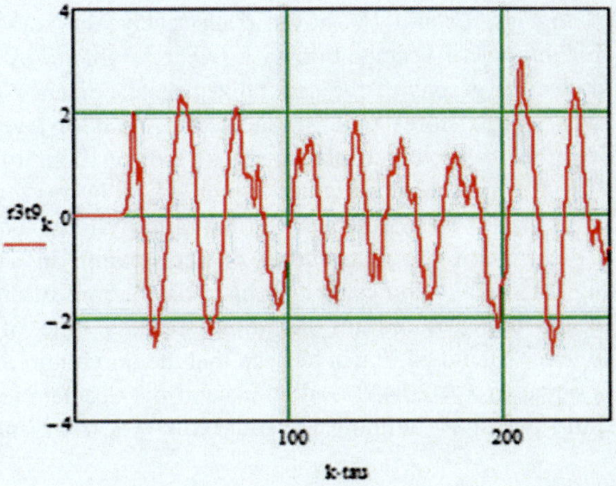

Fig. 16 Acceleration $\ddot{r}$ of the top floor of a twenty-four-story building, m/s^2

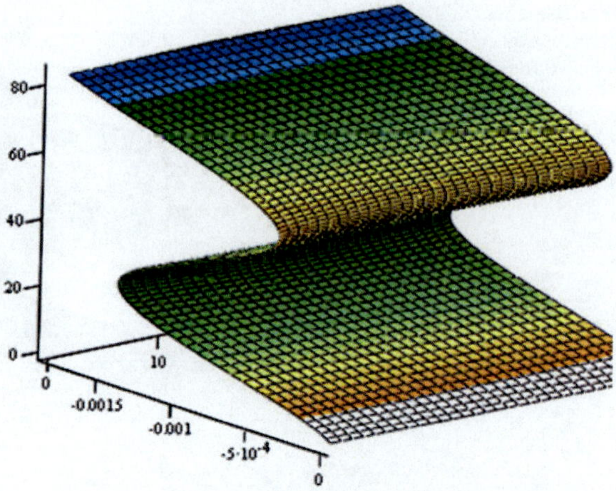

Fig. 17 The waveform of a twenty-eight-story building

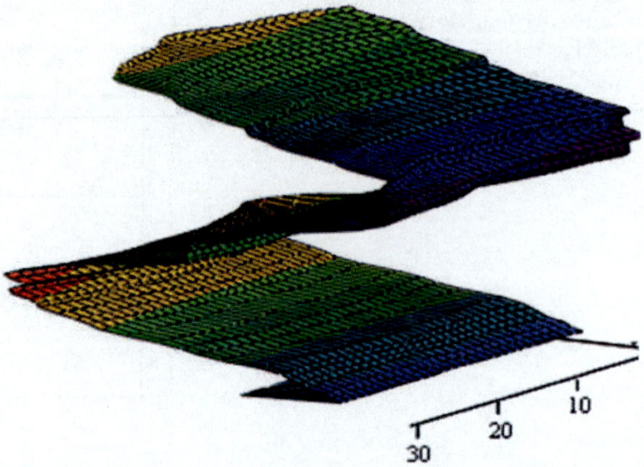

Fig. 18 Plot of acceleration distribution in a twenty-eight-story building

In Figs. 17 and 18 shows, respectively, the waveform and the diagram of the distribution of acceleration in a twenty-eight-storey building. In Figs. 19 and 20 shows the graphs of the generalized displacement r and acceleration in time for a twenty-eight-storey building located at the floor level of the building. As you can see, the maximum displacement of the top floor of the building is $r = -3.4151$ 10^{-3} m, and the acceleration is $\ddot{r} = -4,0074$ m/s^2.

In Fig. 21 shows the waveform of a thirty-two-story building, and Fig. 22 shows a diagram of the distribution of acceleration in a thirty-two-story building. In Figs. 23 and 24 shows the graphs of oscillations of displacements and accelerations of the upper floors for the dimensionless time of a thirty-two-story building. Figures 23 and 24, it can be seen that the maximum displacement of the upper floor is equal to $r = 2.8451 \times 10^{-3}$ m, and the accelerations of the upper floors of the thirty-two-story building are equal to $\ddot{r} = 3,0118$ m/s^2.

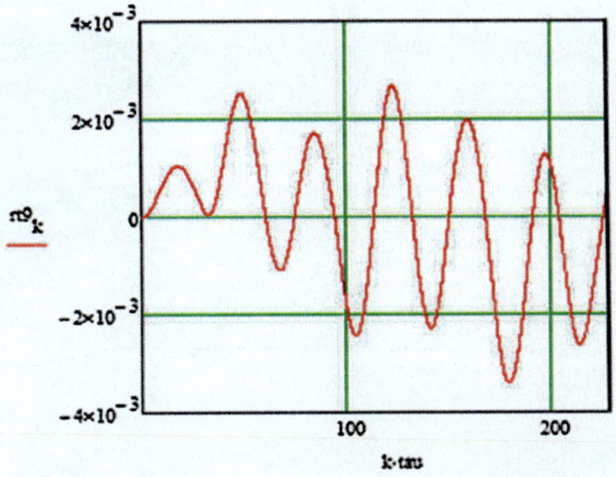

Fig. 19 Moving the top floor of a twenty-eight-story building, m

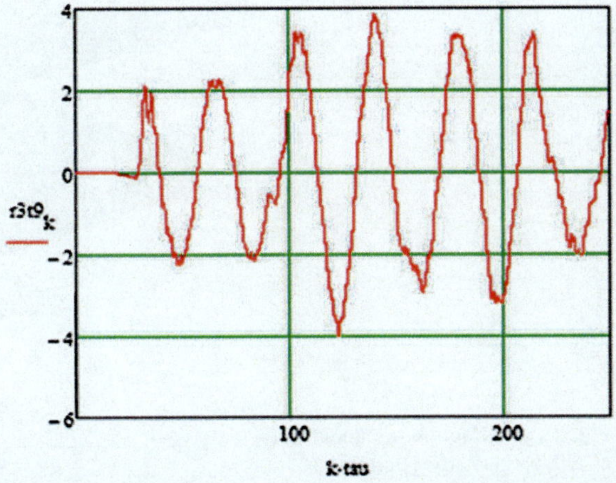

Fig. 20 Acceleration $\ddot{r}$ of the top floor of a twenty-eight-story building, m/s^2

4 Conclusion

Based on the analysis of the numerical results, it can be noted that the maximum displacement values are achieved at the upper level of buildings. The advantage of the proposed methodology for dynamic calculation of buildings based on a lamellar continual model is the reduction of a three-dimensional problem to a two-dimensional one, as well as obtaining the possibility of calculating high accuracy of numerical results for a lamellar structure in the form of a spatial model of multi-storey buildings.

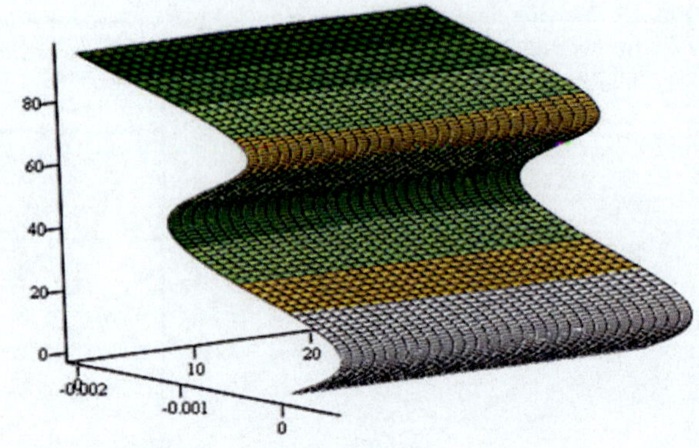

Fig. 21 The waveform of a thirty-two-storey building

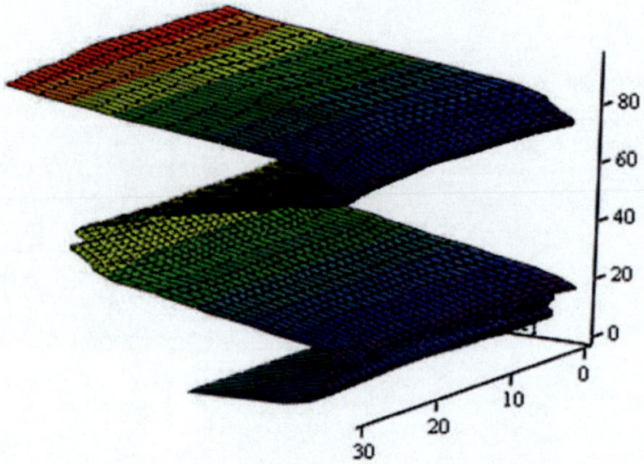

Fig. 22 Diagram of the distribution of acceleration in a thirty-two-storey building

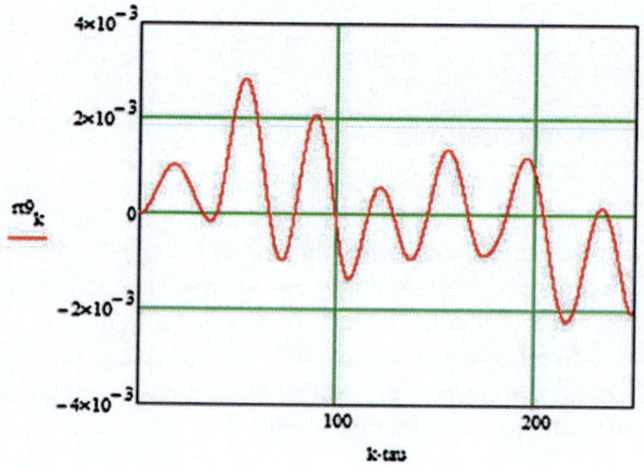

Fig. 23 Moving the top floor of a thirty-two-story building, m

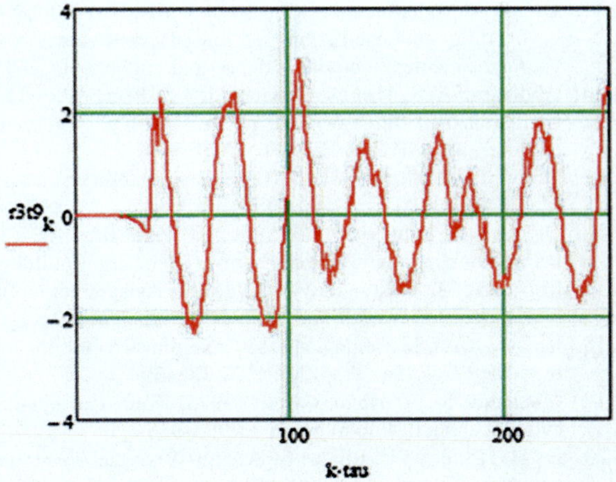

Fig. 24 Acceleration $\ddot{r}$ of the upper floor of a thirty-two-storey building, m/s^2

References

1. Rashidov T, Bekmirzaev D (2015) Seismodynamics of pipelines interacting with the soil. Soil Mech Found Eng 52
2. Rashidov TR, Yuldashev T, Bekmirzaev DA (2018) Seismodynamics of underground pipelines with arbitrary direction of seismic loading. Soil Mech Found Eng 55
3. Vatin NI, Ivanov AY, Rutman YL, Chernogorskiy SA, Shvetsov KV (2017) Earthquake engineering optimization of structures by economic criterion. Mag Civ Eng 76
4. Ватин НИ, Кузнецов ВД, Недвига ЕС (2011) Учет погрешности монтажа при расчете крупнопанельных зданий. Инженерно-строительный журнал. 35–40
5. Usarov MK, Mamatisaev GI (2020) Calculation on seismic resistance of box-shaped structures of large-panel buildings. In: IOP conference series: materials science and engineering. IOP Publishing, p 32041
6. Usarov M, Ayubov G, Mamatisaev G, Normuminov B (2020) Building oscillations based on a plate model. In: IOP conference series: materials science and engineering. IOP Publishing, p 12211
7. Usarov, M., Mamatisaev, G., Ayubov, G., Usarov, D., Khodzhaev, D.: Dynamic calculation of boxed design of buildings. In: IOP Conference Series: Materials Science and Engineering. p. 12186. IOP Publishing (2020).
8. Usarov M, Mamatisaev G, Toshmatov E., Yarashov J (2019) Forced vibrations of a box-like structure of a multi-storey building under dynamic effect. In: Journal of physics: conference series. IOP Publishing, p 12004
9. Yarashov J, Usarov M, Ayubov G (2019) Study of longitudinal oscillations of a five-storey building on the basis of plate continuum model. In: E3S web of conferences.. EDP Sciences, p 4065
10. Toshmatov E, Usarov M, Ayubov G, Usarov D (2019) Dynamic methods of spatial calculation of structures based on a plate model. In: E3S web of conferences. EDP Sciences, p 4072
11. Usarov D, Turajonov K, Khamidov S (2019) Simulation of free vibrations of a thick plate without simplifying hypotheses. In: Journal of physics: conference series. IOP Publishing, p 12115
12. Usarov MK (2015) Buckling of orthotropic plates with bimoments. Mag Civ Eng

13. Usarov M, Salokhiddinov A, Usarov DM, Khazratkulov I, Dremova N (2020) To the theory of bending and oscillations of three-layered plates with a compressible filler. In: IOP conference series: materials science and engineering. IOP Publishing, p 52037
14. Abdikarimov R, Usarov D, Khamidov S, Koraboshev O, Nasirov I, Nosirov A (2020) Free oscillations of three-layered plates. In: IOP conference series: materials science and engineering. IOP Publishing, p 12058
15. Belash TA, Yakovlev AD (2018) Seismic stability of a tsunami-resistant residential buildings. Mag Civ Eng 80
16. Ulybin AV, Zubkov SV, Fedotov SD, Zakrevsky AY (2014) Inspection of pile foundation before constructing additional storeys on existing buildings. Mag Civ Eng
17. Mirsaidov M, Mekhmonov Y (1987) Nonaxisymmetric vibrations of axisymmetric structures with associated masses and hollows (protrusions). Strength Mater 19:424–430
18. Ishmatov AN, Mirsaidov M (1991) Nonlinear vibrations of an axisymmetric body acted upon by pulse loads. Sov Appl Mech 27:388–394
19. Mirsaidov M, Troyanovskii IE (1975) Forced axisymmetric oscillations of a viscoelastic cylindrical shell. Polym Mech 11:953–955
20. Le TQT, Lalin VV, Bratashov AA (2019) Static accounting of highest modes in problems of structural dynamics. Mag Civ Eng 88

Damage and Destruction of Irrigation Canals During Earthquakes and Recommendations for Their Strengthening

Mashrab Akhmedov, Javlon Yarashov, and Doniyor Juraev

Abstract In the complex of water management construction, one of the central links is irrigation canals, through which the required volume of water is transported from the water source to the place of its consumption. The timely delivery and supply of water to its destination largely depends on the safety of the irrigation system. Meanwhile, in areas prone to seismic effects, these types of structures are significantly damaged, and sometimes even destroyed. This can lead to the shut-down of the main system for a long time, causing great material damage. In this work, the types of fastening on slopes are proposed, which ensure the uniformity of operation of all elements of the canal, which contributes to a significant reduction in virtual large deformations expected during strong earthquakes. Analysis of the results of earth canals and their elements behavior, and theoretical and experimental studies showed that, under the same conditions, the canal elements arranged in a chute are more resistant to seismic effects than the elements of the same canal, located in an embankment or in a semi-chute—semi-embankment. To prevent the fracturing of the canal embedded in the chute in non-rocky soils, in conditions of seismicity of 9 points, it is recommended to use concrete, reinforced concrete and asphalt concrete for canal slope protection.

Keywords Earthquake · Damages · Destruction · Irrigation Canals

M. Akhmedov
Academy of Sciences of the Republic of Uzbekistan Institute of Mechanics and Seismic Stability of Structures, Tashkent, Uzbekistan

J. Yarashov (✉) · D. Juraev
Department of Hydraulic Construction, Tashkent Institute of Irrigation and Agricultural Mechanization Engineers, Tashkent, Uzbekistan
e-mail: zhavlon.yarashov@bk.ru

N. Vatin et al. (eds.), *Proceedings of MPCPE 2021*, Lecture Notes in Civil Engineering 182, https://doi.org/10.1007/978-3-030-85236-8_38

1 Introduction

When solving problems related to the issues of seismic resistance of structures, of great interest are data obtained from a survey of the consequences of various destructive earthquakes, which provide reliable information about all kinds of deformations that can appear in structures under seismic impact and post seismic events.

Stable operation of irrigation system is vital for Uzbekistan where over four mln. hectares are irrigated lands. It is has importance for neighbor countries also.

The study of the consequences of a number of destructive earthquakes that occurred in the past worldwide served as an impetus for the deployment of extensive research in the field of seismic stability of linear engineering structures, which includes irrigation canals. This is confirmed by the damage or destruction of irrigation structures during the devastating earthquakes that occurred in 1906 in California, 1908 in Messina, 1923 in Tokyo, 1948 in Fukui (Japan), 1948 in Ashgabat (Turkmenistan), 1954 in California, 1963 in Skopje (Macedonia), 1966 in Tashkent (Uzbekistan), 1976 in Tangshan (China), 1976 in Gazli (Uzbekistan), etc. Because of these and other devastating earthquakes, along with industrial and civil objects, irrigation structures received significant damage and destruction [1–9].

2 Observation Methods

A vivid manifestation of earthquakes consequence for irrigation infrastructure and damages phenomena can be seen, in the report on the aftermath of the Japanese earthquake, 1923 {Formatting Citation}, which describes a case of partial destruction and damage to a water supply canal to Tokyo. The section of the canal, built in 1898, had a length of 4600 m, of which a part of 3800 m long is embedded completely in the embankment, its maximum height reaches 9 m, the channel width at the top is 6 m, at the bottom is 2.4 m and a depth is 1.2 m (Fig. 1).

The following main types of seismic deformations of the channel were observed:

- a slope collapsed near one of the pipelines under the embankment, and the bottom of the channel was washed out for 20 m (Fig. 2);
- wide longitudinal and transverse fractures formed in the canal section of 80 m. For the temporary restoration of the canal, a wooden pipe was installed in this area (Fig. 2);
- on another section of the canal, passing in an embankment of a height of 2.1–2.5 m, the bottom of the canal settled in the canal section of 40 m, which was accompanied by the appearance of cracks (ruptures) in the river bed;
- there was a destruction of the slope paving, made of concrete blocks (Fig. 1);
- 240 sites of transverse cracking were recorded along the entire length of the canal.

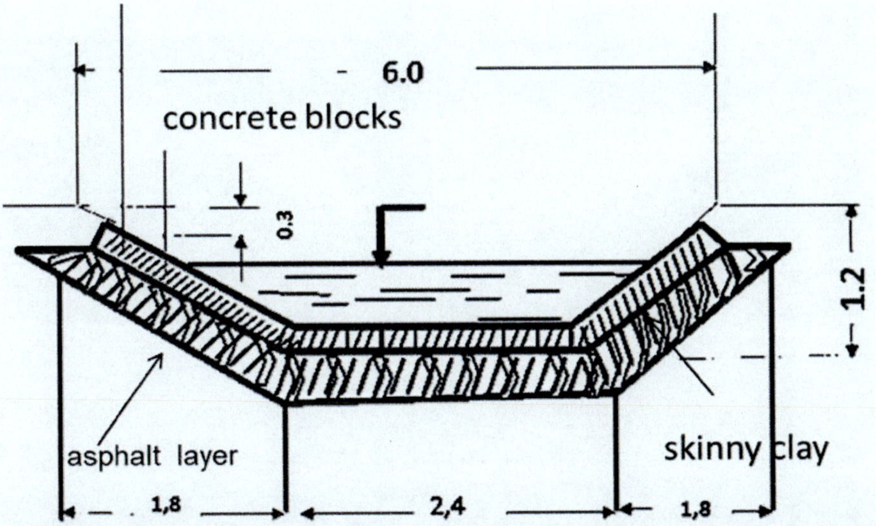

Fig. 1 Cross section before the earthquake [10]

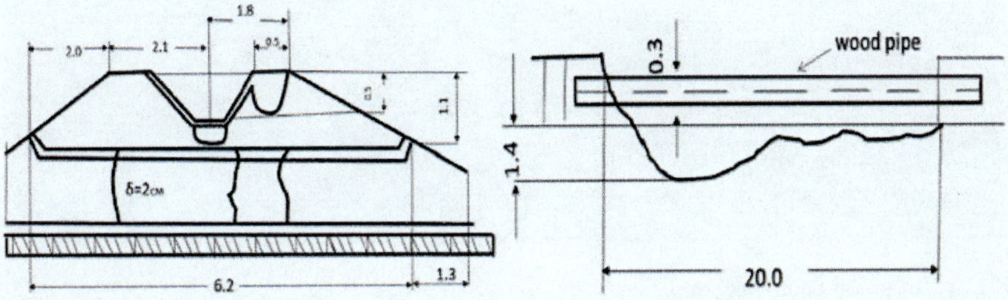

Fig. 2 Channel destruction [10]

Earthquakes also damage irrigation structures using in agriculture—irrigation canals of different types (Figs. 3, 4 and 5). Network of so structures are widespread in irrigated land of Uzbekistan.

Compared to dams and other types of structures, it is much more difficult to protect canals because they are mostly located in weak soils, vulnerable to seismic impact. In addition, irrigation canals during operation are subjected to intense mechanical stresses, as a result of which their bodies and elements receive noticeable or open damage, which is most subjected to destruction during seismic impact. The action of wind and trailing waves is manifested when they collapse onto the slope by hitting the slope face or its fastenings by the water mass transported by the wave. When the waves roll up and roll back, water flows at a significant velocity along the slope surface; pulsating water movements occur in the thickness of permeable elements and along their contact with the slope surface, and in the surface layer of the slope soil.

Fig. 3 Irrigation canal

Fig. 4 Irrigation chute network

Fig. 5 Water for irrigation of fields

As a result of the wave effect, the slopes can deform or, in the presence of fastening, the displacement of the latter can occur, as well as the erosion of the earth slopes or the slope soil, and its uneven settling.

Intensive changes in the water level in the canal or in the slope soil can lead to a difference in levels, filtration of water into the slope soil and into the thickness of the permeable fastening and, as a consequence, to the effect of weighing hydrodynamic pressure on the fastening and on soil particles of the slope. As a result of the weighing load and the occurrence of pore pressure, there is a partial separation of the surface layer or fastening elements from the upper part of the dam, a shift or sliding of the slope soil, and at water filtering under the fastening, soil particles are washed out and the entire slope is prepared for deformation.

Rodents cause significant damage to irrigation canals in Central Asia. By arranging burrows in the body or under the fastening of slopes, they seem to prepare the structure for an accident. Such cases were observed more than once at several irrigation facilities in Uzbekistan and other republics of Central Asia.

A great danger to the elements and structures of the canal is created by the presence of loess and subsiding soils at their base, which contribute to the formation of non-uniform deformation. These deformations increase significantly because of repeated seismic shocks, (insignificant ones) of 4 points or higher. Thus, all kinds of deformations formed in the body and elements of the canal during the period of their operation can reach maximums during destructive earthquakes.

To be convinced of this, consider the formula for relative deformations

$$\varepsilon = v/v_p$$

here v—is the velocity of soil particles vibration; is the velocity of longitudinal wave propagation.

The relative deformation of wet clayey and loess soils for the same seismic intensity is 10–100 times greater than the relative deformations of rocky soils. This explains the appearance of large residual deformations in over-saturated soils.

Besides, during earthquakes, cracks appear in the ground, the level of groundwater rises, and, as a result of ground shaking, eruptions of mud volcanoes may form; dams, inlet and outlet canals, bridges and pipes, drainage and energy systems, etc. can be destroyed. These phenomena were observed during the earthquake on July 6 and August 23, 1954, in Nevada (USA). During these earthquakes of magnitude M = 6.8 Richter scale [10], mud volcanoes arose, which spouted very fine material of blue color.

The embankments that filled the old flow channels were the result of natural sediments or artificial embankments. During the earthquake, the settling of embankments was observed. In Fallon, some cast iron pipelines were bent, which caused water leakage. As an aftermath of the earthquake on August 23, 1954, the ground of the original channel had settled. Sharp cracks were observed, caused by

Fig. 6 Rise of the channel bottom due to an earthquake [11]

the ground motion of the canal banks and adjacent soils, which tended to move down to the canals or rivers. In the Stillwater area, the cracks tended to spread parallel to the directions of canals or rivers. A natural consequence of such ground motion was the raising of the canal bottom, which often led to the rise of the bottom above the initial level of the water surface (Fig. 6).

In the area of Lone Tree and Stillwater, the slopes of the canals settled down from 0.3 to 09.9 m, while the bottom of the canals rose from 0.3 to 0.6 m. In exceptional cases, the bottom of the drainage ditch rose to a height of approximately from 1.5 to 1.8 m.

The damage observed at many locations in the Newlands irrigation system varied in nature, from minor cracks in the Truki Canal tunnels to extensive damage to the canal (distribution canals) and drainage systems. The greatest damage of a separate structure was observed in the drainage dam Coleman Diversion Dam located 1.6 km northwest of Fallon city. The dam was destroyed due to the displacement and fracture formation in the earth fill-in buttresses, which in turn caused the washing out around the structure itself. As a result, there was a partial overturning of the dam, the formation of fractures in it and its settling. The same damage was observed in the earth buttresses of the Rodgers Dam near the town of Lovelock during the earthquake on 23 August 1954.

The damage caused by these earthquakes to the irrigation system structures resulted in the stoppage of water supply in 80% of its service area, approximately 28,300 hectares.

3 Analysis and Recommendations

Analysis of the results of earth canals and their elements behavior, and theoretical and experimental studies showed that, under the same conditions, the canal elements arranged in a chute are more resistant to seismic effects than the elements of the same canal, located in an embankment or in a semi-chute—semi-embankment (Fig. 7). Therefore, when choosing a transverse profile of a canal, especially in the 9-point seismic zone, one should give preference to canals in the chute.

In the presence of fastenings along the canal perimeter, a rigid monolithic fastening is the most unfavorable in canals completely embedded in an embankment with steep slopes and in the event of an earthquake with an intensity of more than 8 points (Fig. 8).

From the analysis of survey data of the consequences of destructive earthquakes, it can be determined that the intensity of the earthquake manifestation is significantly influenced by the engineering-geological, hydrogeological and geomorphological conditions of the area. Seismic microzoning of the territory is conducted in those places in urban and industrial centers, in the locations of hydroelectric

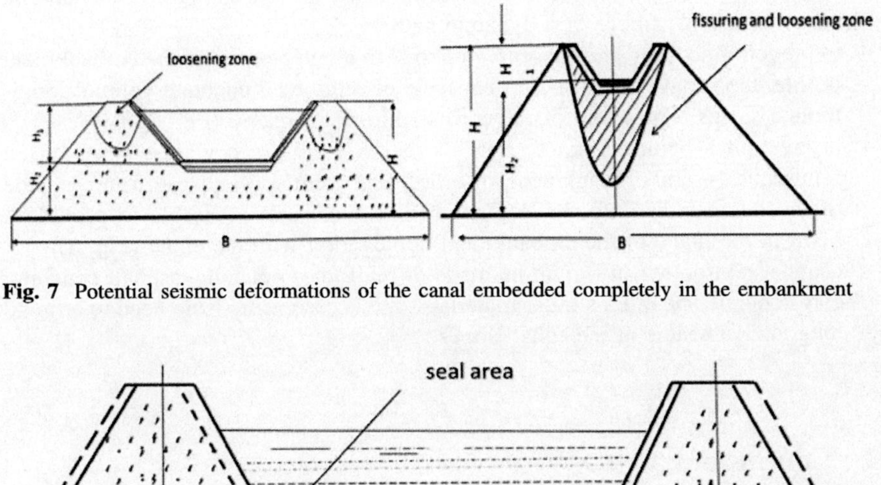

Fig. 7 Potential seismic deformations of the canal embedded completely in the embankment

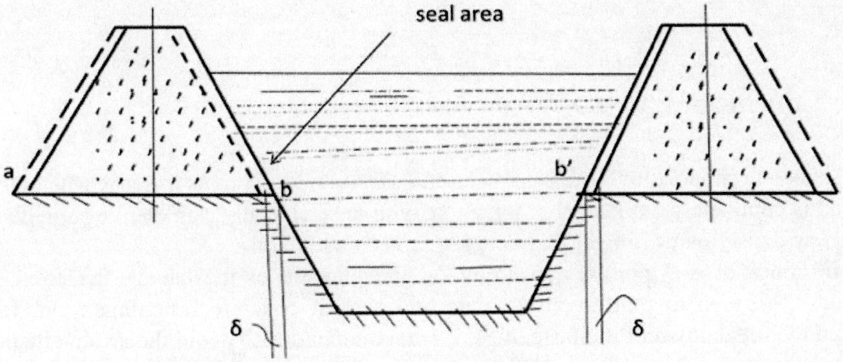

Fig. 8 Potential seismic deformations of the canal embedded in a semi-chute—semi-embankment

facilities, power plants, etc. [12]. Therefore, in vast areas where the canals pass, strong earthquakes are possible but no micro-seismic zoning is conducted.

When making decisions on the canal routes, even in normal conditions, the designers try to avoid the steep slopes, especially the slopes that are characterized by landslide phenomena or if the landslides are expected after the construction of the canal due to water filtration and slope cutting. It is also considered unreasonable to build canals along wetlands, in areas of quicksand and weak soils. In such areas, it is recommended to replace the canal with a chute, tunnel, pipeline or aqueduct. If these conditions are met, then the seismic factor is considered as taken into account. Therefore, these recommendations are mandatory for the areas of 9 points seismicity. The exception is the cases of routing channels along steep slopes when they are built in solid and undisturbed rock in a chute or semi-chute with the arrangement of concrete retaining walls [13].

Before drawing up options for routing and transverse profiles of the canals, it is necessary to conduct seismic microzoning of the canal route territory.

The construction of the canal completely embedded in the embankment of a height of 5–10 m, can be allowed in areas with seismicity of 7–8 points when meeting the anti-seismic measures recommended for earth dams:

- surcharging external slopes with a layer of cohesive soil, gravel or rock fill;
- increasing the slope flatness and checking their stability, taking into account the effect of seismic forces of a dynamic nature;
- to prevent fracturing and loosening of soils in the upper part of the embankment profile, the canal paving should be made of reinforced concrete without longitudinal seams. The width of the curb should be increased (Fig. 7a);
- in areas of 9-point earthquakes in a canal without the paving, erected in the semi-chute—semi-embankment, in order to prevent soil collapse from the side of the internal unreinforced embankment slopes, it is advisable to provide a berm in the plane of the embankment conjugation with the chute (Fig. 8);
- channel reinforcement should be made of reinforced concrete, asphalt concrete, clay concrete and other similar materials, in this case, there is no need to arrange longitudinal seams in the canal slopes.

4 Conclusions

So, concrete supports of canals can be used in zones of 8–9 points in cases when the canal is completely embedded in the chute, and in seismically stable environments, such as dense loams, gravel and coarse-grained sandy soils.

In zones of 8–9 points, depending on the purpose of the canals, preference should be given to reinforced concrete and asphalt concrete fastenings, and, to avoid the breakdown of the fastenings, it is recommended to avoid the arrangement of longitudinal temperature-settling joints, especially in the canal slopes.

To prevent the fracturing of the canal embedded in the chute in non-rocky soils, in conditions of seismicity of 9 points, it is recommended to use concrete, reinforced concrete and asphalt concrete for canal slope protection.

With such fastenings of the slopes, local stepwise thickenings of the paving—the stops—are provided, buried in the soil of the slopes, which cut through the filtering layer of sandy-gravel preparation.

Such types of fastening on slopes ensure the uniformity of operation of all elements of the canal, which contributes to a significant reduction in virtual large deformations expected during strong earthquakes.

References

1. Akhmedov M, Juraev D, Khazratkulov I (2020) Earthquakes and seismic safety of waterworks facilities in the Central Asian region. IOP Conf Ser Mater Sci Eng 883:012027. https://doi.org/10.1088/1757-899X/883/1/012027
2. Mirsaidov M, Sultanov T, Yarashov J, Toshmatov E (2019) Assessment of dynamic behaviour of earth dams taking into account large strains. E3S Web Conf 97:05019. https://doi.org/10.1051/e3sconf/20199705019
3. Mirsaidov MM, Toshmatov ES (2019) Spatial stress state and dynamic characteristics of earth dams. Mag Civ Eng 89:3–15. https://doi.org/10.18720/MCE.89.1
4. Wieland M (2016) Safety aspects of sustainable storage dams and earthquake safety of existing dams. Engineering 2:325–331. https://doi.org/10.1016/J.ENG.2016.03.011
5. Mirsaidov MM, Sultanov TZ, Rumi DF (2013) An assessment of dynamic behavior of the system "structure—Foundation" with account of wave removal of energy. Mag Civ Eng 39:94–105. https://doi.org/10.5862/MCE.39.10
6. Abduvaliev A, Abdulkhayzoda A (2021) Transverse vibrations of an underground cylindrical structure. IOP Conf Ser Mater Sci Eng 1030:012099. https://doi.org/10.1088/1757-899X/1030/1/012099
7. Sultanov KS, Kumakov JX, Loginov PV, Rikhsieva BB (2020) Strength of underground pipelines under seismic effects. Mag Civ Eng 93:97–120. https://doi.org/10.18720/MCE.93.9
8. Bakhodirov AA, Ismailova SI, Sultanov KS (2015) Dynamic deformation of the contact layer when there is shear interaction between a body and the soil. J Appl Math Mech 79:587–595. https://doi.org/10.1016/j.jappmathmech.2016.04.005
9. Sultanov KS, Bakhodirov AA (2016) Laws of shear interaction at contact surfaces between solid bodies and soil. Soil Mech Found Eng 53:71–77. https://doi.org/10.1007/s11204-016-9367-7
10. Davison C (1926) The Japanese Earthquake of 1923. Publications of the imperial earthquake investigation committee in Japanese. No 100 a, 353 pp and 55 pls; No 100 b, 126 pp and 48 pls; No 100 e, 296 pp and 42 pls. Geol Mag. https://doi.org/10.1017/s001675680008420x
11. Steinbrugge KV, Moran DF (1961) Damage caused by the earthquakes of July 6 and August 23, 1954. In: Engineering analysis of the consequences of earthquakes in Japan and the USA, pp 178–193. Moscow
12. Sagdullayeva DA, Maxmudova SA, Adilov FF, Abirov RA, Khazratkulov IO, Nasirov IA (2020) On stability of slopes in mountain zones. Case study. J Phys Conf Ser. https://doi.org/10.1088/1742-6596/1425/1/012016
13. Adilov F, Makhmudova S, Abirov R, Toshmatov E, Maxmudova I (2020) To assessment of stress-strain state in rock continua. IOP Conf Ser Mater Sci Eng 869:072019. https://doi.org/10.1088/1757-899X/869/7/072019.1

Numerical Field Researches of Composite Cross-Section Steel-Reinforced Concrete Rods

Farit Zamaliev

Abstract The first experiments on steel pipes samples filled with concrete were carried out in 1930, Paris. The first steel-reinforced concrete (pipe-concrete) bridge support was patented at the beginning of the twentieth century in Japan after a major earthquake. In Russia, the first tube-concrete structure in 1935 from a package of forty pipes was used in the construction of the bridge, and in 1940 in the construction of a bridge made of monotone pipe concrete. This report is devoted to the analysis of structural solutions of steel–concrete rods and columns, numerical and field researches of composite cross-section compressed rods. Technical solutions, methods and results of numerical and experimental researches have been analyzed using literary sources. On their analysis, a steel–concrete rod based on tubular concrete was selected, which has the inherent positive properties of both reinforced concrete and metal compressed rods. Numerical and field researches of steel–concrete rods, consisting of bent profiles—in the center of which there is a ring, and on four sides—T-bars, filled both in the center and on the sides with concrete, have been carried out. The report contains the methodology and research results, numerical and field tests results comparison. Satisfactory convergence of numerical and field experiments is shown.

Keywords Composite concrete rods · Steel-Reinforced concrete · Structural solutions · Compressed rods · Normal tensions

1 Introduction

The idea of increasing the strength of materials enclosed in steel shells was used by Rabu (France), who made the first experiments (1915) on testing pipes, filled with sand and gravel, in the laboratory of the School of Bridges and Roads. A little bit

F. Zamaliev (✉)
Kazan State University of Architecture and Engineering, Kazan, Russia

N. Vatin et al. (eds.), *Proceedings of MPCPE 2021*, Lecture Notes in Civil Engineering 182, https://doi.org/10.1007/978-3-030-85236-8_39

later, J. Sewell (John Sevel) tested samples of steel pipes filled with concrete in a Paris city laboratory and obtained a 1.25 times increasing of column strength, compared to the pipe and concrete core tested separately. In the V.S. Laura (1934) tests the breaking load for pipe–concrete specimens was 5 times greater than for same cross-section concrete cylinders.

The pipe–concrete rods working in Russia was studied by many scientists: they left a bright mark with their research, which is still relevant today, such scientists as A.A. Gvozdev, L.K. Luksha, V.A. Trull, R.S. Sanzharovsky, L.I. Storozhenko, A. L. Krishan and many others. The first pipe-concrete bridges were built in our country in the 30 s of XX century under the leadership of G.P. Perederiya—across the Neva River in St. Petersburg and under the leadership of V.A. Rosnovsky—across the river Iset near Yekaterinburg in 1940.

Later on in Russia, pipe concrete found application in public and industrial buildings, mainly as columns that require the perception of significant loads.

In KGASU in the 80–90s of the last century, experimental studies of round and square cross-section pipe-concrete rods with different aggregates and insulating gaskets, including steel pipes filled with sand and gravel mixture, different density and strength granite crushed stone, were intensively conducted (see conference proceedings of KISI, KGASA since 1980). In recent years, KGASU received dozens of RF patents for steel–concrete posts, the cross-section of which is formed by a steel profile and concrete, including bent profiles in the center of which there is a ring, and in the periphery there are T-sections, filled with concrete [1–4].

The practical value of composite cross-section rods research, lies in the study of their stress–strain state and identification of structural features of both the cross-section layers and along the compressed element length—the components of such structures, which will make it possible to correctly assess their ultimate bearing capacity and design both reliable and economical structures composite section.

For effective research, an analysis of previously performed research is always necessary. Let's consider the most famous sources published in recent years.

The source [5] is devoted to general issues of calculation and design standards for steel–concrete elements of composite section, including those made of pipe–concrete and steel-reinforced concrete sections based on steel profiles (rigid reinforcement). The article [6] considers the issues of the behavior of steel–concrete pipes under the influence of external influences. Source [7] presents the study's results of circular pipes columns, filled with concrete. In [8], the stress state of circular composite cross-section steel–concrete columns is summarized. Article [9] is devoted to general issues of tubular steel–concrete rods. The source [10] gives the results of experimental studies of thin-walled steel–concrete columns in complex cross-section. In [11], a summary of the investigations results for number of steel pipe concrete columns is given, carried out within the framework of the joint American–Japanese research program for composite and hybrid structures.

Article [12] is devoted to identifying the bearing capacity of concrete tubular rods under uniaxial compression. The source [13] appeared in the basics of design and identification of the limiting state of elements of a composite section, including composite columns of steel–concrete section.

Articles [14, 15] are devoted to the strength of box-shaped steel columns filled with concrete during their buckling.

The article [16] provides behavioral analytical models of the steel square cross-section columns, filled with concrete, during their axial compression. The source [17] is devoted to the hollow profile columns bearing capacity, made of steel and concrete, [18] reflects the analysis of experimental studies and gives recommendations on the design of composite section columns based on the materials of the 1998 congress held in San Francisco (USA). The source [19] summarizes the nonlinear stress–strain state of steel–concrete tubular concrete columns. The article [20] describes the axial stress state of the rods under axial compression.

Article [21] presents the results of research based on fundamental approaches to the behavior of steel–concrete tubular concrete columns. The article [22] describes simplified approaches to the design of pipe-concrete columns, presented at the conference in 2000 in Los Angeles (USA). The article [23] presents the experimental studies results of the pipe–concrete columns behavior for axial compression.

Article [24] presents the results of bearing capacity studies for concrete tubular rods under the eccentric effect of external loads.

When analyzing the stress–strain state of pipe–concrete compressed rods, a steel pipe is a cage for a concrete core; in the source [25], the cage effect is described for a section made of composite materials.

The sources [26–28] describe the features of the stress–strain state of concrete tubular rods when using an internal stressed concrete core. In article [29], the ideas of internally stressed concrete core are extended to the case of square cross section of tubular concrete columns. Article [30] is devoted to the ways of using reinforced concrete rods inside heavily loaded reinforced concrete columns.

In Russia, the features of the work of pipe-concrete rods are described in the books of R.S. Sanzharovsky, V.A. Trull, L.I. Storozhenko, and others.

Analysis of literature sources shows, that research is mainly devoted to pipe–concrete sections, consisting of a steel shell and a concrete core. However, in real structures, circular cross-section pipe concrete, with all its advantages, when coupled with floor slabs, with frame girders, leads to unnecessary costs in conjunction structures. Here, it is desirable to have the usual simple mates, as opposed to mates with the surface of the shells. This is probably why new technical solutions have appeared, patents for which these shortcomings are absent.

In KGASU in 2017–2019, patents were obtained for new technical solutions of composite pipe–concrete rods [1–4]. The purpose of this report is to analyze steel–concrete, pipe-concrete compressed rods and on the basis of the analysis—the choice of an effective steel–concrete rod and its numerical–experimental studies.

2 Methods

Numerical studies of the composite section column model, were carried out using the Ansys software in three stages: at the first stage, a finite element model of the studied column is created, the second stage is the imposition of the necessary physical and boundary conditions on the model, the choice of a method for solving the equations system and solving the FEM equations, the third stage—analysis of calculation results; values of displacements, deformations, tensions, which are presented in the form of tables, graphs, animation. The following types of racks are considered: (a) a steel–concrete rack, consisting of eight bent C-shaped steel profiles, paired with self-tapping screws in a checkerboard pattern in two rows, which are joined with a bent metal pipe and serving as a permanent formwork for the concrete part; (b) a steel–concrete rack consisting of an inner ring to which eight bent channels are attached on self-tapping screws, forming four "T"-shaped sections. Figure 1 presented the model of the steel part of the steel–concrete rack (Fig. 2).

Frictional contact with K = 0.45 was set between the concrete surface and the surface of beams steel part.

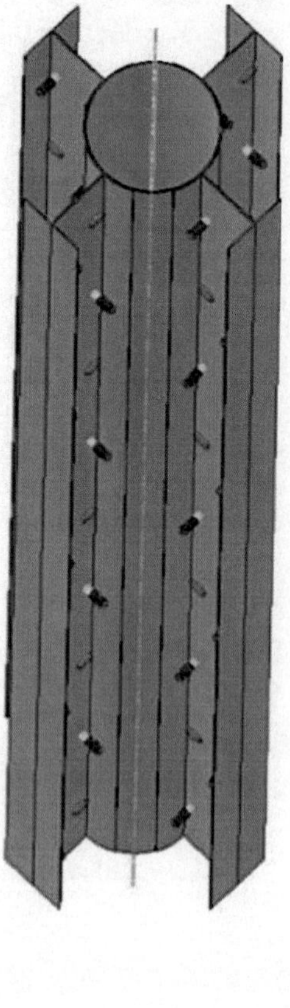

Fig. 1 Modeling the steel parts of a steel concrete pillar

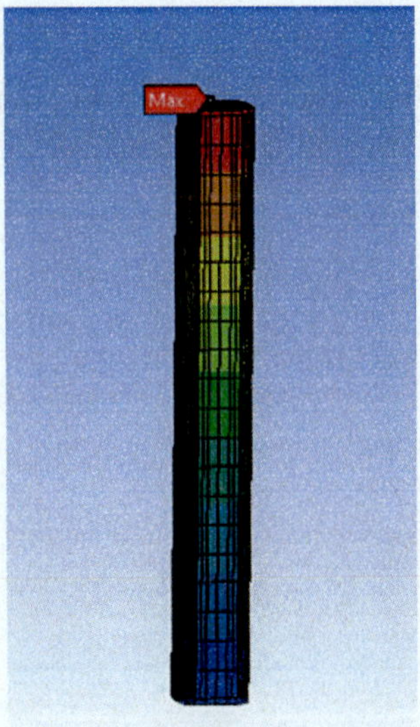

Fig. 2 Deformation mosaic of a steel–concrete rack with self-tapping screws

The contact between the self-tapping screws and the concrete of the beam was set through Rough—this contact implies the absence of friction between the surfaces, with possible separation along the surface. This case corresponds to an infinite coefficient of friction between the bodies being connected.

Next, we carry out the calculation with the above properties and conditions and we obtain a mosaic of deflections and isofield of the steel–concrete rack stresses.

2.1 Numerical Simulation Results and Their Analysis

Let us determine the limiting values of deformations and stresses.

Limit values of deflections and deformations in SP 20.13330.2016 "Loads and Actions", updated edition of SNiP 2.01.07–85.

In accordance with SP 16.13330.2017 "SNiP 2–23-81 * Steel structures" "Standard and design resistances in tension, compression and bending of sheet, broadband universal and shaped steel" for steel in accordance with GOST 27,772 S245 with rolled thickness from 2 to 20 mm, the design resistance of steel to bending according to the yield point is Ry = 240 N/mm^2 = 240 MPa.

According to SP 63.13330.2012 "Concrete and reinforced concrete structures. Basic provisions. Updated edition of SNiP 52-01-2003 (with Amendments No. 1, 2, 3) "clause 5." Requirements for the calculation of concrete and reinforced concrete structures"—strength calculations of concrete and reinforced concrete structures should be made on the basis of the condition that the forces, stresses and deformations in structures from various influences, taking into account the initial stress state (prestress, temperature and other effects) should not exceed the corresponding values established by regulatory documents. The compressive resistance of concrete is represented by stresses not exceeding the design compressive resistance of concrete Rb. For steel–concrete rods, concrete of class B25 (concrete compressive strength class) is used, according to this regulatory document, the calculated value of concrete resistance is Rb = 14.5 MPa.

Consider cross-sections with different thicknesses of a metal sheet with different cross-section fillers (concrete, polymer concrete, and fiber-reinforced concrete).

SSS uprights with 1 mm metal sheet thickness and B15 concrete in the tubular part (Figs. 3 and 4).

SSS racks with 2 mm metal sheet thickness and polymer concrete in the tubular part (Figs. 5 and 6, Table 1).

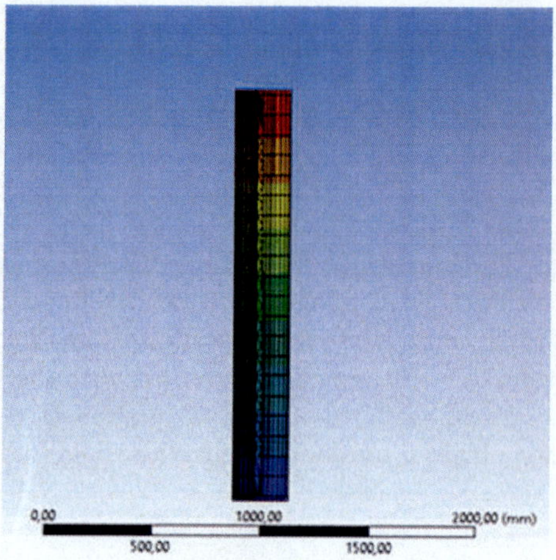

Fig. 3 Deformation mosaic

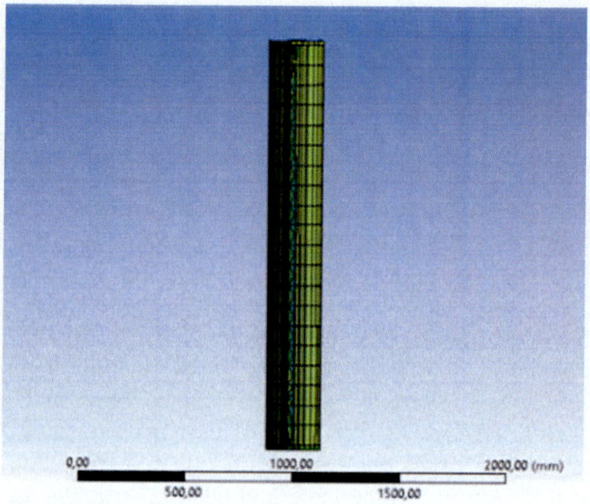

Fig. 4 Isofields of concrete stresses and metal part of the rack

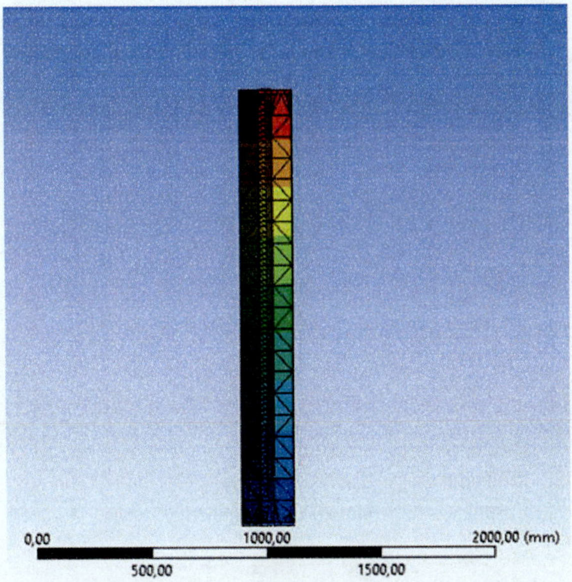

Fig. 5 Deformation mosaic

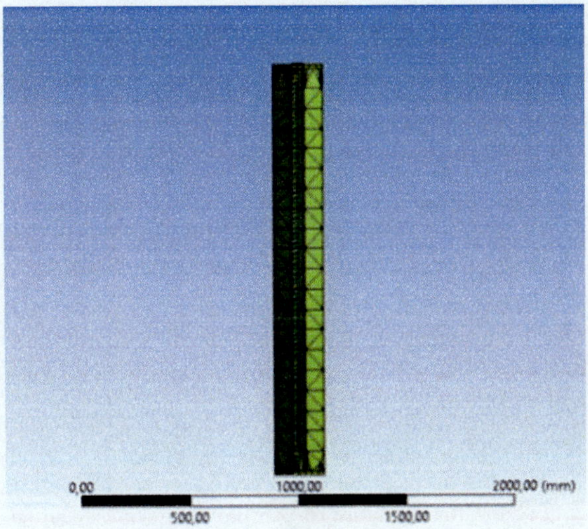

Fig. 6 Isofields of concrete stresses and steel part of the rack

2.2 *Experimental Studies of a Steel–concrete Rack*

Experimental research consisted of several stages, each of which had its own methodological features and in one way or another influenced the final result:

Table 1 Results of deformations and steel and concrete stresses of a steel–concrete rack

Variable rack parameters	Deformations [mm]	Compressive stresses σ [Mpa]	Tensile stresses σ [Mpa]
1 mm, B15	0.829	−61.306	13.551
1 mm, Polymer concrete	0.82807	−61.481	13.561
1 mm, Fiber concrete	0.82804	−61.483	13.407
2 mm, B15	0.84652	−47.501	18.532
2 mm, Polymer concrete	0.84249	−39.487	19.071
2 mm, Fiber concrete	0.84247	−39.357	19.071

1. Production of prototypes.
2. Preparation of the test setup.
3. Preparation of samples for testing.
    3.1. Sticking and insulating resistance strain gauges.
    3.2. Measuring instruments and equipment installation. Setting the specimen in the test position.
4. Research of physical and mechanical properties of concrete and steel.
5. Testing of samples.
6. Processing and analysis of test results.

2.3 *Manufacturing Technology and Characteristics of the Tested Steel–concrete Sample*

Due to the fact that the purpose of the research is to study the strength, deformability and nature of the structure of a compound (composite) section, the test sample had a wide range of features:

1. During developing the tested model of a steel–concrete rack, the structural features used in the construction of civil buildings were taken into account.
2. For the manufacture of samples, the most used steel class (C245) was used, the concrete class was laid according to the calculation—B 25.
3. The geometric dimensions of the steel–concrete rack model were also dictated by the capabilities of the test bases and the laboratory.
4. Concrete and steel parts anchor connection of the steel–concrete rack sample was taken by calculation. The tested steel–concrete column had an octagonal cross-section: the steel part was made of thin-walled cold-formed galvanized profiles (channels) 2000 mm long, paired with self-tapping screws

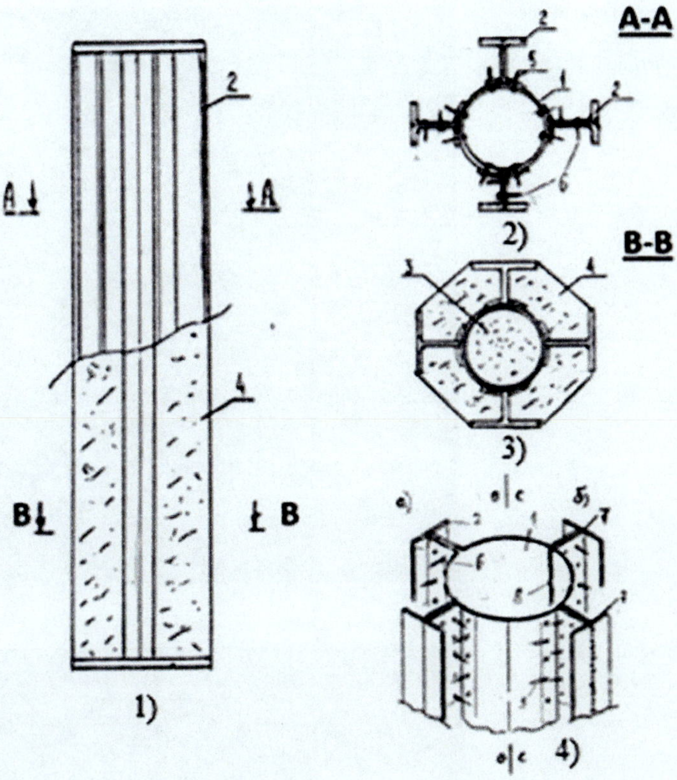

Fig. 7 General view of a steel–concrete rack (steel part in cross-section, perspective view and cross-section of the rack after concrete casting)

4.2 × 41 mm with a press washer; the concrete part had dimensions: length—2000 mm, width—100 mm, height—200 mm. The concrete class for the side faces was taken as B25 (Fig. 7).

The screws were staggered in two rows. The joint work of concrete and steel paired channels was ensured by screwing self-tapping screws into the walls of the profiles in the contact zone (Figs. 8 and 9).

The production of an experimental steel–concrete structure was carried out in the following sequence. The formwork elements were pulled from both end posts.

The concrete parts (center and sides) were concreted in two stages, first—the central tubular part, then—the side segments.

Fresh concrete was cared for in accordance with the requirements of SP 70.13330.2012. To exclude the influence on the test results of the increase in the strength of concrete, as well as changes in the initial stress–strain state of the prototype due to shrinkage, the test of the steel–concrete structure was started after its storage for at least 28 days in laboratory conditions at normal humidity and

Fig. 8 General view of the steel–concrete rack experimental model before concreting

Fig. 9 Concreting the side segments of the steel pillar

Table 2 Mechanical characteristics of the concrete in prototype

№ batch of concrete	On the 28th day			
	R, MPa (kgf/sm^2)	Rb, MPa (kgf/sm^2)	Rbt, MPa (kgf/sm^2)	Eb*10^3, MPa (kgf/sm^2)
1	19, 12 (195)	15, 78 (161)	1, 3 (13,28)	27, 55 (281)
2	25, 1 (256)	20, 49 (209)	1, 62 (16,5)	31, 37 (320)

temperature, and the beams were gaining strength at the concrete plant in the drying chamber. During this time, the strength and deformation properties of concrete have practically stabilized.

The concrete strength and deformation characteristics of the experimental composition were determined by testing auxiliary samples—cubes, made simultaneously with each of the main samples made and stored in similar temperature and humidity conditions.

The compressive strength of concrete was determined by testing cubes with dimensions of 100 × 100 × 100 mm.

Testing of auxiliary concrete samples was carried out in accordance with the methodology of GOST 244.52-80 and GOST 10,180-78. The test results are shown in Table 2.

2.4 Test Procedure for a Steel–Concrete Specimen for Static Loading

Compression testing of a steel–concrete rack was carried out at KGASU in the laboratory—"Department of reinforced concrete and stone structures".

The steel–concrete rack was tested according to the scheme of a free-supported rack with a concentrated force. Rated length of the rack is 2000 mm.

The load was transferred by means of the hydraulic system of the UMM-200 press to the steel–concrete rack.

The samples were preliminarily centered along the physical axis at tensions not exceeding 0.2Rb, ser. Centering was carried out until some of the readings along the edges of the samples differed from the average by no more than 5%.

The load was applied in steps of 0.1 of the expected breaking load with resetting of readings from strain gauges. The value of the unloading load was recorded according to the readings of the force-measuring machine UMM-200. Longitudinal and transverse deformations of concrete were measured along four faces of the samples with resistance strain gauges with a base of 50 mm using an AID 4 device with a switch magazine. T-bars deformations on the steel shelves of the struts composite section were also measured along four faces by strain gauges with a base of 20 mm. To control longitudinal deformations, beakers with a base of 600 mm

Fig. 10 Test stand with testing steel–concrete column

were also installed, equipped with dial indicators (DI) on two mutually perpendicular faces of a composite rack (Fig. 10).

The even distribution of force across the section width, which is especially important in composite structures, was ensured by the setting of rigid metal gaskets in the places, where they rest on the rack.

During the test, the longitudinal and transverse deformations of the concrete and steel in steel–concrete column were measured, as well as the crack opening width. Deformations of concrete and steel were recorded by resistance strain gauges with bases of 50 mm and 20 mm, respectively, through electronic equipment AID-4 with a switch magazine (commutator). Longitudinal deformations of the rack were also recorded according to the readings of the DI, installed on beakers in two planes.

General view and test scheme of a steel–concrete rack, layout of resistance strain gauges for concrete and steel and measuring instruments are shown in Fig. 10.

After checking the operability of the load and force-measuring systems, a trial loading of the steel–concrete rack was carried out with a load not exceeding 20–25% of the permissible destructive load, and was carried out in 2–3 stages. The purpose of the trial loading was to study the behavior of a steel–concrete pillar under load and to first check the correspondence of the values of the measured

Fig. 11 Development of initial cracks in the concrete of the rack

parameters to the expected ones. After the required holding under load, the racks were also unloaded to zero in order. Subsequently, the test was carried out by loading in 0, 1 steps of the expected breaking load. Deformations of concrete and steel, absolute shear deformations, deflections, the nature of the development and formation of cracks were recorded on the steps during holding.

The moment of cracking is determined visually and with a microscope. The breaking load was recorded according to the dial scale readings of the installations testing press and according to the graphs of the given load steps on the screen of a laptop, connected to the press.

Complete physical destruction was taken as the limiting state.

A prototype of a steel–concrete rack was tested with a single short-term static load until complete destruction in order to establish the character of destruction and the patterns of crack development, deformations of concrete and steel during their joint deformation, as part of a single structure.

The prototype of the reinforced concrete support collapsed from the development of cracks in the zone of application of the load.

The first cracks in the concrete of the side segment, which protected the steel composite strut tees, appeared at a load of $0.77N_{des}$ (Figs. 11 and 12).

3 Results and Discussion

Graph 1 shows a strut deformations comparison, according to the experimental data in two planes of the strut section "X" and "Y".

Fig. 12 Destruction of a steel–concrete rack

Graph 1 Deviations by the indicators of beakers

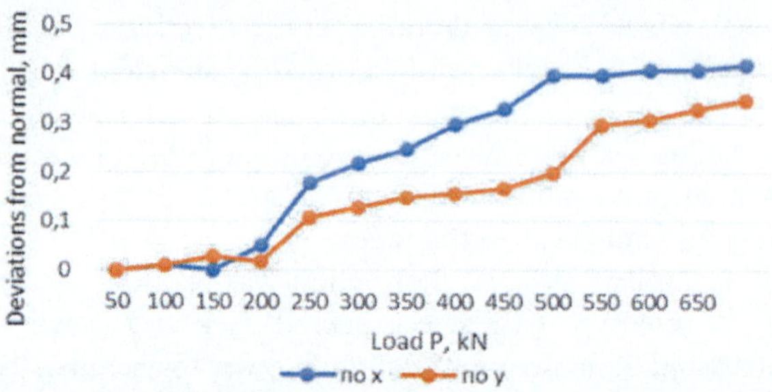

Graph 2 show the stresses of concrete, steel (according to the readings of the strain gages along the section planes "X" and "Y": (a) the tensions of concrete along the rack in the area of application of an external load; (b) the tensions of the concrete in the annular direction of the strut in the area of application of the load; (c) (d) tensions in the rack as measured by vertical and horizontal sensors (along the perimeter of the section) in the middle of the racks length; (e) tensions on the steel profile in the area of application of the load along two planes "X" and "Y"; (f) tensions of the steel part in the middle zone of the length of the strut.

Graph 3 shows the tensions diagrams of racks steel and concrete in the zone of load application and comparison of the experimental data with the numerical results, obtained using the Ansys PC.

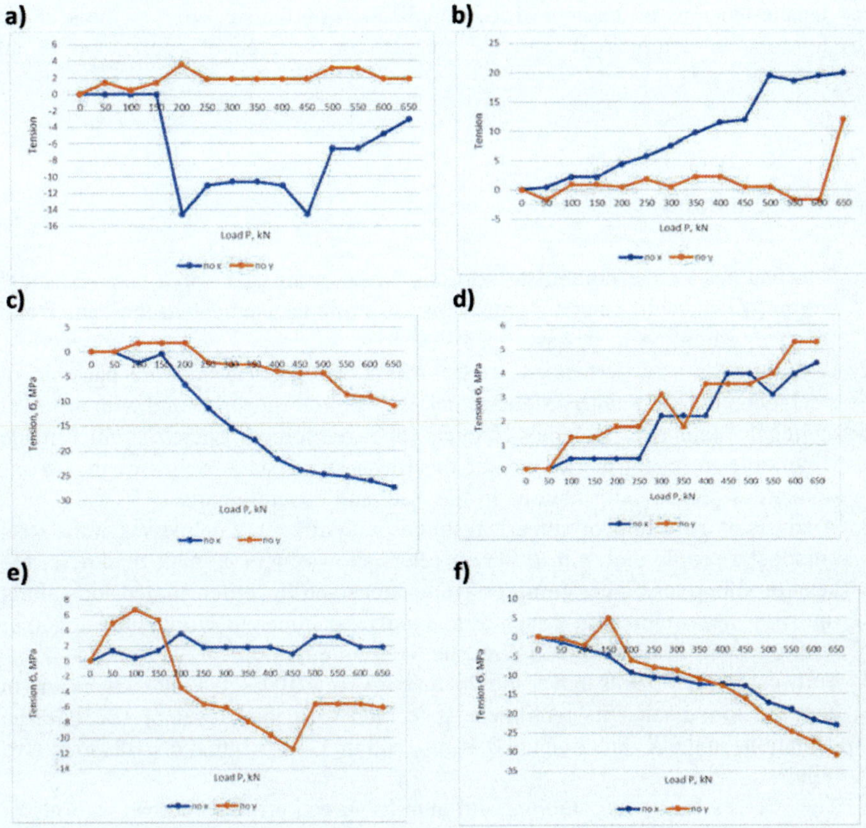

Graph 2 Normal tensions of concrete and steel in steel–concrete rack

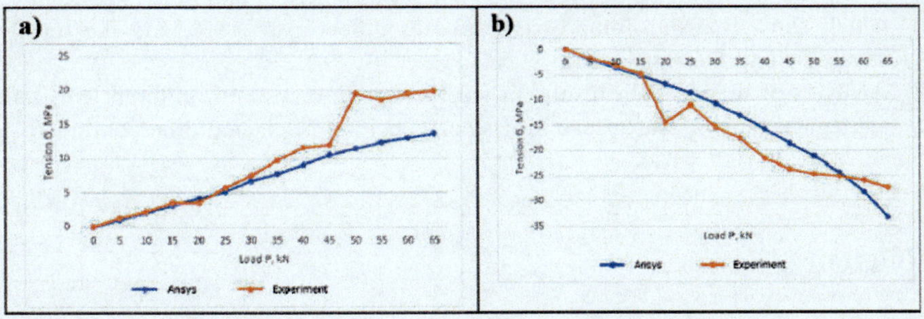

Graph 3 Tensions of concrete and steel according to numerical and physical experiments

(a) tensile tensions in the area of load application on the steel part of the section;
(b) tensions in compressed concrete are also in the area of application of an external load.

4 Conclusions

1. The features of the composite sections stress–strain state of a steel–concrete column filled with concrete, consisting of a tubular element in the center and four tees adjoining to it, have been studied.
2. A calculation was performed on the Ansys Workbench software package of steel–concrete racks with different thicknesses of the steel part and with different aggregates (concrete, polymer concrete, fiber-reinforced concrete), which made it possible to reveal the picture of the stress–strain state, stages of the development of plastic deformations in the steel and concrete parts.
3. Analysis of the results of numerical studies with different thicknesses of the steel composite profile and with different fillers shows: with a small thickness, the steel profile gives higher compressive tensions and lower tensile tensions along the ring, fiber-reinforced concrete and polymer concrete work more together with the shell in tension, and concrete—on compression.
4. Full-scale experiments have shown that up to 0.2Pdes the rack, working in compression, retains its rectilinear state, and to a load level of 0.75Pdesthe deviations in the X direction are 0.4 mm, and in Y—0.3 mm, i.e. e. they are also minor.
5. The stresses in the longitudinal and annular directions in the cross-sections of the rack (on the steel profile and on the concrete) differ depending on the distance from the load application zone; in the load application zone they are more intense than, in the middle of the rack.
6. Comparisons are made between the results of numerical and field experiments, which show that the compressive stresses differ up to 18%, and the tensile stresses along the ring—up to 30%.
7. Analysis of the research results of steel–concrete racks with different concrete aggregates shows, that research of steel–concrete racks should be continued.

References

1. Steel concrete element: US Pat. 174528 Rus. Federation No. 2017114788; publ 19.10.2017, bul. No. 29. 2 sec
2. Steel concrete rack: US Pat. 176508 Rus. Federation No. 2017126965; publ 01/22/2018, bul. Number 3. 2 sec
3. Steel concrete rack: US Pat. 190602 Rus. Federation No. 2019112924; publ 04.07.2019, bul. Number 3. 2 sec

4. Steel concrete rack: US Pat. 191866 Ros. Federation No. 2019114514; publ 26.08.2019, bul. No. 24. 2 sec
5. Eurocode 2 (2011) Design of composite steel and concrete structures—Part 1: General rules for buildings/The European Union Per Regulation 305, p 118
6. Gardner NJ, Jacobson ER (1967) Structural behavior of concrete filled steel tubes. J ACI 64 (7):404–413
7. Georgios G, Lam D (1968) Axial capacity of circular concrete-filled tube columns. J Constr Steel Res 60(2004):104
8. Johanson M (2000) Structural behaviour of circular steel—concrete composite columns. Licentiate thesis, Chalmers University of Technology Div of Concrete Struct Guoteborg, Sweden
9. Johnson RS (1984) Concrete—filled steel tubes. Compos Struct Steel Concrete 1(5):171–177
10. Chen Ju (2010) Wei-Liang Jin experimental intestigatio of thin-walled complex section concrete-filled steel stub columns. Thin-Walled Struct 48(9):718–724
11. Nishiyama I, Morino S, Sakino R, Nakahara H (2002) Summary of research on concrete-filled structural steel tube column system carried out under the US-JAPAN cooperative research program on composite and hybrid structures. Japan, 176 (2002)
12. Kibriya T (2005) Performance of concrete filled steel tubes under uni-axial compression, IV Regional Conference on Civil Engineering Technologi, Joint ASCE/ESIE Conference, Caipo, Egypt
13. Leon RT, Kim DK, Hajjar JF (2007) Limit state response of composite columns and beam-columns part 1. Formulation Design provisions for the 2005 AISC specification. Eng J 341
14. Liang QQ, Uy B, Liew OHK (2007) Strength of concrete-filled steel box columns with buckling effects. Aust J Struct Eng 7(2):145–155
15. Mursi M, Uy B (2003) Strength of concrete filled steel box columns incorporation buckling. Aust J Struct Eng 129(5):626–639
16. Nakahara H, Sacino K, Inai E (1998) Analytical mode for compressive behavior of concrete filled square steel tubular columns. Trans Jpn Concr Inst 20:171–178
17. Roik K, Bergman R, Bode H, Wagenknect C (1975) Tragfahigkeit von aus Betonnierten Hohiprofil-stiitzen aus Bustahl. Eypт-wiss
18. Sakino K, Ninakawa T, Nakahara H, Morno S (1998) Experimental studies and design recommendations of CFT Columns. U.S.-Japan cooperative earthquake research program. Proceedings, Structural Engineers World Congress, San Francisco, pp 169–173
19. Shams M (1997) Non-linear evaluation of concrete-filled steel tubular columns: a Dissertation in partial fulfillment of the requirements for the Degree of Doctor of Philosophy. New Jersey's Institute of Technology 25
20. Schneider SP (1998) Axially loaded concrete-filled steel tubes. J Struct Eng 124:1795–1805
21. Tang C, Zhao B, Zhu H, Shen X (1982) Study of the fundamental structural behavior of concrete filled steel tubular columns. J Build Struct 3(1):13–31
22. Tsuda K, Matsui C, Fujinaga T (2000) Simplified design formula of slender concrete-filled steel tubular beam-columns. In: Proceedings 5th ASCCS conference on composite and hybrid structures 4. Los Angeles, pp 457–464
23. Yamamoto T, Kawaguchi J, Morino S (2002) Experimental study of the size effect on the behavior on concrete filled circular steel tube column under axial compression. J Struct Constr Eng Trans AIJ 561:237–244
24. Zhou C (1982) Investigation on load carrying capacity of concrete-filled steel tubes under eccentric loading. J Harbin Inst Civil Eng 4:29–46
25. Rimshin VI, Kustikova, Yu O (2006) Reinforcement of reinforced concrete structures with clips of composite materials. International scientific and technical conference "Structural physics in the XXI century" M. NIISF RAASN, pp 542–545
26. Krishan AL, Zaikin AI, Melnichuk AS (2010) Calculation of the strength of pipe-concrete columns. Struct Mech Eng Struct Struct 1:20–25

27. Krishan AL, Rimshin VI, Zaikin AI (2014) Strength calculation of compressed reinforced concrete elements with indirect reinforcement. III All-Russian (II International) conference on concrete and reinforced concrete 1, M: 12–16 May 2014, pp 306–314
28. Krishan AL, Krishan MA (2014) To the determination of the bearing capacity of tubular concrete columns of circular cross-section. III All-Russian (II International) conference on concrete and reinforced concrete 1, M: 12–16 May 2014, pp 301–307
29. Krishan AL, Melnichuk AS (2012) Square tube concrete columns. Hous Constr 5:19–20
30. Danel VV (2014) Reinforced concrete elements with tubular concrete elements. III All-Russian (II International) conference on concrete and reinforced concrete 1, M: 12–16 May 2014, pp 251–260

Wood and Steel Rope: A Rational Combination in Floor Beams

Artem Koshcheev, Svetlana Roshchina, Mikhail Lukin, and Nikolai Vatin

Abstract The article is devoted to the experimental study of the method of reinforcing wooden beams with steel wire ropes. The presented reinforcement technology can be used in roof and floor beams of building structures. The process of pulling out the archy glued auxiliary bars made of the double-lay steel wire rope about 8 mm in diameter has been investigated. The investigations have been carried out on the symmetrical double faced samples with local wood reinforcement 100 × 400 × 50 mm in size, with 140 mm in the reinforcement working length. Numerical calculation has revealed the amplitude of the reinforcement trajectory vibration influence on the wooden beam ruggedness. With the reinforcement trajectory amplitude contraction from 35 to 15 mm, the force corresponding to the destruction of the wire rope-wood connection increases from 5 to 23 kN. The researchers have described the manufacturing technique of the control samples, which consists of wooden blanks planing and calibration stages, marking the milling lines, performing milling with the fillet grooved U-shaped cutters, preparation and gluing of the reinforcing elements. The materials of the tests progress photographic evidence have been presented. The results of the experiment have shown that a decrease in the vibration amplitude of the S-shaped form of the reinforcement to 15 mm leads to an increase in the strength of the entire gluing in by 28% in comparison with the rectilinear form. Conclusions on the step-like increase in the ruggedness of the S-shaped gluing in with the decrease in the reinforcement trajectory amplitude have been made.

Keywords Wood · Steel · Timber constructions · Steel rope · Building · Beams

A. Koshcheev (✉) · S. Roshchina · M. Lukin
Vladimir State University Named After Alexander and Nikolay Stoletovs, Vladimir, Russia

N. Vatin
Peter the Great St. Petersburg Polytechnic University, St. Petersburg, Russia

N. Vatin et al. (eds.), *Proceedings of MPCPE 2021*, Lecture Notes in Civil Engineering 182, https://doi.org/10.1007/978-3-030-85236-8_40

1 Introduction

Up to date the worldwide trends in the development of technologies for the construction of buildings and structures have identified several criteria that building structures must meet. The main ones are the environmental friendliness, durability, high strength and performance characteristics, presence of renewable natural resources and recycling materials in production. Wood construction products comport most of these parameters. This is the reason for the wide distribution of this building material throughout the world. The article will focus on wooden beam structures. These structures are used in low-rise and high-rise construction as roof and floor beams. Furthermore, the building load-bearing structure is made of beam elements. Modern methods of wood processing make it possible to produce glued structures with large sections and spans.

A particular niche in the volume of the modern timber structures is occupied by the reinforced wooden structures. Reinforcement in wood can be different. The first works in this direction have proved the efficiency in the application of various types of steel reinforcement in wooden beams and in the interface nodes of building structures [1]. Particular attention has been given to the short-term and long-term strength of such structural elements. The issues of repairing wooden beams in existing structures using steel glued in bars have been investigated [2]. The results of these investigations have shown that short reinforcement bars do not increase the bearing capacity, but they strongly affect the fracture behavior, making fracture plastic and in tact, which affects the structure safety.

The use of prestressing in reinforcement has been the next development direction of the reinforced wooden structures. Performing four-point bending tests on beams carried to failure have been discussed in a series of investigations [3]. Unreinforced, reinforced and prestressed beams have been compared. The cross-sectional ratio of steel to wood has been 0.82%. The results of the reinforced beams have shown that the mechanical strength, load-bearing capacity and rigidness of both unreinforced and prestressed beams have been improved compared to the unreinforced structures. The increase in rigidity has been recorded by 20–30%, and in plasticity—up to 60%. In addition, an increase in the ultimate load—up to 40% has been noted. However, in a number of studies, where the issues of changing the stress-related characteristics of prestressed wooden beams [4–6] were studied, as well as in the analysis of the statistical data on 550 cases of wooden structural damage in Germany [7], there stress reduction of prestressed structures was noted in course of time.

A large number of works in the field of timber reinforcement are devoted to the use of low-grade pine sawn timber in combination with steel reinforcement, FRP, as well as in combination with the use of the glued timber (CLT) [8–12]. According to the research, globally significant volumes of sawn timber are considered to be non-standard and are sold and used at a loss. The necessity of introducing new approaches and methods allowing to increase the strength of structures made of low grade wood has been substantiated. Examples of the cross-laminated timber use with reinforcement (CLT) have been given [13], as a vivid example of construction technology with a good ability to embed large volumes of non-standard timber into the supporting framework of buildings and structures.

Many works are devoted to other types of new engineering wood composites—glued wood reinforced with polymer fibers (FRP) [14–21] and glued laminated lumber with multidirectional wood fibers (LVL) [14, 20, 21], which illustrate several cases of using these composite beams as the basic load-bearing building materials. Their possible disadvantages, expressed by the high fabricating cost and the complexity of manufacture have also been discussed [22]. Comprehensive analysis of 230 damage cases of large span building structures around the world, in which both traditional and similar, advanced methods of wood modification [23] have been applied, has shown that unreinforced timber structures most often suffer from such phenomena as cracking along the grain, which is caused by low or frequent varying moisture content in wood, as well as the effect of systematically varying perpendicular to the fiber tensile stresses. Additionally, in this review of structural failures during the operation, the absence of the above disadvantages in structures with reinforcement due to the presence of elements binding the wood fibers has been noted.

Speaking about failures in the operation of the building structures, special attention should be paid to the issues of brittle fracture of reinforced beams [24, 25]. A number of investigations on the study of the dynamic effects [26] under this nature of destruction [27–32] has revealed the phenomenon of reinforcement tensile properties activation during the transfer of stresses removed as a result of wood fibers destruction from the body of the beam to the reinforcing elements. This sudden process leads to a dynamic shock, which provokes the emergence of the additional stresses in the system and its failure. The magnitude of this stress depends on the ratio of rigidness between the reinforcement and the wood, which breaks down in a brittle manner. It has been revealed that a slight increase in the ability of reinforcement to plastic deformation helps to significantly reduce the growth of stresses due to the dynamic impact and to avoid brittle structure fracture.

The technology of increasing the strength of wooden beams and wooden knots using steel plates [33] has significantly increased the load-bearing capacity of the wooden knot joints, which made it possible to construct high-rise buildings from this material (a vivid example of such reinforcement is the building of the Mjøstårnet hotel that is 85.4 m high on the shore of Lake Mjesa, Norway, or the Brock Commons University Residence Building in Canada (53 m, Fig. 1).

The investigation on this reinforcement method has proved a higher increase in the rigidness of structural components compared to the fiber-reinforced elements (FRP). However, due to the influence of temperature and humidity on the strength and deformation of the adhesive layer between steel and wood, it has been recommended to use additional reinforcement of the connection using such mechanical connectors of materials as bars and bolts. These connectors subsequently protected the structure from the rapid brittle destruction in the event of fire.

In wooden structures much attention is paid to the issues of ruggedness and serviceability of structures, this is especially typical for areas with increased seismic

Fig. 1 Mjøstårnet hotel building (Norway, 85.4 m) and Brock commons university residence building (Canada, 53 m)

activity. Researchers from New Zealand [34] have described in detail several new connections of timber structures based on the steel bars glued onto epoxy resin. The results of testing these nodal structures in modeling seismic loads have shown the importance of increasing the plasticity of deformation in the nodes of building systems [35–37].

In all of the above studies, much attention is paid to the problematic factors that are characteristic of any timber structures. The main disadvantages of such structures are susceptibility to biodamage, low protection against fire and brittle destruction in the event of strong excess of standard operating loads in the seismically active areas of construction [34, 38–41]. These problems are being solved in a wide variety of ways: resistance to biodamage and fire protection increases with the help of wood impregnation, strength can be increased by reinforcing beams and nodes, and brittle fracture can be avoided by reinforcing structures (Fig. 2).

Fig. 2 Brittle destruction of the rafter system from the action of snow load (Russia, Udmurtia)

We have proposed a method for strengthening wooden beams, which can potentially lead to both an increase in strength and a simultaneous increase in the deformation plasticity under the critical loads on a building or structure.

The proposed method of reinforcement consists in the use of elastic elements made of a steel wire rope with a metal or rope core in solid wood, which will be located in the structure tension area along a certain, S-shaped, rational trajectory. Such reinforcement arrangement should ensure the reinforcing bar length increase in the solid wood per linear meter of the structure, which means that it should lead to the effective use of the reinforcing material bearing capacity in the solid wood.

The following experimental data have been obtained:

1. Comparison of reinforcement with all-steel reinforcement and a steel wire rope has shown an increase (Fig. 3) in the connection strength by 19.26 [42].
2. When analyzing the reinforcement optimal trajectory, the most advantageous option in regard to the ruggedness and deformation properties turned out to be 1-line reinforcement with an s-shaped anchorage trajectory (Fig. 4), which should provide a 1.81-fold increase in the wooden beam resistibility in comparison with the unreinforced structure [43].
3. When analyzing the influence of eccentricity when the reinforcement is located on the edge of the beam structure [44], the necessity of covering the reinforced section with solid wood from the outside with the help of overlays has been explained [45].

Before transferring the testing of the proposed type of reinforcement to the in-place testing phase of 6 m beams, it is necessary to study the question of the influence of the steel cable anchoring form in solid wood on the strength characteristics of the adhesive bonding [46]. The purpose of this work is to prove

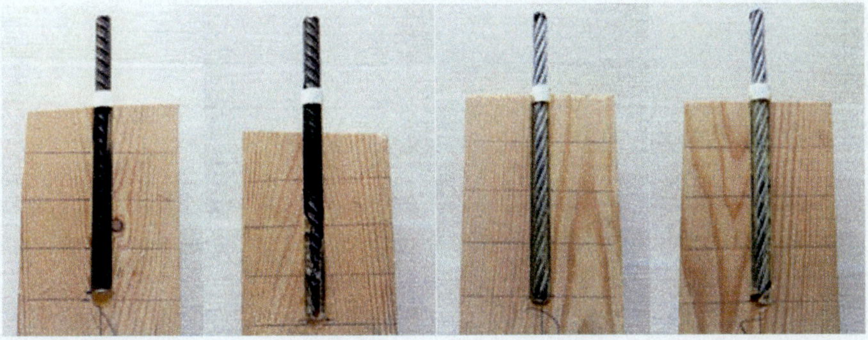

Fig. 3 Samples tested by the pull- out test. Rectilinear trajectory. Reinforcing deformed steel bar and steel wire rope

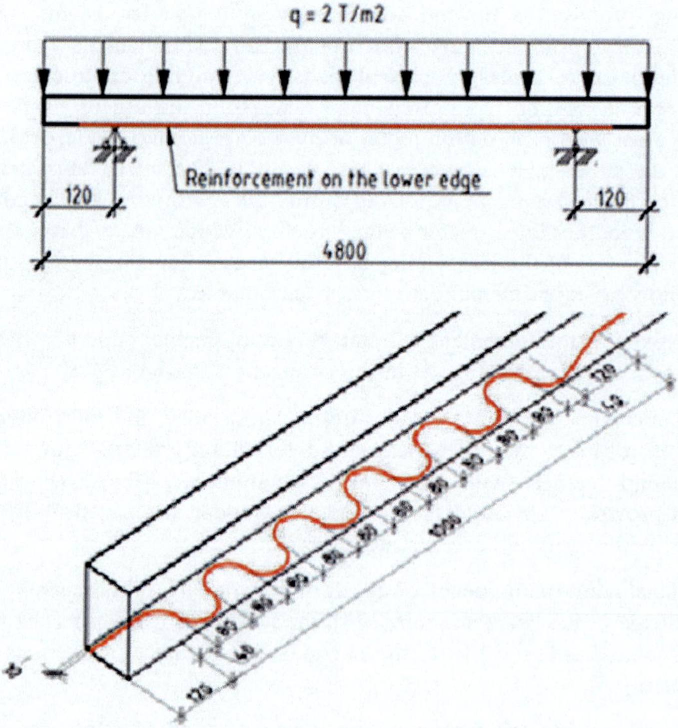

Fig. 4 Design diagram of beams with the archy S-shaped reinforcement

experimentally that the S-shaped trajectory is more advantageous in terms of ruggedness and deformability than rectilinear gluing in a steel cable into wood and to determine the exact geometry of the reinforcement for the manufacture of 6 m beams [47].

2 Materials and Methods

The present study is a test on pulling bars of various shapes from solid wood glued onto an epoxy adhesive composition. 4 series of tests have been performed, 6 samples each.

2.1 Material Specifications

The first series—samples of a rectilinear shape—with eccentricity (displacement of the reinforcing bar to the side plane of the sample) and centered. Samples with eccentricity are open for video production and photographic recording of the dynamics of the sample destruction. The second, third and fourth series (in the quantity of 18 pieces)—samples with 3 different trajectories fluctuation amplitude of the S-shaped reinforcement—15, 25 and 35 mm with respect to the central axis of the solid wood (Table 1).

The base material is wood, grade I pine, from winter cutting forest, dried in a room with a constant humidity of 23%, selected to ensure the minimum number of visible defects.

Reinforcing material—zinc-coated steel wire rope. In all series of samples, a wire rope of 8 mm in diameter has been used, the thickness of the zinc coating has been 50 micron, double-lay cable, 6 bundles of 0.2 mm wire, a flexible steel core.

Adhesive composition—epoxy resin ED-20 type (basic epoxy base, PEPA hardener). To compensate for the arising temperature deformations and increase the strength properties, ground quartz of 0.2 mm fraction has been used as a filling compound. The quartz has been pre-prepared by drying on a gas burner flame to eliminate the effect of moisture on the resin polymerization.

REM-100-A-1 electromechanical tensile testing machine has been the source of the tensile load. Equipped with a two-column power-delivery module and an automated measurement and control system that allows testing according to a given mode thanks to special software—"MTest REM-2.15", which makes it possible to more accurately determine the actual load on the sample [48].

External conditions of the experiment—relative humidity in the room—52%, room temperature—21 °C.

Table 1 Characteristics of the samples series

Name of the samples series	1 series	2 series	3 series	4 series
Geometric dimensions of the base	Sample length—400 mm Sample width—100 mm Sample height—50 mm			
Characteristics of the base material	Wood Pine Grade—1 grade Moisture—23%			
Characteristics of the Reinforcing Material	Steel Wire Rope ISO 2408 standard Outside diameter—8 mm Coating—zinc 50 microns Splice joint—double Core—composite			
Trajectory	Straight	Curved	Curved	Curved
Trajectory amplitude	0 mm	15 mm	25 mm	35 mm

2.2 Creation of Samples for Experimental Tests

Preparing wood for gluing has been the first step in making samples. Wood, previously dried in a chamber to a constant moisture content of no more than 20%, has been calibrated on a thickness planing machine to the correct geometric shape—a parallelepiped with parallel edges. Further, the plane on which the reinforcing element has to be located has been planed with an electric plane with carbide cutting knives at 12,000 rpm [49].

In half of the samples, it has been necessary to ensure centering of the reinforcing element glued into them in the cross section. To do this, these samples, consisting of 2 halves, have been made composite, in one of which a groove for the steel wire rope has been made with a router. To carry out high-quality gluing, an adherence of 2 halves of the workpiece has been necessary, which could only be ensured with the help of a thicknessing machine. In this case, the second half of the workpiece simulates a protective wood layer, which would be made during the manufacture of a 6-m floor beam.

The next step is to make grooves for gluing in the reinforcement. The milling operation has been performed with a Makita 3612C milling cutter, a fillet grooved U-shaped cutter with a diameter of 10 mm. Given the small number of samples, milling has been performed manually, along a predetermined trajectory. The speed of the milling cutter spindle has been 6000 rpm (Fig. 5). This frequency allows you to achieve smooth groove slot sides without the possibility of overheating the wood.

The wire rope must be prepared for gluing. To do this, it has been preliminarily cut into 20 cm samples and degreased with acetone solution. Acetone is a dissolvent for epoxy resins of the ED-20 type and evaporates quickly, therefore it does not affect the adhesion of the resin to steel.

Fig. 5 Milling a groove for the steel wire rope

Fig. 6 Gluing in reinforcing elements into epoxy adhesive

An adhesive composition is being prepared—the epoxy is mixed with a hardener. Within 30 min, the reinforcing elements are glued into the grooves. The adhesive is applied both to the steel wire rope and to the groove until it is completely filled. Further, those elements that will be centered are also treated with an adhesive and glued under a pressure of 0.5–1 MPa using 2 cramps (Fig. 6). The manufactured samples are kept for 10 days until the adhesive composition is completely polymerized.

2.3 Experimental Test Parameters

The source of the tensile load is the REM-100-A-1 electromechanical tensile testing machine. It is equipped with a two-column power-setting module and an automated measurement and control system that allows testing according to a given mode through the use of the special software—"MTest REM-2.15", which makes it possible to more accurately determine the actual load on the sample. External conditions of the running of experiment—relative humidity in the room—52%, room temperature—21 °C. The tensile testing machine traverse speed is 4 mm/min.

3 Results

Numerical calculation of the geometric copies of products has been performed in the Lira 10.10 program as the first stage of the experiment before the production of the experimental samples, which makes it possible to be convinced of the

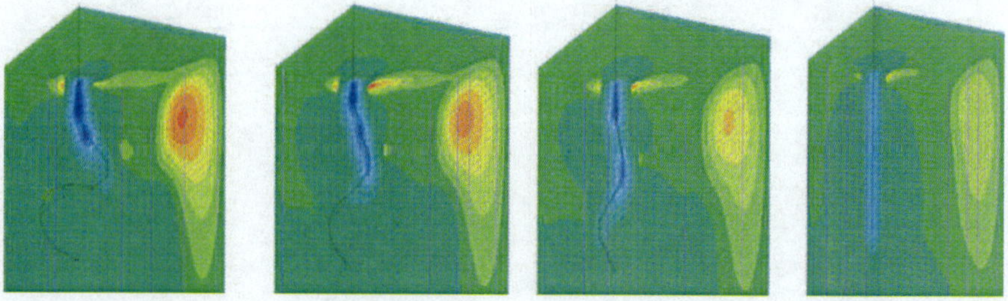

Fig. 7 Virtual experiment on pulling out archy wire ropes of various shapes. Results on the shear stresses in solid wood (red—positive, blue—negative, green—neutral)

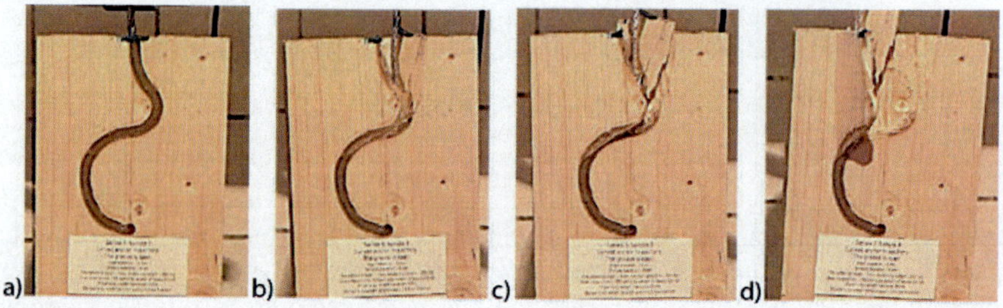

Fig. 8 Experiment on the tensile testing machine. The sample deformation with the maximum amplitude (35 mm) with increasing load. A—is the initial state of the sample; B—is the beginning of fracture, corresponding to the ultimate strength; C—is the 1st wave destruction; D—the beginning of the second wave destruction

discrepancies in the degree of the steel wire rope loading in the solid wood along its length. The numerical experiment clearly demonstrates that with a decrease in the amplitude of the S-shaped reinforcement trajectory, the steel wire rope begins to work together with wood along its entire length (Fig. 7 and 8). This is especially clearly observed in the analysis of the shear stresses in wood and tensile stress diagrams in the steel wire rope (Table 2) (Fig. 9).

4 Discussion

Numerical results of the calculations and analytical data obtained during testing have shown how strongly the form of reinforcement affects the stress-related characteristics of the study samples. Analyzing the graphical results, the following key moments of the tests should be noted:

Table 2 Characteristics of the samples series

Name of the specimens series	Ultimate tensile strength, kN	Deformation, mm
1 Series—rectilinear glued-in piece	17.46	10.1
1 Series—archy glued-in piece, amplitude 15 mm	25.13	12.4
3 Series—archy glued-in piece, amplitude 25 mm	19.02	8.8
4 archy glued-in piece, amplitude 35 mm	7.77	6.2

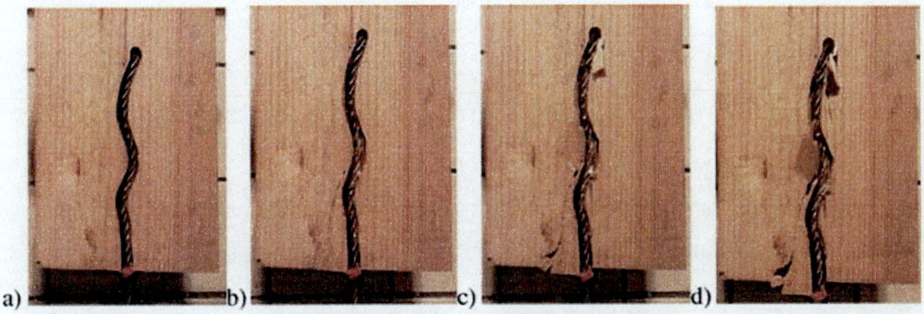

Fig. 9 Experiment on the tensile testing machine. The sample deformation with the maximum amplitude (15 mm) with increasing load. A—is the initial state of the sample; B—is the beginning of fracture corresponding to the ultimate strength; C—is the course of destruction, the simultaneous nature of destruction is visible along the entire length of the sample; D—is the complete destruction of the sample

Firstly, unstable failure is characteristic of the straight form of the wire rope anchorage in solid wood, as evidenced by a pronounced impact at the end of the deformation and instantaneous rupture of the sample.

Secondly, the S-shaped form of reinforcement behaves very ambiguously. At the beginning of the experiment, it would be logical to assume that the more the amplitude of the trajectory is, the more the strength of the connection we will get, for the reason that this shape will be located on a larger plane of the solid wood. However, that's not the case. The experiment has shown that the more the reinforcement trajectory bends, the more the degree of sample deformation in time is. A strongly curved reinforcing element is pulled out of the solid gradually, therefore, only a part of it participates in pulling resistance in each time unit.

Therefore, the reinforcement trajectory with a small amplitude of the reinforcing cable equal to 2 diameters is more effective in terms of strength and rigidity. In this case, a larger (than with direct gluing) solid wood will be swung into action, however, the wire rope will resist pulling out along its full length at the same time, which will lead to gain in strength.

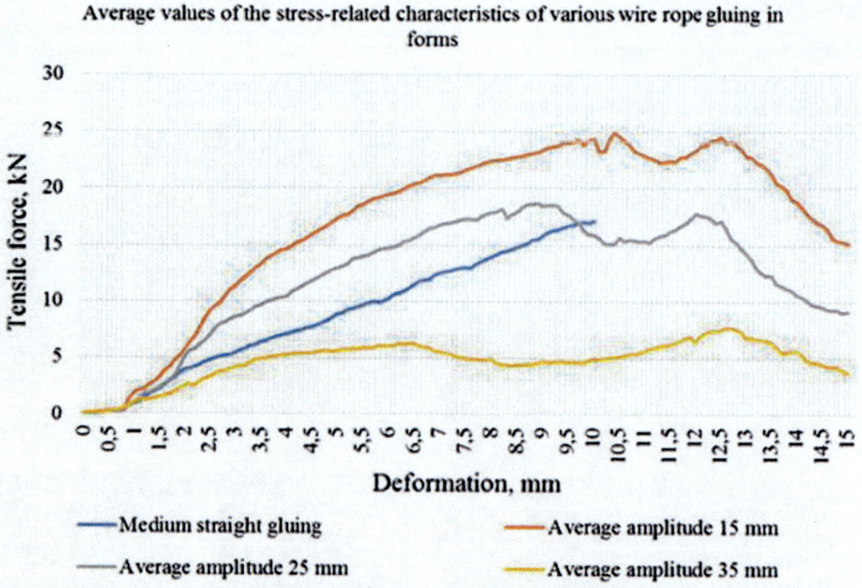

Fig. 10 Diagrams of stress-related characteristics for the control series of samples

Diagrams of the stress-related characteristics obtained from the tensile testing machine confirm these conclusions (Fig. 10). A gradual increase in strength with the S-shaped trajectory amplitude attenuation has been clearly visible. If we compare the most successful shape (15 mm wave amplitude) with the rectilinear reinforcement, the resulting increase in strength in the zone of rupture load will be 24–38%.

It is important to note the graphs upward shifting inclination angle in the horizontal axis—this indicates an increase in the sample rigidity at the beginning of the deformation.

The investigation has proved the consistency of the hypothesis about the advantages of the archy shape trajectory shape located along the S-shaped line in the solid wood, in terms of stress-related characteristics.

However, the wave amplitude (or width) drastically affects the strength. The S-shaped trajectory with the reinforcing bar of 2 diameters width turned out to be most effective. It is that trajectory that has been adopted by the author for use in wooden beams with this type of reinforcement, and will be used in further tests.

The advantage of the resulting trajectory shape is the ability to reduce the steel wire rope anchorage length on the beam support zones—which will lead to an increase in the ruggedness of the structure while maintaining the amount of metal and wood that is required for its manufacture.

5 Conclusions

Visual analysis of the conducted tests results allows us draw a conclusion that the shape of the reinforcing elements arrangement in wood has a direct effect on the building structure stress-related characteristics. The correct choice of the reinforcement form makes it possible to achieve a more efficient use of the bearing capacity of the reinforcing element in wooden beams.

Our research has shown that S-wave reinforcement form of the wooden beams stretched zone becomes effective only when the wave amplitude is approximately 2 diameters of the reinforcing bar for a wire rope of 8 mm in diameter. These results have been tracked in the process of visual and numerical analysis of the experimental data. A decrease in the vibration amplitude of the S-shaped form of reinforcement from 35 to 15 mm leads to an increase in the strength of the entire gluing in by a factor of 3.23, from 7.77 to 25.13 kN, which is 28% stronger in comparison with the straight position. But, with a decrease in the trajectory amplitude, the sample deformation plasticity decreases. Therefore, we see no rationale for further amplitude contraction, since the deformation plasticity under the action of the destructive loads is a positive quality of the building structures and it increases their operating abilities, expanding the range of their application. In future we will use the obtained optimum shape of the reinforcement for carrying out full-scale tests of full size 6 m wooden beams with a section of 75 by 120 mm.

Acknowledgements The reported study was funded by RFBR, project number 19-38-90062.

References

1. Labudin B, Popov E, Stolypin D, Sopilov V (2019) The wood composite ribbed panels on mechanical joints. In: E3S Web of Conferences. https://doi.org/10.1051/e3sconf/20199102021
2. Orlando N, Taddia Y, Benvenuti E, Pizzo B, Alessandri C (2019) End-repair of timber beams with laterally-loaded glued-in rods: experimental trials and failure prediction through modelling. Constr Build Mater. https://doi.org/10.1016/j.conbuildmat.2018.11.045
3. De Luca V, Marano C (2012) Prestressed glulam timbers reinforced with steel bars. Constr Build Mater. https://doi.org/10.1016/j.conbuildmat.2011.11.016
4. Kreher K, Natterer J, Natterer J (2004) Timber-glass-composite girders for a hotel in Switzerland. Struct Eng Int J Int Assoc Bridg Struct Eng. https://doi.org/10.2749/101686604777963964
5. McConnell E, McPolin D, Taylor S (2014) Post-tensioning of glulam timber with steel tendons. Constr Build Mater. https://doi.org/10.1016/j.conbuildmat.2014.09.079
6. Winter W, Tavoussi K, Pixner T, Parada FR (2012) Timber-steel-hybrid beams for multi-storey buildings. In: World Conference on Timber Engineering 2012, WCTE 2012
7. Frese M, Blaß HJ (2011) Statistics of damages to timber structures in Germany. Eng Struct. https://doi.org/10.1016/j.engstruct.2011.02.030
8. Karelskiy AV, Zhuravleva TP, Labudin BV (2015) Load-to-failure bending test of wood composite beams connected by gang nail. Mag Civ Eng. https://doi.org/10.5862/MCE.54.9

9. Steiger R, Serrano E, Stepinac M, Rajčić V, O'Neill C, McPolin D, Widmann R (2015) Strengthening of timber structures with glued-in rods. Constr Build Mater. https://doi.org/10.1016/j.conbuildmat.2015.03.097
10. Tavoussi K, Winter W, Pixner T, Kist M (2010) Steel reinforced timber structures for multi storey buildings. In: 11th World Conference on Timber Engineering 2010, WCTE 2010
11. Borri A, Corradi M, Speranzini E (2013) Reinforcement of wood with natural fibers. Compos Part B Eng. https://doi.org/10.1016/j.compositesb.2013.04.039
12. Buell TW, Saadatmanesh H (2005) Strengthening timber bridge beams using carbon fiber. J Struct Eng. https://doi.org/10.1061/(ASCE)0733-9445(2005)131:1(173)
13. Cherry R, Manalo A, Karunasena W, Stringer G (2019) Out-of-grade sawn pine: a state-of-the-art review on challenges and new opportunities in cross laminated timber (CLT). https://doi.org/10.1016/j.conbuildmat.2019.03.293
14. Sun X, He M, Li Z (2020) Novel engineered wood and bamboo composites for structural applications: State-of-art of manufacturing technology and mechanical performance evaluation. https://doi.org/10.1016/j.conbuildmat.2020.118751
15. De La Rosa García P, Escamilla AC, Nieves González García M (2013) Bending reinforcement of timber beams with composite carbon fiber and basalt fiber materials. Compos Part B Eng. https://doi.org/10.1016/j.compositesb.2013.07.016
16. Gentile C, Svecova D, Rizkalla SH (2002) Timber beams strengthened with GFRP bars: development and applications. https://doi.org/10.1061/(ASCE)1090-0268(2002)6:1(11)
17. Gribanov AS, Rimshin VI, Roshchina SI (2019) Experimental investigations of composite wooden beams with local wood modification. In: IOP conference series: materials science and engineering. https://doi.org/10.1088/1757-899X/687/3/033039
18. Gribanov AS, Roshchina SI, Popova MV, Sergeev MS (2018) Laminar polymer composites for wooden structures. Mag Civ Eng. https://doi.org/10.18720/MCE.83.1
19. Raftery GM, Harte AM (2011) Low-grade glued laminated timber reinforced with FRP plate. Compos Part B Eng. https://doi.org/10.1016/j.compositesb.2011.01.029
20. Raftery GM, Whelan C (2014) Low-grade glued laminated timber beams reinforced using improved arrangements of bonded-in GFRP rods. Constr Build Mater. https://doi.org/10.1016/j.conbuildmat.2013.11.044
21. Buka-Vaivade K, Serdjuks D, Goremikins V, Pakrastins L, Vatin NI (2018) Suspension structure with cross-laminated timber deck panels. Mag Civ Eng. https://doi.org/10.18720/MCE.83.12
22. Lisyatnikov MS, Glebova TO, Ageev SP, Ivaniuk AM (2020) Strength of wood reinforced with a polymer composite for crumpling across the fibers. In: IOP conference series: materials science and engineering. https://doi.org/10.1088/1757-899X/896/1/012062
23. Dietsch P, Winter S (2018) Structural failure in large-span timber structures: a comprehensive analysis of 230 cases. Struct Saf. https://doi.org/10.1016/j.strusafe.2017.11.004
24. Ilyichev V, Emelianov S, Kolchunov V, Bakaeva N (2018) Principles of urban areas reconstruction ensuring safety and comfortable living conditions. In: IOP conference series: materials science and engineering. https://doi.org/10.1088/1757-899X/463/3/032011
25. Rimshin VI, Labudin BV, Melekhov VI, Orlov AO, Kurbatov VL (2018) Improvement of strength and stiffness of components of main struts with foundation in wooden frame buildings. ARPN J Eng Appl Sci
26. Roshchina S, Lukin M, Lisyatnikov M (2020) Compressed-bent reinforced wooden elements with long-term load. https://doi.org/10.1007/978-3-030-42351-3_7
27. Dietsch P, Kreuzinger H (2020) Dynamic effects in reinforced beams at brittle failure—evaluated for timber members. Eng Struct. https://doi.org/10.1016/j.engstruct.2019.110018
28. Savin SY, Kolchunov VI, Emelianov SG (2018) Modelling of resistance to destruction of multi-storey frame-connected buildings at sudden loss of bearing elements stability. In: IOP conference series: materials science and engineering. https://doi.org/10.1088/1757-899X/456/1/012089

29. Labudin BV, Popov EV, Nikitina TA (2019) Notes for calculated resistance to tension for laminated wood. In: IOP conference series: materials science and engineering. https://doi.org/10.1088/1757-899X/687/3/033028
30. Roshchina S, Sergeev M, Lukin M, Strekalkin A (2018) Reconstruction of fixed fertilizer folders in the vladimir region. In: IOP conference series: materials science and engineering. https://doi.org/10.1088/1757-899X/463/4/042011
31. Lisyatnikov MS, Shishov II, Sergeev MS, Hisham E (2020) Precast monolithic coating of an industrial building based on variable-height beam-slabs. In: IOP conference series: materials science and engineering. https://doi.org/10.1088/1757-899X/896/1/012064
32. Travush VI, Gordon VA, Kolchunov VI, Leontiev EV (2019) Dynamic deformation of a beam at sudden structural transformation of foundation. Mag Civ Eng. https://doi.org/10.18720/MCE.91.12
33. Nowak T, Jasieńko J, Kotwica E, Krzosek S (2016) Strength enhancement of timber beams using steel plates—Review and experimental tests. https://doi.org/10.12841/wood.1644-3985.150.06
34. Buchanan AH, Fairweather RH (1993) Seismic design of glulam structures. Bull New Zeal Natl Soc Earthq Eng. https://doi.org/10.5459/bnzsee.26.4.415-436
35. Dražić J, Vatin N (2016) The influence of configuration on to the seismic resistance of a building. Procedia Eng. https://doi.org/10.1016/j.proeng.2016.11.788
36. Kasal B, Pospisil S, Jirovsky I, Heiduschke A, Drdacky M, Haller P (2004) Seismic performance of laminated timber frames with fiber-reinforced joints. Earthq Eng Struct Dyn. https://doi.org/10.1002/eqe.368
37. Labudin BV, Popov EV, Sopilov VV (2019) Stability of compressed sheathings of wood composite plate-ribbed structures. In: IOP conference series: materials science and engineering. https://doi.org/10.1088/1757-899X/687/3/033041
38. Lukin M, Prusov E, Roshchina S, Karelina M, Vatin N (2021) Multi-Span composite timber beams with rational steel reinforcements. Buildings 11:1–12. https://doi.org/10.3390/buildings11020046
39. D'Agata G, D'Aveni A (2017) Post-tensioned timber structures: new perspectives. Constr Build Mater. https://doi.org/10.1016/j.conbuildmat.2017.07.031
40. Varenik KA, Varenik AS, Sanzharovsky RS, Labudin BV (2019) Wood moisture accounting in creep equations. In: IOP conference series: materials science and engineering. https://doi.org/10.1088/1757-899X/656/1/012054
41. Wanninger F, Frangi A (2016) Experimental and analytical analysis of a post-tensioned timber frame under horizontal loads. Eng Struct. https://doi.org/10.1016/j.engstruct.2016.01.029
42. Koshcheev AA, Roshchina SI, Naichuk AY, Vatin NI (2020) The effect of eccentricity on the strength characteristics of glued rods made of steel cable reinforcement in solid wood. In: IOP conference series: materials science and engineering. https://doi.org/10.1088/1757-899X/896/1/012059
43. Koshcheev AA, Roshchina SI, Aleksiievets V, Labudin BV (2020) Local deformation and strength characteristics of S-shaped reinforcement in wood. In: IOP conference series: materials science and engineering. https://doi.org/10.1088/1757-899X/896/1/012060
44. Lukin MV, Roshchina SI, Gribanov AS, Naychuk AY (2020) Stress-strain state of wooden beams with external reinforcement. In: IOP conference series: materials science and engineering. https://doi.org/10.1088/1757-899X/896/1/012066
45. Lukin MV, Roshchina SI, Smirnov EA, Shunqi M (2020) Strengthening of the operated wooden floor beams with external rigid reinforcement. In: IOP conference series: materials science and engineering. https://doi.org/10.1088/1757-899X/896/1/012065
46. Popova M, Sergeev M, Lukina A, Shunqi M (2020) Strength and deformability of lightweight metal trusses with elements from cut I-beams. In: IOP conference series: materials science and engineering. https://doi.org/10.1088/1757-899X/896/1/012061

47. Gonshakov NG, Gonshakov AG, Aleksiievets II (2020) Reinforcement of brick structures with carbon fiber. In: IOP conference series: materials science and engineering. https://doi.org/10.1088/1757-899X/896/1/012032
48. Gribanov AS, Strekalkin AA, Kudryatseva AA, Zdralovic N (2020) CFRP composites for strengthening wooden structures. In: IOP conference series: materials science and engineering. https://doi.org/10.1088/1757-899X/896/1/012114
49. Romanovich A, Kleshcunov Y, Vlasov A (2019) On potentiality and practicability of installing flooring suspended in geodesic domes by means of cable system. In: IOP conference series: materials science and engineering. https://doi.org/10.1088/1757-899X/687/3/033025

Roofing of an Industrial Building with Variable Height Rafters and Wooden Decking

Ivan Shishov, M. V. Lukin, and M. S. Sergeev

Abstract Single storey production buildings are used in many industries. Typically, these are frame-type buildings: columns clamped in foundations, truss structures that overlap spans and support flooring of ribbed slabs 6 or 12 m long. In the latter case, a truss structure is added to provide support for the rafter in the gap between the columns.

Often there is a need to increase the usable area of the workshop—the spans of the building or the pitch of the transverse axes, but the structures of the reinforced concrete pavement become cumbersome, laborious to manufacture, transport and install, and their construction height significantly increases the height of the building. It is believed that for buildings with spans of 30 m, it is better to use metal coatings, but it is not indisputable, and, of course, it is necessary to have such a coating made of reinforced concrete.

Keywords Buildings · Beam · Reinforced concrete · Industrial buildings · Coating

1 Introduction

Reinforced concrete structures are quite widespread in both civil and industrial construction [1–6].

In works [5, 7–9], the racks of the rafter frame are made of two cross walls, some of which are directed along the frame and impart the greatest rigidity to it, others are perpendicular and support the ribbed plates. Between the ends of the slabs, monolithic concrete is laid, partially including the slabs in the girder, which makes it possible to reduce the height of the coating.

The influence of seismic effects on the deformation of rod structures and industrial buildings is considered in works [10–15].

I. Shishov · M. V. Lukin (✉) · M. S. Sergeev
Vladimir State University Named After Alexander and Nikolay Stoletovs, Gorky Street 87, Vladimir 600000, Russia

N. Vatin et al. (eds.), *Proceedings of MPCPE 2021*, Lecture Notes in Civil Engineering 182, https://doi.org/10.1007/978-3-030-85236-8_41

In works [16–20] an experimental study of a precast-monolithic coating and calculation of its elements taking into account the physical nonlinearity of concrete and reinforcement deformation are given.

In [21], a precast-monolithic coating of a one-story industrial building with spans of 18, 24, 30 m and an increased step of the transverse axes is presented.

One-story industrial buildings with spans up to 30 m are being built for many industries [22–24]. The elements of the coating are rather cumbersome, require prestressing of the reinforcement, laborious to manufacture, transport and install; the coating turns out to be high and increases the height of the building [25–28].

Reinforced concrete coatings for industrial buildings are usually used with spans up to 30 m. In a typical solution, they are made in the form of flooring of ribbed slabs on rafter beams, trusses or arches resting on columns or under-rafter structures. The coverage is high; with spans of 30 m, it increases the height of the building to 4 m; structures have a high labor intensity of manufacture, transportation and installation. The task arises to eliminate these shortcomings, and, in addition, it is desirable to have a reinforced concrete coating for spans of 30 m. All this gives rise to the search for new solutions.

We suggest the following. Discard ribbed plates. For buildings with spans of 30 m, rafter beams with an I-beam cross-section with a width of the upper belt of 1 m (Fig. 1) should be placed with a step of 3 m, and the distances between them should be covered with shields made of boards 40 mm thick, which will bear the load and serve as part of the thermal insulation layer.

The rafter beams are pivotally supported at the ends. Bending moments vary from zero at the supports to maximum values in the middle of the span. Beams with stepwise variable heights are offered: in the middle part, the height and rigidity of the beam are maximum, and in the direction to the supports—several steps with a decrease in these values. For example, we took a beam in five steps: one in the middle and two in the support zones (Fig. 1).

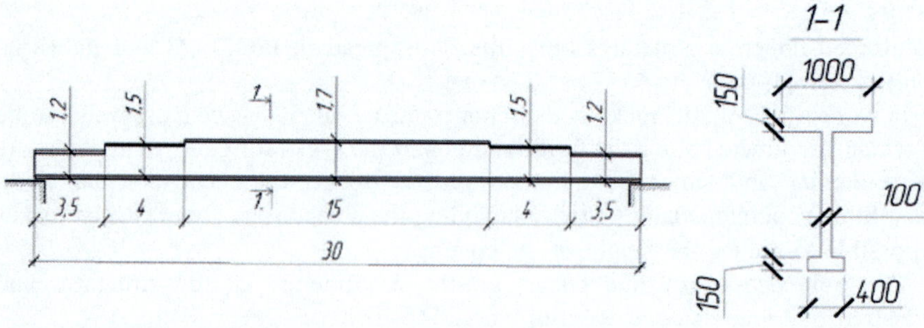

Fig. 1 Rafter beam of stepped variable height with a span of 30 m

2 Methods

Strength calculation is performed according to the limit state on the basis of the following prerequisites [29–34]:

- In a compressed zone, the stresses in concrete at all points are the same and equal to the design resistance of concrete R_b.
- The concrete resistance of the tensioned zone is assumed to be zero.
- The stresses in the reinforcement in the tensile zone are equal to the design resistance of the reinforcement R_s.

The design load from snow and dead weight of the pavement elements is 0.0209 MN/m.

The beam is made of concrete of class B40 and reinforced with bar reinforcement of class A500C. The heights of steps 1, 2, 3 (in the middle of the span and in the support zones) are taken respectively: 1.7, 1.5, 1.2 m. Reinforcement by steps was determined based on strength calculation:

$$A_s = 0.003563.14\emptyset18, \ \mu = 0.0219$$
$$A_s = 0.003054.12\emptyset18, \ \mu = 0.0214$$
$$A_s = 0.002036.8\emptyset18, \ \mu = 0.0181$$

The calculation for deflections is made taking into account the nonlinearity of deformation of reinforced concrete, based on the following prerequisites [35–39]:

- Hypothesis of flat sections: when the beam is bent, its cross sections rotate relative to the neutral line, remaining flat;
- The concrete of the compressed zone deforms non-linearly in accordance with the recommended three-line diagram;
- Stretched zone reinforcement of class A500C is deformed in accordance with the recommended norms of a two-line diagram;
- The work of the concrete in the tensioned zone is taken into account indirectly using the coefficient ψ_s increasing the modulus of elasticity of the reinforcement to the value $\frac{E_s}{\psi_s}$,

$$\psi_s = 1 - 0.8\frac{M_{crc}}{M}, \quad (1)$$

where: M_{crc}—the moment of cracking; M—moment from external load.

The moment of cracking is determined taking into account the inelastic deformations of concrete in the tensile zone [40–44] based on the following assumptions (1):

- Hypothesis of flat sections;
- In the compressed zone, concrete works elastically at an initial modulus of elasticity E_b;
- In a stretched zone, stresses in concrete increase elastically to the value of the design tensile strength of concrete for the limiting states of the second group $R_{bt,ser}$, with further deformation remain constant;
- The deformation of the most stressed fiber of the stretched zone reaches the limit value for a short-term load $\varepsilon_{bt,u} = 0.00015$;
- The reinforcement is elastically deformed.

Cross-section of a beam at a moment equal to M_{crc}, is divided into elastic and plastic zones with deformation along the boundary (Fig. 2).

$$\varepsilon_{b1} = R_{bt,\ ser}/E_b \tag{2}$$

The rate of change of deformations along the height of the plastic zone y and the entire section can be found by the formula, after which it is easy to find deformations and stresses at any point of the section and the magnitude of the moment:

$$u = \left(\varepsilon_{bt,u} - \varepsilon_{b1}\right)/y \tag{3}$$

The stiffness of the beam is determined from the deformation ε_b the most stressed concrete fiber in the compressed zone. It is required to determine the height of the compressed zone and the bending moment corresponding to this deformation —the sum of moments relative to the neutral line of the cross-section of the stresses in the concrete of the compressed zone and the forces in the reinforcement of the tensioned and compressed zones. The task is complicated by the fact that the compressed zone has a T-section. The following technique is proposed: with x the plot of deformations is tripled within the height of the compressed zone; a stress diagram is constructed in accordance with the deformation diagram and the adopted concrete deformation diagram; resultant stress in the concrete of the compressed

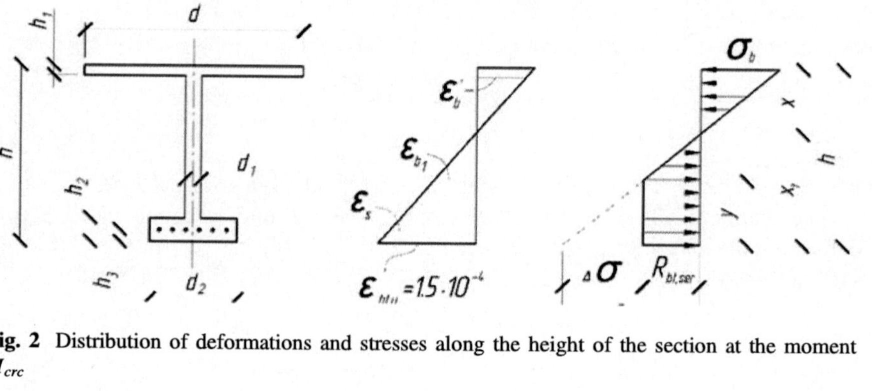

Fig. 2 Distribution of deformations and stresses along the height of the section at the moment M_{crc}

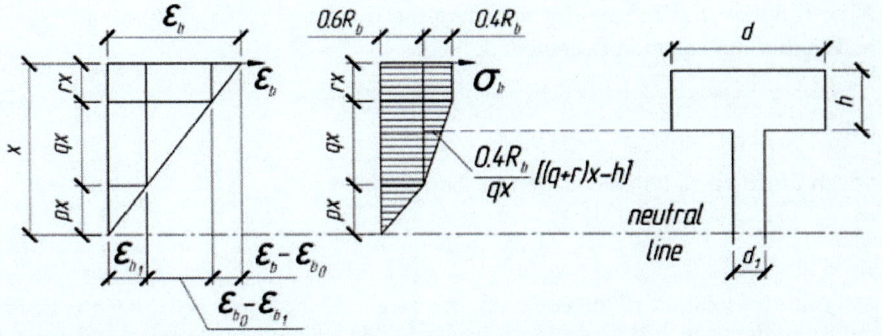

Fig. 3 Distribution of deformations and stresses along the height of the compressed zone of the section

zone x N_b calculated in parts $N_{b,i}$, which are distinguished taking into account the T-shape of the section so that for each of them it is easy to determine the distance Z_i to the neutral line of the cross-section of the beam. Figure 3 shows an example for the case when $\varepsilon_{b0} < \varepsilon_b < \varepsilon_{b2}$ and the height of the plastic zone of the height of the compressed flange of the section = 0.002, = 0.0035 are the parameters of the concrete deformation diagram.

The designations are accepted:

$$p = \frac{\varepsilon_{b1}}{\varepsilon_b};\ q = \frac{\varepsilon_{b0} - \varepsilon_{b1}}{\varepsilon_b};\ r = \frac{\varepsilon_b - \varepsilon_{b0}}{\varepsilon_b}. \tag{4}$$

According to Fig. 3:

$$N_{b1} = \frac{1}{2}0.6R_b p_x d_1;\ z_1 = \frac{2}{3}p_x;$$

$$N_{b2} = 0.6R_b q_x d_1;\ z_2 = p_x + \frac{q_x}{2};$$

$$N_{b3} = \frac{1}{2}0.4R_b q_x d_1;\ z_3 = p_x + \frac{2}{3}q_x;$$

$$N_{b4} = R_b r_x d;\ z_4 = x - \frac{r_x}{2};$$

$$N_{b5} = -R_b(r_x - h_1);\ z_5 = x - \frac{r_x + h_1}{2}$$

Reinforcement forces are written as:

$N_s = E_s A_s \varepsilon_s = \frac{E_s A_s}{\psi_s}\varepsilon_b \frac{h_0 - x}{x}$—for the extended area;

$N'_s = E_s A'_s \varepsilon'_s = E_s A'_s \varepsilon_b \frac{x-a'}{x}$—for the compressed zone.

Equilibrium equation is compiled.

$$N_b + N'_s - N_s = 0, \quad (5)$$

which turns out to be an equation of the form

$$ax^2 + bx + c = 0 \quad (6)$$

From the solution of the equation, the height of the compressed zone is determined x, after which the bending moment is determined M, curvature of a curved beam axis $k = \varepsilon_b / x$ and the stiffness of the beam $g = M/k$.

3 Results and Discussion

The calculation of the rafter beam begins with the fact that for each of its steps the deformations of the extreme fiber of the compressed zone from to with a step of 0.00001 are sequentially set and the corresponding values of the bending moment, deformation of the curvature of the bent axis and the stiffness of the beam are calculated and entered into the table. $\varepsilon_b = 0.00001 \cdot \varepsilon_b = 0.00350 \cdot \varepsilon_b$.

Further, the method of elastic solutions is applied in combination with the method of finite differences. In each elastic solution, the beam is divided along its length into small parts, along the boundaries of which points are applied $j = 1, 2, 3...n$.

The deflections of the beam at these points are taken as the main unknowns; they are determined from the solution of the system of equilibrium equations of small parts, highlighted in the vicinity of each point. The deflections are used to determine the curvatures of the curved axis of the beam, and from them—the lines in the table, from where the stiffnesses for the next elastic solution are taken, and individually for each point $j : j = 1, 2, 3...n$. The solution shows stable convergence.

The rafter beam for a span of 30 m was divided in the solution into 30 parts 1.0 m long (Fig. 4).

Results of the solution: beam deflections, bending moments, stiffness, height of the compressed zone, deformation of the most stressed fiber of concrete in the compressed zone are shown in Fig. 4.

It is interesting to note that the beam does not have excessive reserves either in strength or in rigidity. The reinforcement of the steps is determined by the maximum moments, the change of which within the step is small. The calculation for deformations shows a deflection in the middle of the span of 0.103 m with the maximum permissible value determined by the standards of 0.109 m.

Such a coating can also be used for a span of 36 m. The rafter beams have three steps: an average height of 1.7 m with a length of 15 m and extreme heights of

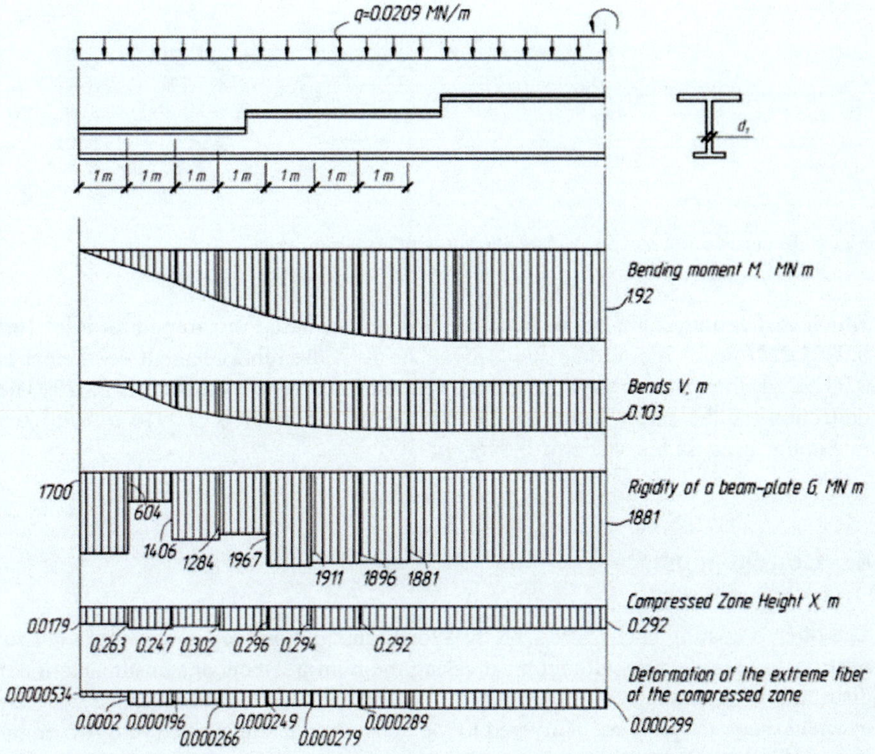

Fig. 4 Stress—deformed state of the rafter beam of stepwise variable height for a span of 30 m

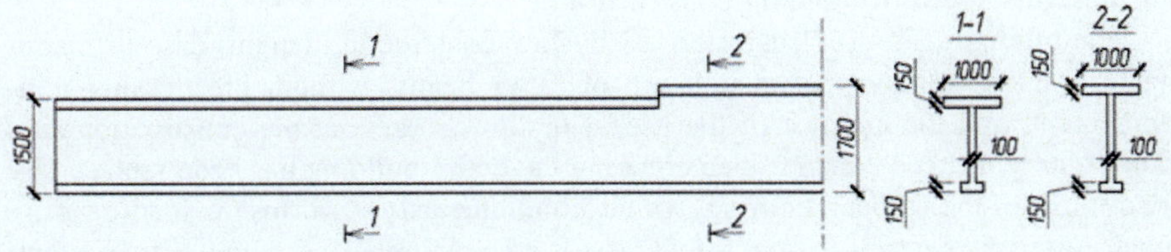

Fig. 5 Rafter beam of stepped variable height with a span of 36 m

1.5 m with a length of 10.5 m each (Fig. 5). Beams are often located—with a step of 1.5 m, the gaps between the upper shelves, covered with wooden boards, 0.5 m. reinforcement 0.032. The deflection of the rafter beam in the middle of the span is 0.118 m with the maximum permissible value of 0.120 m.

With a span of 24 m (Fig. 6), the beams also have three steps: 1.3 m high for the middle and 1.1 m for the extreme ones (Fig. 2), 15 and 4.5 m long, respectively.

Beams are located with a step of 6 m; the distance between the upper shelves of 5 m is covered by shields made of planks connected in a quarter, the thickness of

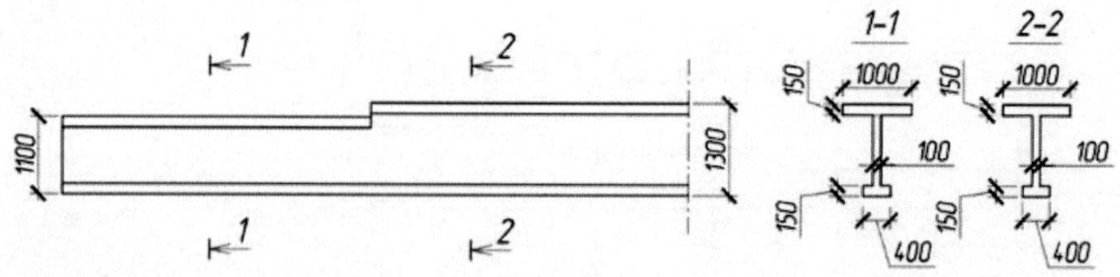

Fig. 6 Rafter beam of stepped variable height with a span of 24 m

which is determined by calculation. Reinforcement based on strength at a load of 0.0322 MN/m for the middle step 18Ø18 A500C, the reinforcement coefficient is 0.0312, for the extreme 14Ø18 A500C, the reinforcement coefficient is 0.029. The deflection of the rafter beam in the middle of the span is 0.0914 m with the maximum permissible value of 0.0960 m.

4 Conclusions

The offered coating is characterized by a low construction height. The rafter beams are easy to manufacture without prestressing the reinforcement and ensure a rational distribution of concrete and reinforcement along the length in accordance with the moment diagram. They are delivered to the construction site divided into two equal parts, which are then rigidly connected along stretched and compressed zones in the middle of the span—in a section with a maximum moment and zero shear force. A small number of assembly units and a small amount of assembly welding increase the speed of building construction.

The offered coating is characterized by low construction height, ease of manufacture, transportation and installation of rafter beams without prestressing reinforcement, and the absence of the use of in-situ concrete at the construction site. The solid reinforced concrete slab covering the entire building has been replaced by two-thirds of the wooden flooring, which simultaneously performs the load-bearing, enclosing and heat-insulating functions. A small number of crane lifts, a small amount of assembly welding, the absence of monolithing of joints between ribbed plates increase the speed and quality of the coating device, the profile of which, given by the rafter beams, facilitates the drainage device.

References

1. Modin A, Lukin M, Vlasov A, Hisham E (2020) Energy-efficient indicators of panel housing mass construction in the climatic conditions of central Russia. In: IOP conference series: materials science and engineering. https://doi.org/10.1088/1757-899X/896/1/012063
2. Rimshin VI, Kuzina ES, Shubin IL (2020) Analysis of the structures in water treatment and sanitation facilities for their strengthening. J Phys Conf Ser. https://doi.org/10.1088/1742-6596/1425/1/012074
3. Bednarz L, Bajno D, Matkowski Z, Skrzypczak I, Leśniak A (2021) Elements of pathway for quick and reliable health monitoring of concrete behavior in cable post-tensioned concrete girders. Materials (Basel) 14. https://doi.org/10.3390/ma14061503
4. Shubin LI, Rimshin VI, Truntov PS, Suleymanova LA (2020) Evaluation method of technical condition façade systems. In: IOP conference series: materials science and engineering. https://doi.org/10.1088/1757-899X/896/1/012023
5. Lisyatnikov MS, Shishov II, Sergeev MS, Hisham E (2020) Precast monolithic coating of an industrial building based on variable-height beam-slabs. In: IOP conference series: materials science and engineering. https://doi.org/10.1088/1757-899X/896/1/012064
6. Roschina SI, Lukin MV, Lisyatnikov MS, Sergeyev MS (2017) Reconstruction of coating by a single-stage adjustment of a lind-fitting factory in the city of vyazniki
7. Lee S-J, Eom T-S, Yu E (2021) Investigation of diagonal strut actions in masonry-infilled reinforced concrete frames. Int J Concr Struct Mater 15. https://doi.org/10.1186/s40069-020-00440-x
8. Roshchina S, Ezzi H, Shishov I, Lukin M, Sergeev M (2017) Evaluation of the deflected mode of the monolithic span pieces and preassembled slabs combined action. In: IOP conference series: earth and environmental science. https://doi.org/10.1088/1755-1315/90/1/012075
9. Lukin MV, Popov MV, Lisyatnikov MS (2020) Short-term and long-term deformations of the lightweight concrete. In: IOP conference series: materials science and engineering. https://doi.org/10.1088/1757-899X/753/3/032071
10. Pan X, Qiao S, Li H, Gao H, Ma H, Yao L (2021) Seismic behaviour of a new type of external hoop assembled joint for RC Frame-added layer hybrid structure. Structures 32:170–193. https://doi.org/10.1016/j.istruc.2021.03.016
11. Zameeruddin M, Sangle KK (2021) Performance-based seismic assessment of reinforced concrete moment resisting frame. J King Saud Univ Eng Sci 33:153–165. https://doi.org/10.1016/j.jksues.2020.04.005
12. Belleri A, Torquati M, Riva P (2014) Seismic performance of ductile connections between precast beams and roof elements. Mag Concr Res 66:553–562. https://doi.org/10.1680/macr.13.00092
13. Park S, Mosalam KM (2013) Simulation of reinforced concrete frames with nonductile beam-column joints. Earthq Spectra 29:233–257. https://doi.org/10.1193/1.4000100
14. Liel AB, Haselton CB, Deierlein GG, Baker JW (2009) Incorporating modeling uncertainties in the assessment of seismic collapse risk of buildings. Struct Saf 31:197–211. https://doi.org/10.1016/j.strusafe.2008.06.002
15. Priestley MJN (1997) Displacement-based seismic assessment of reinforced concrete buildings. J Earthq Eng 1:157–192. https://doi.org/10.1080/13632469708962365
16. Krauthammer T, Assadi-Lamouki A, Shanaa HM (1993) Analysis of impulsively loaded reinforced concrete structural elements-I. Theory. Comput Struct 48:851–860. https://doi.org/10.1016/0045-7949(93)90507-A
17. Hago AW, Al-Jabri KS, Alnuaimi AS, Al-Moqbali H, Al-Kubaisy MA (2005) Ultimate and service behavior of ferrocement roof slab panels. Constr Build Mater 19:31–37. https://doi.org/10.1016/j.conbuildmat.2004.04.034

18. Koksal HO, Erdogan A (2021) Stress–strain model for high-strength concrete tied columns under concentric compression. Structures 32:216–227. https://doi.org/10.1016/j.istruc.2021.02.063
19. Hassan MAJ, Izzet AF, Oukaili NK (2021) Structural performance under monotonic static loading of reinforced concrete gable roof beams with multiple web openings. Int J Civ Eng 19:421–440. https://doi.org/10.1007/s40999-020-00578-5
20. Roshchina SI, Gonshakov AG, Gonshakov NG (2020) Taking into consideration the work of monolithic reinforced concrete floor with shaped steel profiles in two directions. In: IOP conference series: materials science and engineering. https://doi.org/10.1088/1757-899X/753/4/042084
21. Popova M, Sergeev M, Lukina A, Shunqi M (2020) Strength and deformability of lightweight metal trusses with elements from cut I-beams. In: IOP conference series: materials science and engineering. https://doi.org/10.1088/1757-899X/896/1/012061
22. Zamorowski J, Gremza G (2019) Modelling of roof bracings of single-storey industrial buildings. In: SDSS 2019—international colloquium on stability and ductility of steel structures
23. Chen F, Jin Z, Wang E, Wang L, Jiang Y, Guo P, Gao X, He X (2021) Relationship model between surface strain of concrete and expansion force of reinforcement rust. Sci Rep 11. https://doi.org/10.1038/s41598-021-83376-w
24. Trekin NN, Kodysh EN, Kelasiev NG, Shmakov SD, Chaganov AB, Terehov IA (2020) The improvement of protection methods from the progressive collapse of one-storey industrial buildings. J Phys Conf Ser. https://doi.org/10.1088/1742-6596/1425/1/012050
25. Varlamov A, Kostyuchenko Y, Rimshin V, Kurbatov V (2020) Diagrams of concrete behavior over time. In: IOP conference series: materials science and engineering. https://doi.org/10.1088/1757-899X/896/1/012085
26. Castañeda-Saldarriaga DL, Alvarez-Montoya J, Martínez-Tejada V, Sierra-Pérez J (2021) Toward structural health monitoring of civil structures based on self-sensing concrete nanocomposites: a validation in a reinforced-concrete beam. Int J Concr Struct Mater 15. https://doi.org/10.1186/s40069-020-00451-8
27. Merkulov S, Rimshin V, Akimov E, Kurbatov V, Roschina S (2020) Regulatory support for the use of composite rod reinforcement in concrete structures. In: IOP conference series: materials science and engineering. https://doi.org/10.1088/1757-899X/896/1/012022
28. Slowik O, Novák D, Novák L, Strauss A (2021) Stochastic modelling and assessment of long-span precast prestressed concrete elements failing in shear. Eng Struct 228. https://doi.org/10.1016/j.engstruct.2020.111500
29. Khamidulina D, Rimshin V, Varlamov A, Nekrasov S (2019) Power and energy characteristics of concrete. In: E3S web of conferences. https://doi.org/10.1051/e3sconf/201913503057
30. Akbar J, Ahmad N, Alam B (2020) Response modification factor of haunch retrofitted reinforced concrete frames. J Perform Constr Facil 34. https://doi.org/10.1061/(ASCE)CF.1943-5509.0001525
31. Gonshakov NG, Gonshakov AG, Aleksiievets II (2020) Reinforcement of brick structures with carbon fiber. In: IOP conference series: materials science and engineering. https://doi.org/10.1088/1757-899X/896/1/012032
32. Karpenko NI, Rimshin VI, Eryshev VA, Shubin LI (2020) Deformation models of concrete strength calculation in the edition of Russian and Foreign norms. In: IOP conference series: materials science and engineering. https://doi.org/10.1088/1757-899X/753/5/052043
33. Vasiliev AS, Plekhanova EA (2020) To the multi-hollow reinforced concrete floor panels' calculation according to the two groups of the limit states. In: IOP conference series: materials science and engineering. https://doi.org/10.1088/1757-899X/913/2/022033
34. Das D, Ayoub A (2021) Mixed formulation for geometric and material nonlinearity of shear-critical reinforced concrete columns. Eng Struct 229. https://doi.org/10.1016/j.engstruct.2020.111587

35. Hashemi SS, Sadeghi K, Javidi S, Malakouti M (2020) Analysis of RC deep beams considering the shear deformations and bar-concrete interaction. Period Polytech Civil Eng 65:99–108. https://doi.org/10.3311/PPci.16192
36. Hartung B, Krebs A (2004) An extension of the technical bending theory | Erweiterung der Technischen Biegelehre—Teil 1: Aufstellung der Bedingungsgleichungen mit Überprüfung der Lösbarkeit. Beton- und Stahlbetonbau 99:378–387. https://doi.org/10.1002/best.200490045
37. Gunnin BL, Rad FN, Furlong RW (1977) A general nonlinear analysis of concrete structures and comparison with frame tests. Comput Struct 7:257–265. https://doi.org/10.1016/0045-7949(77)90044-X
38. Hieu Dinh N, Kim J-C, Lee S-J, Choi K-K, Park H-G (2020) Modeling parameters for the nonlinear analysis of reinforced concrete beams under cyclic loading. Eng Struct 216. https://doi.org/10.1016/j.engstruct.2020.110715
39. Pochinok V, Tamov M, Greshkina E (2020) Nonlinear shear analysis of I-shaped beams with arbitrary inclination angles of shear reinforcement. In: IOP conference series: materials science and engineering. https://doi.org/10.1088/1757-899X/869/5/052057
40. Mercedes L, Escrig C, Bernat-Masó E, Gil L (2021) Analytical approach and numerical simulation of reinforced concrete beams strengthened with different frcm systems. Materials (Basel) 14. https://doi.org/10.3390/ma14081857
41. Fomin SL, Bondarenko YV, Butenko SV, Plakhotnikova IA (2019) DBN Concrete and reinforced concrete structures intended to operate under elevated and high temperatures. In: IOP conference series: materials science and engineering. https://doi.org/10.1088/1757-899X/708/1/012047
42. Fomin S, Izbash Y, Plakhotnikova I, Butenko S, Shemet R (2017) Improvement of the mathematical model of the diagram of deformation of the compressed composite steel and concrete structures. In: MATEC web of conferences. https://doi.org/10.1051/matecconf/201711602012
43. Bondarenko Y, Spirande K, Iakymenko M, Mol'Skyj M, Oreshkin D (2017) Study of the stress-strain state of compressed concrete elements with composite reinforcement. In: MATEC web of conferences. https://doi.org/10.1051/matecconf/201711602008
44. Fu C, Zhu Y, Tong D (2021) Stiffness assessment of cracked reinforced concrete beams based on a fictitious crack model. KSCE J Civ Eng 25:516–528. https://doi.org/10.1007/s12205-020-2056-0
45. Li Z-X, Zhong B, Shi Y, Ding Y, Hao Y (2019) A computationally efficient numerical model for progressive collapse analysis of reinforced concrete structures. Int J Prot Struct 10:330–358. https://doi.org/10.1177/2041419619854768

Investigation of the Stress–Strain State of Wooden Beams with Rational Reinforcement with Composite Materials

S. Roshchina, A. Gribanov, M. Lukin, D. Chibrikin, and Mei Shunqi

Abstract The purpose of this work is to study the stress–strain state of wood-glued beams reinforced with composite materials. To achieve this goal, a comparative analysis of the types of polymer composites was carried out and a numerical calculation of beam structures with a span of 4.5 m and a cross section of 100 × 200 mm was performed in the PC "LIRA 10.10". The rational variant of the reinforcing material-carbon fiber is chosen: the tensile strength 2000 MPa, the modulus of elasticity 95,000 GPa, the density 400 g/m^2, the equivalent thickness 0.14 mm, the elongation 2.3%. The data obtained as a result of the calculation clearly illustrates that the design load on the reinforced beam is 2 times greater than that on the non-reinforced one. For the reinforced beam, the calculated load was 1800 kg/m, for the non-reinforced one 900 kg/m. At the same time, the ultimate strength of the reinforced beam decreased by 40% relative to the non-reinforced one. A method for calculating the reinforcement of wooden elements with composite material based on the geometric characteristics of the cross section is presented. The possibility of using polymer composite materials for strengthening beam structures is theoretically justified and confirmed by calculation.

Keywords Building construction · Construction materials · Strength · Timber

1 Introduction

Wood has traditionally been used as the main building material in the construction of homes for many centuries in many countries. Currently, there are various building materials that use woodworking products with various adhesives-binders

S. Roshchina (✉) · A. Gribanov · M. Lukin · D. Chibrikin
Vladimir State University Named After Alexander and Nikolay Stoletovs, Vladimir, Russian Federation

M. Shunqi
School of Mechanical Engineering and Automation, Wuhan Textile University, Wuhan, China

N. Vatin et al. (eds.), *Proceedings of MPCPE 2021*, Lecture Notes in Civil Engineering 182, https://doi.org/10.1007/978-3-030-85236-8_42

as a basis. Wood materials have ceased to represent pure wood in their structure, becoming composite materials [1–8].

In the modern foreign practice of construction in recent years, there are more and more examples of the introduction of wood as a material for load-bearing structures. The main feature of this material is the excellent ratio between strength and density, which is four times higher than steel. The use of wood is associated with the following main reasons. The first reason is that wood is a relatively inexpensive building material and easy to use. The second reason is the development of technologies for creating building materials based on wood, as well as wood protection products. The third reason is the environmental friendliness and renewable nature of this natural resource. The topic of environmental friendliness of building materials throughout its entire life cycle in Western countries is given great attention [9–12]. There are other reasons for using wood as the main material in construction, which are pointed out by architects: the possibility of factory production of many parts and assemblies in the assembly, high assembly speed, no need to use a large number of heavy equipment on the construction site, etc. [12–32].

However, its ability to lift a small load, combined with exposure to different environmental conditions compared to other building materials, significantly limits its range of application. Increasing the mechanical strength of the wood elements used in construction is a goal that, once achieved, will allow for an increase in the use of this material with noticeable economic consequences. The use of composite materials, such as carbon fiber, as reinforcement for wood elements under bending loads, such as beams, requires attention to reinforcement methods for various carbon fiber element layouts. Each of the amplification options can potentially lead to different results.

Architectural solutions and technological processes often limit the dimensions of the load-bearing wooden elements, which leads to the search for an effective method to increase the load-bearing capacity of the structure without changing its appearance. Requirements for the preservation of the external appearance of the structure are also imposed during the reconstruction of buildings of historical and cultural value.

One of the ways to solve the problem is to strengthen the structures. The most common wooden structures are bent elements and the issues of strengthening these elements are relevant. Typical defects for the bent elements are bioreflections, large knots in the stretched zone, poor-quality toothed spikes, non-project inserts, delaminations and longitudinal shrinkage cracks. In this regard, one of the urgent tasks is to develop methods for restoring and improving the load-bearing capacity and rigidity of bent wooden structures.

In structures operated in a chemically aggressive environment, as well as in special structures that require radio transparency and increased dielectric characteristics, the corresponding requirements are imposed on wood structures and materials used for reinforcement.

Polymer composites have high strength characteristics, low volume weight and are resistant to aggressive environmental influences. These advantages and a wide variety of composites in appearance and shape allow us to consider them as promising materials for reinforcement.

The main reason limiting the use of polymer composites for reinforcement of wooden structures is the lack of regulatory documents for the calculation and design of reinforcement both in Russia and abroad. This is due to the small amount of research in this area compared to reinforced concrete and steel structures. This indicates that the topic has not been sufficiently studied. Thus, research in this area is relevant.

The subject of the study is the stress–strain state of wood-glued beams reinforced with composite materials.

To achieve this goal the following tasks were set:

1. To analyze the world experience of development in the reinforcement of wooden and wood-glued structures.
2. Perform theoretical studies of a wooden beam structure with the justification of the physical and mathematical model of an anisotropic material.
3. Develop a method of engineering calculation of wooden beams reinforced with carbon fiber.

2 Methods

The calculation was carried out taking into account the physical nonlinearity of the wood material. The non-linearity in the software package is taken into account when setting the finite elements (plates) of stiffness.

Choosing a physical and mathematical model and ensuring the accuracy of the simulation is considered one of the most important research tasks. Modeling always involves making assumptions of varying degrees of importance. At the same time, the basic requirements for models must be met: adequacy, accuracy, versatility and reasonable economy.

The method of reinforcement proposed by the authors consists in strengthening the side faces of the reference zones. Numerical studies were carried out on wooden floor beams with standard cross-sectional dimensions of 100 × 200 mm and a length of 4.5 m. The material is pine. Reinforced with carbon unidirectional tape of the FibARM.

The design, as well as the type of cross-section of one of the variants of reinforcement of wooden beams with the location of reinforcement in the side faces of the support zones along the main compressive stresses, are shown in the form of graphic material (Fig. 1).

Based on the results of the calculations, for further analysis and evaluation of the effectiveness of reinforcement, the isofields of displacements and stresses are derived and graphs are constructed that characterize the operation of the structure. Marking of beams: B-1-wooden composite beam, B-2-wooden beam.

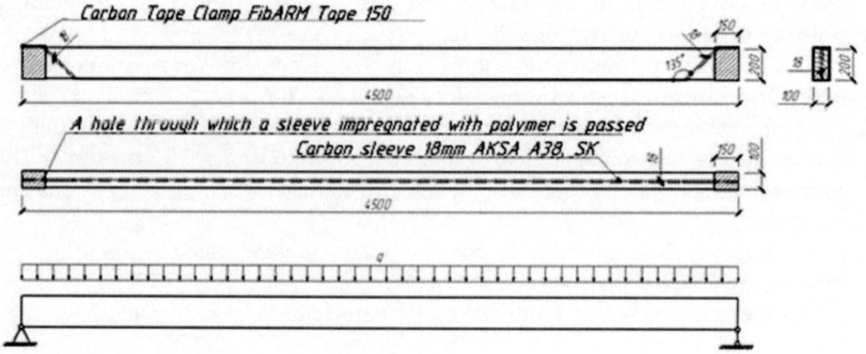

Fig. 1 General view of the design scheme of the beam

3 Results and Discussion

Results of calculations of the values of the physical and mathematical model of a wooden beam (Fig. 2). Results of calculations of the values of the physical and mathematical model of a wood composite beam (Fig. 3).

The data obtained as a result of the calculation clearly illustrates that the design load on the reinforced beam is 2 times greater than that on the non-reinforced one. For the reinforced beam, the calculated load was 1800 kg/m, for the non-reinforced one 900 kg/m. At the same time, the ultimate strength of the reinforced beam decreased by 40% relative to the non-reinforced one. Analysis (Fig. 4).

4 Conclusions

Based on the results of the research, the following conclusions can be drawn:

1. The rational variant of the reinforcing material-carbon fiber is chosen: the tensile strength—2000 MPa, the modulus of elasticity—95000 HPa, the density-400 g/m^2, the equivalent thickness—0.14 mm, the elongation—2.3%.
2. The data obtained as a result of the calculation clearly illustrates that the design load on the reinforced beam is 2 times greater than that on the non-reinforced one. For the reinforced beam, the calculated load was 1800 kg/m, for the non-reinforced one 900 kg/m. At the same time, the ultimate strength of the reinforced beam decreased by 40% relative to the non-reinforced one.
3. A method for calculating the reinforcement of wooden elements with composite material based on the geometric characteristics of the cross section is presented.
4. The possibility of using polymer composite materials for strengthening beam structures is theoretically justified and confirmed by calculation.

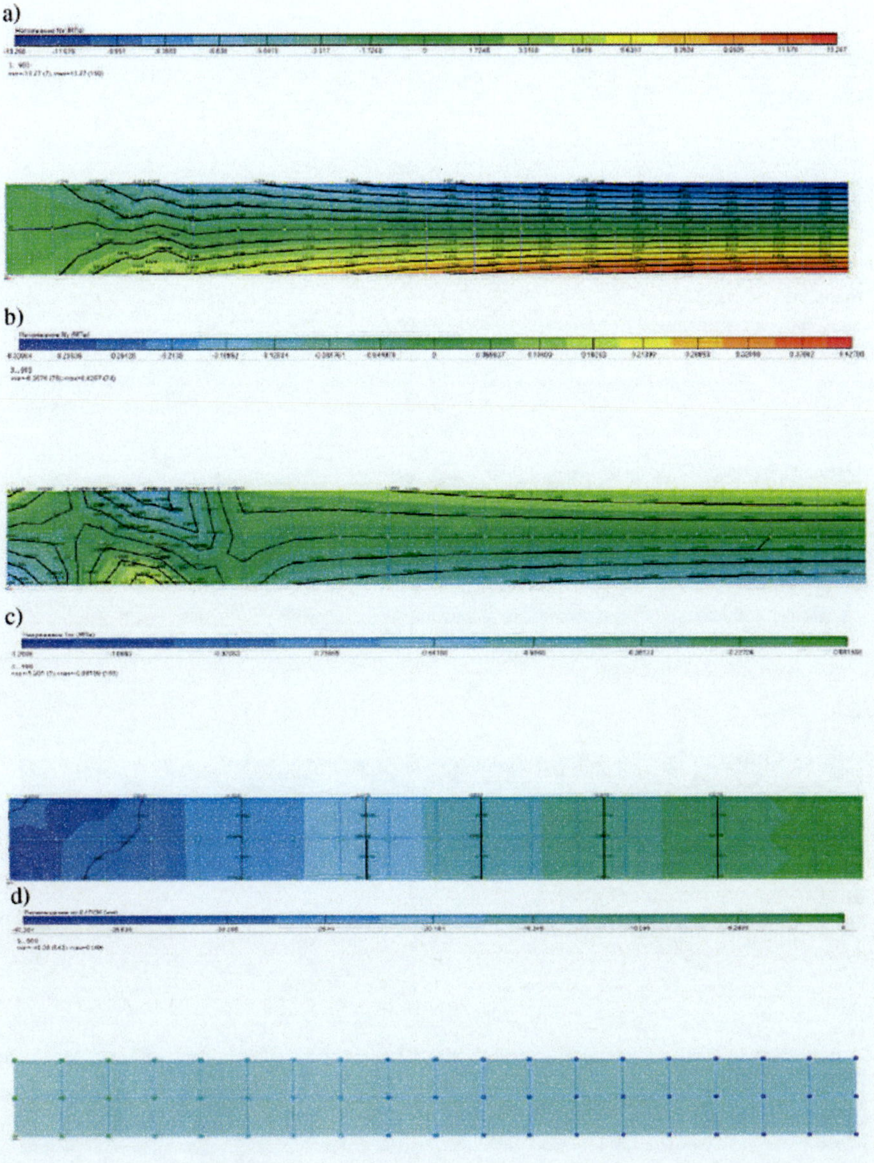

Fig. 2 Test results of calculations of the values of the physical and mathematical model of a wooden beam: **a** isofield Nx (MPa) at half of the beam span; **b** isofield Ny (MPa) at the half span of the beam; **c** isofield Txz (MPa) (tangential stresses); **d** mosaic of node movements along the Z-axis for a half-beam structure

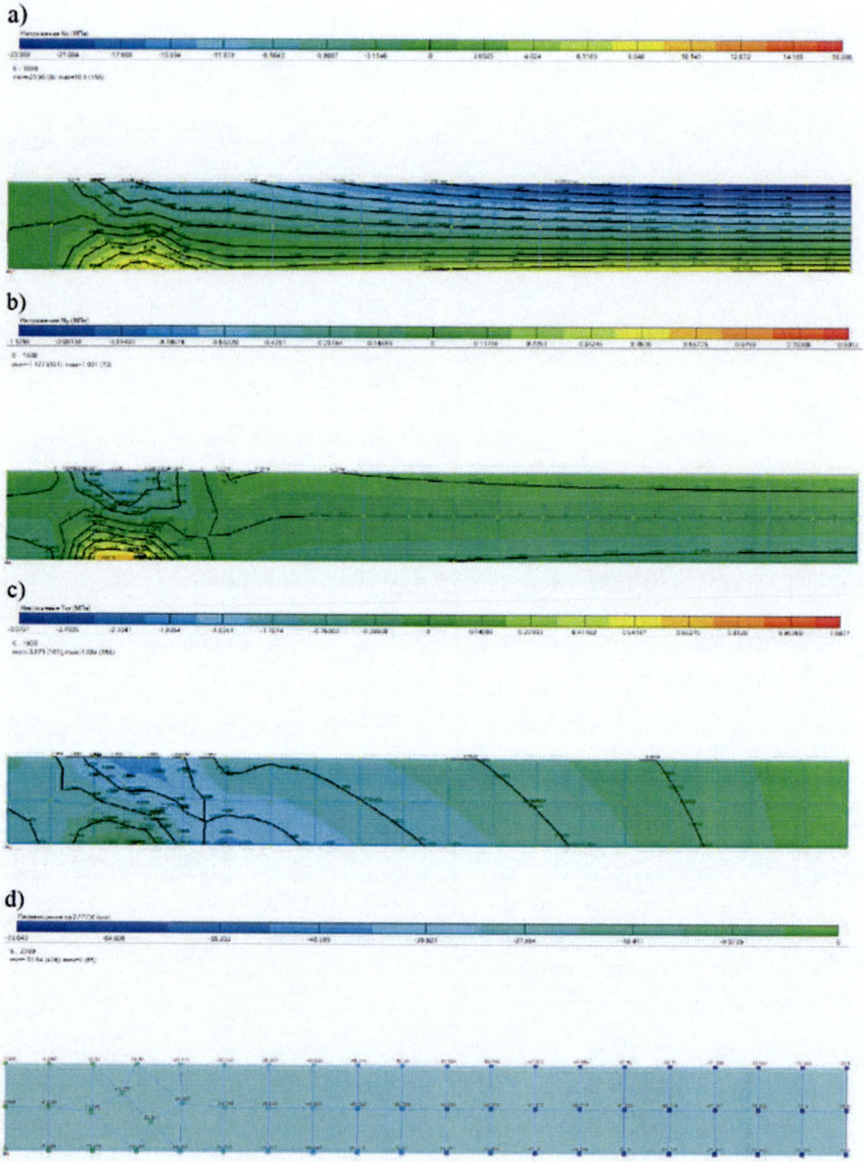

Fig. 3 Test results of calculations of the values of the physical and mathematical model of a wood composite beam: **a** isofield Nx (MPa) at half of the beam span; **b** isofield Ny (MPa) at the half span of the beam; **c** isofield Txz (MPa) (tangential stresses); **d** mosaic of node movements along the Z-axis for a half-beam structure

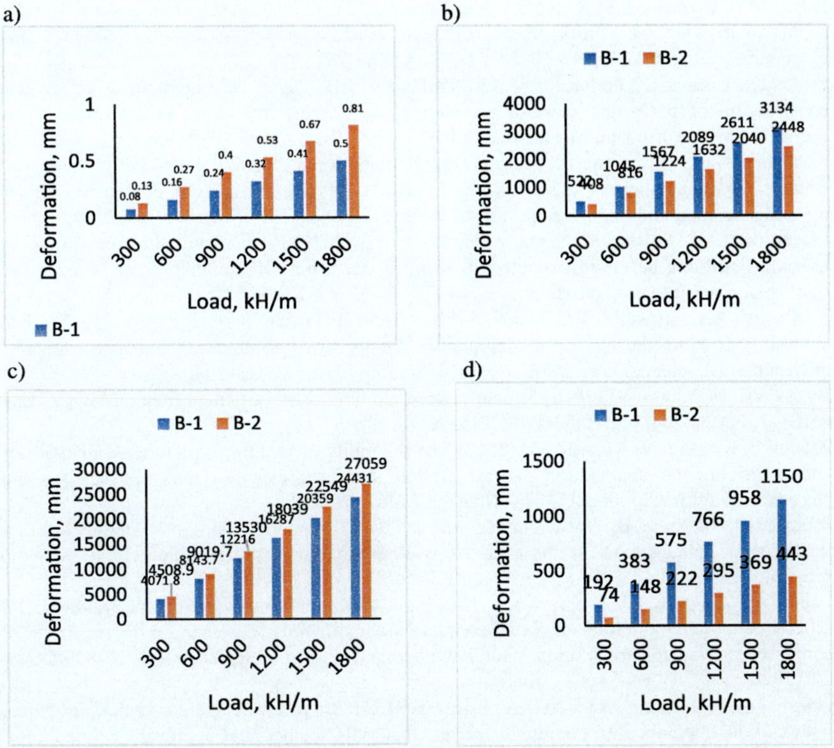

Fig. 4 Analysis of results: **a** diagram "Load—deformation" based on the results of results of calculations of the values of the physical and mathematical models; **b** diagram "Load—Tension T_{xz}" based on the results of results of calculations of the values of the physical and mathematical models; **c** diagram "Load—Tension N_x" based on the results of results of calculations of the values of the physical and mathematical models; **d** diagram "Load—Tension N_y" based on the results of results of calculations of the values of the physical and mathematical models

References

1. Gribanov AS, Roshchina SI, Naichuk AY, Melekhov VI (2020) Wooden beams with local wood modification. In: IOP conference series: materials science and engineering. https://doi.org/10.1088/1757-899X/896/1/012067
2. Turkovskij SB, Pogorel'tsev AA, Eknados'yan IL (2003) Selection of design scheme of lens-shaped trusses from adhesive wood. Stroit Mater 18–20
3. Labudin BV, Popov EV, Nikitina TA (2019) Notes for calculated resistance to tension for laminated wood. In: IOP conference series: materials science and engineering. https://doi.org/10.1088/1757-899X/687/3/033028
4. Lisyatnikov MS, Glebova TO, Ageev SP, Ivaniuk AM (2020) Strength of wood reinforced with a polymer composite for crumpling across the fibers. In: IOP conference series: materials science and engineering. https://doi.org/10.1088/1757-899X/896/1/012062

5. Lukin MV, Roshchina SI, Gribanov AS, Naychuk AY (2020) Stress-strain state of wooden beams with external reinforcement. In: IOP conference series: materials science and engineering. https://doi.org/10.1088/1757-899X/896/1/012066
6. Preobrazhenskaya IP, Pogorel'tsev AA, Turkovskij SB (2003) Development of design and construction of potassium chloride storehouse with framework from prefabricated wood frames with size of 63 m. Stroit Mater 14–16
7. Koshcheev AA, Roshchina SI, Aleksiievets V, Labudin BV (2020) Local deformation and strength characteristics of S-shaped reinforcement in wood. In: IOP conference series: materials science and engineering. https://doi.org/10.1088/1757-899X/896/1/012060
8. Lisyatnikov MS, Shishov II, Sergeev MS, Hisham E (2020) Precast monolithic coating of an industrial building based on variable-height beam-slabs. In: IOP conference series: materials science and engineering. https://doi.org/10.1088/1757-899X/896/1/012064
9. Varlamov AA, Rimshin VI, Tverskoi SY (2018) Planning and management of urban environment using the models of degradation theory. In: IOP conference series: earth and environmental science. https://doi.org/10.1088/1755-1315/177/1/012040
10. Byzov VE, Melekhov VI (2016) Structural sawn timber: resource enhancement. Mag Civ Eng 65:67–76. https://doi.org/10.5862/MCE.65.5
11. Kuzina E, Rimshin V, Kurbatov V (2018) The reliability of building structures against power and environmental degradation effects. In: IOP conference series: materials science and engineering. https://doi.org/10.1088/1757-899X/463/4/042009
12. Jocovic M, Melovic B, Vatin N, Murgul V (2014) Modern business strategy customer relationship management in the area of civil engineering. https://doi.org/10.4028/www.scientific.net/AMM.678.644
13. Labudin BV, Popov EV, Sopilov VV, Nikitina TA, Karelsky AV, Bardin IN, Vakhlov DG (2020) Structural solutions of wood-concrete floors using shear-resistant joints. In: IOP conference series: materials science and engineering. https://doi.org/10.1088/1757-899X/896/1/012028
14. Gonshakov NG, Gonshakov AG, Aleksiievets II (2020) Reinforcement of brick structures with carbon fiber. In: IOP conference series: materials science and engineering. https://doi.org/10.1088/1757-899X/896/1/012032
15. Varenik KA, Varenik AS, Sanzharovsky RS, Labudin BV (2019) Wood moisture accounting in creep equations. In: IOP conference series: materials science and engineering. https://doi.org/10.1088/1757-899X/656/1/012054
16. Naichuk AY (2020) Causes of defects in timber arches of the buildings covering and methods of strengthening. In: IOP conference series: materials science and engineering. https://doi.org/10.1088/1757-899X/896/1/012042
17. Labudin BV, Popov EV, Sopilov VV (2019) Stability of compressed sheathings of wood composite plate-ribbed structures. In: IOP conference series: materials science and engineering. https://doi.org/10.1088/1757-899X/687/3/033041
18. Petrichenko M, Vatin N, Nemova D, Kharkov N, Staritcyna A (2014) EOR (oil recovery enhancement) technology using shock wave in the fluid. https://doi.org/10.4028/www.scientific.net/AMM.627.297
19. Lukin M, Prusov E, Roshchina S, Karelina M, Vatin N (2021) Multi-Span composite timber beams with rational steel reinforcements. Buildings 11:1–12. https://doi.org/10.3390/buildings11020046
20. Usmanov R, Vatin N, Murgul V (2014) Experimental research of a highly compacted soil beds. https://doi.org/10.4028/www.scientific.net/AMM.633-634.1082
21. Myuller OD, Melekhov VI, Malygin VI (2015) Elastoplastic deformation of fine-grain media. Russ Eng Res 35:911–918. https://doi.org/10.3103/S1068798X15120151
22. Gorpinchenko VM, Pogorel'tsev AA, Ecknadosyan IL (2005) Large-scale tests provided for a block, consisting of two wooden lens-type roof trusses of the sports complex "Strogino." Promyshlennoe i Grazhdanskoe Stroit 38 (2005)

23. Labudin BV, Nikitina TA, Popov EV, Varenik KA, Novoselova VI (2020) Assessment of the anisotropic wood strength on local crushing. In: IOP conference series: materials science and engineering (2020). https://doi.org/10.1088/1757-899X/939/1/012039
24. Romanovich A, Kleshcunov Y, Vlasov A (2019) On potentiality and practicability of installing flooring suspended in geodesic domes by means of cable system. In: IOP conference series: materials science and engineering. https://doi.org/10.1088/1757-899X/687/3/033025
25. Sergeev M, Rimshin V, Lukin M, Zdralovic N (2020) Multi-span composite beam. In: IOP conference series: materials science and engineering. https://doi.org/10.1088/1757-899X/896/1/012058
26. Rimshin VI, Labudin BV, Melekhov VI, Orlov A, Kurbatov VL (2018) Improvement of strength and stiffness of components of main struts with foundation in wooden frame buildings. ARPN J Eng Appl Sci 13:3851–3856
27. Nagruzova LP, Shibaeva GN, Saznov KV, Kubanychbek Kyzy A (2019) Technology and manufacture of reinforced concrete structures with application of silica fume for multi-storey house-building in the Republic of Khakassia. In: IOP conference series: materials science and engineering. https://doi.org/10.1088/1757-899X/597/1/012029
28. Labudin B, Popov E, Stolypin D, Sopilov V (2019) The wood composite ribbed panels on mechanical joints. In: E3S Web of Conferences. https://doi.org/10.1051/e3sconf/20199102021
29. Kuntysh VB, Melekhov VI, Margolin GA, Vaisman VE (1988) Test-stand for determining the quality of contact in bimetallic finned tubes. Chem Pet Eng 24:643–646. https://doi.org/10.1007/BF01274195
30. Roshchina S, Lukin M, Lisyatnikov M (2020) Compressed-bent reinforced wooden elements with long-term load. https://doi.org/10.1007/978-3-030-42351-3_7
31. Popova M, Sergeev M, Lukina A, Shunqi M (2020) Strength and deformability of lightweight metal trusses with elements from cut I-beams. In: IOP conference series: materials science and engineering. https://doi.org/10.1088/1757-899X/896/1/012061
32. Nagruzova L, Saznov K, Aytbu KK (2019) Thermal efficient panels on a wooden frame for quickly erectable low-rise buildings. In: E3S Web of Conferences. https://doi.org/10.1051/e3sconf/201911001023

Stress–Strain State of a Wood-Glued Three-Span Beam with Layer-By-Layer Modification

M. Sergeev, A. Lukina, N. Zdralovic, and D. Reva

Abstract One of the priority areas of research is the creation of lightweight composite bendable structures based on wood. Composite multi-span wood-glued beams with the proposed reinforcement scheme are an experimental design and to date their operation is poorly studied. The essence of the proposed design is to form a glued beam in which fiberglass is placed in the adhesive seam between the boards. To perform the calculations, we set the model of a beam with three spans. The load is assumed to be evenly distributed over the entire length of the beam. A comparison of the structure to assess its rationality was carried out with an unreinforced glued beam with the same design scheme. In the course of the study, it was found that, in contrast to glued beams, the strength of composite beams with layer-by-layer modification increases by 14–17%, and the deformability decreases by 4–7%. The destruction of the beams occurs along normal sections, which eliminates the possibility of chipping and splitting in the supporting sections, i.e. ensures the reliability of the structures for the action of shear forces in the supporting sections, thereby increasing the reliability of the structure against collapse.

Keywords Composite beam · Layer-by-layer modification · Three-span beam

1 Introduction

Glued wooden structures in our time are used mainly in the construction of large sports facilities and in some cases in the construction of bridges. Taking into account the requirements of fire protection standards, they are also used for

M. Sergeev (✉) · A. Lukina · D. Reva
The Department of Building Constructions, Institute of Architecture of Construction and Energy, Vladimir State University Named After Alexander and Nikolay Stoletovs, Gorky Street 87, Vladimir 600000, Russian Federation
e-mail: sergeevmichael@inbox.ru

N. Zdralovic
Fylkesmannen I Troms Og Finnmark, Damsveien 1, 9815 Vadsø, Norway

N. Vatin et al. (eds.), *Proceedings of MPCPE 2021*, Lecture Notes in Civil Engineering 182, https://doi.org/10.1007/978-3-030-85236-8_43

industrial buildings, especially for warehouses and buildings with a chemically aggressive environment, the use of metal and reinforced concrete in which is associated with high costs for their anti-corrosion protection. One of the important advantages of wooden structures is the aesthetic component, which plays an increasingly important role in the selection of future building structures [1–5].

The scientific novelty of the article lies in the study of layer-by-layer modified glued beams, the use of which is possible both in new construction and in reconstruction. Composite wood-glued beams are an experimental design and to date their work is poorly studied [6–9].

The scientific nature of the article is traced in the study of the stress–strain state of glued beams with layer-by-layer reinforcement with fiberglass by studying the results of a numerical experiment performed in a computational complex.

2 Methods

The study is based on a glued multi-span beam, in which a layer of fiberglass is glued between each board in the package. The epoxy resin ED-20 modified by the addition of carbon nanotubes is used as an adhesive. The compound with the inclusion of carbon nanotubes allows to achieve a higher conversion, and, consequently, to obtain a more regular and frequent grid of chemical cross-links than in the original systems. Such pre-cured compositions will have a higher glass transition temperature, a higher elastic modulus, a greater tensile strain and, as a result, a higher ultimate strength. The strength of the ED-20 epoxy matrix containing carbon nanotubes increases by 6–8% during cold curing [10–13].

As a design scheme for the numerical experiment and further testing of the models, a three-span continuous beam with a single layer of fiberglass reinforcement in each seam between the boards in the beam package is adopted. The cross section of the beam for numerical modeling and further experimental research is 80×160 (h) mm. Boards for the set of cross-sections are accepted for 19 mm, in the amount of 8 pieces, which eventually allows you to dial a cross-section with a height of 160 mm. Thus, the number of layers of fiberglass is 7.

The reinforcement coefficient is set experimentally, at the first stage laying one layer of fiberglass between each plate of the recruited package of boards. In the future, the reinforcement coefficient can be varied by increasing the number of layers of fiberglass. Sizing with tape is made along the entire length of the beams without breaking (Fig. 1).

To compare the results and evaluate the feasibility of the proposed design, we will take a beam of a similar cross-section as a reference, but without layer-by-layer reinforcement with fiberglass.

Modern methods of calculating wooden structures allow us to accurately assess both the load-bearing capacity and deformability for any cross-sections and at any stage of work. When loading wood structures with an external load, three

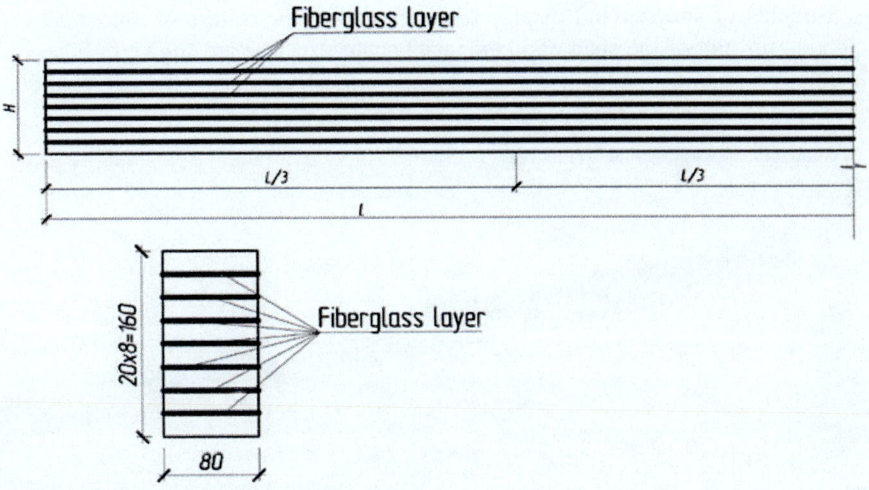

Fig. 1 Diagram of the beam in question

characteristic and successive stages of the stress–strain state are clearly manifested: conditionally elastic, elastic–plastic, and destruction.

The design and calculation of composite beam structures in the normative literature is carried out under the assumption of the elastic work of materials using the coefficients of reduction to wood of the geometric characteristics of the cross-sections of these beams [14–21]. This calculation method is applied and uses a number of assumptions, which in turn allow us to obtain compact mathematical expressions for finding the required parameters.

The finite element method is used to calculate structures in the elastic–plastic and plastic stages of the stress–strain state, as well as to visualize transient physical processes in structural elements at various loading stages. Reliable calculation results are obtained using a physically nonlinear model with the use of actual deformation diagrams of the materials used and taking into account the duration of the applied loads [21–30].

The analysis of the stress–strain state (SSS) of structures was carried out sequentially at the conditionally elastic stage, then at the elastic–plastic stage, and at the failure stage using the finite element method in the Lira software package. The calculation is carried out by the method of successive loads in the linear and physically-nonlinear formulation of the problem.

The design scheme of the structure is adopted in the form of a pivotally supported continuous three-span beam. The spans are taken at 2.0 m, the beam is loaded with a uniformly distributed load along the entire length. For the possibility of further full-scale tests on the models, a numerical experiment is performed on the cross-section sizes corresponding to the dimensions of the model beams that are available for full-scale tests.

Isofields of stresses and displacements based on the results of numerical calculation for half of the span of a composite beam structure are shown in Figs. 2, 3 and 4.

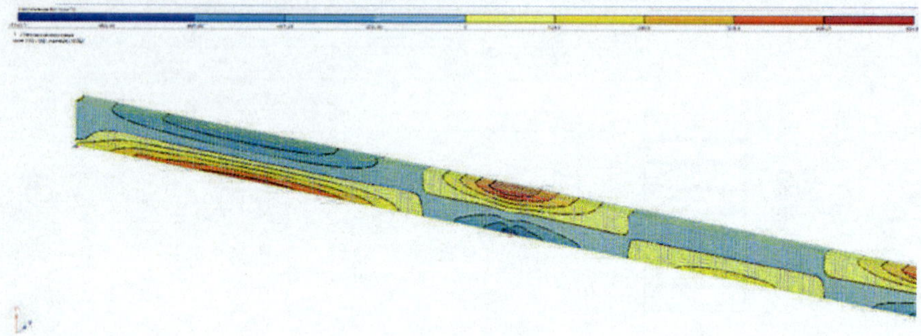

Fig. 2 Normal stress isofield σ_x, MPa for a layer-by-layer modified beam (half length)

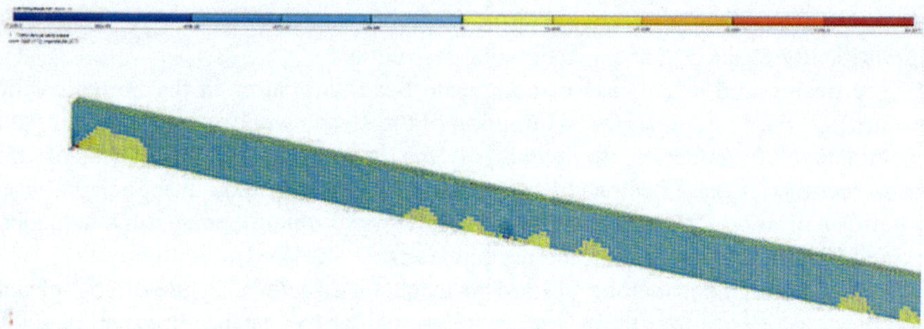

Fig. 3 Normal stress isofield σ_z, MPa for a layer-by-layer modified beam (half length)

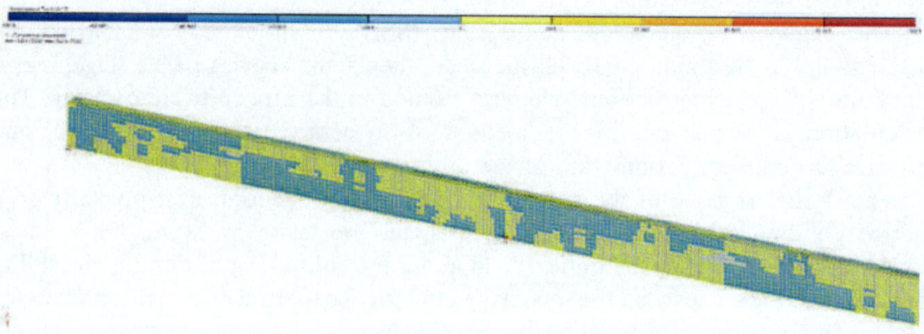

Fig. 4 Isofields of tangent stresses τ_{xy} MPa for a layer-by-layer modified beam (half length)

3 Results and Discussion

As a result of modeling, it is established that the destruction of layer-by-layer modified beams is plastic in nature, it can be stated that the destruction begins with crumpling in the compressed zone, after which stress concentrations are formed in the stretched zone. The separation of fiberglass from wood and its rupture does not occur. The destruction of composite beams occurs only in normal cross-sections, which eliminates the possibility of destruction of the proposed structure of beams from chipping and splitting in the supporting areas, i.e. provides reliable operation of structures under the action of shear forces in the support sections, thereby increasing the reliability of the structure against collapse.

In contrast to simple glued beams, the strength of layer-by-layer modified composite beams increases by 14–17%, and the deformability decreases by 4–7%.

The reliability of the results obtained is ensured by the correctness of the tasks set, the use of hypotheses and assumptions accepted in structural mechanics; modern research tools using a certified tool base; methods of conducting numerical experiments using computational programs.

4 Conclusion

By evaluating structural and technological indicators such as cross-sectional dimensions and mounting weight, it is possible to evaluate the effectiveness of load-bearing building structures, as well as by technical and economic indicators, expressed in this case in the level of consumption of basic materials, factory cost, the cost of structures in the case, the reduced costs and operational suitability. The effectiveness of composite wooden structures, namely, the proposed structure can be attributed to them, in comparison with traditional ones, is beyond doubt: reducing the cross-section of reinforced elements allows you to reduce the total volume of the structure and, as a result, the cost of enclosing structures and heating, reducing the size and weight of the elements makes it possible to more effectively solve the issues of storage, transportation and installation of structures.

Based on the results obtained during the numerical study, it can be argued about the feasibility and effectiveness of the considered method of strengthening wooden beams. According to the conducted numerical experiment, it was found that, in contrast to simple glued beams, the strength of composite beams with layer-by-layer modification increases by 14–17%, and the deformability decreases by 4–7%.

References

1. Augeard E, Michel L, Ferrier E (2018) Experimental and analytical study of the mechanical behavior of heterogeneous glulam–concrete beams and panels assembled by a specific treatment of wood. Constr Build Mater 191:812–825. https://doi.org/10.1016/j.conbuildmat.2018.10.038
2. Gribanov A, Glebova T, Roschina S (2020) Restoration of destructive wood in supporting zones of wooden beams. In: Lecture Notes in Civil Engineering. https://doi.org/10.1007/978-3-030-42351-3_14
3. Gribanov AS, Roshchina SI, Popova MV, Sergeev MS (2018) Laminar polymer composites for wooden structures. Mag Civ Eng. https://doi.org/10.18720/MCE.83.1
4. Moses DM, Prion HGL (2004) Stress and failure analysis of wood composites: a new model. Compos Part B Eng. https://doi.org/10.1016/j.compositesb.2003.10.002
5. Málaga-Chuquitaype C, Ilkanaev J (2018) Novel digitally-manufactured wooden beams for vibration reduction. Structures 16:1–9. https://doi.org/10.1016/j.istruc.2018.08.003
6. Labudin BV, Nikitina TA, Popov EV, Varenik KA, Novoselova VI (2020) Assessment of the anisotropic wood strength on local crushing. In: IOP conference series: materials science and engineering. https://doi.org/10.1088/1757-899X/939/1/012039
7. Roshchina S, Ezzi H, Shishov I, Lukin M, Sergeev M (2017) Evaluation of the deflected mode of the monolithic span pieces and preassembled slabs combined action. In: IOP conference series: earth and environmental science. https://doi.org/10.1088/1755-1315/90/1/012075
8. Lu XZ, Chen JF, Ye LP, Teng JG, Rotter JM (2009) RC beams shear-strengthened with FRP: Stress distributions in the FRP reinforcement. Constr Build Mater 23:1544–1554. https://doi.org/10.1016/j.conbuildmat.2008.09.019
9. Koshcheev AA, Roshchina SI, Aleksiievets V, Labudin BV (2020) Local deformation and strength characteristics of S-shaped reinforcement in wood. In: IOP conference series: materials science and engineering. https://doi.org/10.1088/1757-899X/896/1/012060
10. Roschina SI, Lukina AV, Sergeev MS, Vlasov AV, Gribanov AS (2016) Restoration of wooden constructions by impregnation of polymer composition on the example of industrial buildings of light and textile industry. Izv. Vyss. Uchebnykh Zaved. Seriya Teknol. Tekst. Promyshlennosti
11. Kliger IR, Haghani R, Brunner M, Harte AM, Schober K-U (2016) Wood-based beams strengthened with FRP laminates: improved performance with pre-stressed systems. Eur J Wood Wood Prod 74:319–330. https://doi.org/10.1007/s00107-015-0970-5
12. Roschina SI, Lisyatnikov MS, Lukin MV, Popova MV (2018) Technology of strengthening the supporting zones of the glued-wood beaming structure with the application of nanomodified prepregs. In: Materials science forum. https://doi.org/10.4028/www.scientific.net/MSF.931.226
13. Labudin BV, Popov EV, Tyurikova TV, Nikitina TA, Ruslanova AV (2020) Experimental studies of the strength and rigidity of screw connections of covers with wooden ribs. In: IOP conference series: materials science and engineering. https://doi.org/10.1088/1757-899X/753/4/042063
14. Lukin M, Prusov E, Roshchina S, Karelina M, Vatin N (2021) Multi-span composite timber beams with rational steel reinforcements. Buildings. https://doi.org/10.3390/buildings11020046
15. Merkulov S, Rimshin V, Akimov E, Kurbatov V, Roschina S (2020) Regulatory support for the use of composite rod reinforcement in concrete structures. In: IOP conference series: materials science and engineering. https://doi.org/10.1088/1757-899X/896/1/012022
16. Lisyatnikov MS, Glebova TO, Ageev SP, Ivaniuk AM (2020) Strength of wood reinforced with a polymer composite for crumpling across the fibers. In: IOP conference series: materials science and engineering. https://doi.org/10.1088/1757-899X/896/1/012062

17. Borri A, Corradi M (2011) Strengthening of timber beams with high strength steel cords. Compos Part B Eng 42:1480–1491. https://doi.org/10.1016/j.compositesb.2011.04.051
18. Lukin MV, Popov MV, Lisyatnikov MS (2020) Short-term and long-term deformations of the lightweight concrete. IOP Conf Ser Mater Sci Eng 753:032071. https://doi.org/10.1088/1757-899X/753/3/032071
19. Gutkowski R, Balogh J, Rogers C, Saribeiro R (2002) Laboratory tests of deep composite wood-concrete beam and deck specimens. In: Proceedings, annual conference—Canadian society for civil engineering, pp 1617–1625
20. Roshchina S, Lukin M, Lisyatnikov M (2020) Compressed-bent reinforced wooden elements with long-term load. In: Lecture Notes in Civil Engineering. https://doi.org/10.1007/978-3-030-42351-3_7
21. Lukin MV, Roshchina SI, Gribanov AS, Naychuk AY (2020) Stress-strain state of wooden beams with external reinforcement. In: IOP conference series: materials science and engineering. https://doi.org/10.1088/1757-899X/896/1/012066
22. Popova M, Sergeev M, Lukina A, Shunqi M (2020) Strength and deformability of lightweight metal trusses with elements from cut I-beams. In: IOP conference series: materials science and engineering. https://doi.org/10.1088/1757-899X/896/1/012061
23. Lukin MV, Roshchina SI, Smirnov EA, Shunqi M (2020) Strengthening of the operated wooden floor beams with external rigid reinforcement. In: IOP conference series: materials science and engineering. https://doi.org/10.1088/1757-899X/896/1/012065
24. Piao C, Cai Z, Stark NM, Monlezun CJ (2014) Dimensional stability of wood-plastic composites reinforced with potassium methyl siliconate modified fiber and sawdust made from beetle-killed trees. Eur J Wood Wood Prod 72:165–176. https://doi.org/10.1007/s00107-013-0736-x
25. Modin A, Lukin M, Vlasov A, Hisham E (2020) Energy-efficient indicators of panel housing mass construction in the climatic conditions of central Russia. In: IOP conference series: materials science and engineering. https://doi.org/10.1088/1757-899X/896/1/012063
26. Bajno D, Bednarz Ł, Nowak T (2013) Problems relating to assessment, repair and restoration of wooden roof structures in historic buildings, as exemplified by two case studies in southern Poland. https://doi.org/10.4028/www.scientific.net/AMR.778.888
27. Roshchina S, Sergeev M, Lukin M, Strekalkin A (2018) Reconstruction of Fixed fertilizer folders in the Vladimir region. In: IOP conference series: materials science and engineering. https://doi.org/10.1088/1757-899X/463/4/042011
28. Lisyatnikov MS, Roshchina SI, Chukhlanov VY (2020) The use of cenospheres for the production of spheroplastics with high dielectric characteristics, obtained from ash of thermal power plant operating on solid fuel. In: IOP conference series: earth and environmental science. https://doi.org/10.1088/1755-1315/421/7/072005
29. Biscaia HC, Chastre C, Cruz D, Viegas A (2017) Prediction of the interfacial performance of CFRP laminates and old timber bonded joints with different strengthening techniques. Compos Part B Eng 108:1–17. https://doi.org/10.1016/j.compositesb.2016.09.097
30. Roshchina S, Lukin M, Lisyatnikov M, Koscheev A (2018) The phenomenon for the wood creep in the reinforced glued wooden structures. In: MATEC Web of Conferences. https://doi.org/10.1051/matecconf/201824503020

Work of Wood-Composite Beams in Panel Floors of Prefabricated Buildings

M. Sergeev, A. Lukina, Mei Shunqi, T. Glebova, and A. Kryukov

Abstract In the study, the design of a panel wooden floor with wood-composite beams was considered. Such overlappings make it possible to perform quick installation on site because they are supplied with large-assembled elements, and the production of panels in an equipped workshop allows you to comply with the technology at a lower cost than on the construction site. The proposed design of the shield has dimensions of 6.0 × 3.0 m and is made in the form of a beam floor. The load-bearing beams are made of wood-composite, with a cross-section of 10 × 15 (h) cm with a step of 0.5 m, with sizing of the lower edge zone with glass wool on an epoxy oligomer, and the upper and lower linings are made of approximately chipboard sheets with a thickness of 16 mm. To analyze the effectiveness, a numerical study of similar structures made of wood without modifications was carried out. After analyzing the stress–strain state of structures at various stages of loading, it was found that, unlike wooden boards, the strength of boards with composite beams increases by 22–27%, and the deformability decreases by 7–10%.

Keywords Wood-composite beam · Wooden structures · Panel floors · Prefabricated buildings

1 Introduction

The development of wooden structures is inherently associated with the emergence of new structural elements with high performance parameters. Under this definition, the creation of buildings from modern modular elements that allow you to observe

M. Sergeev (✉) · A. Lukina · T. Glebova · A. Kryukov
The Department of Building Constructions, Institute of Architecture of Construction and Energy, Vladimir State University Named After Alexander and Nikolay Stoletovs, Gorky Street 87, Vladimir 600000, Russian Federation
e-mail: sergeevmichael@inbox.ru

M. Shunqi
School of Mechanical Engineering and Automation, Wuhan Textile University, Wuhan, China

N. Vatin et al. (eds.), *Proceedings of MPCPE 2021*, Lecture Notes in Civil Engineering 182, https://doi.org/10.1007/978-3-030-85236-8_44

the manufacturing technology and high speed of installation on the construction site is suitable. The Covid-19 pandemic gave another impetus to individual low-rise construction, which is what modular wooden structures are focused on [1–8].

The main indicators of the effectiveness of load-bearing building structures are both structural and technological indicators, and technical and economic indicators. The effectiveness of composite wooden structures in comparison with traditional ones is beyond doubt: reducing the cross-section of reinforced elements reduces the volume of the building and, consequently, the cost of enclosing structures and heating; reducing the width of the cross-section of the elements makes it possible to use lumber of non-deficient width; reducing the size and weight of the elements makes it possible to more effectively solve the issues of storage, transportation and installation of structures.

The scientific novelty of the article lies in the study of wooden floor boards made on the basis of wood-composite beams. The use of shields is possible in prefabricated buildings of the frame type, or in stone buildings of individual development. Panel floors with composite beams are an experimental design and to date their work is poorly studied.

The scientific nature of the article is traced in the study of the stress–strain state of a wood-composite floor panel by studying the results of a numerical experiment performed in a software calculation complex and searching for optimal design parameters.

2 Methods

The study is based on a wooden floor board, the main load-bearing elements of which are wooden beams modified with several layers of fiberglass at the lower edge zone, on an epoxy oligomer ED-20. The fiberglass tape is applied along the entire length of the beam without breaking. As a design scheme, a plate supported on two sides was taken. The load was set evenly distributed over the entire plane of the shield.

The proposed design of the shield has dimensions of 6.0 × 3.0 m and is made in the form of a beam floor. The load-bearing beams are made of wood composite, with a cross section of 10 × 15 (h) cm and laid in increments of 0.5 m. The modification of the beams consists in sizing the lower edge zone with glass wool on an epoxy oligomer. The upper and lower paneling are made of approximately 16 mm thick chipboard sheets, and mineral wool insulation is laid in the space between the beams to create sound insulation of the panel. The epoxy oligomer ED-20 was used as an adhesive composition in two variations: for modifying the beams with fiberglass with the inclusion of carbon nanotubes in the compound, for gluing the sheaths—in the form of a two-component resin. Modification of the composition of ED-20 was carried out and studied.

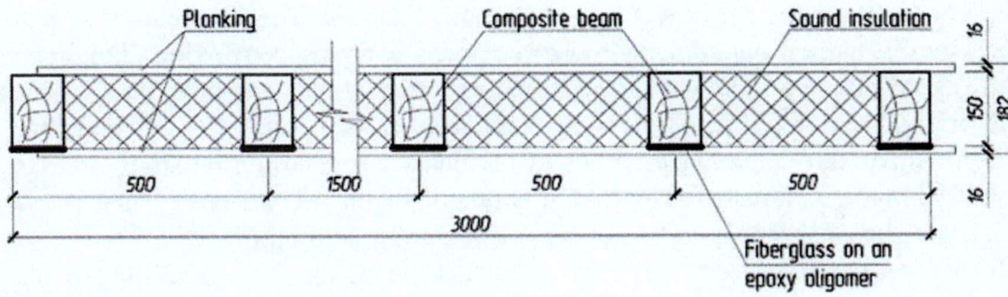

Fig. 1 Diagram of the shield under consideration

The optimal coefficient of reinforcement for such a beam structure was established in studies [9–15] and is $\mu = 1.0\ldots3.0\%$, which is achieved with different cross-sections of beams by a different number of glued layers (Fig. 1).

To analyze the effectiveness, a numerical study of similar panel structures with wooden beams without modifications was carried out.

The current methods of calculating wooden structures allow us to estimate their load-bearing capacity and deformability with sufficient accuracy for any cross-sections and at any stage of work. When loading wooden structures with an external load, three characteristic and successive stages of the stress–strain state are clearly manifested: conditionally elastic, elastic–plastic, and destruction.

The stage of conditional elastic work is characterized by the amount of deformations that do not exceed the limit values of elastic deformations of wood and rebar. When unloading reinforced elements in this stage, the residual deformations are absent or so insignificant that they can be ignored. Due to the fact that even at low stresses, the linear relationship between the stresses and deformations of wood is somewhat violated, the first stage of the stress–strain state can only be considered as conditionally elastic. The stage of elastic–plastic work is characterized by the appearance of noticeable plastic deformations in the compressed wood fibers, and then in the compressed reinforcement. In the compressed part of the cross-section, a plastic zone is formed, which spreads with increasing load into the depth of the cross-section. There is a redistribution of forces, as a result of which the neutral layer is shifted towards the stretched fibers. The deformations of the stretched wood fibers increase to the value of the proportionality limit. When the element is unloaded at this stage, significant residual deformations appear. The failure stage is characterized by a significant increase in the deformability of the reinforced element with a small increase in the load. Plastic deformations get the maximum development. There is a destruction of the element, the nature of which depends on the type of reinforcement [16–23].

The paper presents a detailed analysis of the stress–strain state (VAT) of structures at all stages of operation using the finite element method in the Lira software package. The calculation is performed in a linear and physically-nonlinear formulation of the problem by the method of successive loads. The reinforcement is made of fiberglass on an epoxy matrix with the inclusion of carbon nanotubes along the entire length of the lower edge zone in 4 layers. The compound with the inclusion of carbon nanotubes allows to achieve a higher conversion, and, consequently, to obtain a more regular and frequent grid of chemical cross-links than in the original systems. Such pre-cured compositions will have a higher glass transition temperature, a higher modulus of elasticity, a greater tensile strain and, as a result, a higher ultimate strength. The strength of the ED-20 epoxy matrix containing carbon nanotubes increases by 6–8% during cold curing.

The design scheme of the structure is adopted in the form of a hinged plate supported on two sides. The span is assumed to be 6.0 m, the plate is loaded with a uniformly distributed load over the entire area. The numerical experiment is performed on the cross-section dimensions corresponding to the dimensions of the shield model, which are available for full-scale tests. Isofields of stresses and displacements based on the results of numerical calculation are presented on the example of a single beam, the design of the sheaths is not conditionally shown. The beam stresses in the shield are shown in Figs. 2, 3 and 4.

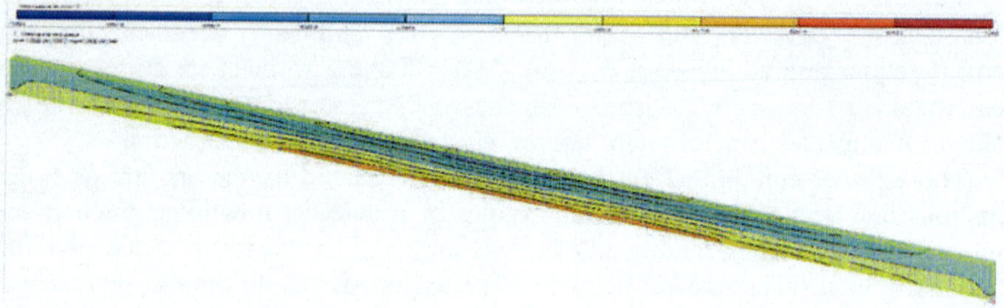

Fig. 2 Normal stress isofield σ_x, MPa of the beam as part of the panel floor

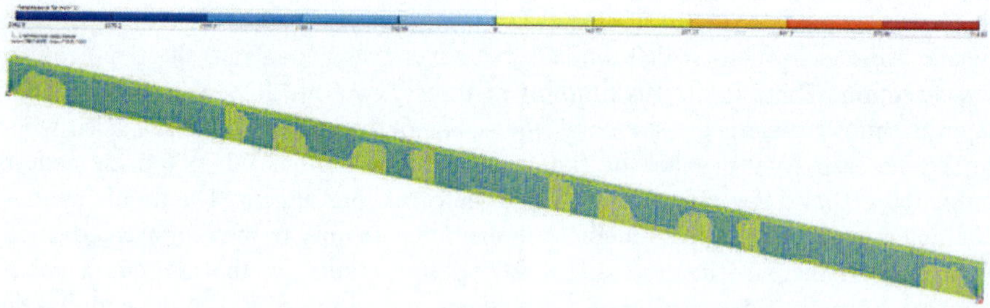

Fig. 3 Normal stress isofield σ_z, MPa of the beam as part of the panel floor

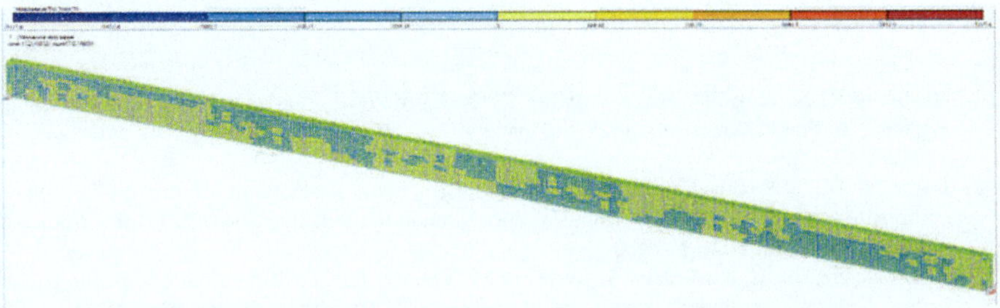

Fig. 4 Isofields of tangent stresses τ_{xy} MPa of the beam as part of the panel floor

3 Results and Discussion

During the tests, it was found that the destruction of shields with composite beams is plastic in nature, in contrast to shields with solid wood beams. As a result of modeling, it is established that the destruction of shields with composite beams is plastic in nature, it can be stated that the destruction begins with crumpling in the compressed zone, after which stress concentrations are formed in the stretched zone in the places of defects. The separation of fiberglass from wood and its rupture does not occur. Destruction occurs only in normal cross-sections.

In contrast to wooden boards, the strength of boards with composite beams increases by 22–27%, and the deformability decreases by 7–10%.

The reliability of the results obtained is ensured by the correctness of the tasks set, the use of hypotheses and assumptions accepted in construction mechanics; modern research tools using a certified tool base; methods for conducting numerical experiments using computational programs.

4 Conclusion

The construction of panel housing allows you to build buildings in a short time, and the main manufacturing processes are organized on equipped factory sites. The panel in question is part of such a home kit.

After analyzing the stress–strain state of the structure of the floor panel with composite beams at various stages of loading, it was found that, in contrast to wooden boards, the strength of boards with composite beams increases by 22–27%, and the deformability decreases by 7–10%. During the tests, it was found that the destruction of shields with composite beams is plastic in nature, in contrast to shields with solid wood beams. T. k. the manufacturing process takes place in the factory, for the production of composite beams of the shield, it is proposed to use the method of vacuum infusion, which is proven, which has a positive effect on the technical and economic indicators.

References

1. Vasiljevs R, Serdjuks D, Buka-Vaivade K, Podkoritovs A, Vatin N (2020) Load-carrying capacity of timber-concrete composite panels. Mag Civ Eng 93:60–70. https://doi.org/10.18720/MCE.93.6
2. Demertzi M, Silvestre JD, Durão V (2020) Life cycle assessment of the production of composite sandwich panels for structural floor's rehabilitation. Eng Struct 221. https://doi.org/10.1016/j.engstruct.2020.111060
3. Lukin M, Prusov E, Roshchina S, Karelina M, Vatin N (2021) Multi-span composite timber beams with rational steel reinforcements. Buildings. https://doi.org/10.3390/buildings11020046
4. Gonshakov NG, Gonshakov AG, Aleksiievets II (2020) Reinforcement of brick structures with carbon fiber. In: IOP conference series: materials science and engineering. https://doi.org/10.1088/1757-899X/896/1/012032
5. Roschina S, Gribanov A, Lukin M, Lisyatnikov M, Strekalkin A (2018) Calculation of wooden beams reinforced with polymeric composites with modification of the wood compression area. In: MATEC web of conferences.https://doi.org/10.1051/matecconf/201825104029
6. Woschitz R, Deix K, Huber C, Kampitsch T (2021) Entwicklung neuartiger Holz-Betonverbunddecken in Fertigteilbauweise.https://doi.org/10.1002/bate.201800097
7. Roshchina SI, Gonshakov AG, Gonshakov NG (2020) Taking into consideration the work of monolithic reinforced concrete floor with shaped steel profiles in two directions. In: IOP conference series: materials science and engineering.https://doi.org/10.1088/1757-899X/753/4/042084
8. Kuzina E, Rimshin V (2019) Strengthening of concrete beams with the use of carbon fiber. https://doi.org/10.1007/978-3-030-19868-8_90
9. Lisyatnikov MS, Roshchina SI, Chukhlanov VY (2020) The use of cenospheres for the production of spheroplastics with high dielectric characteristics, obtained from ash of thermal power plant operating on solid fuel. In: IOP conference series: earth and environmental science.https://doi.org/10.1088/1755-1315/421/7/072005
10. Pacchioli S, Pozza L, Trutalli D, Polastri A (2021) Earthquake-resistant CLT buildings stiffened with vertical steel ties. J Build Eng 40. https://doi.org/10.1016/j.jobe.2021.102334
11. Koshcheev AA, Roshchina SI, Naichuk AY, Vatin NI (2020) The effect of eccentricity on the strength characteristics of glued rods made of steel cable reinforcement in solid wood. In: IOP conference series: materials science and engineering.https://doi.org/10.1088/1757-899X/896/1/012059
12. Sergeev M, Rimshin V, Lukin M, Zdralovic N (2020) Multi-span composite beam. In: IOP conference series: materials science and engineering.https://doi.org/10.1088/1757-899X/896/1/012058
13. Aleksiievets VI, Aleksiievets II, Ivaniuk AM, Roshchina SI (2020) Load-carrying capacity of bolted joints of timber structures under static loading. In: IOP conference series: materials science and engineering.https://doi.org/10.1088/1757-899X/896/1/012043
14. Romanov PG, Sivtsev PV (2021) Deformation features of cross-glued wooden panel structures for northern construction. In: IOP conference series: earth and environmental science.https://doi.org/10.1088/1755-1315/625/1/012018
15. Sergeev MS, Gribanov AS, Roschina SI (2020) The stress strain state of composite multi-span beams. In: IOP conference series: materials science and engineering. https://doi.org/10.1088/1757-899X/753/3/032068
16. Gribanov AS, Strekalkin AA, Kudryatseva AA, Zdralovic N (2020) CFRP composites for strengthening wooden structures. In: IOP conference series: materials science and engineering.https://doi.org/10.1088/1757-899X/896/1/012114
17. Byzov VE, Melekhov VI (2016) Structural sawn timber: resource enhancement. Mag Civ Eng 65:67–76. https://doi.org/10.5862/MCE.65.5

18. Gribanov AS, Roshchina SI, Naichuk AY, Melekhov VI (2020) Wooden beams with local wood modification. In: IOP conference series: materials science and engineering.https://doi.org/10.1088/1757-899X/896/1/012067
19. Kukk V, Kers J, Kalamees T (2021) Field measurements and simulation of an massive wood panel envelope with ETICS. Wood Mater Sci Eng 16:27–34. https://doi.org/10.1080/17480272.2020.1712738
20. Lisyatnikov MS, Shishov II, Sergeev MS, Hisham E (2020) Precast monolithic coating of an industrial building based on variable-height beam-slabs. In: IOP conference series: materials science and engineering.https://doi.org/10.1088/1757-899X/896/1/012064
21. Koshcheev AA, Roshchina SI, Lukin MV, Lisyatnikov MS (2018) Wooden beams with reinforcement along a curvilinear trajectory. Mag Civ Eng. https://doi.org/10.18720/MCE.81.19
22. Lilge DS, Haselein CR, Santini EJ (2021) Performance evaluation of wood-cement panels manufactured with two forest species in combination with rice husk | Avaliação do desempenho de painéis cimento-madeira fabricados com duas espécies florestais em combinação com casca de arroz. Cienc Florest 31:20–42. https://doi.org/10.5902/1980509810859
23. Roschina SI, Lisyatnikov MS, Koshcheev AA (2019) Technical- and- economic efficiency of reinforced wooden structures. IOP Conf Ser Mater Sci Eng 698:022005. https://doi.org/10.1088/1757-899X/698/2/022005

Rigid Joint of Bent Glued Laminated Timber Structures Using Inclined Glued-In Rods

Anatoliy Naichuk, Aleksandr Pogoreltsev, Igor Demchuk, Andrii Ivaniuk, and Svetlana Roshchina

Abstract The study of the stress–strain state of a bent rigid joint was carried out by solving a volumetric problem of the theory of elasticity using a software package based on the finite element method, which makes it possible to take into account the anisotropy of the mechanical properties of wood, the physical nonlinearity of materials, and the contact failure between the element surfaces. Experimental studies were carried out by the method of four-point static loading of a full-size specimen of a rigid joint, which in its geometric parameters and material properties corresponded to the FE-model. As a result of the performed studies, it was found that the height of the compression joint zone depends on the strain capacity of the embedded parts connection, the level of stresses in the connecting plates and the elastic characteristics of timber. The strain capacity of the embedded parts connection in the compressive zone is greater than the strain capacity of the embedded parts connection in the tensile zone. The distribution of stresses along the length of the inclined reinforcing bars is characterized by great unevenness. The coefficient of non-uniformity of normal stresses between the inclined reinforcing bars of the compressive zone connection is in the range of $0.61 \div 0.38$, and in the tensile zone —$0.7 \div 0.51$.

A. Naichuk
Brest Technical University, Brest, Belarus

A. Pogoreltsev
Central Research Institute of Building Structures Named After V.A. Kucherenko, JSC Research Center "Construction", Moscow, Russian Federation

I. Demchuk
RUE "Institute Belniis"—Scientific and Technical Centre, Arkhangelsk, Belarus

A. Ivaniuk
National University of Water and Environmental Engineering, Rivne, Ukraine

S. Roshchina (✉)
Vladimir State University Named After Alexander and Nikolay Stoletovs, Vladimir, Russian Federation
e-mail: rsi3@ mail.ru

N. Vatin et al. (eds.), *Proceedings of MPCPE 2021*, Lecture Notes in Civil Engineering 182, https://doi.org/10.1007/978-3-030-85236-8_45

Keywords Timber · Glued laminated timber · Glued-in rods · Structure · Joint · Strength · Load · Stress

1 Introduction

Construction of the large-span buildings and civil engineering works using glued laminated timber structures often demands additional jointing of elements performed at the construction site. The necessity for enlargement joints is most often due to the impossibility of manufacturing a lengthy structure or transporting such structural element to the construction site. The choice of the appropriate construction for the jointing unit depends on both the values of the forces and exposure conditions of the structure. The most common joints used in such units are those with the use of dowel pins, screws and glued-in rods. The first two types of joints are widely used both in foreign and domestic practice in the construction of large-span buildings, as well as girders of road and pedestrian bridges. As for the joints of timber structures with glued-in rods, their research and widespread use in large-span building structures began at the end of the last century in the former USSR, then in Russia and Belarus, and since the 2000s it expanded abroad [1–18].

Any given type of joint used to connect the elements of timber structures has both its advantages and disadvantages. One of the disadvantages of the dowelled connections is their relatively low passive fire protection concerning structural elements, which, as a rule, have significant cross-sectional dimensions in large-span structures. Besides, such joints undergo considerable flexibility changes over time that can significantly affect the distribution of forces between structural elements. For the joints using the glued-in rods, which perceive the action of shear forces, their ductility is much lower in comparison the dowelled or screw connections, and the bearing capacity with the same diameters and length of the embedment is higher [15, 19].

An analysis of the works [7–14] allows noting that in foreign construction practice, glued-in rods are mainly used to strengthen structures and increase the resistance of the base platform to compression of wood perpendicular to the grain, as well as the platforms in a place where concentrated forces are transferred. To reinforce the base platforms, threaded steel rods are mainly used. As for the joints of structural elements using glued-in rods, there are only a few examples [9]. Despite the significant amount of the research carried out [7–14], the bonding technique using glued-in rods has a rather limited area of application in foreign construction practice. A limiting factor in the use of joints with glued-in rods is the increased requirements for quality control in the process of glued rods into structural elements.

Over the past 30 years, more than 300 buildings and structures have been built on the territory of the Russian Federation and the Republic of Belarus (warehouses for mineral fertilizers with spans up to 90 m, pedestrian bridges, roofs for water parks, ice arenas, athletics and equestrian arenas, markets, stadiums with spans up to 100 m, buildings with the use of three-hinged frames with spans up to 40 m) [1–6, 16–18]. The joints of the timber structural elements were made with the usage of glued-in rods (Fig. 1). As a result of monitoring the existed buildings and civil engineering works, no damage was found in the joining connections that confirms their sufficient reliability and safety [5, 6, 16, 18]. It should be noted that the steel reinforcing bars glued into pre-drilled holes with a two-component epoxy adhesive were used as the rods -connectors and cement was used as filler.

The accumulated forty years of experience in the design and exploitation of joints with glued-in rods and the continued improvement of the production technology made it possible to create a system of joints, and the obtained results of the study allowed developing calculation models that were included in the regulatory documents [20, 21]. According to [20, 21] the optimal angle of inclination of the glued-in rods concerning the direction of the fibers of the connected elements is an angle from 300 to 450 in the multiple bar joints. The value of this angle depends on the type of stress–strain state of the connected elements. It should be noted that for joints with glued-in rods, improving their construction solutions, manufacturing technology, and increasing fire resistance are still relevant. Moreover, the improvement of design models that allow taking into account the type of stress–strain state of the connected structural elements is still an actual problem.

Fig. 1 Different type of roofs made of glued laminated timber with joints using glued-in rods [6]. **a** and **b** potassium chloride warehouse of St. Petersburg seaport 63 × 300 × 45 m (2003); **c** public center of Lipki holiday home in the Moscow region with a diameter of up to 100 m with multi-span continuous beams (1985); **d** oval dome of a sports arena in St. Petersburg with a diameter 48 × 68 m (2018)

The purpose of this paper is to present an improved design model of a rigid joint with glued-in rods. To achieve this goal, experimental and theoretical studies of a rigid joint consisted of unidirectional inclined glued-in steel rods designed according to [21] were carried out. During the research, the stress–strain state relations of timber in the area of location of inclined glued-in reinforcing bars, the ductility of the connection, the depth of the compression zone, the width of the joint opening were registered.

Theoretical studies included the determination of the stress–strain state of the elements of a designed rigid bent joint using a software package based on the finite element method. The FE-software takes into account the anisotropy of the mechanical properties of timber, the physical nonlinearity of deformation of materials and contact failure. Experimental studies were carried out by the method of four-point static loading of a specimen of a rigid joint, which is similar in its geometric parameters and properties of materials to the design model adopted in the performance of numerical calculations. The results of the experimental studies were compared with the data obtained in the process of numerical calculations.

2 Construction of the Joint

The full-size specimen of a rigid joint was designed in accordance with the requirements of [20, 21] and consisted of two rectilinear elements LWB1 and LWB2 and cross-section b × h = 200 × 1700 mm. Glued laminated timber conforms to the strength of class GL 24c. The total length of the specimen L = 12,900 mm was assigned under the condition of excluding the effect of applied concentrated loads on the behaviour of unidirectional inclined glued reinforcing bars Ib1 located in the rigid joint zone (Fig. 2).

The connection of timber elements LWB1 and LWB2 is made by means of two metal plates Sp1 and Sp2, welded to the embedded parts MD1 and MD2 (Fig. 3).

Each anchored plate was a steel welded T-cross section element connected to three unidirectional inclined reinforcing bars Ib1 (Fig. 3). The diameter of the rods Ib1 was 22 mm and the length was 1154 mm. The inclined bars Ib1 were connected to wall 2 of the embedded part MD1 or MD2 utilizing electric welding. The thickness and width of flange 2 of the T-bar of anchored plates MD1 and MD2 were 20 × 70 mm respectively, and the height and thickness of wall 1 were 60 × 20 mm respectively. The angle of inclination of the reinforcing bars Ib1 was 30° relative to the longitudinal axis of the T-anchored plates MD1 or MD2. Anchored plates MD1 and MD2 were connected to the elements LWB1 and LWB2 by gluing the inclined reinforcing bars Ib1 into the pre-drilled holes. Rebar holes Ib1 were drilled in the elements LWB1 and LWB2 at an angle of 30° to the fiber

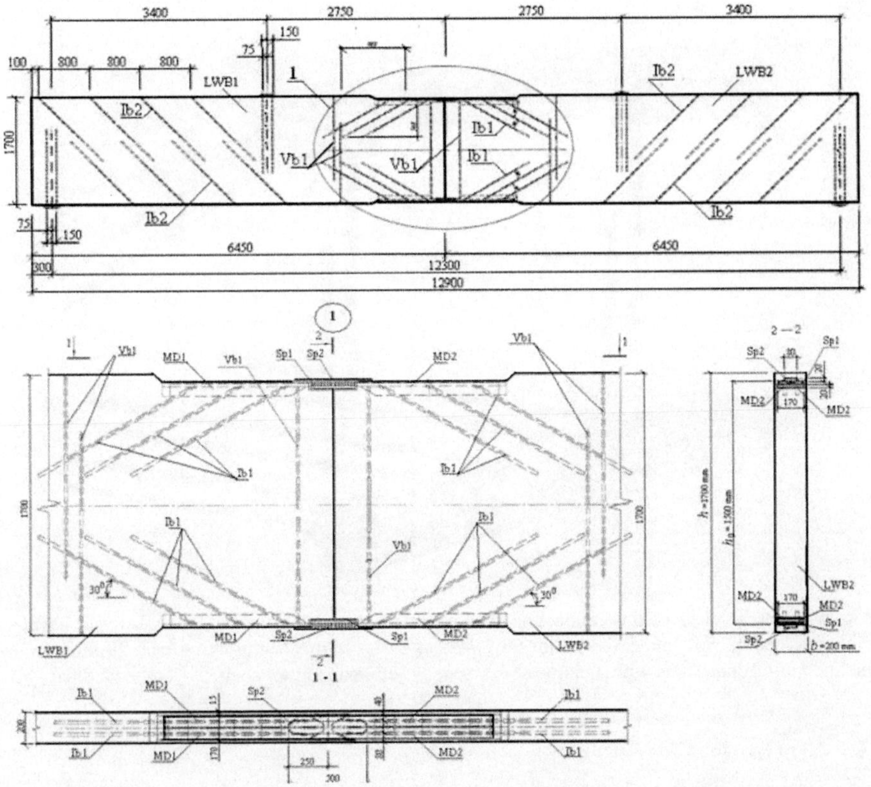

Fig. 2 Scheme of a rigid joint specimen made using unidirectional inclined glued-in rods. LWB1 and LWB2—glued laminated timber elements; Sp1 and Sp2—connecting steel plates; MD1 and MD2—anchored plates; Vb1—vertical reinforcing bars; Ib1 and Ib2—inclined reinforcing bars

direction. The Ib1 rods were glued in using a two-component epoxy adhesive. Anchored plates MD1 and MD2 were installed both on the lower and upper edges of the elements LWB1 and LWB2 (two anchored plates were installed along the width of the section of each timber element). To prevent the possible a cracking in the timber of LWB1 and LWB2 at the level of the end of the tensile inclined bars Ib1 and the neutral joint axis, caused by the timber tension perpendicular to the fibers, vertical bars Vb1 were installed (Fig. 2).

Two vertical bars Vb1 with a diameter 20 mm were installed at the edge of the fracture of the inclined bars Ib1 and two vertical bars Vb1 at the contacting ends of the elements LWB1 and LWB2 (Figs. 2 and 3).

Reinforcing bars Vb2, glued in at an angle 450 (Fig. 2) with a diameter 22 mm in the elements LWB1 and LWB2, were installed to increase the shear resistance of timber elements to shear at supports during testing of a rigid joint.

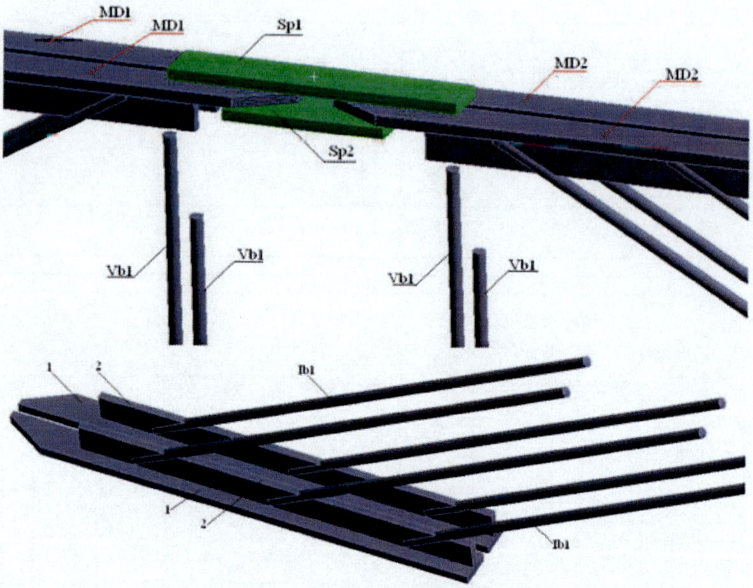

Fig. 3 General view of the connection of embedded parts MD1 and MD2. MD1 and MD2—embedded parts; Sp1 and Sp2—connecting plates; Vb1—vertical reinforcing bars 20 mm; Ib1—inclined reinforcing bars with a diameter 20 mm; 1—embedded part shelf; 2—embedded part wall

3 Methods

The object of the study is a rigid joint designed by the requirements [21]. The ultimate value of the applied load Fmax = 473 kN is determined based on the material strength of the connecting plates Sp1 and Sp2 under tension. The research was carried out in two stages. At the first stage depending on the value of the applied load F the stress–strain state of the connected elements was determined using FE-modeling. At the second stage, the ductility of the connected elements and the actual value of the joint bearing capacity Rsup were determined by the static loading method.

3.1 Finite Element Modeling

The study of the stress–strain state of a rigid joint was carried out by solving a volumetric problem of the theory of elasticity using a software package based on the finite element method (FEM), which allows taking into account the anisotropy of mechanical properties of wood, physical nonlinearity of materials, contact failure between the surfaces of the elements LWB1, LWB2 and embedded parts MD1,

MD2, as well as the surfaces of the elements LWB1 and LWB2. The coefficient of friction between the contacting surfaces of the elements was taken to be zero. The calculations were performed for two variants of the joint models (Fig. 4). A distinctive feature of the model of the first version from the second one was that in the first version there were no vertical reinforcing bars Vb1 (Fig. 4).

In the FE-model, timber was modeled with an orthotropic material, reinforcing bars and anchored plates—with an isotropic material made of steel. The properties of materials are shown in Table 1.

In the process of solving the problem, the deformation characteristics of the materials of the joint elements were corrected in accordance with the diagrams shown in Fig. 5.

One of the supports of the joint design model was modeled as movable, and the second was pinned with displacement restrictions in the direction of the Z axis. The concentrated forces F was transmitted through the distribution plates during loading.

3.2 Bending Test

A full-size specimen of the joint (Fig. 6), equivalent in shape, size and materials used for FE-model was made for testing.

In the specimen production, GL24c glued laminated timber was used. The inclined Ib1 and vertical bars Vb1 were made of reinforcing steel of class B500,

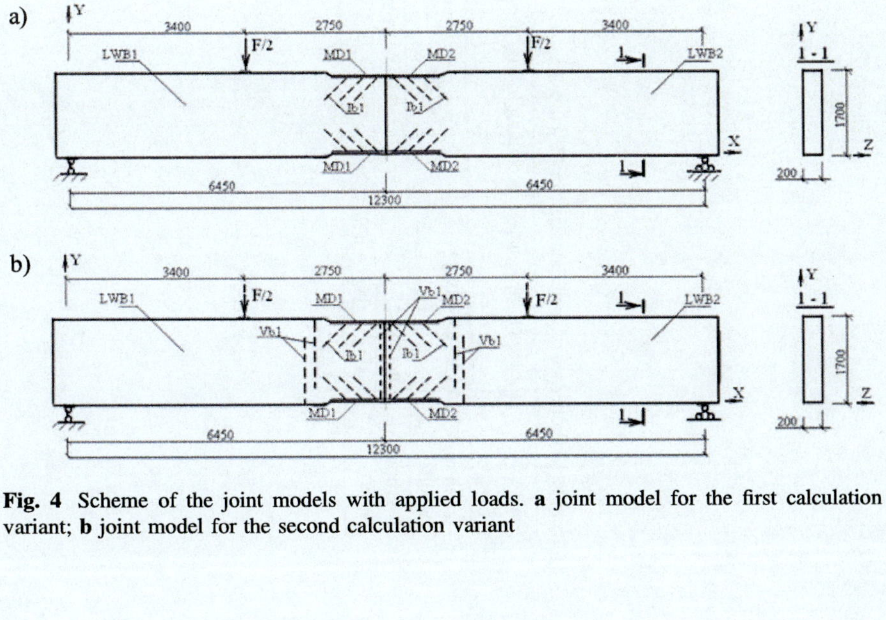

Fig. 4 Scheme of the joint models with applied loads. **a** joint model for the first calculation variant; **b** joint model for the second calculation variant

Table 1 Properties of materials used in FE-modeling

Material	Element designation according to Fig. 3	Elastic modulus values, MPa			Shear modulus values, MPa			Poisson's ratios		
		E_x	E_y	E_z	G_{xy}	G_{yz}	G_{xz}	υ_{xy}	υ_{yz}	υ_{xz}
Orthotropic (wood)	LWB1, LWB2	11,000	400	300	650	67	690	0.43	0.5	0.5
Isotropic (steel)	Ib1, Vb2	235,000	235,000	235,000	90,385	90,385	90,385	0.3	0.3	0.3
	MD1, MD2, Sp1 и Sp2	200,000	200,000	200,000	76,923	76,923	76,923	0.3	0.3	0.3

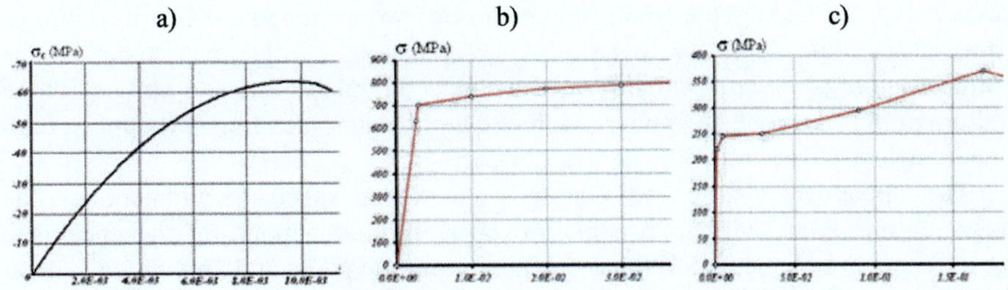

Fig. 5 Stress–strain relations for materials. **a** for wood when compressed along the grain [23]; **b** for reinforcing steel B500 [24]; **c** for steel [25]

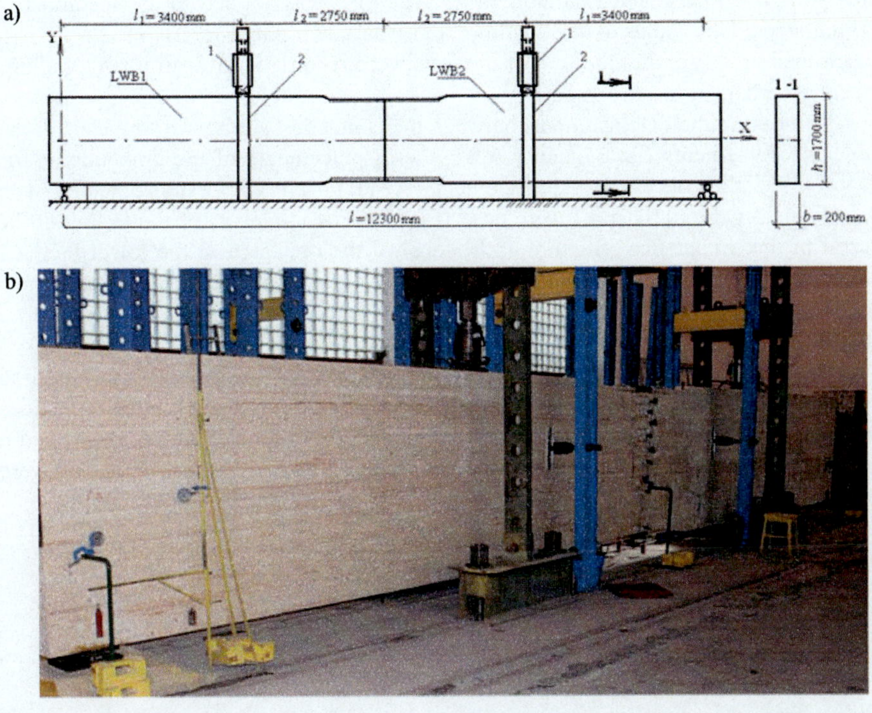

Fig. 6 Scheme and general view of testing the specimen joint. LWB1 and LWB2 – glued laminated timber elements; **a** hydraulic jacks DG100; **b** steel frames anchored in the load-bearing floor

embedded parts MD1, MD2, connecting plates Sp1 and Sp2 were made of steel 345. The moisture content of the glued laminated timber was 12%.

At the first stage of testing, the maximum value of the load applied to the specimen was Fmax = 473 kN. After reaching the load value Fmax = 473 kN and fixing deformations for all instruments, the tested specimen was unloaded. At the second stage of testing, the connecting plates Sp1 were strengthened with additional steel plates. The necessity to perform strengthening of the joint is due to the elimination of plastic deformations in the connecting plates Sp1 and Sp2 before the failure of the specimen by timber. At this stage of testing, loading was applied until the complete failure of the joint was obtained.

The specimen testing was carried out using special equipment of the load-bearing floor, which provides the possibility of supporting the specimen, unfastening the elements LWB1 and LWB2 from the plane and applying a vertical load F to it (Fig. 6).

Loading with a vertical load F was carried out using two hydraulic jacks of the DG100 brand, connected to one pumping station. The value of the applied force F was controlled according to the readings of a force meter installed in the system of the pumping station and hydraulic jacks. The load was applied to the specimen in equal steps. The value of the loading step was taken equal to $\Delta F = 62.4$ kN. At each loading stage, the duration of the tests were recorded. All instrument readings were recorded in the test report.

To measure the displacements between the contacting surfaces of the ends of the connected elements LWB1 and LWB2, the displacements of the embedded parts MD11, MD2 concerning the elements LWB1, LWB2, extensometers with mechanical gage (0.01 mm) were used (Fig. 7). The strains ε_y of the timber at the level of the timber fibre fracture at the level of the breakage of the lower inclined rods (Fig. 7b, d), were measured using electrical gages with a 20 mm base, glued across the grain on both sides of the elements LWB1 and LWB2 connected to a digital measuring complex.

The strains ε_x of the timber located in the contact zone (Fig. 7c) were measured using electrical gages glued parallel to the grain along with the height on both sides of the elements LWB1 and LWB2 at the line of their contact and connected to a digital measuring complex. The strains ε_x in steel plates Sp1 and Sp2 (Fig. 7e) were obtained similarly. The duration of the tests was recorded using a stopwatch.

4 Construction of the Joint

4.1 Results of the Numerical Investigation

In the process of numerical investigations, the load F varied from 157 to 1400 kN (Fig. 8).

The following parameters were determined depending on the load value F:

- distribution of normal stresses σx along with the height of the cross-section 2-2, which is located at the contacting ends of the elements LWB1 and LWB2;
- the width of the joint opening δ_x;
- distribution of normal stresses σ_{y1} along the length of the inclined bars Ib1;
- distribution of normal stresses $\sigma_{s,x}$ in the connecting plates Sp1 and Sp2;

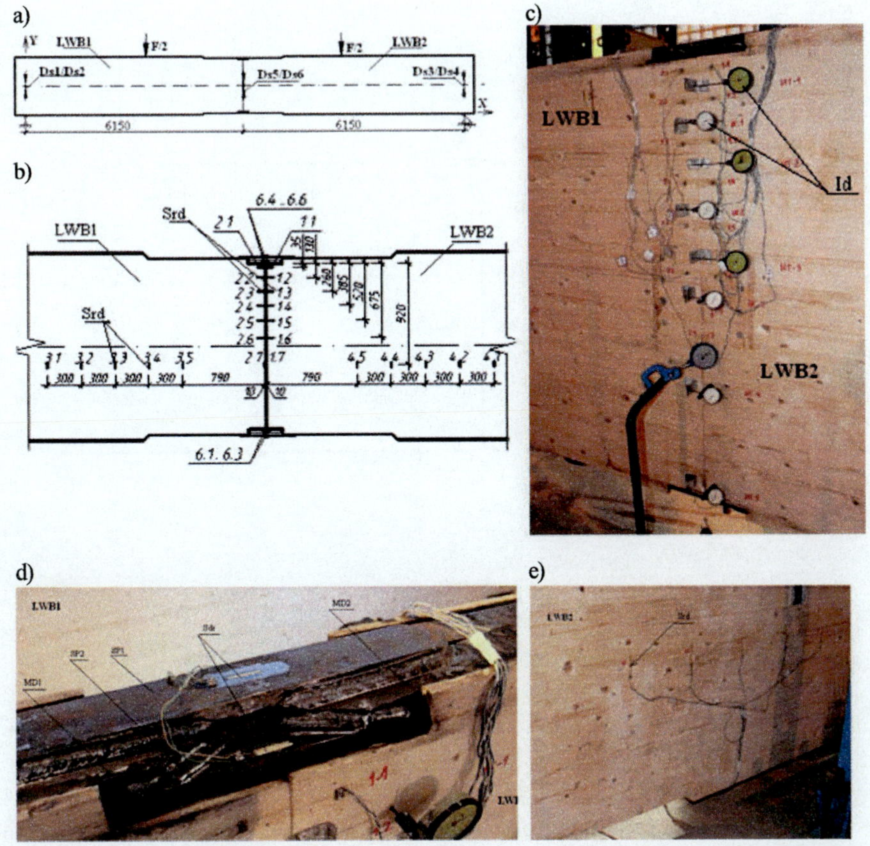

Fig. 7 Scheme and general view of the devices arrangement. **a** Ds deflection meters placement diagram; **b** layout of the Srd strain gauges in the area of the contact surfaces of the elements LWB1 and LWB2, and in the area of the fracture of the lower inclined rods; **c** general view of the location of measuring instruments in the joint area; **d** general view of the location of the Srd strain gauges in the area of the lower inclined rods fracture; **e** general view of the location and strain gauge Srd sensors of the connecting plates Sp1 and Sp2

- distribution of normal stresses σ_y acting across the timber fibre in the fracture zone of the inclined bars.

Analyzing the diagrams of normal stresses σ_x distribution, acting in timber along with the height of the contact zone (Fig. 8b), it can be noted that the nature of their distribution depends on the load value F. In the range F = 0 ÷ 1000 кN, the timber situated in the compressive zone of the section is in the stage of linear elastic deformation. With the load F > 1000 кN, on a part of the height of the compressive zone, an area appears where the timber is in a nonlinear deformation region (Fig. 8c). The depth of the nonlinear deformation region of timber depends on the

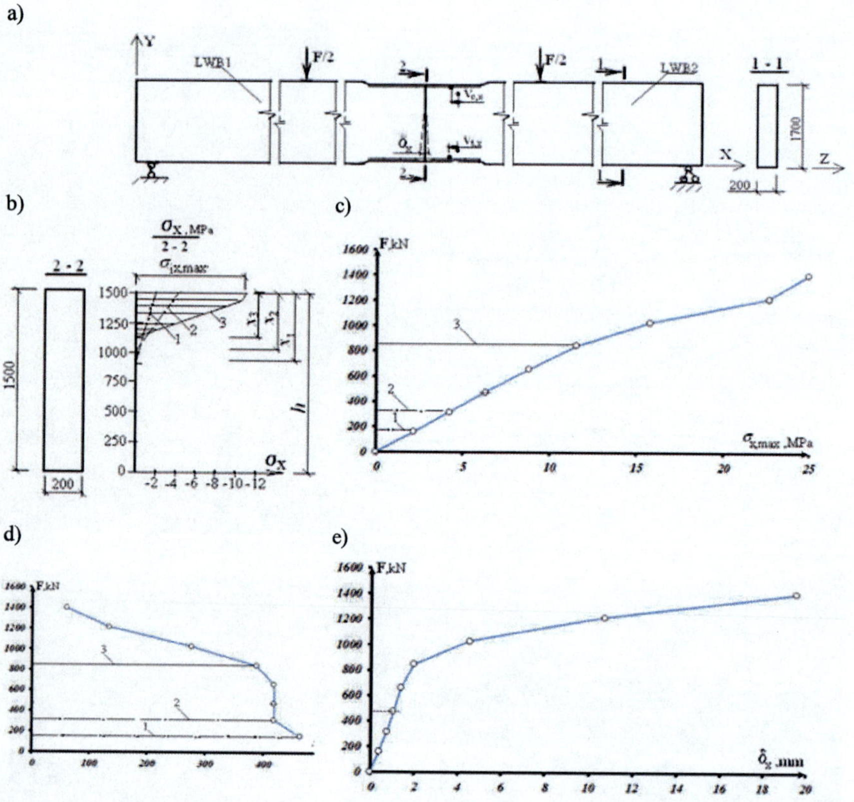

Fig. 8 Stress–strain state of the bent joint. **a** loading scheme; **b** diagram of normal stresses σx in section 2-2; **c** dependence of maximum normal σx,max stresses in section 2-2 on the load value; **d** dependence of the height of the compressive zone x on the load value; **e** dependence of the width δx of the joint opening on the load value; 1—under load F = 157 кN; 2—at load F = 315.3 кN; 3—under load F = 843.8 кN

ductility of the connection of the anchored plates MD1 and MD2, the stress level $\sigma_{y1,max}$ in the inclined bars, as well as the stress level $\sigma_{s,x}$ in the connecting plates.

Considering the relation of the depth of the compressed zone x on the load value F (Fig. 8d), it can be noted that in the interval F = 315, 3 kN ÷ 658 kN—neutral axis depth x remains constant and equal to 418 mm, and with under load F = $\geq$658 kN—the value x decreases. A decrease in the depth of the compressive zone x can be explained by the occurrence of plastic strains in the connecting plate Sp1 of the tension zone of the joint, which leads to an increase in the width of the joint opening δ_x (Fig. 8e). As for the plastic deformations of the connecting plate Sp1 located in the compressed zone of the joint, they arise under load F = 844 kN, i.e. larger than in a tensile plate (Fig. 9b). This is because part of the compressive force is absorbed by the timber compressive zone. The values of the forces in the connecting plate Sp1 and the timber of the compressed zone depends on the ductility of the connection of the anchored plates MD1 and MD2 and the frontal stop between the surfaces of the timber elements LWB1 and LWB2.

As a result of the calculations performed, the values of strain capacity (ductility) c of the connection of the embedded plates MD1 and MD2 listed in Table 2. Analyzing the data in Table 2, we can conclude that strain capacity (ductility) c of the connection of the embedded plates MD1 and MD2 located in the compressed zone is higher than in the tensile zone $c_c > c_t$. With increasing load F, the difference between c_c and c_t decreases. This is because with increasing the load F, most of the compressive force is absorbed by the timber of the compressive zone, as well as by the redistribution of forces between the inclined bars Ib1 in the case when stresses in one of the connecting rods is $\sigma_{y1,max} > 700$ MPa. As for the state tension of the inclined reinforcing bars Ib1, the most loaded are the bars located at a distance from the contact surfaces of the elements LWB1 and LWB (Fig. 9b, c). Based on the distribution of normal stresses σ_{y1}, one can notice a significant unevenness of their distribution both along the length of the rods and between them in the joint. The maximum values σy1 are reached at the beginning of the rod (in the zone of connection with the embedded part MD1 or MD2), and at the end of the rod—they approach zero. It was found that the inclined rods of the compressive and tensile zones perceive forces of different value.

To calculate the ununiformity distribution of stresses $\sigma_{y1,max}$ between the rods, a conditional coefficient k was introduced. The value of the coefficient k was determined as the ratio of the stress $\sigma_{y1,max}$ in the considered rod to the stress of the

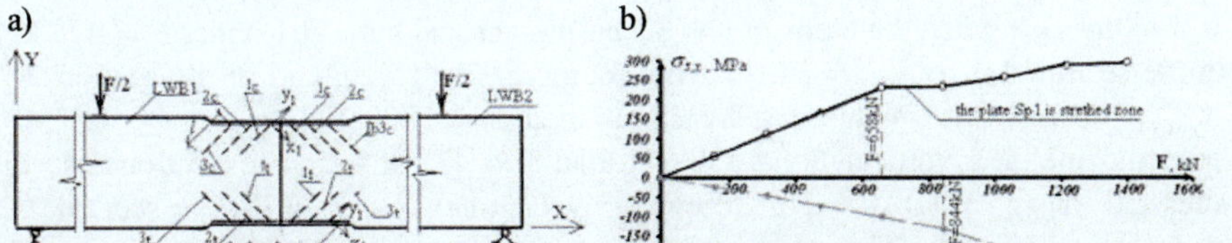

Fig. 9 Stress–strain state of the bent joint. **a**—loading scheme; **b**—diagram of normal stresses σx in section 2-2; **c** dependence of maximum normal σx,max stresses in section 2-2 on the load value; **d** dependence of the height of the compressive zone x on the load value; **e** dependence of the width δx of the joint opening on the load value; 1—under load F = 157 кN; 2—at load F = 315.3 кN; 3—under load F = 843.8 кN

Table 2 Value of strain capacity c of the connection of the embedded parts MD1 and MD2

Load F, kN	In the joint compressive zone		In the joint tensile zone	
	Offset $v_{x,c}$ of the parts MD1 MD2, mm	Strain capacity c_c of the parts MD1and MD2, mm/kN·10^{-4}	Offset $v_{x,c}$ of the parts MD1 MD2, mm	Strain capacity c_t of the parts MD1and MD2, mm/kN·10^{-4}
157.7	0.03	3.65	0.06	3.26
315	0.05	3.05	0.11	2.92
473	0.07	2.84	0.14	2.48
658	0.08	2.32	0.17	2.13
844	0.10	2.18	0.20	1.98
1029	0.14	2.21	0.22	1.98
1215	0.17	1.99	0.22	1.98
1400	0.19	1.96	0.23	1.81

most loaded joint rod. For the rods of the tensile zone under load F = 157.7 kN, the conditional coefficient of ununiformity between the second 2t and the third 3t rods is $k_{t,2/3} = \sigma_{y1,max,2t}/\sigma_{y1,max,3t} = 0.70$, between the first 1t and the third 3t – $k_{t,1/3} = 0.51$, and for the compressive zone—$k_{c,2/3}$ = 0,63 and $k_{c,1/3} = 0.38$. Analyzing the values of the coefficients $k_{t,i/3}$ and $k_{c,i/3}$, it can be argued that the redistribution of forces between the rods of both the tensile and compressive zones will occur under stresses $\sigma_{y1,max}$ = >700 MPa in the most loaded rod of the joint (Fig. 5b).

Analyzing the stress state of the timber belonging to volume V1 (Fig. 10), it can be noted that it is in a complex state, where the action σ_y takes place.

For the case when the element LWB2 has no vertical bars Vb1 under F = 473 kN, in the section 1-1, $\sigma_{y,max} = -0.11$ MPa (compressive zone), and in the section 2-2, $\sigma_{y,max} = 0.12$ MPa. With an increase in load $F \geq 800$ kN, $\sigma_{y,max} \geq 1.5$ MPa with the inclined vertical bars and the load F = 473 kN, in the section 1-1, the stresses $\sigma_{y,max} = -0.10$ MPa (compressive zone), and in the section 2-2, $\sigma_{y,max} = -0.13$ MPa. Thus, the usage of vertical bars Vb1 in the areas where the inclined bars Ib1 are located is necessary to absorb the tensile stress of the timber across the fiber. Despite the insignificant values of stresses $\sigma_{y,max}$, causing the timber perpendicular tension, these rods should be considered as structural elements and are not subject to computation. When utilizing these rods, the length of their mutual overlapping must be at least 9d, where d is the rod diameter.

4.2 *Results of the Numerical Investigation*

As a result of processing the data obtained during testing the specimen, depending on the value of the load F, the relations for the development of the following

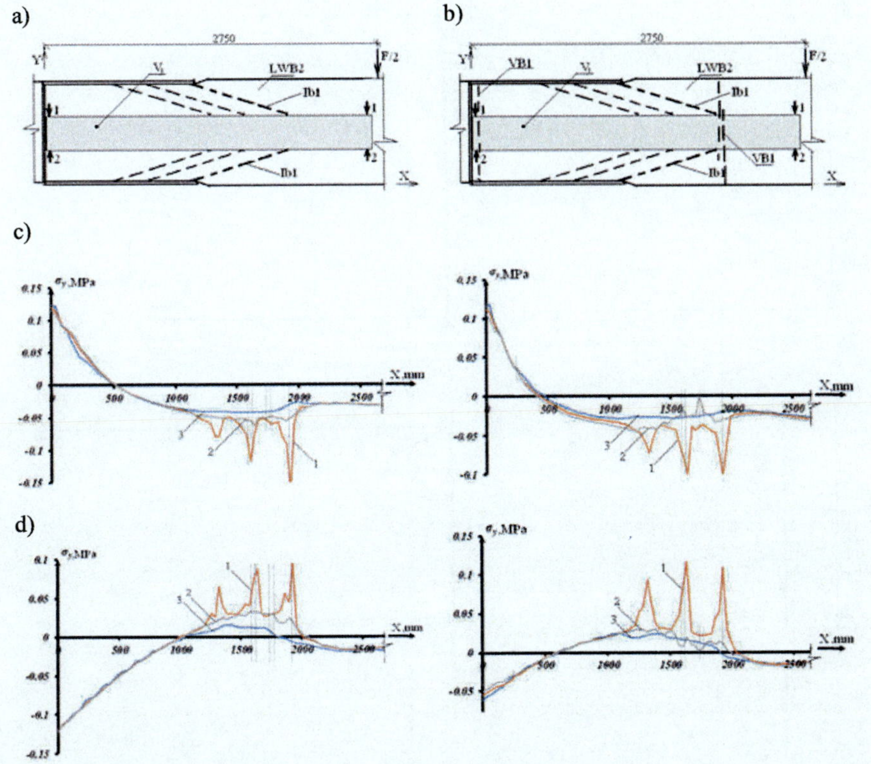

Fig. 10 Distribution of normal stresses σy acting across the grain in sections of volume V1 under load F = 473 kN. **a** joint diagram in the absence of vertical bars Vb1; **b** joint diagram with the presence of vertical bars Vb1; **c** in section 1-1; **d** in section 2-2; 1—along the line of a row of inclined rods; 2—in the middle of the section width; 3—on the element surface

parameter were determined: shear strains vx of the anchored plates MD1 and MD2; normal stresses $\sigma s,x$ in the connecting plates Sp1 and Sp2 located both in the compressive and tensile zones of the joint; the depth of the compressed zone x; the maximum compression stress $\sigma x,max$ in the timber of the compressive zone and the width of the joint opening δx (Fig. 11).

Analyzing the relations shown in Fig. 10, it can be noted that the materials of all elements of the joint under a load F $\leq$ 473 kN were in the linear-elastic region. As for the experimental values of the width δ_x of the joint opening, the maximum stresses $\sigma x,max$, in the contact zone of the elements LWB1 and LWB2, the stresses $\sigma s,x$ in the connecting plates Sp1 and Sp2 of the tension zone, shear deformations $v_{x,t}$ and $v_{x,c}$ of the anchored plates MD1 and MD2—they turned out to be lower than the corresponding values obtained from FEM, and the height of the compressive zone x and stresses $\sigma s,x$ in the connecting plates Sp1 and Sp2 of the compressed zone were higher. So, with the load value F = 473 kN, the values of the

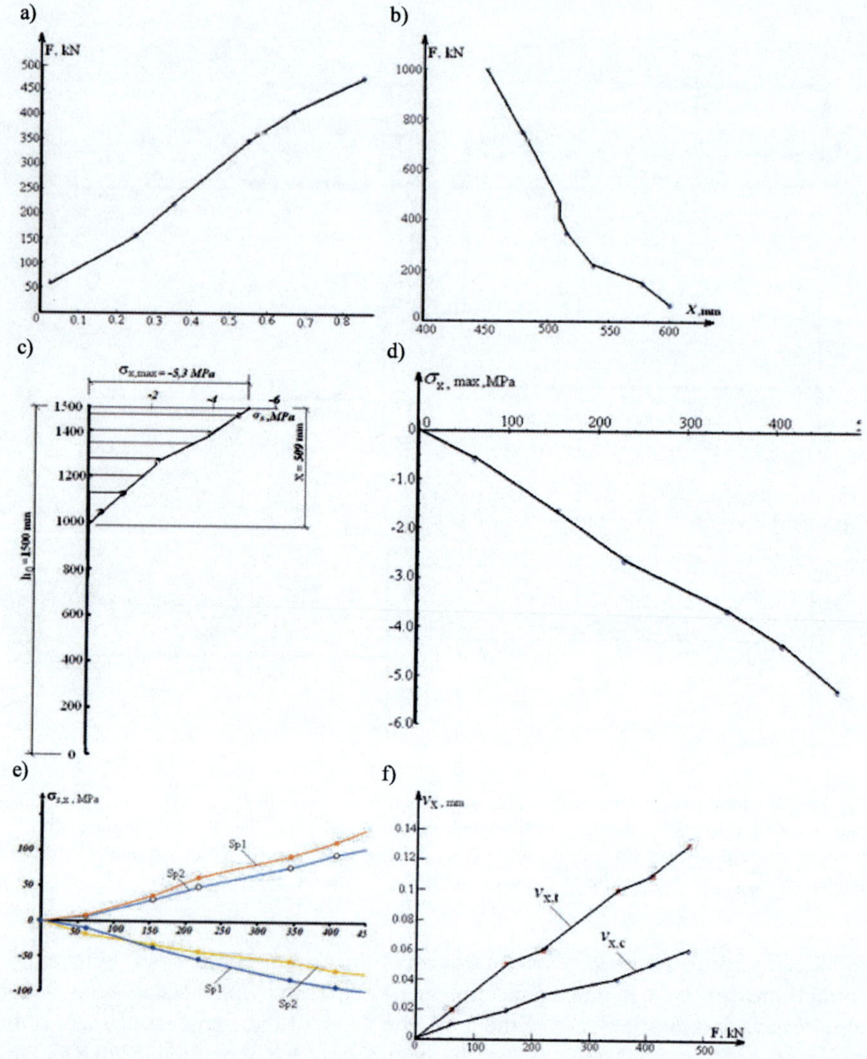

Fig. 11 Relationships of change in the controlled parameters of the joint on the load value F, identified during testing the specimen. **a** change in the joint opening width $\boldsymbol{\delta_x}$; **b** the height of the tensile x zone. **c** distribution of normal stresses σ_x along the height x of the tensile zone under F = 473 kN; **d** compressive stress $\sigma_{x,max}$, acting in the wood of the tensile joint zone along the grain; **e** stresses $\sigma_{sx,}$ acting in the steel connecting plates Sp1 and Sp2; **f** change in deformations $v_{x,t}$ and $v_{x,c}$ shear of the embedded parts MD1 and MD2, respectively, in the tensile and compressive zones of the joint

parameters obtained during the tests turned out to be lower than the FEM values: by 18% for the width δ_x of the joint opening and compressive stresses $\sigma_{x,max}$; by 15% for stresses $\sigma_{s,x}$ in the tensile plate Sp1; by 14% for shear strains $v_{x,t}$ of the plates MD1 and MD2 in the tensile zone and by 7% for shear strains vx,c in the compressive zone. At the same time, the values of stresses $\sigma_{s,x}$ in the compressed plate Sp1 are higher by 38%, and the height value x of the compressive zone is higher by 22%. The revealed difference in the values is explained by the fact that during the specimen production, the contacting surfaces of the plates MD1 and MD2 were connected with the elements LWB1 and LWB2 when the glue was squeezed out of the holes during the immersion of the inclined reinforcing bars Ib1. Therefore, the values of shear strains $v_{x,t}$ and $v_{x,c}$ of the embedded parts MD1 and MD2 connection with the wood of the elements LWB1 and LWB2 decreased, which led to an increase in the stiffness of the connection of anchored plates with the wood. This is confirmed by an increase in stresses $\sigma_{s,x}$ in the compressed plate Sp1, as well as in the depth x of the tensile zone.

After strengthening the connecting plates Sp1 with steel plates, the specimen was brought to failure under the breaking load F_t = 1003 kN. The time the specimen to destruction was t_1 = 7200 s. The failure of the specimen was brittle and occurred in two stages.

At the first stage, the timber of the lower lamella fractured at the plates MD1 and MD2, where, due to the change in the cross-section, there was a concentration of all stress components. With a further increase in the load, there was a successive rupture of the toothed joints of the lamellas, and then chipping of the timber along the grain at the level of the fracture of the inclined tensile rods (Fig. 12).

It should be noted that no crushing of the timber in the contact zone between the surfaces of the elements LWB1 and LWB2 was detected. This allows concluding that the timber in the compression zone works in the linear-elastic region of deformation until the specimen failure. From this conclusion, it follows that for such structural solutions of the joint, the transition of timber in the compression zone to the nonlinear region of deformation may be due to the presence of plastic deformations in the connecting plates Sp1 and Sp2 of the tensile zone and the strain

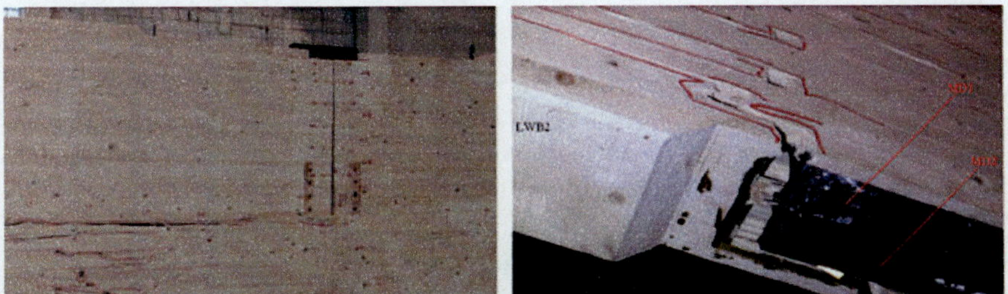

Fig. 12 General view of the failure of the specimen joint

ductility of the connection of the anchored plates MD1 and MD2 with the elements LWB1 and LWB2.

The value of the actual load-bearing capacity Rsup of the tested specimen according to [26] was determined by the formula (1)

$$R_{\text{sup},v} = \frac{F_t}{\gamma_v}, \tag{1}$$

where $R_{sup,v}$ is the value of bearing capacity with brittle nature of failure [27], kN;

F_t is the ultimate load value, kN;

γ_v is the partial factor determined by the formula (2).

$$\gamma_v = 1,64(1,94 - 0,116 \lg t), \tag{2}$$

here t is the reducing time, s, corresponding to the action of constant load F_t, determined by the formula [28]:

$$t = \frac{t_1}{38,2}, \tag{3}$$

where t_1 is the time of increasing load from zero to the ultimate value F_t, s.

Using the formulae (1)–(3), the actual value of the load-bearing capacity of a bent rigid joint was $R_{sup,v}$ = 364,9 kN that is higher than the calculated design value of the load-bearing capacity Fd, determined according to [20, 21, 29, 30]. When determining F_d according to [21] the following assumptions were made:

The timber of the compression joint zone works linear;

- stresses in the compression zone of the joint do not exceed the calculated values of the timber compressive strength parallel to the fibre;
- the force in the compression joint zone is perceived by the wood only, and in the tensile one—by the inclined rods;
- the hypothesis of the plain section is valid;
- the equality of deformations and constancy of the timber modules and the material of the connecting plates of the joint is observed;
- the strain capacity of the embedded parts is constant.

As the results of the performed studies have shown, the main disadvantages of the used model are that wood is considered as an elastic material, the change in the tensile strength of the embedded parts and its change in time is not taken into account, as well the perception of the force in the compression zone not only by timber but also by the inclined rods [31]. Therefore, there is partial use of the timber strength, the bearing capacity of the joints of the anchored plates and, as a consequence, a decrease in the calculated design value of the joint load-bearing capacity F_d and overspending of materials.

5 Conclusions

Based on the results of numerical and experimental studies of a rigid joint specimen, it should be concluded:

- the discrepancies between the data obtained as a result of numerical calculations and test are most typical of the value of the compression zone depth, which amounted to 22%. This difference is because, during the specimen production, there was a partial adhesive connection of the anchored plates surfaces and the timber in the process of glueing the inclined reinforcing bars. This circumstance led to a decrease in the strain capacity (ductility) of the anchored plates and an increase in the depth of the compression zone. In the FE-model, the coefficient of friction between the surfaces of the anchored plates and the timber was taken to be zero;
- the depth of the compression joint zone is not a constant value and depends on the strain capacity (ductility)of the anchored plates and the level of stresses in the connecting plates. With the load increases, the values of strain capacity and the depth of the compression zone decrease. The strain capacity (ductility) of the anchored plates in the compression zone is higher than that of the embedded parts in the tensile zone;
- the distribution of stresses along the length of the inclined reinforcing bars is characterized by the high rate of non-uniformity. The coefficient of non-uniformity of the distribution of normal stresses between the inclined reinforcing bars of the connection for the embedded parts of the compression zone is in the range $0.61 \div 0.38$, and for the reinforcing bars of the embedded parts of the tensile zone—$0.7 \div 0.51$. Therefore, when calculating the bearing capacity of the anchored plates, the anchoring length of the inclined reinforcing bars should be taken with the account of the coefficients of non-uniformity distribution of stresses between the rods;
- to prevent the destruction of the joint from timber tension across the grain at the contacting surfaces of the connected ends of the elements and in the fracture zone of the inclined rods, vertical rods must be installed along with the entire height of the cross-section;
- the calculated design value of the joint load-bearing capacity should be determined taking into account the ductility diagrams of the anchored plate connections of the compression and tensile zones, depending on the stress value and the time of its action, as well as the stress–strain relations of the materials used. The calculation is performed using an iterative method.

References

1. Anon (1973) Wood fights back to hold its own in railroad bridges
2. Sergeev M, Rimshin V, Lukin M, Zdralovic N (2020) Multi-span composite beam. IOP Conf Ser Mater Sci Eng. https://doi.org/10.1088/1757-899X/896/1/012058
3. Gorpinchenko VM, Pogorel'tsev AA, Ecknadosyan IL (2005) Large-scale tests provided for a block, consisting of two wooden lens-type roof trusses of the sports complex "Strogino." Promyshlennoe i Grazhdanskoe Stroit 38
4. Turkovskij SB, Pogorel'tsev AA, Eknados'yan IL (2003) Selection of design scheme of lens-shaped trusses from adhesive wood. Stroit Mater 18–20
5. Dilik T, Erdinler S, Kurtoglu A (2012) Edge glued wood panel technology and an assessment on the development of edge glued wood panel industry. Am J Appl Sci 9:1625–1635. https://doi.org/10.3844/ajassp.2012.1625.1635
6. Turkovskij SB, Pogorel'tsev AA (2001) Wooden structures with rigid joints in structures with corrosive medium. Promyshlennoe i Grazhdanskoe Stroit 10–13
7. Zhou H-Z, Zhu E-C, Zhou G-C (2013) Investigation of stress perpendicular to grain and cracking of curved glulam beams. Jianzhu Cailiao Xuebao/J Build Mater 16:913–918. https://doi.org/10.3969/j.issn.1007-9629.2013.05.031
8. Parida G, Johnsson H, Fragiacomo M (2013) Provisions for ductile behavior of timber-to-steel connections with multiple glued-in rods. J. Struct. Eng. (United States) 139:1468–1477. https://doi.org/10.1061/(ASCE)ST.1943-541X.0000735
9. Tlustochowicz G, Serrano E, Steiger R (2011) State-of-the-art review on timber connections with glued-in steel rods. Mater Struct Constr 44:997–1020. https://doi.org/10.1617/s11527-010-9682-9
10. Steiger R, Serrano E, Stepinac M, Rajčić V, O'Neill C, McPolin D, Widmann R (2015) Strengthening of timber structures with glued-in rods. Constr Build Mater 97:90–105. https://doi.org/10.1016/j.conbuildmat.2015.03.097
11. Steiger R, Gehri E, Widmann R (2007) Pull-out strength of axially loaded steel rods bonded in glulam parallel to the grain. Mater Struct Constr 40:69–78. https://doi.org/10.1617/s11527-006-9111-2
12. Serrano E (2001) Glued-in rods for timber structures - a 3D model and finite element parameter studies. Int J Adhes Adhes 21:115–127. https://doi.org/10.1016/S0143-7496(00)00043-9
13. Rossignon A, Espion B (2008) Experimental assessment of the pull-out strength of single rods bonded in glulam parallel to the grain. https://doi.org/10.1007/s00107-008-0263-3
14. Negrão J, Martins J (2021) Bondline strength of glued-in threaded rods in new and ancient chestnut (Castanea sativa Mill.) wood. Constr Build Mater 267. https://doi.org/10.1016/j.conbuildmat.2020.120989
15. Koshcheev AA, Roshchina SI, Naichuk AY, Vatin NI (2020) The effect of eccentricity on the strength characteristics of glued rods made of steel cable reinforcement in solid wood. IOP Conf Ser Mater Sci Eng. https://doi.org/10.1088/1757-899X/896/1/012059
16. Naichuk AY (2020) Causes of defects in timber arches of the buildings covering and methods of strengthening. IOP Conf Ser Mater Sci Eng. https://doi.org/10.1088/1757-899X/896/1/012042
17. Gribanov AS, Roshchina SI, Naichuk AY, Melekhov VI (2020) Wooden beams with local wood modification. IOP Conf Ser Mater Sci Eng. https://doi.org/10.1088/1757-899X/896/1/012067
18. Gribanov AS, Strekalkin AA, Kudryatseva AA, Zdralovic N (2020) CFRP composites for strengthening wooden structures. IOP Conf Ser Mater Sci Eng. https://doi.org/10.1088/1757-899X/896/1/012114
19. Ribeiro AS, de Jesus AMP, Lima AM, Lousada JLC (2009) Study of strengthening solutions for glued-laminated wood beams of maritime pine wood. Constr Build Mater 23:2738–2745. https://doi.org/10.1016/j.conbuildmat.2009.02.042

20. Liu W, Yang H (2019) Research progress on modern timber structures | 现代木结构研究进展. Jianzhu Jiegou Xuebao/J Build Struct 40:16–43. https://doi.org/10.14006/j.jzjgxb.2019.02.002
21. Azinović B, Serrano E, Kramar M, Pazlar T (2018) Experimental investigation of the axial strength of glued-in rods in cross laminated timber. Mater Struct Constr 51. https://doi.org/10.1617/s11527-018-1268-y
22. Wang C-J, Wang C-J, Xu J, Zhang P, Shan D-B, Guo B, Wang Z-L (2016) Tensile deformation behaviors of pure nickel fine wire with a few grains across diameter. Trans Nonferrous Met Soc China (English Ed. 26:1765–1774). https://doi.org/10.1016/S1003-6326(16)64287-5
23. Ohtani T (2008) The effects of mechanical parameters of the stress-strain diagram on wood abrasion. Wear 265:1557–1564. https://doi.org/10.1016/j.wear.2008.02.007
24. Zhang J, Lounis Z (2009) Nonlinear relationships between parameters of simplified diffusion-based model for service life design of concrete structures exposed to chlorides. Cem Concr Compos 31:591–600. https://doi.org/10.1016/j.cemconcomp.2009.05.008
25. Zhang LL, Song XG, Zhang JJ (2014) Performance based optimized design of steel structure. https://doi.org/10.4028/www.scientific.net/AMR.919-921.199
26. Rinaldin G, Fragiacomo M, Amadio C (2017) On the accuracy of the N2 inelastic spectrum for timber structures. Soil Dyn Earthq Eng. https://doi.org/10.1016/j.soildyn.2017.05.026
27. Buka-Vaivade K, Serdjuks D, Goremikins V, Pakrastins L., Vatin, N.I.: Suspension structure with cross-laminated timber deck panels. Mag. Civ. Eng. (2018). https://doi.org/10.18720/MCE.83.12.
28. Karelskiy AV, Zhuravleva TP, Labudin BV (2015) Load-to-failure bending test of wood composite beams connected by gang nail. Mag Civ Eng. https://doi.org/10.5862/MCE.54.9
29. Gravit MV, Serdjuks D, Bardin AV, Prusakov V, Buka-Vaivade K (2019) Fire design methods for structures with timber framework. Mag Civ Eng. https://doi.org/10.18720/MCE.85.8
30. Pyatikrestovsky KP, Travush VI (2017) A complex analysis of stress-strain state of ribbed wooden structures with anisotropic sheathings. Mater Phys Mech
31. Saknite T, Serdjuks D, Goremikins V, Pakrastins L, Vatin NI (2016) Fire design of arch-type timber roof. Mag Civ Eng. https://doi.org/10.5862/MCE.64.3

The Strength of Wood-Reinforced Polymer Composites in Tension at an Angle to the Fibers

Mikhail Lisyatnikov, Anastasiya Lukina, Danila Chibrikin, and Boris Labudin

Abstract A method of reinforcing beams on a support with a polymer composite based on fiberglass and an epoxy matrix with the inclusion of carbon nanotubes is proposed. To prove the effectiveness of amplification and to determine the quantitative and qualitative characteristics, experimental studies were carried out on standard wooden samples with and without reinforcement. Before the experiment, its planning was carried out in order to achieve maximum measurement accuracy with a minimum number of experiments and preserve the statistical reliability of the results. As a result of planning, a multifactor formula was obtained that is intended to determine the strength of a sample depending on any combination of three variable factors: the reinforcement coefficient, the quantitative component of carbon nanotubes, and the curing temperature of the composite. The tests were carried out under normal conditions. As a result of tensile tests at an angle to the fibers of the prototypes, it was found that the tensile strength of reinforced samples compared to wooden ones without reinforcement increases 2 times, and the destruction is of a plastic nature.

Keywords Timber · Building · Beams · Construction materials · Reinforcement of wooden beams · Strengthening

1 Introduction

One of the types of wooden and wood-glued structures are beams [1–6]. The nature of the work of beam elements is bending. The criteria for wood performance in this case can be reduced to the theory of classical destruction [7]. Accordingly, in the

M. Lisyatnikov (✉) · A. Lukina · D. Chibrikin
Vladimir State University Named After Alexander and Nikolay Stoletovs,
Vladimir, Russian Federation

B. Labudin
Northern (Arctic), Federal University Named After M.V. Lomonosov,
Arkhangelsk, Russian Federation

N. Vatin et al. (eds.), *Proceedings of MPCPE 2021*, Lecture Notes in Civil Engineering 182, https://doi.org/10.1007/978-3-030-85236-8_46

vast majority of cases, the loss of load-bearing capacity or structural failure [8, 9] occurs in the middle of the span under the influence of normal stresses in the stretched and compressed zone of the beam. In this regard, the bulk of solutions for strengthening beams are aimed at reducing the influence of normal stresses in the middle of the span, and strengthening the support zone is mistakenly neglected.

Under the influence of a uniformly distributed load [10], the following stress conditions occur in the support zone of beam structures: chipping along the fibers (shear); crumpling across the fibers (surface compression from the support reaction); stretching at an angle to the fibers.

Based on this, defects and foci of the first signs of failure [11–14] often occur outside the zone of maximum stress. The nature of destruction in the form of separation on the main platforms when chipping will be characteristic of high wooden beams of overlap and coating, glued beams-inserts in cross-beam systems. The strength of the beam [15, 16] at the supports must be evaluated as a bent anisotropic element using the criteria for the strength of wood in a complex stressed state.

It is also worth noting that the destruction of the beam structure on the support is also possible when the wood is destroyed due to unsatisfactory operating conditions.

The feasibility of evaluating glued wooden beams at an angle to the fibers also occurs in the vicinity of the application of concentrated forces, including increased support reactions, in areas of steep undercutting, especially at stretched edges, on curved sections with a bend that reduces the curvature of the element, etc.

After analyzing all of the above, we can conclude that the complex strenuous work of wood [17] due to anisotropy of the structure, as well as the need in certain cases to strengthen the support zones of beam structures.

To strengthen beam structures on supports, classical methods of reinforcement are used: restoration of the section with overlays, installation of a prosthesis, gluing steel rods, etc.

There are also methods of strengthening [18–24] support zones of beam structures, involving the use of polymers, enclosing in the body of wood reinforcing compounds, or pasting the support zones with tapes and nets made of carbon fiber. In addition, one of the perspective directions of strengthening of structures may be the use of new lightweight composite materials, metal-based [25–27].

We have developed a method of strengthening, consisting in the device of the cage on the support part of the beam, made of fiberglass impregnated with an epoxy composition (ED-20), with the addition of carbon nanotubes (CNTS). The number of clip layers varies from 1 to 5. The use of adhesive compositions [28, 29] based on fiberglass and nanotechnologies in strengthening the supporting zones of wood-glued beam structures requires special research, including experimental studies.

The object of research is wood reinforced with a polymer composite made of fiberglass, epoxy resin and carbon nanotubes.

The subject of research is the tensile strength of wood reinforced with a polymer composite at an angle to the fibers.

The purpose of the study is to carry out experimental studies to determine the tensile strength of wood reinforced with a polymer composite at an angle to the fibers.

The following tasks were set:

1. Set the maximum strength indicators and the destructive load of experimental wooden samples and samples with surface reinforcement.
2. To identify the dependence of the strength limit of experimental samples on the coefficient of reinforcement, the concentration of carbon nanotubes and the curing temperature of the composite.

Establish the nature of destruction [30–33] of the tested wooden and reinforced samples.

2 Methods

Determination of the strength characteristics of wooden and reinforced specimens under tension at an angle to the fibers was carried out on a tensile testing machine REM-100-A-1.

Before conducting an experiment, it is necessary to plan it [34]. The main goal of experiment planning is to achieve maximum measurement accuracy with a minimum number of experiments performed and maintain the statistical reliability of the results.

The methodical grid of experiments to determine the rational strengthening of the support section of a glued beam is based on a combination square developed for three primary factors of influence, each of which consists of three options. Ed-20 epoxy-Diane resin was chosen as a permanent factor. The coefficient of reinforcement, the quantitative composition of CNTS and the curing temperature of the polymer are variable factors.

To identify the influence of each of the variable factors, you need to set at least three different values or options for it. With the minimum number of experiments, the most evenly cover the entire area of the table of possible combinations of influencing factors [35, 36]. To do this, we can further develop the idea of the so-called "Latin square". It is suggested to look for the dependence of the result on three factors and plan the experiment so that there are no repeated combinations in any row or column. Figure 1 shows one of the possible plans for this combination of three factors, each of which can take three values.

This planning of the experiment allows us to narrow the number of experiments from 27 to 9 with an acceptable decrease in the accuracy of the dependencies of secondary strength and deformability factors on the primary factors: "r" (reinforcement coefficient), "c" (the quantitative component of CNT), "t" (the curing temperature of the composite).

Coefficient. reinforcements, layers		1			3			5		
Number of CNTS, %		0.1	0.3	0.5	0.1	0.3	0.5	0.1	0.3	0.5
Curing temperature, °C	20			№1		№4		№7		
	40		№2		№5					№8
	60	№3					№6		№9	

Fig. 1 Scheme design of experiments for three factors

Factor "r"—takes into account the effect of the composite clip layers on the strength of the reinforced wood-glued beam, changes 1, 3, 5 layers.

Factor "c"—takes into account the influence of the concentration of carbon nanotubes in the adhesive composition on the strength of the reinforced wood-glued beam, changes 0.1, 0.3, 0.5%.

Factor "t"—takes into account the effect of the curing temperature of the composite on the strength of the reinforced wood-glued beam, varies 20, 40, 60 °C.

It is necessary to find empirical formulas that would cover the influence of these factors. The study of the strength properties of experimental models was carried out on samples with surface reinforcement. In the course of research, the values of deformation of samples from the applied load were recorded.

As a result, we obtained the following empirical dependencies, found from the graphs of the influence of each amplification factor:

$$f(r) = 0.506 \cdot \ln a + 0.949 \tag{1}$$

$$f(c) = -0.0087 \cdot u^2 + 0.0731 \cdot u + 0.8827 \tag{2}$$

$$f(t) = 0.0061 \cdot t^2 + 0.0186 \cdot t + 0.9889 \tag{3}$$

Thus, based on the found partial equations of each variable, an empirical dependence is obtained that takes into account the influence of each design factor on the strength of the prototypes:

$$P = \Delta_{av} \cdot f(r) \cdot f(c) \cdot f(t) \tag{4}$$

The resulting multi-factor formula is designed to determine the strength depending on any combination of three variable factors. Substituting the appropriate parameters of factors, we get the corresponding value of the destructive load. The reinforcement with the best strength characteristics was used for the experiment.

Tests of prototypes were carried out with a short-term load [37]. Samples for experimental studies were performed according to the norms. Geometric dimensions were measured using an electronic caliper. The samples were weighed on a laboratory electronic balance. The moisture content of wood was measured by a moisture meter and was 12% with a possible error of ±2%. The temperature and humidity of the environment were determined by a portable weather station. The air temperature in the room was in the range of 18–22 °C, and the relative humidity was 50–60%.

All tests were carried out in series, the strength indicators of wooden and reinforced samples were determined according to the variable parameters of the methodical grid. The coefficient of reinforcement was changed by the number of layers of adhesive composite on the corresponding wooden samples. The CNT concentration varied by the percentage of mass parts of nanotubes to the total weight fraction of the adhesive composition. The curing temperature of the reinforced samples was at room temperature of 20 °C ± 1 °C, additional heating of the samples was performed in a drying test chamber "heat/cold/moisture".

Tensile testing of the samples at an angle to the fibers was carried out in accordance with the requirements of the standards. Thickness (a) and width (b) were measured in the middle of the sample length with an error of no more than 0.1 mm. The shape and dimensions of the sample are shown in Fig. 2. The fiber direction angle is taken as 27° ± 2° according to numerical studies and engineering calculation.

The prototypes were placed in the grippers so that the sample surfaces were adjacent to the corrugated gripping surfaces, and the part of each head adjacent to the rounding remained free for about 10 mm, and the tensile load coincided with the longitudinal geometric axis of the sample.

The test speed was such that the sample collapsed within 1.5–2 min from the moment of loading. It was allowed to load the sample uniformly at a rate of (2500 ± 500) N/min or to carry out the test at a loading rate of the crosshead of the REM-100-A-1 testing machine—4 mm/min.

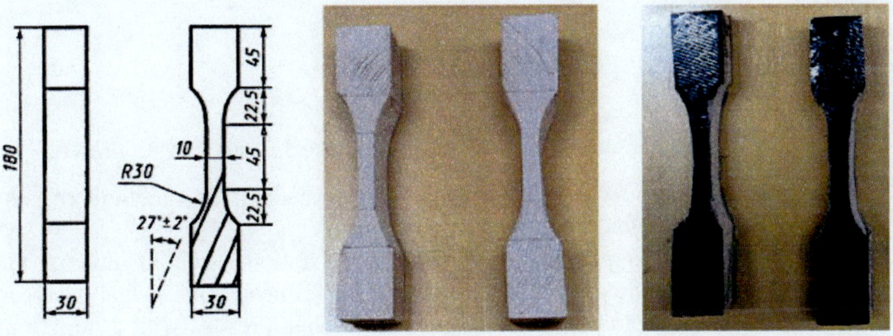

Fig. 2 Tensile test piece at an angle to the fibers

The tests were conducted in the laboratory of building constructions named after V. Yu Shuko on the basis of the Vladimir state University. The wood type is pine. Designation of test samples: W—wooden sample; C1, 3, 5 (c-t)—wooden sample with surface reinforcement composite in 1, 3 and 5 layers, respectively (CNT concentration—curing temperature).

A series of tests was performed on samples in the following sequence:

1. Before testing, the samples were weighed, their humidity and geometric dimensions were measured in three planes.
2. Tensile specimens were fixed in special clamps of a tensile testing machine at the required distance.
3. Spend stretching of the samples.
4. According to the data on the laptop monitor from the tensile testing machine, the values of the breaking load F (N) and the corresponding deformations of the sample were recorded.
5. We performed the construction of diagrams to visualize the results of the experiment.

3 Results and Discussion

The General view of the prototypes at the time and after testing is shown in Fig. 3.

During the tests, the dependences "load—normal tensile stresses" (Fig. 4) and "load—relative deformation" (Fig. 5) were built. Data on test results are summarized in the Table 1.

Thus, when testing prototypes, the rational concentration of CNT—0.3% and the curing temperature of the composite equal to 60 °C were determined.

Experimentally, it was found that the destruction of wooden samples was of a brittle nature, in contrast to samples with surface reinforcement, which were characterized by a plastic nature of destruction. Also, when the reinforced samples were destroyed, the separation of the composite from the wood was not detected.

4 Conclusions

As a result of experimental studies, the following conclusions can be drawn.

1. The installation and tool base for testing are presented. The method of conducting an experimental study is described.
2. The experiment was planned to determine the rational composition of the composite. A methodical grid of three-factor experiment experiments has been developed. This research scheme allows you to obtain results with a margin of error, which reduces the number of tests performed.

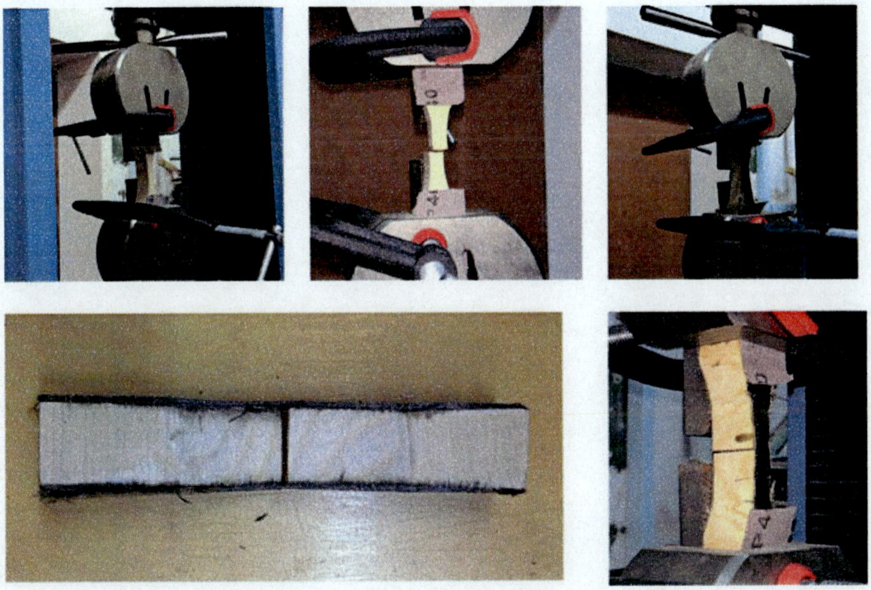

Fig. 3 General view of the test and tested prototypes in tension at an angle to the fibers

Fig. 4 Dependence "load—normal tensile stresses" during tensile tests at an angle to the fibers

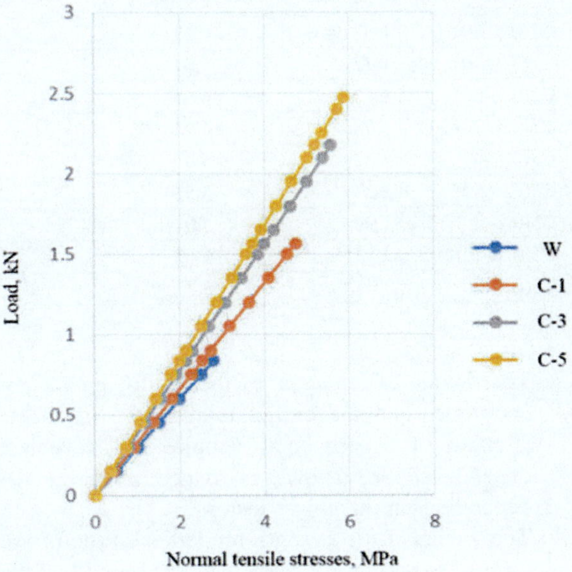

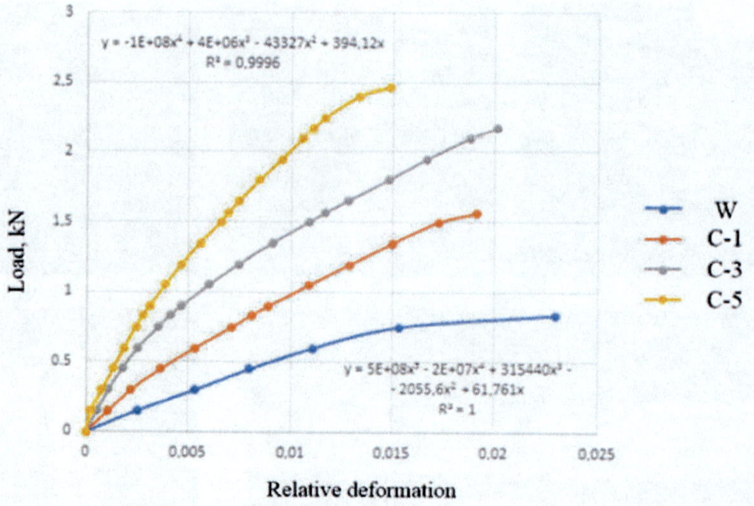

Fig. 5 Dependence "load—relative deformation" during tensile tests at an angle to the fibers

Table 1 Test results of experimental wooden and reinforced specimens

Sample marking	Breaking load, kN	Redistribution of strength, MPa	Increasing the strength of, %
W	0.84	2.79	–
C1(0.5–20)	1.57	4.19	50
C1(0.3–40)	1.77	4.51	63
C1(0.1–60)	1.72	4.67	60
C3(0.3–20)	1.98	4.92	77
C3(0.1–40)	2.02	5.0	80
C3(0.5–60)	2.27	5.53	97
C5(0.1–20)	1.96	5.01	76
C5(0.5–40)	2.32	5.62	100
C5(0.3–60)	2.47	5.88	107

3. The degree of influence of the reinforcement coefficient, the concentration of nanotubes and the curing temperature of the CNT-based composite on the strength of experimental samples was revealed. We obtained an empirical dependence that allows us to determine the destructive load of the sample depending on the above factors.
4. The values of the strength properties of the prototypes were determined on the basis of experimental studies. It was found that the tensile strength of reinforced specimens in comparison with wooden specimens increases by a factor of 2 during tensile testing at an angle to the fibers.

References

1. Ling Z, Liu W, Yang H, Chen X (2018) Modelling of glued laminated timber joints with glued-in rod considering bond-slip location function. Eng Struct. https://doi.org/10.1016/j.engstruct.2018.08.098
2. Babiak M, Gaff M, Sikora A, Hysek Š (2018) Modulus of elasticity in three- and four-point bending of wood. https://doi.org/10.1016/j.compstruct.2018.07.113
3. Lisyatnikov MS, Glebova TO, Ageev SP, Ivaniuk AM (2020) Strength of wood reinforced with a polymer composite for crumpling across the fibers. In: IOP conference series: materials science and engineering. https://doi.org/10.1088/1757-899X/896/1/012062
4. Lukin M, Prusov E, Roshchina S, Karelina M, Vatin N (2021) Multi-span composite timber beams with rational steel reinforcements. Buildings. https://doi.org/10.3390/buildings11020046
5. Nagruzova L, Saznov K, Aytbu KK (2019) Thermal efficient panels on a wooden frame for quickly erectable low-rise buildings. In: E3S web of conferences. https://doi.org/10.1051/e3sconf/201911001023
6. Lokaj A, Vavrusova K, Mikolasek D (2017) Analysis of behavior and carrying capacity of glued timber-timber joints loaded with bending moment. Procedia Eng. https://doi.org/10.1016/j.proeng.2017.05.336
7. Gribanov AS, Roshchina SI, Naichuk AY, Melekhov VI (2020) Wooden beams with local wood modification. In: IOP conference series: materials science and engineering. https://doi.org/10.1088/1757-899X/896/1/012067
8. Naychuk AY (2013) Estimation of load-bearing capacity and stiffness of timber beams with through-thickness cracks. In: Advanced materials research. https://doi.org/10.4028/www.scientific.net/AMR.778.361
9. Karelskiy AV, Zhuravleva TP, Labudin BV (2015) Load-to-failure bending test of wood composite beams connected by gang nail. Mag Civ Eng. https://doi.org/10.5862/MCE.54.9
10. Sergeev M, Rimshin V, Lukin M, Zdralovic N (2020) Multi-span composite beam. In: IOP conference series: materials science and engineering. https://doi.org/10.1088/1757-899X/896/1/012058
11. Jelušič P, Kravanja S (2018) Flexural analysis of laminated solid wood beams with different shear connections. Constr Build Mater. https://doi.org/10.1016/j.conbuildmat.2018.04.102
12. Gribanov AS, Strekalkin AA, Kudryatseva AA, Zdralovic N (2020) CFRP composites for strengthening wooden structures. In: IOP conference series: materials science and engineering. https://doi.org/10.1088/1757-899X/896/1/012114
13. Lisyatnikov MS, Shishov II, Sergeev MS, Hisham E (2020) Precast monolithic coating of an industrial building based on variable-height beam-slabs. In: IOP conference series: materials science and engineering. https://doi.org/10.1088/1757-899X/896/1/012064
14. Koshcheev AA, Roshchina SI, Naichuk AY, Vatin NI (2020) The effect of eccentricity on the strength characteristics of glued rods made of steel cable reinforcement in solid wood. In: IOP conference series: materials science and engineering. https://doi.org/10.1088/1757-899X/896/1/012059
15. Gribanov AS, Roshchina SI, Popova MV, Sergeev MS (2018) Laminar polymer composites for wooden structures. Mag Civ Eng. https://doi.org/10.18720/MCE.83.1
16. Gudkov AB, Dyomin AV, Gribanov AV (2015) Postural control characteristics in elderly women with fallers. Adv Gerontol. = Uspekhi Gerontol
17. Zmijewski T, Wojtowicz-Jankowska D (2017) Timber —material of the future—examples of small wooden architectural structures. In: IOP conference series: materials science and engineering. https://doi.org/10.1088/1757-899X/245/8/082019

18. Lukin MV, Roshchina SI, Smirnov EA, Shunqi M (2020) Strengthening of the operated wooden floor beams with external rigid reinforcement. In: IOP conference series: materials science and engineering. https://doi.org/10.1088/1757-899X/896/1/012065
19. Ando D, Umemura K (2021) Bond structures between wood components and citric acid in wood-based molding. Polymers (Basel) 13:1–9. https://doi.org/10.3390/polym13010058
20. Merkulov S, Rimshin V, Akimov E, Kurbatov V, Roschina S (2020) Regulatory support for the use of composite rod reinforcement in concrete structures. In: IOP conference series: materials science and engineering. https://doi.org/10.1088/1757-899X/896/1/012022
21. Roschina SI, Lisyatnikov MS, Lukin MV, Popova MV (2018) Technology of strengthening the supporting zones of the glued-wood beaming structure with the application of nanomodified prepregs. In: Materials science forum. https://doi.org/10.4028/www.scientific.net/MSF.931.226
22. Lisyatnikov MS, Roshchina SI, Chukhlanov VY, Ivaniuk AM (2020) Repair compositions based on methyl methacrylate modified with polyphenylsiloxane resin for concrete and reinforced concrete structures. In: IOP conference series: materials science and engineering. https://doi.org/10.1088/1757-899X/896/1/012113
23. Vatin NI, Ivanov AY, Rutman YL, Chernogorskiy SA, Shvetsov KV (2017) Earthquake engineering optimization of structures by economic criterion. Mag Civ Eng 76:67–83. https://doi.org/10.18720/MCE.76.7
24. Buka-Vaivade K, Serdjuks D, Goremikins V, Pakrastins L, Vatin NI (2018) Suspension structure with cross-laminated timber deck panels. Mag Civ Eng. https://doi.org/10.18720/MCE.83.12
25. Aborkin A, Khorkov K, Prusov E, Ob'edkov A, Kremlev K, Perezhogin I, Alymov M (2019) Effect of increasing the strength of aluminum matrix nanocomposites reinforced with microadditions of multiwalled carbon nanotubes coated with TiC nanoparticles. Nanomaterials. 9. https://doi.org/10.3390/nano9111596
26. Prusov ES, Panfilov AA, Kechin VA (2017) Role of powder precursors in production of composite alloys using liquid-phase methods. Russ J Non-Ferrous Met 58:308–316. https://doi.org/10.3103/S1067821217030154
27. Prusov E, Deev V, Rakhuba E (2019) Aluminum matrix in-situ composites reinforced with Mg <inf>2</inf> Si and Al <inf>3</inf> Ti. In: Materials today: proceedingsm pp 386–391. https://doi.org/10.1016/j.matpr.2019.01.001
28. Sergeev MS, Gribanov AS, Roschina SI (2020) The stress strain state of composite multi-span beams. In: IOP conference series: materials science and engineering. https://doi.org/10.1088/1757-899X/753/3/032068
29. Rimshin VI, Labudin BV, Melekhov VI, Orlov A, Kurbatov VL (2018) Improvement of strength and stiffness of components of main struts with foundation in wooden frame buildings. ARPN J Eng Appl Sci 13:3851–3856
30. Franke S, Franke B, Harte AM (2015) Failure modes and reinforcement techniques for timber beams-State of the art. Constr Build Mater. https://doi.org/10.1016/j.conbuildmat.2015.06.021
31. Ponomarev AN, Rassokhin AS (2016) Hybrid wood-polymer composites in civil engineering. Mag Civ Eng. https://doi.org/10.5862/MCE.68.5
32. Labudin B, Popov E, Stolypin D, Sopilov V (2019) The wood composite ribbed panels on mechanical joints. In: E3S web of conferences. https://doi.org/10.1051/e3sconf/20199102021
33. Negrão J, Martins J (2021) Bondline strength of glued-in threaded rods in new and ancient chestnut (Castanea sativa Mill.) wood. Constr Build Mater 267. https://doi.org/10.1016/j.conbuildmat.2020.120989
34. Roshchina S, Lukin M, Lisyatnikov M (2020) Compressed-bent reinforced wooden elements with long-term load. In: Lecture notes in civil engineering. https://doi.org/10.1007/978-3-030-42351-3_7

35. Kamar NT, Hossain MM, Khomenko A, Haq M, Drzal LT, Loos A (2015) Interlaminar reinforcement of glass fiber/epoxy composites with graphene nanoplatelets. Compos Part A Appl Sci Manuf. https://doi.org/10.1016/j.compositesa.2014.12.010
36. Gravit M, Simonenko Y, Larionov A (2019) Method single burning item (SBI) for fire hazard of wood constructions. In: E3S web of conferences. https://doi.org/10.1051/e3sconf/20199102049
37. Naichuk AY (2020) Causes of defects in timber arches of the buildings covering and methods of strengthening. In: IOP conference series: materials science and engineering. https://doi.org/10.1088/1757-899X/896/1/012042

Analysis of the Modern "International Harmonized Format" of the Theory of Concrete Creep

Rudolf Sanzharovsky, Frieder Sieber, and Tatyana Ter-Emmanuilyan

Abstract The purpose of the study is to analyze the main directions of the considered theory of concrete creep. This theory is the basis for calculating buildings and structures, taking into account the time factor, when considering the long-term resistance of reinforced concrete structures. Methods. Using the rules of mathematics, principles of mechanics and the results of solid experiments to reveal: the fallacy of the superposition principle, which ultimately leads to the construction of incorrect kernels of integral equations; illegal substitution of instantaneous non-linear concrete properties, creep properties (minute creep); non-stationary instant elastic properties of concrete, transforming into a model of Maxwell's theory of creep, leading to an error of up to 300%; only linear creep is considered, leading to an error of 200–500%.; the principle of "algebraization" of integral equations naturally leads to significantly contradictory results in the calculations of structures. Results. We have found that the basics of the theory under consideration grossly violate the rules of mathematics, the principles of mechanics and the results of solid experiments.

Keywords Theory of concrete creep · Principle of imposition · Instant elastic deformations · Long-term resistance of reinforced concrete

1 Introduction

The theory under consideration, according to its developers, is coordinated and promoted by international institutions (USA, Europe, Asia) for standardization within the framework of the "global harmonization scenario". It is positioned:

R. Sanzharovsky (✉)
ENU After L.N. Gumilev, 010000 Nur-Sultan, Kazakhstan

F. Sieber
Leibniz- Institut Fur Interdisziplinare Studien (LIFIS), 12489 Berlin, Germany

T. Ter-Emmanuilyan
Russian University of Transport, 127994 Moscow, Russia

N. Vatin et al. (eds.), *Proceedings of MPCPE 2021*, Lecture Notes in Civil Engineering 182, https://doi.org/10.1007/978-3-030-85236-8_47

internationally—a new advanced format for calculating of concrete creep for various structures and also of mathematical theory. The leaders also report on the theory that it "was derived from debate and collaboration between different schools and academics at a global level with the assistance of international standards organizations over the past 40 years." The following comprehensively testifies to the failure of the theory: the presence of a system of gross mathematical errors; violations of the principles and rules of classical mechanics and Eurocodes; inconsistencies with well-known experimental data; negative results of design practices, including global experience in designing unique structures by RAMBOLL structures, United Kingdom [1].

Let us to emphasize the first that we are analyzing no any abstract scientific theory, but the main directions of the theory underlying in the design of the modern construction and it is harming of the world level. This theory is the basis for calculating buildings and structures, taking into account the time factor, when considering the long-term resistance of reinforced concrete structures.

2 Methods

The leading scientists point out: the unique properties of reinforced concrete make it today the main building material used in the construction of residential, public and industrial buildings, as well as in the development of underground space in cities, in transport construction, bridges and tunnels; construction has turned into a nature-altering factor in its impact, due to its scale in terms of the tasks being implemented.

Among them:

2.1 *The Principle of Superposition*

The foundation of the theory—its superposition principle—violates the rules for differentiating functions. This violation is accompanied by the incorrect justification that "the principle of superposition is inherent in Voltairre's theory." As a result, erroneous kernels of integral equations are created. The principle of superposition is the basis of both the modern scientific creep theory of concrete, which is called the "world harmonized format" by foreign scientists, and the developments "in recent decades of international standardization institutions … for recommendations, norms and technical guidance documents" [2–4]. These works also indicate that McHenry [5] in USA (1943) "substantiated this trend by experimental studies of the creep of hermetic specimens using the principle of superposition which is characteristic for the theory of Volterra."

We give the basic law of creep of concrete in the original notation [2]:

$$\varepsilon_\sigma(\mathrm{t}) = \sigma(\mathrm{t_0})\mathrm{J}(\mathrm{t},\mathrm{t_0}) + \int_{\mathrm{t_0}}^{\mathrm{t}} \mathrm{J}(\mathrm{t},\mathrm{t}')\mathrm{d}\sigma(\mathrm{t}'), \tag{1}$$

where $\varepsilon_\sigma(t)$ is the complete strain from stress σ(t);

$J(t,t') = \frac{1}{E_c(t')} + \frac{\varphi(t,t')}{E_c(t')}$—compliance function;

$E_c(t')$ is nonstationary modulus of elasticity;

$\varphi(t,t')$ is nonstationary creep characteristic considering ageing.
In scientific publications (1) is usually integrated by parts, thus obtaining

$$\varepsilon_\sigma(\mathrm{t}) = \frac{\sigma(\mathrm{t})}{\mathrm{E_c}(\mathrm{t})} - \int_{\mathrm{t_0}}^{\mathrm{t}} \sigma(\mathrm{t}')\frac{\partial'}{\partial \mathrm{t}}\left[\frac{1}{\mathrm{E_c}(\mathrm{t}')} + \frac{\varphi(\mathrm{t},\mathrm{t}')}{\mathrm{E_c}(\mathrm{t}')}\right]\mathrm{dt}' \tag{1b}$$

The last part of the integral (1b).

$$\int_{t_0}^{t} \frac{\varphi(t,t')}{E_c(t')} d\sigma(t'),$$

describing the development of creep deformations with σ(t') variables is demonstrated. The principle of Boltzmann's linear superposition, corresponding to the stationary properties of creep of the material, is copied with the name of the principle of superposition; that is, substitution of the fundamental property of concrete occurs in this case. This substitution, on the one hand, leads to the loss of three components in the basic law (1a), caused by the rate of change of the coefficient of compliance.

$$\sigma(t')\frac{1}{E_c(t')}\frac{\partial\varphi(t,t')}{\partial t} + \sigma(t')\frac{1}{E_c(t')}\frac{\partial\varphi(t,t')}{\partial t'} - \sigma(t')\varphi(t,t')\frac{\dot{E}_c(t')}{E_c^2(t')},$$

at that they are comparable in importance to the remaining term. These losses cause significant discrepancies between theory and experiments, described in the scientific literature. They lead to the incorrect expression of the creep kernel, even within the framework of the non-existent linear creep theory of concrete. The principle of superposition distorts this linear theory, causing the appearance of additional non-existent bodies. The number of such bodies depends on the form of the function $\varphi(t,t')$, which describes the non-stationary creep characteristic in the basic law (1). We write this function in the well-known, widely used in scientific literature, as

$$\frac{\varphi(t,t')}{E(t')} = \frac{\varphi_{\infty}(t')\left[1 - e^{-\gamma(t-t')}\right]}{E_c(t')} \tag{2}$$

where $\varphi_{\infty}(t')$ is a function considering aging of concrete.

In the famous monograph of Prokopovich I.E. the creep characteristic $\varphi(t,t')$ of foreign scientists is designated as $\overline{C}(t,\tau)$ these are identical values.

In the case of (2), the basic law (1a) forms four superfluous (fictitious) bodies: two bodies of the Voigt type and two viscous elements connected in series with each other. The deformations of these bodies are equal.

$$\varepsilon_{1\phi}(t) = \int_{t_0}^{t} \sigma(t')\frac{1}{\eta_{1\phi}(t')}e^{-\gamma(t-t')}dt', \eta_{1\phi}(t') = \frac{E_c(t')}{\phi_{\infty}(t')}; \tag{3}$$

$$\varepsilon_{2\phi}(t) = \int_{t_0}^{t} \sigma(t')\frac{1}{\eta_{2\phi}(t')}dt', \eta_{2\phi}(t') = \frac{E_c^2(t')}{\dot{E}_c(t')}\frac{1}{\varphi_{\infty}(t')}; \tag{4}$$

$$\varepsilon_{3\phi}(t) = \int_{t_0}^{t} \sigma(t')\frac{1}{\eta_{3\phi}(t')}e^{-\gamma(t-t')}dt', \eta_{3\phi}(t') = -\frac{E_c^2(t')}{\dot{E}_c(t')}\frac{1}{\varphi_{\infty}(t')}; \tag{5}$$

$$\varepsilon_{4\phi}(t) = \int_{t_0}^{t} \sigma(t')\frac{1}{\eta_{4\phi}(t')}dt', \eta_{4\phi}(t') = -\frac{E_c^2(t')}{\dot{\varphi}_{\infty}(t')}, \tag{6}$$

where $\eta 1\phi, \ldots, \eta 4\phi$ are viscosity coefficients or coefficients of internal resistance of fictitious bodies; moreover, the bodies (8) of the Voigt and (9) of the viscous element expand when compressed.

Creep deformations (3)–(6), caused by the influence of the superposition principle on the classical connection (1c), are fiction; they are also summarized with short-term fictitious deformation

$$\varepsilon_{5\phi}(t) = -\int_{t_0}^{t} \sigma(t')\frac{\partial}{\partial t'}\frac{1}{E_c(t')}dt' \tag{7}$$

$$\varepsilon_{\sigma\phi}(t) = \sum_{i=1}^{5} \varepsilon_{i\phi}(t)$$

and introduce large errors in the total strain $\varepsilon\sigma$ (t), determined by the creep law (1b).

This revealed fact of a significant erroneous complication of the theory, caused by the principle of superposition, shows the inconsistency of the judgments of leading scientists currently expressed about the mythical advantages and benefits of this principle, evaluating it with the exact opposite: “and, on the other hand, this hypothesis greatly simplifies the phenomenological theory of creep and makes it simpler and more accessible for use in engineering calculations”; “as applied to linear creep deformations, the superposition principle was first used by L. Boltzmann (1874), but only recently it was proved (Persoz B.) for non-linear creep deformations”.

2.2 *Substitution of Instantaneous Nonlinear Properties of Concrete*

An erroneous substitution of instantaneous nonlinear concrete properties on creep properties (minute creep) is performed, which leads to the appearance of resistance forces proportional to acceleration, creates a violation of the principle of independence of the action of forces (Fourth Axiom), distorts the theory of calculation of reinforced concrete.

In the framework of the requirements of Eurocode 2 to the diagram of instantaneous deformation of concrete [6] (Fig. 1), it is necessary to recognize the error of the creep theory, the removal of plastic deformation εн from the total instantaneous deformation ε_i and its transfer into the category of creep deformation $\varepsilon_c(t)$: plastic deformation ε_n develops about 1–2 min (Aleksandrovsky, Bazant), and creep deformation $\varepsilon_c(t)$ lasts for years; the rate of increase of nonlinear deformations шs up to 2000 times the rate of increase of creep deformations (in 1 day); growth rate and time of elastic ε_e and nonlinear deformations ε_n have the same order; an error is the separation of these deformations by splitting the total quantity ε_i in violation of the Eurocode 2 rules.

Plastic instantaneous deformation ε_n is endowed with the name of fast-flowing or minute creep; total deformation of the usual $\varepsilon_c(t)$ and fast-flowing creep ε_n is sought using a creep measure.

$$C(t,\tau) = C_{\text{оп}}(t,\tau) + C_{\text{бп}}(t,\tau)$$

presented in the form of two functions for ordinary and for fast-flowing creep. Such a technique artificially creates unnecessary mathematical difficulties, and a violation of the principle of independence of the action of forces that is fundamental in mechanics arises; ridiculous results also arise in the design calculations.

The mathematical complexity consists in the necessity of constructing an unnecessary integral, followed by defects in the principle of superposition,

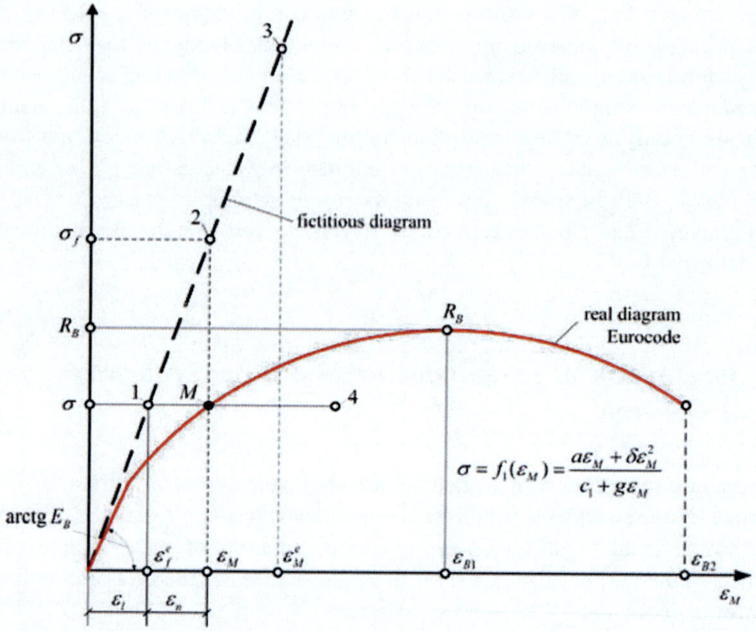

Fig. 1 The distortion of the σ-ε diagram of concrete

$$\varepsilon_{\text{н}}(t) = \int_{\tau_1}^{t} \sigma(\tau) \frac{\partial}{\partial \tau} C_{\text{бп}}(t, \tau)$$

whereas εн is easily found from the Sargin formula, other equations describing instant diagrams, for example, from Emperger's parabola $\varepsilon_{\text{н}} = B_2 \sigma^2$or from the dependence proposed by NIIZHB [6]

$$\varepsilon_{\text{н}} = \frac{\sigma^4}{ER_{\text{пр}}^3}\left(0{,}1 + \frac{24}{2 + \rightleftarrows R_{\text{пр}}}\right)$$

Comparing these formulas with each other, we see the fallacy of the integral form, designed to find the fast-flowing creep, its artificiality.

Let us give an instructive example showing the absurdity of the results obtained using fast-flowing creep deformations. Consider the longitudinal bending of the compressed rack in the interval of one day after loading, when, in the main, only fast-flowing creep has time to appear. A long-term critical force in accordance with the well-known decisions of Rzhanitsyn, Rabotnov, Shesterikov, Prokopovich [7–11], is equal to $P_{\text{Д}} = \frac{\pi^2 HI}{e^2}$, where $H = \frac{E}{1+\phi_{\text{бн}}}$, where φбн—characteristic of fast-flowing creep. This critical force tends to infinity with a length l → 0, Fig. 2, what is rejected by both experiments and common sense.

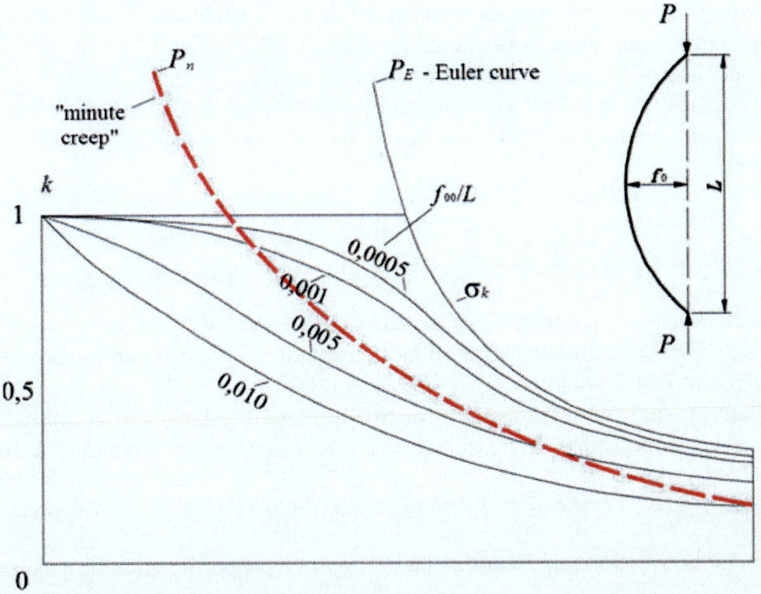

Fig. 2 The schedule of calculation of compressed-curved concrete structures with an initial deflection

If instantaneous nonlinear deformations are not added to creep deformations, then we have a tangential-modular (or reduced-modular) critical force with a finite value as $l \to 0$.

Note that the renaming of plastic deformations ε_n (Fig. 1) in the creep deformation $\varepsilon_c(t)$ and their uniform mathematical description

$$\varepsilon(t) = \frac{\sigma(t)}{E(t)} - \int_{\tau_1}^{t} \frac{\sigma(u)}{E(u)} L_E(t, u) du$$

in the record of function LE(t,u) leads to distortion of the results of experimental research on concrete creep problems in all countries of the world [2]. As a result of such mixing, creep deformations mistakenly acquire initial "vertical segments", distorting the values of creep deformations (up to 50%), distracting concrete creep researchers and misleading experts in the theory of reinforced concrete.

The erroneous assumption of "fast-flowing creep", "minute creep" and "vertical segments" has distorted the direction of the development of the theory of creep of reinforced concrete. The introduction of this assumption in the norm is detrimental to reinforced concrete construction.

Writing a concrete creep measure in the form of such a sum not only leads to mathematical complication of the creep theory, but also violates the principle of independence of the action of Newtonian mechanics.

For clarity, we consider a simple and instructive case. We will write down the measure of creep in the form proposed by Aleksandrovsky S.V. [12] (in his notation)

$$C(t,\tau) = A_3\left[1 - e^{-\gamma(t-\tau)}\right] + A_4\left[1 - e^{-\alpha(t-\tau)}\right] \tag{8}$$

where $A_3 = \psi(\infty) = const$; $A_4 = \Delta(\infty) = const$; $\alpha > \; > \gamma > 0$.

"The presence of the second term in the formula ... provides an initial steep rise in creep curves for small t-τ".

Differentiating the integral Eq. (1a) two times in t, taking into account (9), we obtain the second order differential (E = const) equation corresponding to that.

$$\varepsilon E + (\gamma + \alpha)E\dot{\varepsilon} + \gamma\alpha E\varepsilon = \sigma + [(\gamma + \alpha) + EA_3\gamma + EA_4\alpha]\dot{\sigma} + [1 + EA_3 + EA_4]\gamma\alpha\sigma.$$

From this equation it is clear that there is a force proportional to the acceleration.

$$\sigma = \frac{E}{(1 + EA_3 + EA_4)\gamma\alpha}\ddot{\varepsilon}(t).$$

The remaining forces are proportional to $\varepsilon, \dot{\varepsilon}, \dot{\sigma}, \ddot{\sigma}$, insignificant.

In Newtonian mechanics, the presence of forces proportional to acceleration, indicates violation of the principle of independence of action of forces, and the impossibility of using expression (8) for concrete creep in practical problems, with variable forces σ(t). We will come to the same result if we use many other formulas to describe the creep measure in the form of two or more terms (Yashin, Mac-Henry, Prokopovich, Ulitsky, etc.).

2.3 *Nonstationary Instantaneous Elastic Properties of Concrete*

Nonstationary instantaneous elastic properties of concrete are transformed into a model of Maxwell's creep theory, leading to an error of up to 300%.The substitution, unnoticed by scientists, distorts the Hooke elastic model, erroneous here, Fig. 1; it attaches to the classical linear connection $\frac{\sigma(t)}{E(t)}$ a non-existent and unreal body of a viscous fluid, with Newton's linear viscosity coefficient $\eta(t') = \frac{E_c^2(t')}{\dot{E}_c(t)}$.

$$\varepsilon_y(t) = \frac{\sigma(t_0)}{E_c(t_0)} + \int\limits_{t_0}^{t} \frac{1}{E_c(t')} d\sigma(t') = \frac{\sigma(t)}{E_c(t)} - \int\limits_{t_0}^{t} \sigma(t') \frac{\partial}{\partial t'} \frac{1}{E_c(t')} dt' \tag{9}$$

Formula (9) represents the first terms in (1a), (1b), and demonstrates the transformation of a classical nonstationary elastic body into Maxwell’s viscoelastic medium.

The essence of the second substitution follows from the principle of superposition, the fundamental principle in the construction of the law of creep (1a). The principle of superposition, being a kind of catachresis (abuse), simultaneously combines two concepts that are incompatible in meaning: stationarity and non-stationarity of the mechanical properties of concrete. Borrowing the Boltzmann scheme, the principle of superposition borrows the nonstationary of the corresponding material properties of this scheme, that is, rejects the fundamental non-stationary linear properties of concrete 6, replacing them with stationary properties. The principle of superposition is applied in non-stationary linear properties (1b), under the conditions of the fundamental meaning of this non-stationarity.

The mathematical essence of the error arises from the second substitution in the values of deformations of concrete, detected as follows.

The rate of elastic deformation is

$$\dot{\varepsilon}_y(t') = \dot{\sigma}(t') \frac{1}{E_c(t')} + \sigma(t') \frac{\partial}{\partial t'} \frac{1}{E_c(t')}.$$

Integrating, we obtain

$$\varepsilon_y(t) - \varepsilon_y(t_0) = \int\limits_{t_0}^{t} \frac{1}{E_c(t')} d\sigma(t') + \int\limits_{t_0}^{t} \sigma(t') \frac{\partial}{\partial t'} \frac{1}{E_c(t')} dt'.$$

Integrating the first term in parts, we find

$$\varepsilon_y(t) - \varepsilon_y(t_0) = \frac{\sigma(t)}{E_c(t)} - \frac{\sigma(t_0)}{E_c(t_0)} - \int\limits_{t_0}^{t} \sigma(t') \frac{\partial}{\partial t'} \frac{1}{E_c(t')} dt' + \int\limits_{t_0}^{t} \sigma(t') \frac{\partial}{\partial t'} \frac{1}{E_c(t')} dt'.$$

Hence the short-term deformation is equal to

$$\varepsilon_y(t) = \frac{\sigma(t)}{E_c(t)}; \tag{10}$$

it is also seen that the first term under the integral sign (1a) is excessive, and the use of superposition principle.

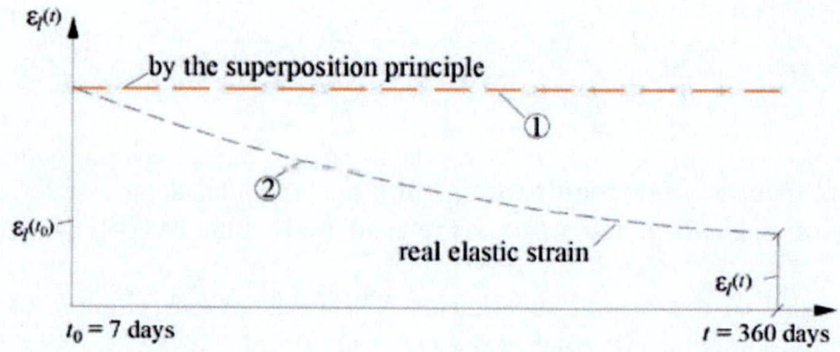

Fig. 3 Comparison of εy(t0) and εy(t)

$$\varepsilon_y(t) = \frac{\sigma(t_0)}{E_c(t_0)} - \int_{t_0}^{t} \frac{1}{E_c(t')} d\sigma(t') = \frac{\sigma(t)}{E_c(t)} - \int_{t_0}^{t} \sigma(t') \frac{\partial}{\partial t'} \frac{1}{E_c(t')} dt', \tag{11}$$

in (1a) and (1b) is deeply mistaken.

Let us make a numerical estimate of the error arising in determining the instantaneous elastic deformation distorted by the principle of superposition. Using $\sigma(t) = \sigma_0 = const$ in (10), (11) we obtain $\varepsilon_y(t) = \frac{\sigma_0}{E_c(t)}$ and $\varepsilon_y(t_0) = \frac{\sigma_0}{E_c(t_0)} = const.$ Comparison of these deformations is shown in Fig. 3.

Curve 2 in Fig. 3 corresponds to the VNIIG data on the change of the elastic modulus Ec(t) with time. Errors in the value of the elastic deformation reach, at t = 360 days, ≈ 300%.

2.4 *Only Linear Creep is Considered.*

Only linear creep is considered. Famous scientists S.V. Aleksandrovsky and P.I. Vasil'yev emphasize (based on unique experiments): "there is no linear creep"; the error is 200–500%. "Creep deformations are non-linear from the lowest loading levels, … no linear creep area … exists." So testify the founders of the theory Gvozdev AA, Arutyunyan N.Kh., Aleksandrovsky S.V., Vasilyev P.I., Fig. 4 [13].

As we known In the theory of linear creep of concrete, which is a world theory, the Volterra integral equations representing the creep of concrete with its unsteady and non-linear properties have fictitious nuclei that violate the prescribed mathematical order of their construction: as a result, concrete forms an erroneous set of fictitious forces that incorrectly form creep deformations.

The classical Abel case allows us to emphasize an important essence: the substitution or insignificant change of only one parameter of the integral equation distorts the essence of the theory.

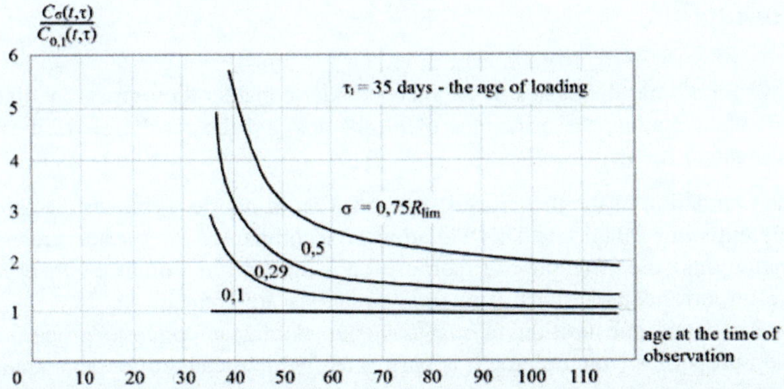

Fig. 4 Change in the ratio of specific creep deformations at different initial stress levels Cσ(t,τ) to specific creep deformations at initial stress level C0,1(t,τ)

In the theory of reinforced concrete creep, the substitution of ten fundamental properties of structural concrete, described in detail in [8], is carried out; substitution of each of them distorts the essence of the theory and leads to gross calculation errors. Let us give numerical estimates of the two substitutions indicated above under numbers 1 and 2: 1—substitution of non-stationary short-term deformations—leads to an error of up to 300% [14]; 2—substitution of non-linear creep deformations by nonexistent linear ones—changes their values by a factor of 2–5 [15], Fig. 4. The nonexistent linear creep property is used instead. From the data in Fig. 1 shows that the error from such a substitution is up to +400%, at. If we take as a basis the average curve corresponding to its experimental parameters, then the error from such a distortion will be from +200% to −200%.

2.5 *The Principle of "Algebraization" of Integral Equations*

The principle of "algebraization" of integral equations certainly leads to significantly contradictory results in the calculations of structures. This was repeatedly accented by N.Kh. Harutyunyan [16]. In the theory, "the relationship between stresses and strains is established by a formula based on a linear relationship between stresses and strains and on the principle of superposition."

$$\varepsilon^*(t) = \frac{\sigma^*(t)}{E(t)} - \int_{\tau_1}^{t} \sigma^*(\tau)\frac{\partial\delta(t,\tau)}{\partial\tau}d\tau, \quad (12)$$

where $\delta(t,\tau) = \frac{1}{E(\tau)} + C(t,\tau)$;

C(t,τ)—a measure of creep.

3 Results

We have found that the basics of the theory under consideration grossly violate the rules of mathematics, the principles of mechanics and the results of solid experiments. Among them:

1. The foundation of the theory—its superposition principle—violates the rules for differentiating functions. This violation is accompanied by the incorrect justification that "the principle of superposition is inherent in Voltairre's theory." As a result, erroneous kernels of integral equations are created.
2. An erroneous substitution of instantaneous nonlinear concrete properties on creep properties (minute creep) is performed, which leads to the appearance of resistance forces proportional to acceleration, creates a violation of the principle of independence of the action of forces (Fourth Axiom), distorts the theory of calculation of reinforced concrete.
3. Nonstationary instantaneous elastic properties of concrete are transformed into a model of Maxwell's creep theory, leading to an error of up to 300%.
4. Only linear creep is considered. Famous scientists S.V. Aleksandrovsky and P.I. Vasil'yev emphasize (based on unique experiments): "there is no linear creep"; the error is 200–500%.
5. The principle of "algebraization" of integral equations certainly leads to significantly contradictory results in the calculations of structures. This was repeatedly accented by N.Kh. Harutyunyan.

In addition, there are a number of incorrect actions for a simplified calculation of the creep of reinforced concrete structures: this results in unreliability or uneconomic structures.

4 Conclusions

The analyzed theory of reinforced concrete in certain aspects is unscientific. It was pointed at different times out by authoritative scientists: Skramtaev B.G., Keldysh V. M., Nikitin G.V., Rzhanitsyn A.R., Geniev G.A., Drozdov P. The problem of the unscientific nature of the theory of concrete creep is indicated by the negative results of design practice, including the world experience in the design of unique structures with RAMBOLL structures (Great Britain) [1]. President of fib Gordon Clarke warns: "accurately predicting the impact of creep … is highly controversial". We have established the reasons for the unscientific nature of this theory—among them are mathematical errors and violation of the principles of classical mechanics [14, 15, 17–19]; we also developed a new nonlinear theory of concrete creep, which has not yet been published, supplementing the general theory [20].

References

1. Klark G (2014) Challenges of high buildings. Industry. Engineering Newspaper 5(11–12)
2. Chiorino MA (2014) Analysis of structural effects of time—dependent behavior of concrete: an internationally harmonized format. In: Concrete and reinforced concrete—glance at future. III All Russian (International) conference on concrete and reinforced concrete, vol 7, pp 338–350. Moscow
3. fib, Model code for concrete structures 2010. Ernst & Sohn (2013)
4. Chiorino MA (2011) Analysis of creep and shrinkage effects on concrete structures, final draft (Chairm. of Edit. Team), ACI Com-mittee 209
5. Mc.Henr (1943) A new aspect of creep in concrete and its application to design. In: Proc. ASTM, V.S.
6. Creep and shrinkage of concrete and reinforced concrete structures. Problems and prospects of development/GOSSTROY of the USSR; NIIZB. Construction publishing, Moscow (1976)
7. Prokopovich IE (1963) The influence of long-term processes on the stressed and deformed state. Gosstroyizdat, Moscow
8. Rzhanitsyn AR (1968) Creep theory. Construction publishing, Moscow
9. Prokopovich I.E, Ulitsky I.I (1963) On the theory of the use of concrete. Izvestiya VUZov. Constr Archit (10):23–27
10. Rabotnov YuN (1977) Elements of hereditary mechanics of solids. Nauka, Moscow
11. Leaderman H (1943) Elastic and creep properties of filamentous and other high polymers. Textile Foundation, Washington
12. Aleksandrovsky SV (1973) Calculation of concrete and reinforced concrete structures for changes in temperature and humidity, taking into account the use of. Construction publishing, Moscow
13. Creep and shrinkage of concrete and reinforced concrete structures. State problems and prospects of development/Gosstroy of the Ukrainian SSR; NIIZB. Construction publishing, Moscow (1976)
14. Sanzharovskij RS, Manchenko MM, Gadzhiev MA, Musabaev TT, Ter-Emmanuil'yan TN, Varenik KA (2019) System of stability of the modern theory of long-term resistance of reinforced concrete and prevention of designers. Constr Mech Eng Struct Struct 1(15):3–24
15. Sanzharovsky RS, Ter-Emmanuilyan TN, Manchenko MM (2018) Superposi-tion principle as the fundamental error of the creep theory and standards of the rein-forced concrete. Struct Mech Eng Constr Buildings 14(2):92–104
16. Harutyunyan N (1952) Some questions of the theory of creep. Construction publishing, Moscow
17. Concrete and reinforced concrete-a look into the future. Scientific papers of the III All-Russian (II International) Conference on Concrete and Reinforced Concrete. Moscow (2014)
18. Sanzharovskij RS, Manchenko MM (2016) Errors in the theory of the use of reinforced concrete and modern standards. Constr Mech Eng Struct Struct (3):25–32
19. Sanjarovskiy R, Ter-Emmanuilyan T, Manchenko M (2015) Creep of concrete and its instantaneous nonlinear deformation in the calculation of structures. Concretization 10:238–247
20. Sanzharovskij RS (1984) Stability of elements of building structures during use. LSU

Strength and Deformability of Reinforced Wooden Beams of Variable Stiffness

Mikhail Lisyatnikov, Tatyana Glebova, Kazimir Rusak, and Andrii Ivaniuk

Abstract The method of designing reinforced wooden beams of variable stiffness with group reinforcement is described. The dependence of the breakage of reinforcement in the span on the reinforcement coefficient is given and the influence of changes in the stiffness of structures on the deformability is determined. Theoretical stress diagrams are analyzed, indicating the possibility of hazardous stress concentrations in reinforced beams of variable stiffness. Diagrams M and Q, as well as schemes of theoretical breakage of reinforcement in beams were constructed. The results of determining the influence of the reinforcement coefficient on the place of breakage of reinforcement in the span of beams are displayed graphically. The dependences of the anchorage length on the diameter and reinforcement coefficient, the relative length of the anchorage of the rods on the span of the reinforced beams have been constructed. Defined deflection of reinforced wooden beams of variable stiffness. A further direction of research is indicated, which consists in determining local stresses in zones of change in the stiffness of beams.

Keywords Timber · Building · Beams · Reinforcement of wooden beams

1 Introduction

To save materials of steel and epoxy compound, it became necessary to create the most efficient reinforced wooden structures [1–4]. Which can be obtained by rational placement of reinforcement in the span in accordance with the diagram of the greatest bending moments, i.e. create structures of variable stiffness by breaking the reinforcement in the span. But as shown by preliminary calculations performed by Boldyunum V.F. [5], in such structures it is possible that a dangerous stress

M. Lisyatnikov (✉) · T. Glebova · K. Rusak
Vladimir State University Named After Alexander and Nikolay Stoletovs,
Vladimir, Russian Federation

A. Ivaniuk
National University of Water and Environmental Engineering, Rivne, Ukraine

N. Vatin et al. (eds.), *Proceedings of MPCPE 2021*, Lecture Notes in Civil Engineering 182, https://doi.org/10.1007/978-3-030-85236-8_48

concentration in the wood occurs at the ends of the reinforcing bars (Fig. 1). This is indicated by the dependence describing the distribution of shear stresses along the length of the reinforcement [6, 7]:

$$\tau_{\alpha} = -\frac{q_x}{S_e \psi}\left\{1 - \left[1 - \frac{\alpha}{2}\left(l^2 - b^2\right)\right]\frac{Sh\alpha x}{\alpha x Ch\alpha b}\right\} \quad (1)$$

The designations are taken from the statics condition.

Analysis of expression (1) shows that the magnitude of shear stresses is significantly influenced by the place of breakage of reinforcement in the span [8]. Based on this, it becomes necessary to theoretically determine the optimal place of breakage of reinforcement in the span of reinforced wooden structures [9–11], i.e. so that the resulting structures of variable stiffness would have the same or similar strength and stiffness indicators as structures of constant stiffness [12, 13].

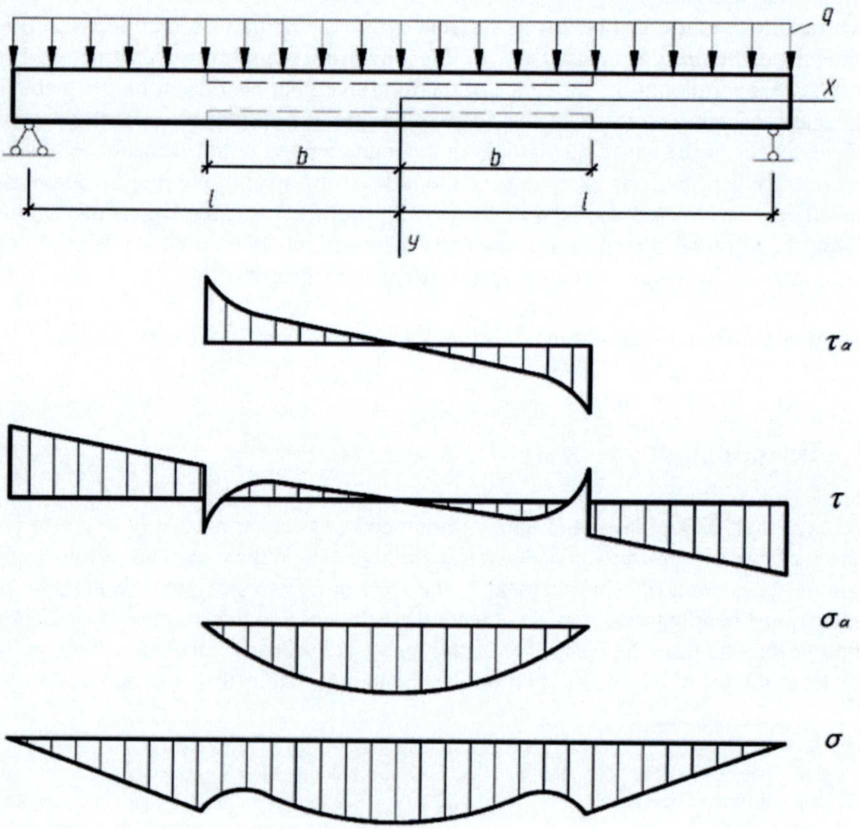

Fig. 1 Theoretical diagrams indicating the possibility of hazardous stress concentrations in reinforced beams of variable stiffness

2 Methods

It is more expedient to determine the place of breakage of reinforcement using a simpler method for calculating reinforced structures—according to the given sections [14–16].

The place of the theoretical breakage of reinforcement in bent structures is determined from the condition that the wood perceives a certain part of the bending moment $M_{\max}$, acting on the complex structure (Fig. 2), i.e.:

$$M_x = W_w R_b \tag{2}$$

For a single-span element loaded with a uniformly distributed load—q, the value of the bending moment [17] at any point of the beam section is equal to:

$$M_x = \frac{q_x(l - x)}{2} \tag{3}$$

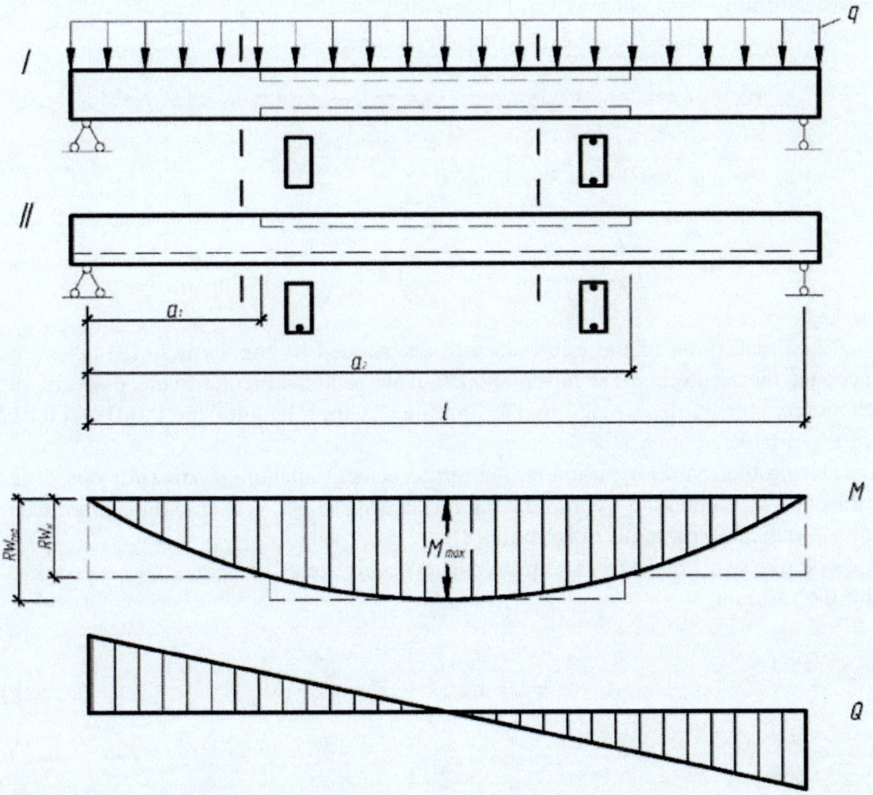

Fig. 2 Diagrams M, Q and circuit breakage theoretical reinforcement girders

Solving Eqs. (2) and (3) together, we determine the distance from the support to the place of theoretical breakage of the reinforcement:

$$a_{1,2} = \frac{l}{2} \pm \sqrt{\left(\frac{l}{2}\right)^2 - \frac{2R_b W_w}{q}} \tag{4}$$

The value of the calculated uniformly distributed load is found from the condition of the strength of the reinforced beams [18]:

$$\begin{aligned} M_{\max} &= W_{red} R_b = W_0(1+3\eta\mu)R_b \\ M_{\max} &= \frac{ql^2}{8} \end{aligned} \tag{5}$$

From identity (5) we determine the load:

$$q = \frac{8W_0(1+3\eta\mu)R_b}{l^2} \tag{6}$$

Substituting the load into Eq. (4), we get:

$$a_{1,2} = \frac{l}{2}\left(1 \pm \sqrt{1 - \frac{R_b W_w}{R_b W_0(1+3\eta\mu)}}\right) = \frac{l}{2}\left(1 \pm \sqrt{1 - \frac{M_w}{M_{\max}}}\right) \tag{7}$$

Let us assume that $W_w = W_0$ than:

$$a_{1,2} = \frac{l}{2}\left(1 + \sqrt{1 - \frac{1}{1+3\eta\mu}}\right) = \frac{l}{2}\left(1 \pm \sqrt{\frac{3\eta\mu}{1+3\eta\mu}}\right) \tag{8}$$

The break point of the reinforcement determined by the formula (8) takes into account the reinforcement termination, i.e. its establishment for the place of the theoretical break, determined by the formula (7), for a length from 0.0012 to 0.017 of the calculated span [19].

During the search experiment, the reinforcement anchorage zone for one series of beams was increased by introducing the safety factor $C = 2$ into the denominator of the radical expression of formula (8).

The place of breakage of reinforcement in this series of beams was determined by the formula:

$$a_{1,2} = \frac{l}{2}\left(1 \pm \sqrt{1 - \frac{1}{C(1+3\eta\mu)}}\right) \tag{9}$$

Analytical formulas for determining the theoretical break point of the compressed zone reinforcement in beams [20–22] with the tensile zone reinforcement along the entire span were derived using a similar method.

We solve jointly Eq. (3) with the equation describing the maximum bending moment perceived by an asymmetrically reinforced section of the beam, we get:

$$M_x = W^c_{red} R_b \tag{10}$$

Determine the location of the breakage of the reinforcement in the compressed zone:

$$a^c_{1,2} = \frac{l}{2} \pm \sqrt{\left(\frac{l}{2}\right)^2 - \frac{2R_b W^c_{red}}{q}} \tag{11}$$

Substituting the load determined from the strength condition of the symmetrically reinforced section [23, 24] by formula (6), we obtain:

$$a^c_{1,2} = \frac{l}{2}\left(1 \pm \sqrt{1 - \frac{W^c_{red}}{W_{red}}}\right), \tag{12}$$

where W^c_{red}—reduced moment of resistance of the compressed zone relative to the neutral axis (13) located at a distance h_c(14) from the upper edge of the beam [25]:

$$W^c_{red} = \frac{I_{red}}{h_c} \tag{13}$$

$$h_c = \frac{h_{n.a.}(1 + 2\eta\mu_{as})}{2(1 + \eta\mu_{as})} \tag{14}$$

The moment of inertia of an asymmetrically reinforced section [26] is determined by the formula:

$$I_{red} = \frac{bh^3_{n.a.}(1 + 4\eta\mu_{as})}{12(1 + \eta\mu_{as})} \tag{15}$$

Substituting expressions (14) and (15) into formula (13), we obtain the reduced moment of resistance of the compressed zone:

$$W^c_{red} = \frac{bh^2_{n.a.}(1 + 4\eta\mu_{as})}{6(1 + 2\eta\mu_{as})} \tag{16}$$

Let us assume that the design height of a symmetrically reinforced section is equal to the height of the section with an asymmetrical arrangement of reinforcement, i.e. $h_0 = h_{0as}$. In this case, the reinforcement coefficient for a section with an asymmetrical arrangement of reinforcement will be half that of a symmetrical one, i.e. $\mu_{as} = \mu/2$. Then the place of breakage of the reinforcement in the compressed zone, taking into account the anchoring, should be determined by the formula:

$$a^{c}_{1,2} = \frac{l}{2}\left(1 \pm \sqrt{1 - \frac{1+2\eta\mu}{(1+\eta\mu)(1+3\eta\mu)}}\right), \tag{17}$$

when substituting the reinforcement coefficient $\mu = F_a/bh_o$—sections with symmetrically located reinforcement.

3 Results and Discussion

The place of breakage of reinforcement, determined by formula (17), takes into account the insertion of reinforcement beyond the place of theoretical break, determined by formula (11) for the anchorage length from 0.0018 to 0.025 of the calculated span of the beams. The anchorage length was obtained under the condition that during the operation of the beams in the zone of the maximum bending moment, the calculated strength characteristics of the materials: steel and wood are fully used. In this case, several factors affect the anchorage zone and the location of the breakage of the reinforcement: the reinforcement coefficient, the diameter of the reinforcement and the span of the beams.

The influence of the reinforcement coefficient on the place of breakage of reinforcement in the span of beams is determined by expressions (7)–(9), (17) and is shown in Fig. 3.

The dependence of the anchorage zone—l_{an} on the diameter of the reinforcement —d and the reinforcement coefficient is shown in Fig. 4. It is obtained on the basis of the assumptions made when deriving formulas (8) and (17). In the first case, due to the fact that the height of the unreinforced section of the beam is greater than the calculated one by the diameter of the bar $h = h_0 + d$, and in the second case, the calculated height of the asymmetrically reinforced section is greater than the calculated height of the symmetrically reinforced section $0.5d$ on the reinforcement.

The influence of the span of beams on the relative length of the embedment of rods $\lambda_{an} = l_{an}/d$ is shown in Fig. 5.

It should be noted that in practice the anchorage zone turns out to be much larger than that obtained by formulas (8) and (17). Since the main criterion in the design and operation of wooden beams, in most cases, is their deformability.

Consequently, the design load, as a rule, for such structures is taken to be significantly less than that determined by formula (6), from the condition of the strength of symmetrically reinforced beams. Therefore, the strength characteristics of materials, steel and wood, in the zone of action of the maximum bending moment, are not fully used.

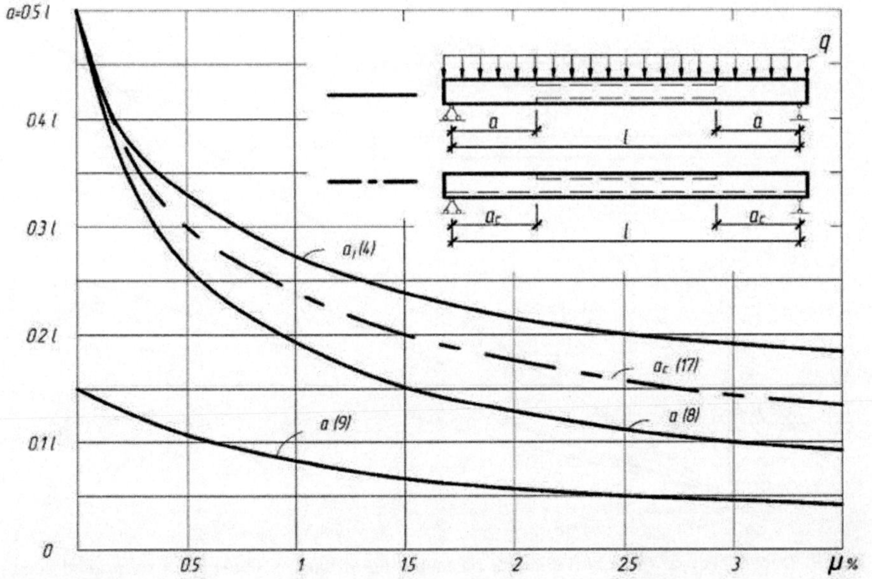

Fig. 3 Graph for determining the theoretical breakage location of reinforcement in beams of variable stiffness

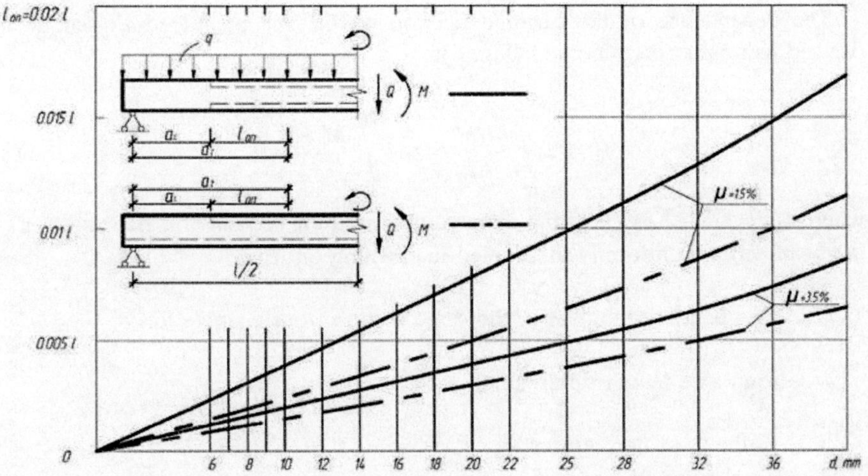

Fig. 4 Dependence of anchorage length on diameter and reinforcement coefficient

It is known that an increase in the anchorage zone leads to an increase in the strength and rigidity of the structure, and a change in the stiffness of beams, due to the breakage of reinforcement in the span, leads to an increase in deflection.

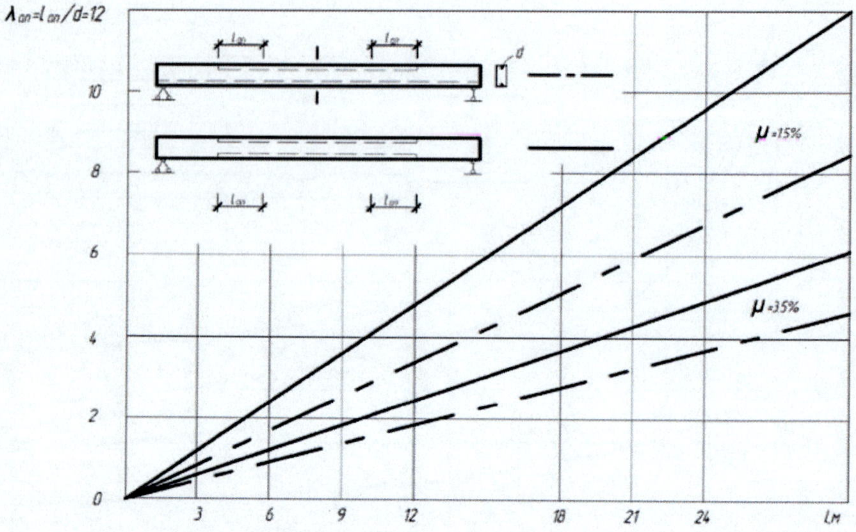

Fig. 5 Dependence of the relative length of anchoring of bars on the span of reinforced beams

Therefore, it became necessary to theoretically determine the deformability of beams of variable stiffness, loaded, as a rule, during operation with an evenly distributed load—q^n all over the span—l.

The dependence of the beam deflection on the accepted loading scheme is derived using the force method (Fig. 6):

$$f_0 = 2\int_0^a \frac{M_x M_{x_1}}{EI_w} dz + \int_a^{l/2} \frac{M_x M_{x_1}}{EI_{red}} dz, \quad (18)$$

where $M_x = \frac{q^n zl}{2} - \frac{q^n z^2}{2}$—bending moment in the current section of a given section of the beam from a uniformly distributed load acting on it;

$M_{x_1} = \frac{z}{2}$—bending moment in the same section from a single load;

dz—beam axle length differential;

EI_w—stiffness of unreinforced area;

EI_{red}—stiffness of the reinforced section.

We transform expression (18):

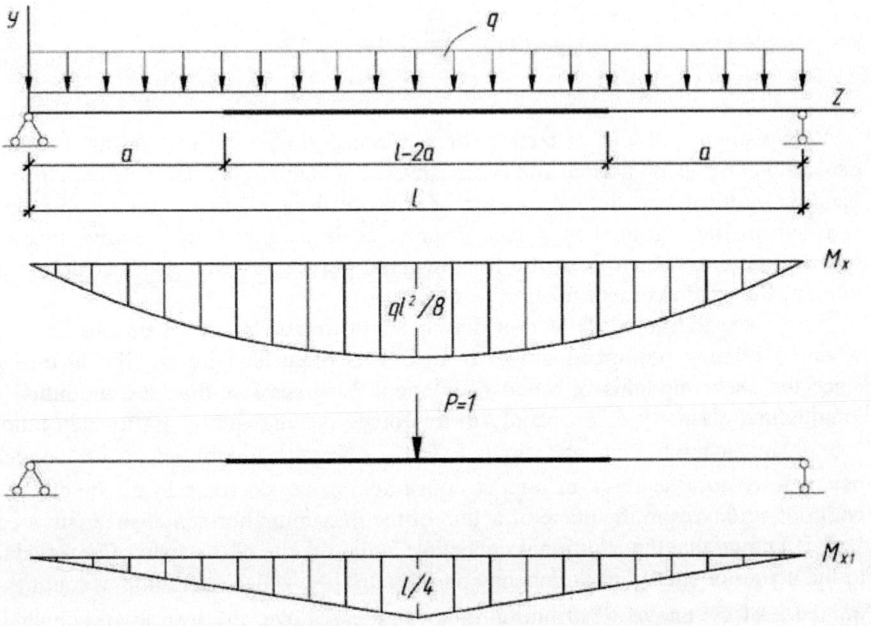

Fig. 6 Design scheme for determining displacements (deflections) of reinforced beams of variable stiffness

$$f_0 = 2\int\limits_0^a \left(\frac{q^n l z^2}{4} - \frac{q^n z^3}{4}\right)\frac{dz}{EI_w} + 2\int\limits_a^{l/2}\left(\frac{q^n l z^2}{4} - \frac{q^n z^3}{4}\right)\frac{dz}{EI_{red}} = \frac{2}{EI_w}\left|\frac{q^n l z^3}{12} - \frac{q^n z^4}{16}\right|_0^a$$
$$+\frac{2}{EI_{red}}\left|\frac{q^n l z^3}{12} - \frac{q^n z^4}{16}\right|_a^{l/2} = \frac{2}{EI_w}\left(\frac{q^n l a^3}{12} - \frac{q^n a^4}{16}\right) + \frac{2}{EI_{red}}\left(\frac{q^n l^4}{96} - \frac{q^n l^4}{256} - \frac{q^n l a^3}{12} + \frac{q^n a^4}{16}\right)$$
$$= \frac{5q^n l^4}{384EI_{red}} - \frac{4q^n l a^3 - 3q^n a^4}{24}\left(\frac{1}{EI_w} - \frac{1}{EI_{red}}\right)$$

where I_w—the moment of inertia of an unreinforced section, in the case of reinforcement of a stretched zone along the entire span, and compressed on a part of the length is taken equal to:

$$I_{red.as} = \frac{bh^3_{n.a.}(1+4\eta\mu_{as})}{12(1+\eta\mu_{as})} \tag{19}$$

I_{red}—reduced moment of inertia of the section in the zone of action of the maximum bending moment:

$$I_{red} = \frac{bh_0^3}{12}(1 + 3\eta\mu) \quad (20)$$

It is known that the deflection of reinforced timber beams during bending generally depends on normal and shear stresses, i.e. consists of deformations from a bending moment, causing the rotation of cross-sections relative to the neutral axis, and deformations from shear forces, manifested in parallel shears relative to each other of the cross-sections of the base material—wood, as well as shear deformations of the reinforcement relative to wood.

As a rule, additional deflection due to lateral forces is not taken into account when calculating reinforced timber beams. This often leads to significant errors, since the shear modulus of wood G is about 20 times less than the modulus of longitudinal elasticity E. Especially in reinforced beams with height-to-span ratios $\frac{h}{l} > \frac{1}{17}$ and a reinforcement ratio μ = 2.5%, since the reinforcement of wooden beams leads to a decrease in lumber consumption, i.e. decrease in the height and width of structures in the plane of action of the maximum horizontal shear stresses.

It is known that the additional deflection in the middle of the span of beams due to the action of shear forces depends on the ratio $\frac{\tau_{\max}}{\tau_{mid}}$, which takes into account the influence of the uneven distribution of shear stresses over the section and depends on the shape of the cross section. With plastic deformations of beams, the ratio $\frac{\tau_{\max}}{\tau_{mid}}$ becomes a variable value, depending on the value of the elastic-stressed volume of the wood. Consequently, with an increase in the bending moment, the effect of shear forces on the total relative deflection of the beams in the elastic–plastic stage of work decreases. Therefore, the greatest influence of shear forces on the deflection of reinforced beams affects the conditionally elastic stage of work, when the presence of variable bending stiffness due to the flexibility of the steel-wood adhesive bond is only slightly manifested. Consequently, the additional deflection from the flexibility of the glue joint steel–wood, at the stage of conditionally elastic work of the reinforced wooden beams, can be neglected. However, when a constant or long-term load is applied in time, an increase in deflections of reinforced wooden structures is observed, which is caused by a change in the elastic modulus of wood and the effect of shear deformations of the adhesive bond of reinforcement with wood. Such an increase in deflections of complex structures must be taken into account when determining the total deflection by introducing a coefficient K_T, that takes into account the level of loading of reinforced structures.

Hence, the total deflection of reinforced wooden beams should be determined in accordance with the norms by the formula:

$$f = \frac{f_0}{k}\left[1 + C\left(\frac{h}{l}\right)^2\right]K_T \quad (21)$$

Formula (21) takes into account the effect of shear deformation from shear forces by the coefficient—C, at $k = 1$ for beams with a constant section height.

Here, f_0—is the deflection of reinforced wooden beams of constant or variable stiffness without taking into account shear deformations, is determined by the formula (18). The coefficient K_T in addition to the loading level, also takes into account the redistribution of stresses in time between the reinforcement and the wood, which leads to an increase in the normal stresses in the reinforcement and the glue seam and to a decrease in them in the wood. In beams with breakage of reinforcement in the span, the redistribution of stresses from wood to steel leads first to an increase in local stresses in the wood, and then to a decrease in them due to a change in the elastic modulus of wood over time.

Hence, it becomes necessary to determine local stresses in zones of change in the stiffness of beams, which is accepted as a priority task for further research.

4 Conclusions

As a result of the research carried out, the following conclusions can be drawn:

1. The theoretical definition of the optimal place of breakage of reinforcement in the span of reinforced wooden structures (beams of variable stiffness) has been carried out.
2. The influence of the reinforcement coefficient on the place of breakage of reinforcement in the span of beams has been determined.
3. Revealed the dependence of anchoring on the diameter of the reinforcement and the coefficient of reinforcement of complex structures.
4. The influence of the span of beams on the relative length of the embedment of the rods has been established.
5. The influence of the change in the stiffness of the reinforced structure on the deformability is considered.
6. A task for further research has been formulated.

References

1. Turkovskij SB, Pogorel'tsev AA (2001) Wooden structures with rigid joints in structures with corrosive medium. Promyshlennoe i Grazhdanskoe Stroit 10–13
2. Buka-Vaivade K, Serdjuks D, Goremikins V, Pakrastins L, Vatin NI (2018) Suspension structure with cross-laminated timber deck panels. Mag Civ Eng. https://doi.org/10.18720/MCE.83.12
3. Lukin M, Prusov E, Roshchina S, Karelina M, Vatin N (2021) Multi-span composite timber beams with rational steel reinforcements. Buildings. https://doi.org/10.3390/buildings11020046

4. Göldi M, Sell J, Strässler H (1979) The shearing strength of glue-joints of preimpregnated wood—contribution to the development of weatherproof gluelams | Scherfestigkeit der Klebverbindung von vorimprägniertem Holz—Beitrag zur Entwicklung wetterbeständigen Brettschichtholzes. https://doi.org/10.1007/BF02607422
5. Roshchina S, Lukin M, Lisyatnikov M (2020) Compressed-bent reinforced wooden elements with long-term load. In: Lecture notes in civil engineering. https://doi.org/10.1007/978-3-030-42351-3_7
6. Gravit MV, Serdjuks D, Bardin AV, Prusakov V, Buka-Vaivade K (2019) Fire design methods for structures with timber framework. Mag Civil Eng. https://doi.org/10.18720/MCE.85.8
7. Aleksiievets VI, Aleksiievets II, Ivaniuk AM, Roshchina SI (2020) Load-carrying capacity of bolted joints of timber structures under static loading. In: IOP conference series: materials and science engineering. https://doi.org/10.1088/1757-899X/896/1/012043
8. Gorpinchenko VM, Pogorel'tsev AA, Ecknadosyan IL (2005) Large-scale tests provided for a block, consisting of two wooden lens-type roof trusses of the sports complex "Strogino." Promyshlennoe i Grazhdanskoe Stroit 38
9. Koshcheev AA, Roshchina SI, Naichuk AY, Vatin NI (2020) The effect of eccentricity on the strength characteristics of glued rods made of steel cable reinforcement in solid wood. In: IOP conference series: materials and science engineering. https://doi.org/10.1088/1757-899X/896/1/012059
10. Haghani R, Al-Emrani M (2012) A new design model for adhesive joints used to bond FRP laminates to steel beams—part a: background and theory. Constr Build Mater. https://doi.org/10.1016/j.conbuildmat.2012.02.051
11. Lukin MV, Roshchina SI, Smirnov EA, Shunqi M (2020) Strengthening of the operated wooden floor beams with external rigid reinforcement. In: IOP conference series: materials and science engineering. https://doi.org/10.1088/1757-899X/896/1/012065
12. Ando D, Umemura K (2021) Bond structures between wood components and citric acid in wood-based molding. Polymers (Basel) 13:1–9. https://doi.org/10.3390/polym13010058
13. Parida G, Johnsson H, Fragiacomo M (2013) Provisions for ductile behavior of timber-to-steel connections with multiple glued-in rods. J Struct Eng (United States) 139:1468–1477. https://doi.org/10.1061/(ASCE)ST.1943-541X.0000735
14. Roschina SI, Lisyatnikov MS, Koshcheev AA (2019) Technical- and- economic efficiency of reinforced wooden structures. In: IOP conference series: materials and science engineering. https://doi.org/10.1088/1757-899X/698/2/022005
15. Turkovskij SB, Pogorel'tsev AA, Eknados'yan IL (2003) Selection of design scheme of lens-shaped trusses from adhesive wood. Stroit Mater 18–20
16. Roschina S, Gribanov A, Lukin M, Lisyatnikov M, Strekalkin A (2018) Calculation of wooden beams reinforced with polymeric composites with modification of the wood compression area. In: MATEC web of conferences. https://doi.org/10.1051/matecconf/201825104029
17. De Luca V, Marano C (2012) Prestressed glulam timbers reinforced with steel bars. Constr Build Mater 30:206–217. https://doi.org/10.1016/j.conbuildmat.2011.11.016
18. Koshcheev AA, Roshchina SI, Lukin MV, Lisyatnikov MS (2018) Wooden beams with reinforcement along a curvilinear trajectory. Mag Civil Eng. https://doi.org/10.18720/MCE.81.19
19. Babiak M, Gaff M, Sikora A, Hysek Š (2018) Modulus of elasticity in three- and four-point bending of wood. https://doi.org/10.1016/j.compstruct.2018.07.113
20. Lisyatnikov MS, Glebova TO, Ageev SP, Ivaniuk AM (2020) Strength of wood reinforced with a polymer composite for crumpling across the fibers. In: IOP conference series: materials and science engineering. https://doi.org/10.1088/1757-899X/896/1/012062
21. Steiger R, Gehri E, Widmann R (2007) Pull-out strength of axially loaded steel rods bonded in glulam parallel to the grain. Mater Struct Constr 40:69–78. https://doi.org/10.1617/s11527-006-9111-2
22. Riberholt H (1986) Glued bolts in Glulam

23. Lukin MV, Roshchina SI, Gribanov AS, Naychuk AY (2020) Stress-strain state of wooden beams with external reinforcement. In: IOP conference series: materials and science engineering. https://doi.org/10.1088/1757-899X/896/1/012066
24. Preobrazhenskaya IP, Pogorel'tsev AA, Turkovskij SB (2003) Development of design and construction of potassium chloride storehouse with framework from prefabricated wood frames with size of 63 m. Stroit Mater 14–16
25. Ling Z, Liu W, Yang H, Chen X (2018) Modelling of glued laminated timber joints with glued-in rod considering bond-slip location function. Eng Struct. https://doi.org/10.1016/j.engstruct.2018.08.098
26. Tlustochowicz G, Serrano E, Steiger R (2011) State-of-the-art review on timber connections with glued-in steel rods. Mater Struct Constr 44:997–1020. https://doi.org/10.1617/s11527-010-9682-9

Reinforced Concrete Shallow Shell of Negative Double Gaussian Curvature Built on the Basis of a Four-Lobed Hyperbolic Paraboloid

Mikhail Lukin, Marina Popova, Dmitry Reva, and Rustamkhan Abdikarimov

Abstract The stress–strain state of a flat reinforced concrete shell of negative double Gaussian curvature is numerically investigated. The studies were carried out on a shell built on the basis of a four-lobed hyperbolic paraboloid with a plan view of 80.0 × 80.0 m. It was experimentally proved that the strength of the shell is provided with the following reinforcement: two meshes made of B500 class reinforcement over the shell area, A500 class reinforcement at a distance of 20 m from shell angle—at the corners of the gipar petals, with A500 class rods in the contour edges. The non-displacement of the corners of the gipar is ensured by tightening from the bundles of double lay reinforcing ropes of the LK-RO type of the structure. The deflection of the shell is within the permissible value. The high efficiency of the calculation by the finite element method for evaluating the criteria of the bearing capacity of the shells up to their destruction has been proved.

Keywords Building · Reinforced concrete · Hyperbolic paraboloid · Shell · Stresses

1 Introduction

The main structural elements of the shell are a thin surface, side elements, and diaphragms.

The main characteristic of the classification of shells according to the work of the structure is spacer or no spacer. The purpose of the shells is certainly different, in this regard, they can be in the form of coverings, awnings or canopies and other structures. The structure can be supported along the surface, along the line, at individual points, as well as in a combination of these options. Different sources

M. Lukin (✉) · M. Popova · D. Reva
Vladimir State University Named After Alexander and Nikolay Stoletovs, 87 Gorky St., Vladimir 600005, Russian Federation

R. Abdikarimov
Tashkent Institute of Finance, Amir Temur avenue, 60A, Tashkent, Uzbekistan

N. Vatin et al. (eds.), *Proceedings of MPCPE 2021*, Lecture Notes in Civil Engineering 182, https://doi.org/10.1007/978-3-030-85236-8_49

have completely different values of overlapping spans [1–4]. It can be said unequivocally that in the modern world spans can reach 300 m or more. Depending on the Gaussian curvature, three types of surfaces are distinguished: parabolic (zero Gaussian curvature), hyperbolic (negative Gaussian curvature), and elliptic (positive Gaussian curvature).

A thin-walled shell of double curvature is formed by moving a parabolic curve in two directions. The shell, resting on four corners, creates very difficult working conditions for the structure [5–8]. Therefore, its contour must be supported by diaphragms—rigid supporting structures, the role of which can be played by arches, segmental trusses, and walls with rounded fronts. The diaphragms in this case will support the shell and work together with it, perceiving the shear forces developing along the contour [9–12].

Depending on the ratio of the lift boom (h) to the covered span (l), the shells are divided into ragged (h/l = 1/1 ÷ 1/4), shallow (h/l = 1/5 ÷ 1/6) and forming a sail arch (l = h).

Gentle shells of positive Gaussian curvature, made of reinforced concrete, are capable of covering rooms in the range from 18 to 100 m. Prefabricated-monolithic shell structures are widespread in construction [12, 13], the modules of which have a thickness of 3–5 cm and are reinforced with diagonal and contoured ribs. The given thickness of such a coating is from 8 to 15 cm.

Shells in the form of a hyperbolic paraboloid on square and rectangular plans are used to cover industrial, public and warehouse buildings [14–19]. The dimensions of the overlapping plan can vary from 10 to 70 m and more. Single-leaf and multi-leaf gipars can be used in the coatings.

In the practice of design and construction from monolithic reinforced concrete, the use of a fragment of the surface of a gipar (petal) on a square (rhombic) plan is often found [20–26]. Both single surfaces and their combination can be used.

Calculations and experience show that prefabricated shells of double curvature in comparison with flat systems in coatings with a span of 30–36 m can significantly reduce the consumption of concrete (25–30%), steel (15–20%), as well as the total cost of construction. With an increase in spans, these advantages of double curvature shell vaults increase, but at the same time, the labor intensity and cost of installation increase [27–31].

When the space is covered with a flat double-curved shell, the building volume of the building is significantly reduced, which is economically beneficial in comparison with domed and expanded structures. Gentle shells of positive Gaussian curvature in terms of material consumption are 25–30% more economical than cylindrical shells. The shell of the hyperbolic paraboloid type makes it possible to simplify the formwork device and the workpiece of the reinforcement, in comparison with shells of positive Gaussian curvature.

To facilitate orientation in an increasingly complex environment: when moving around the city, transport hubs, streets and crowded buildings, one has to turn to the visual structuring of the directions of movement, which cannot do without the absence of visual obstacles, without the unification of space. With which today arches, frames, shells and other types of structures cope best of all, of which shells show themselves to be the most effective in all respects [32–36]. Thus, related to the study of the calculation of reinforced concrete shells is an urgent scientific task.

Taking into account the above, for further work, it was decided to take as a basis a covering of four petals of a hyperbolic paraboloid with a contour consisting of straight lines, on a rectangular plan, supported along the perimeter on rectangular columns and walls. All edges and all folds of the received coating are inclined.

The object of the study was the construction of a covering made of a four-lobed parabolic paraboloid.

The subject of the research was the determination of the theoretical and computational substantiation of the constructive solution of the covering from a four-lobed parabolic paraboloid.

The aim of the study was to improve the shells by scientifically substantiated provision of strength, stiffness and performance characteristics.

At the same time, the tasks were set:

Select a physical model and create a mathematical model of a shallow shell of double curvature. Conduct numerical studies of the operation of a shallow shell of double curvature. To assess the bearing capacity and deformability of the shell.

To develop recommendations for improving a shallow shell of double curvature, aimed at reducing material consumption and increasing their economic efficiency.

2 Methods

To study the issues of calculating reinforced concrete shells of double curvature, we set the following initial data [37, 38]:

- the coating consists of four petals of a hyperbolic paraboloid with a contour consisting of straight lines, and is a shallow shell of negative Gaussian curvature;
- base dimensions (see Fig. 1):

$$2 \cdot a = 2 \cdot 40 = 80\,\text{m}\ on\ 2 \cdot b = 2 \cdot 40 = 80\,\text{m}, \tag{1}$$

- the walls and columns are the supports for the covering.
- class B40 shell concrete.

The section is shown diagonally across the building due to its more informational content (see Fig. 2). A general view of the building under consideration, covered with a reinforced concrete shell of a four-lobed hyperbolic paraboloid, is shown in Fig. 3.

Arrow for lifting the shell:

$$f = \frac{1}{8} \cdot L = \frac{80}{8} = 10\,\text{m}, \tag{2}$$

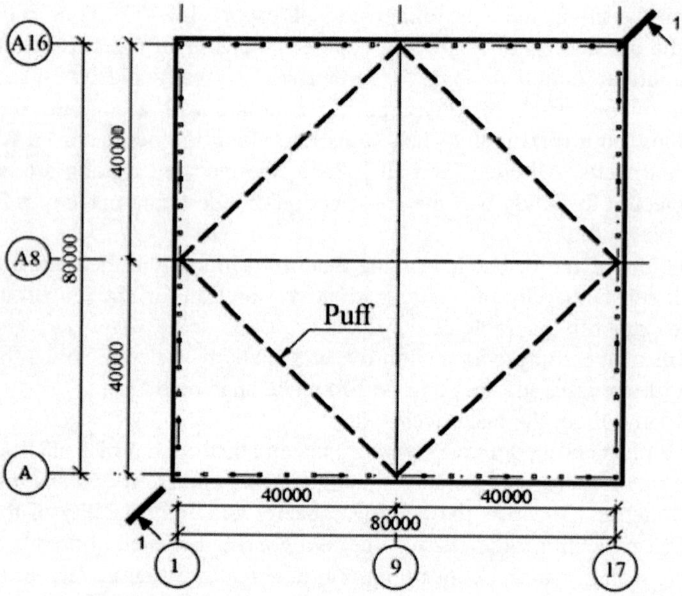

Fig. 1 Plan of a building covered with a reinforced concrete shell

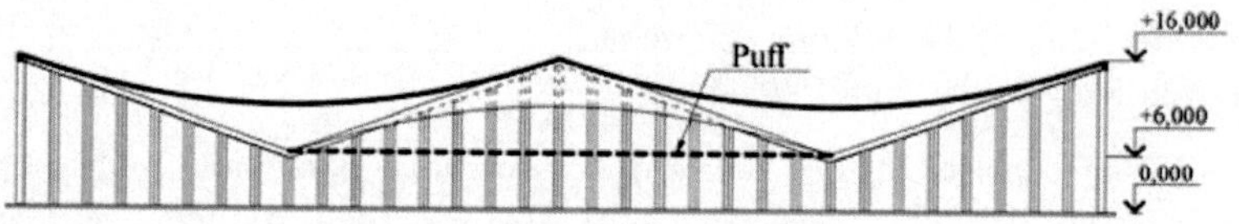

Fig. 2 Diagonal section of the building (section 1-1)

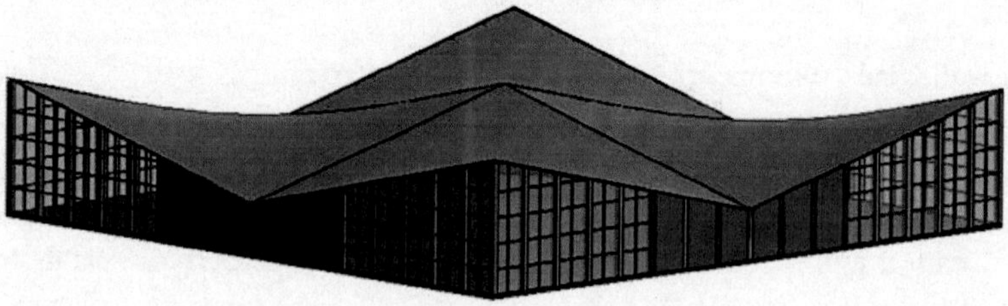

Fig. 3 General view of a building covered with a reinforced concrete shell

Sheath thickness:

$$\frac{1}{570} \cdot L = \frac{80}{570} = 0.14\,\mathrm{m}, \tag{3}$$

We accept the height of the contour element:

$$H_{kon} = \frac{80}{120} = 0.67\,\mathrm{m}, \tag{4}$$

We take the height of the contour edges 70 cm. The width of the contour element is taken as 1/5–1/2 of the height of the edge. Thus, we take the width of the contour edge b_{fin} = 70/2 = 35 cm.

Before starting the calculation, the following hypothesis is accepted: there are no shear deformations in the middle surface of a thin-walled prismatic rod. This hypothesis assumes that the linear elements normal to the middle surface remain direct normals to the deformable middle surface of the shell [37, 39–41].

Calculations of reinforced concrete spatial structures should be made taking into account the presence of cracks and inelastic deformations in concrete and reinforcement according to limit states should be performed for all effects on structures or their elements at the stage of manufacturing, transportation, erection and operation [42–45], including: monolithic structures should be calculated for strength and crack resistance during unburdening.

Determination of forces and deformations from various influences in structures and in the systems of buildings and structures formed by them should be carried out according to the methods of structural mechanics, as a rule, taking into account the physical and geometric nonlinearity of the work of structures. Efforts and deformations are allowed to be determined on the assumption of elastic work of reinforced concrete elements with subsequent correction of the calculation results to take into account the influence of physical nonlinearity of their work.

Shear deformations affect mainly the stiffness of the bar and practically do not affect the axial normal and tangential stresses of the cross section. Only the deformations of the elongations along the generatrix are taken into account. The middle surface of the shell is defined by the z and s coordinates. The first one sets the position of any point along the generatrix, and the second one—the position on the contour line of the cross section.

Forces and displacements are recommended to be determined based on the assumption of elastic work of the material according to the moment theory, taking into account the actual stiffness of the contour elements and ridge beams.

The shells can be supported along a contour by a series of struts. Depending on the relative position of the shells, under the action of a uniformly distributed load, compressive or tensile forces arise in the contour elements and ring beams. The calculation can be made approximately according to the momentless theory.

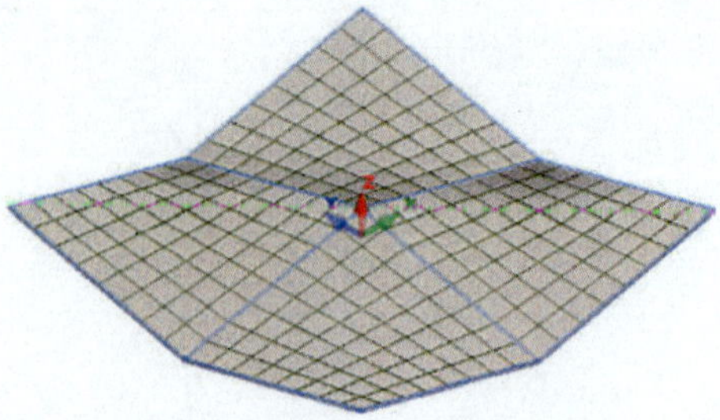

Fig. 4 General view of the shell with contoured edges (isometry)

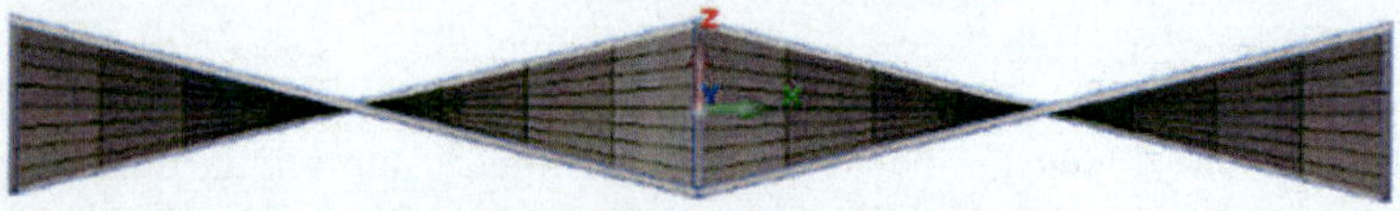

Fig. 5 General view of the shell with contoured edges (side view)

In the calculations, it is necessary to take into account the influence of the known initial distortions of the shell shape. The moment of loss of stability from the main momentless state to the bending and momentary state.

The general view of the obtained coverage of four petals of a hyperbolic paraboloid in isometric view is shown in Fig. 4, side view—in Fig. 5.

To calculate the shell of negative Gaussian curvature, the Lira-CAD computational complex was used.

Based on the general shell model, a computational model is created, which will then be converted into a computational complex.

The design model of finite elements "bar" and "plate" is shown in Fig. 6. The top view is shown in Fig. 7.

The main characteristics of the sections (types of stiffness) are shown in Table 1.

The calculation was carried out for the following load cases:

- loading 1—static (dead weight);
- load 2—static (External constant load);
- load 3—static (External short-term load).

3 Results and Discussions

Under the action of loads, displacements occur in the shell along the "Z" axis—deflections shown in Fig. 8. In the shell, forces M_x, M_y, Q_x, Q_y, N_x, N_y, and τ_x appear. Isofields of stresses arising in plates are shown in Figs. 9, 10 and 11.

Fig. 6 Design model (Isometric)

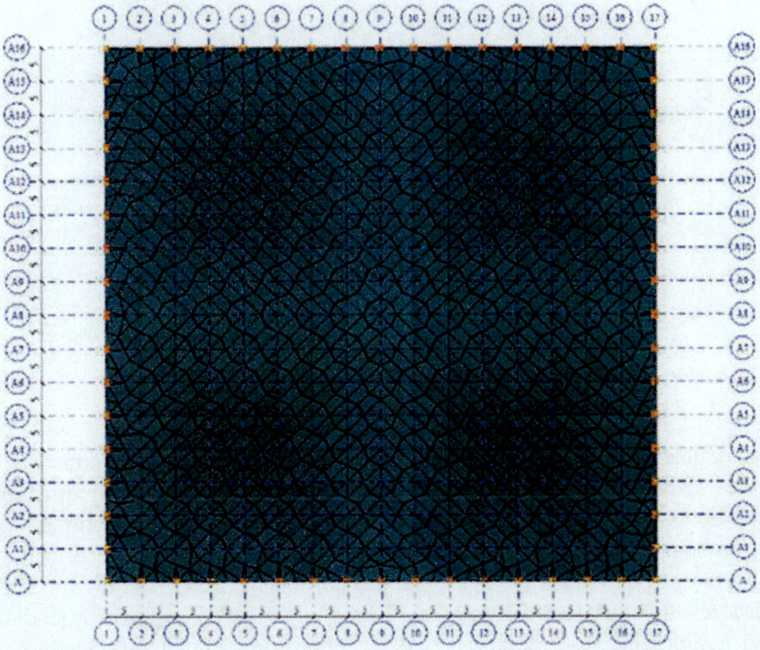

Fig. 7 Design scheme (Top view)

The Figure shows that the largest displacements occur in the central part of the shell. The maximum allowable deflection of building structures of coatings, open for review, with a span of more than 36 m, is 1/300 of a span. The graph of the change in the deflection in the shell is shown in Fig. 12.

Table 1 Main characteristics of cross-sections

Name	E (kN/m^2)	R$_O$ (kN/m^3)	EF (kN)	q (kN/m)
Balk 35 × 70	35,990,405.50	24.52	8,817,649.00	6.01
Plate H 14	35,990,405.50	24.52	5,038,656.77	3.43
Element 310	8,172,173.19	76.98	8,172,173.19	4.97

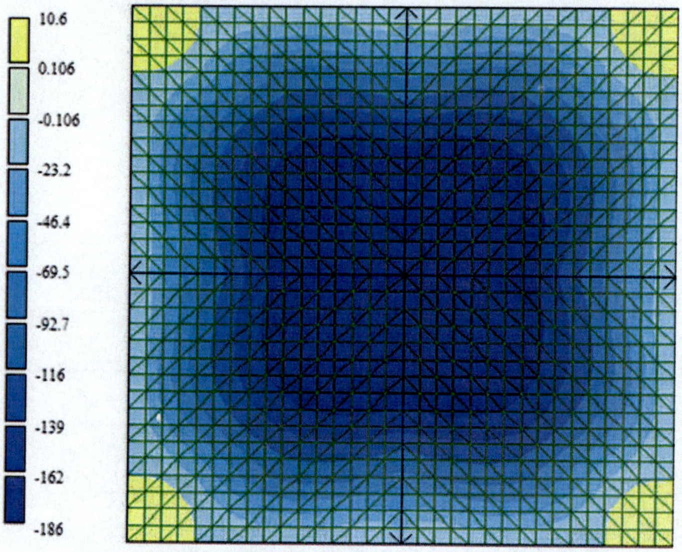

Fig. 8 Isofields of displacements along the "Z" axis, mm

The deflection contour curves resemble slab deflection. In contrast to freely supported slabs, the deflections arising in the gipar-coverings are significantly less than the deflections arising in the slabs. The difference is about 40 times.

The voltage change from the Mx moments occurs in a similar way (Fig. 13).

The graph of the stress change from the normal force N_x is shown in Fig. 14.

FEM results must be carefully evaluated in light of experimental data and general engineering considerations. To be more confident in the results of FEM calculations, it is advisable to vary the model using a different number of elements or alternative types of elements. For some types of elements, this task can be solved by the program itself. The software package implements H- and P-methods of mesh modification, which allow you to automatically obtain a given level of accuracy of the results.

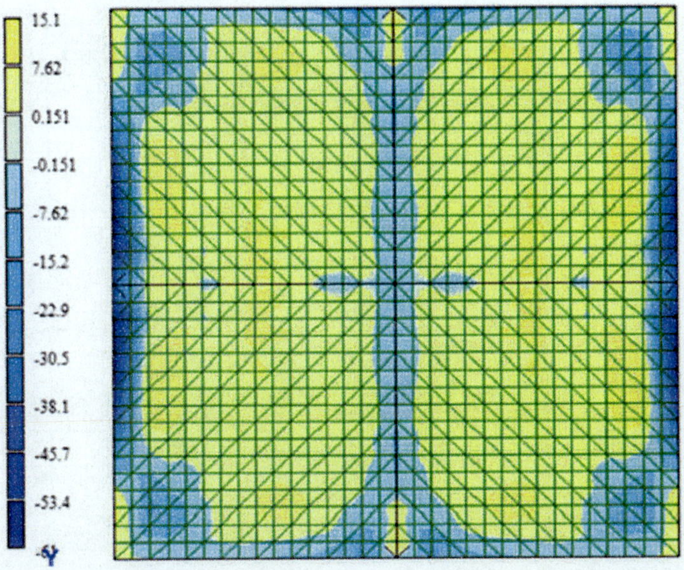

Fig. 9 Isofields of stresses in Mx, (kN * m)/m

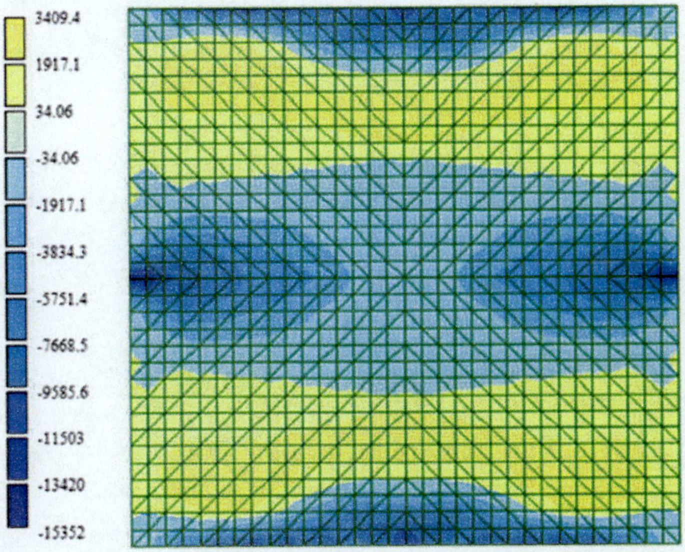

Fig. 10 Isofields of stresses by Nx, kN/m^2

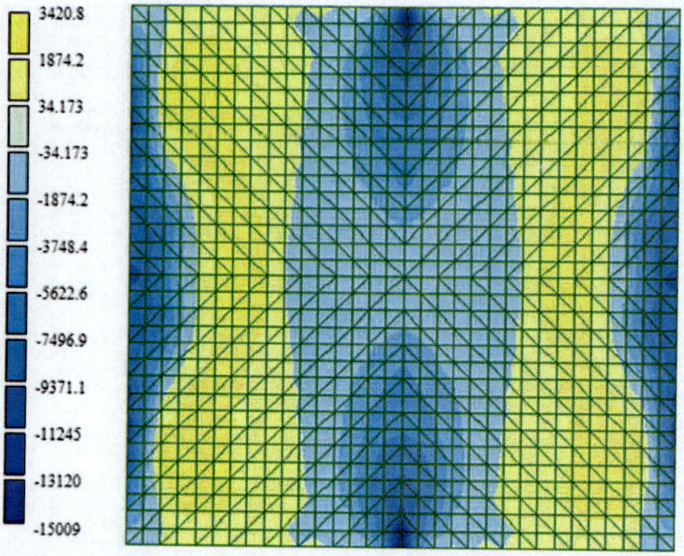

Fig. 11 Isofields of stresses according to Ny, kN/m^2

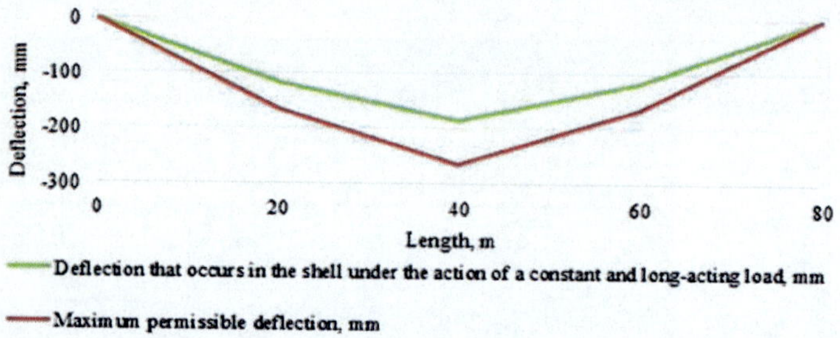

Fig. 12 Graph of changes in the deflection in the shell

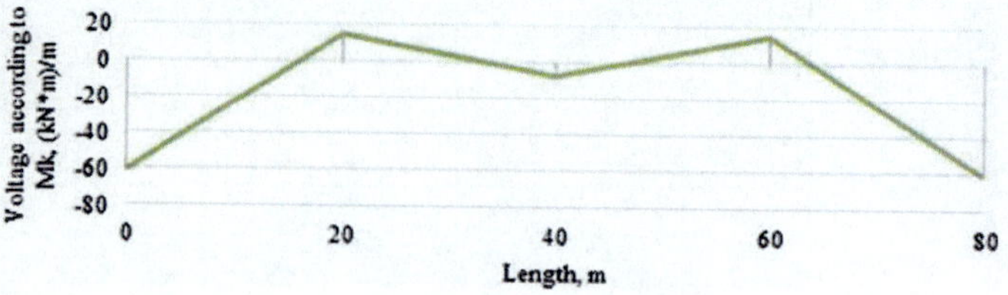

Fig. 13 Change in stresses from moments on M_x in the shell

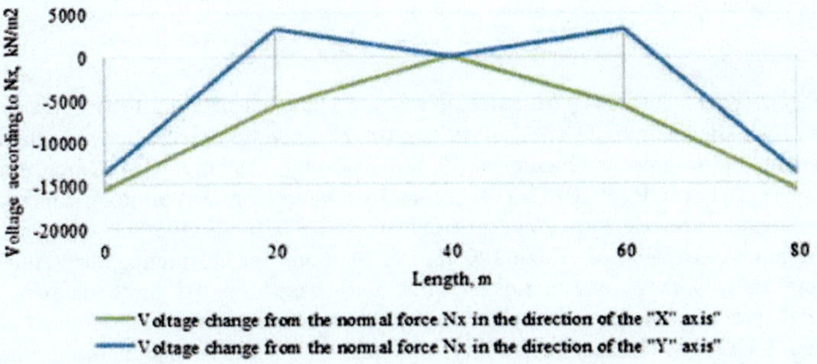

Fig. 14 Graph of stress change from normal force N_x

4 Conclusions

Based on the results of studies carried out to analyze the stress–strain state of shallow shells of double Gaussian curvature by a numerical method, the following conclusions can be drawn:

1. When the space is covered with a shallow shell of double curvature, the building volume of the building is significantly reduced, which is economically beneficial in comparison with domed and expanded structures. Gentle shells of double Gaussian curvature save material consumption, in contrast to cylindrical shells, by 25–30%.
2. The bearing capacity of a reinforced concrete pavement on a square plan of 80 × 80 m is provided with the following geometrical characteristics: shell thickness 140 mm, cross-section of contour edges 700 (h) × 350 (b) mm; the length and width of the half-spans of the shell are 40 × 40 m.
3. The strength of the shell is ensured with the following reinforcement: two meshes made of B500 class reinforcement with a diameter of 10 mm with a cell of 100 × 100 mm along the shell area, A500 class reinforcement with a diameter of 25 mm with a pitch of 70 mm at a distance of 20 m from the shell corner - at the corners of the gipar petals, twelve rods of class A500 with a diameter of 28 mm in contour edges.
4. The non-displacement of the corners of the gipar is ensured by puffs from bunches of double-lay reinforcing ropes of the LK-RO type of construction 6 × 36 (1 + 7 + 7/7 + 14) + 7 × 7 (1 + 6) in an amount of 24 pieces.
5. The deflection of the shell, which is 186 mm, is within the permissible value, which is 267 mm with a span of 80 m.
6. The high efficiency of the calculation of shallow shells with the use of numerical experimental studies in computational software systems has been proved.

References

1. Hartung B, Krebs A (2004) An extension of the technical bending theory | Erweiterung der Technischen Biegelehre - Teil 1: Aufstellung der Bedingungsgleichungen mit Überprüfung der Lösbarkeit. Beton- und Stahlbetonbau 99:378–387. https://doi.org/10.1002/best.200490045
2. Rimshin V, Truntov P (2020) Calculation and strengthening of reinforced concrete floor slab by composite materials. https://doi.org/10.1007/978-3-030-37919-3_43
3. Telichenko V, Rimshin V, Kuzina E (2018) Methods for calculating the reinforcement of concrete slabs with carbon composite materials based on the finite element model. In: MATEC web of conferences. https://doi.org/10.1051/matecconf/201825104061
4. Sprague TS (2013) "Beauty, versatility, practicality": The rise of hyperbolic paraboloids in post-war America (1950–1962). Constr Hist 28:165–184
5. Lukin MV, Popov MV, Lisyatnikov MS (2020) Short-term and Long-Term Deformations of the Lightweight Concrete. In: IOP conference series: materials science and engineering. https://doi.org/10.1088/1757-899X/753/3/032071
6. Wang W, Teng S (2007) Modeling cracking in shell-type reinforced concrete structures. J Eng Mech 133:677–687. https://doi.org/10.1061/(ASCE)0733-9399(2007)133:6(677)
7. Wang W, Teng S (2008) Finite-element analysis of reinforced concrete flat plate structures by layered shell element. J Struct Eng 134:1862–1872. https://doi.org/10.1061/(ASCE)0733-9445(2008)134:12(1862)
8. Chapman GP, Roeder AR (1970) Effects of sea-shells in concrete aggregates
9. Lukin M, Prusov E, Roshchina S, Karelina M, Vatin N (2021) Multi-Span composite timber beams with rational steel reinforcements. Buildings 11:1–12. https://doi.org/10.3390/buildings11020046
10. Lukin MV, Roshchina SI, Gribanov AS, Naychuk AY (2020) Stress-strain state of wooden beams with external reinforcement. In: IOP conference series: materials science and engineering. https://doi.org/10.1088/1757-899X/896/1/012066
11. Merkulov S, Rimshin V, Akimov E, Kurbatov V, Roschina S (2020) Regulatory support for the use of composite rod reinforcement in concrete structures. In: IOP conference series: materials science and engineering. https://doi.org/10.1088/1757-899X/896/1/012022
12. Lisyatnikov MS, Shishov II, Sergeev MS, Hisham E (2020) Precast monolithic coating of an industrial building based on variable-height beam-slabs. In: IOP conference series: materials science and engineering. https://doi.org/10.1088/1757-899X/896/1/012064
13. Roschina SI, Lukin MV, Lisyatnikov MS, Sergeyev MS (2017) Reconstruction of coating by a single-stage adjustment of a lind-fitting factory in the city of vyazniki
14. Noh H-C, Park T (2011) Response variability of laminate composite plates due to spatially random material parameter. Comput Methods Appl Mech Eng 200:2397–2406. https://doi.org/10.1016/j.cma.2011.03.020
15. Noh HC (2005) Ultimate strength of large scale reinforced concrete thin shell structures. Thin-Walled Struct 43:1418–1443. https://doi.org/10.1016/j.tws.2005.04.004
16. De Bolster E, Cuypers H, Van Itterbeeck P, Wastiels J, De Wilde WP (2009) Use of hypar-shell structures with textile reinforced cement matrix composites in lightweight constructions. Compos Sci Technol 69:1341–1347. https://doi.org/10.1016/j.compscitech.2008.10.028
17. Scholzen A, Chudoba R, Hegger J (2015) Thin-walled shell structures made of textile-reinforced concrete: part I: structural design and construction. Struct Concr 16:106–114. https://doi.org/10.1002/suco.201300071
18. Lisyatnikov MS, Roshchina SI, Chukhlanov VY, Ivaniuk AM (2020) Repair compositions based on methyl methacrylate modified with polyphenylsiloxane resin for concrete and reinforced concrete structures. In: IOP conference series: materials science and engineering. https://doi.org/10.1088/1757-899X/896/1/012113
19. Min CS (1997) Ultimate behavior of RC hyperbolic paraboloid saddle shell. Struct Eng Mech 5:507–521. https://doi.org/10.12989/sem.1997.5.5.507

20. Nagruzova L, Saznov K, Aytbu KK (2019) Thermal efficient panels on a wooden frame for quickly erectable low-rise buildings. In: E3S web of conferences. https://doi.org/10.1051/e3sconf/201911001023
21. Nagruzova LP, Shibaeva GN, Saznov KV, Kubanychbek Kyzy A (2019) Technology and manufacture of reinforced concrete structures with application of silica fume for multi-storey house-building in the Republic of Khakassia. In: IOP conference series: materials science and engineering. https://doi.org/10.1088/1757-899X/597/1/012029
22. Ramasubramani R, Gunasekaran K (2021) Sustainable alternate materials for concrete production from renewable source and waste. Sustainability 13:1–33. https://doi.org/10.3390/su13031204
23. Gunasekaran K, Ramasubramani R, Annadurai R, Prakash Chandar S (2014) Study on reinforced lightweight coconut shell concrete beam behavior under torsion. Mater Des 57:374–382. https://doi.org/10.1016/j.matdes.2013.12.058
24. Kabir AF, Scordelis AC (1979) Analysis of RC shells for time dependent effects. Bull Int Assoc Shell Spat Struct 20–1:3–13
25. Alexander SDB, Simmonds SH (1992) Punching shear tests of concrete slab-column joints containing fiber reinforcement. ACI Struct J 89:425–432
26. Radwańska M, Waszczyszyn Z (1995) Buckling analysis of a cooling tower shell with measured and theoretically-modelled imperfections. Thin-Walled Struct 23:107–121. https://doi.org/10.1016/0263-8231(95)00007-Z
27. Lourenço PB, Figueiras JA (1995) Solution for the design of reinforced concrete plates and shells. J Struct Eng (United States) 121:815–823. https://doi.org/10.1061/(ASCE)0733-9445(1995)121:5(815)
28. Gunnin BL, Rad FN, Furlong RW (1977) A general nonlinear analysis of concrete structures and comparison with frame tests. Comput Struct 7:257–265. https://doi.org/10.1016/0045-7949(77)90044-X
29. Vasiliev AS, Plekhanova EA (2020) To the multi-hollow reinforced concrete floor panels' calculation according to the two groups of the limit states. In: IOP conference series: materials science and engineering. https://doi.org/10.1088/1757-899X/913/2/022033
30. Trekin NN, Kodysh EN, Kelasiev NG, Shmakov SD, Chaganov AB, Terehov IA (2020) The improvement of protection methods from the progressive collapse of one-storey industrial buildings. J Phys: Conf Ser. https://doi.org/10.1088/1742-6596/1425/1/012050
31. Khamidulina D, Rimshin V, Varlamov A, Nekrasov S (2019) Power and energy characteristics of concrete. In: E3S web of conferences. https://doi.org/10.1051/e3sconf/201913503057
32. Lukin MV, Roshchina SI, Smirnov EA, Shunqi M (2020) Strengthening of the operated wooden floor beams with external rigid reinforcement. In: IOP conference series: materials science and engineering. https://doi.org/10.1088/1757-899X/896/1/012065
33. Roshchina S, Sergeev M, Lukin M, Strekalkin A (2018) Reconstruction of fixed fertilizer folders in the Vladimir region. In: IOP conference series: materials science and engineering. https://doi.org/10.1088/1757-899X/463/4/042011
34. Roschina SI, Lisyatnikov MS, Lukin MV, Popova MV (2018) Technology of strengthening the supporting zones of the glued-wood beaming structure with the application of nanomodified prepregs. https://doi.org/10.4028/www.scientific.net/MSF.931.226
35. Lin Y, Chen Z, Guan D, Guo Z (2021) Experimental study on interior precast concrete beam-column connections with UHPC core shells. Structures 32:1103–1114. https://doi.org/10.1016/j.istruc.2021.03.087
36. Aleksiievets VI, Aleksiievets II, Ivaniuk AM, Roshchina SI (2020) Load-carrying capacity of bolted joints of timber structures under static loading. In: IOP conference series: materials science and engineering. https://doi.org/10.1088/1757-899X/896/1/012043
37. Pochinok V, Tamov M, Greshkina E (2020) Nonlinear shear analysis of I-shaped beams with arbitrary inclination angles of shear reinforcement. In: IOP conference series: materials science and engineering. https://doi.org/10.1088/1757-899X/869/5/052057

38. Karpenko NI, Rimshin VI, Eryshev VA, Shubin LI (2020) Deformation models of concrete strength calculation in the edition of Russian and foreign norms. IOP Conf Ser Mater Sci Eng. https://doi.org/10.1088/1757-899X/753/5/052043
39. Mercedes L, Escrig C, Bernat-Masó E, Gil L (2021) Analytical approach and numerical simulation of reinforced concrete beams strengthened with different from systems. Materials (Basel) 14. https://doi.org/10.3390/ma14081857
40. Fomin SL, Bondarenko YV, Butenko SV, Plakhotnikova IA (2019) DBN Concrete and reinforced concrete structures intended to operate under elevated and high temperatures. Conf Ser Mater Sci Eng. https://doi.org/10.1088/1757-899X/708/1/012047
41. Li Z-X, Zhong B, Shi Y, Ding Y, Hao Y (2019) A computationally efficient numerical model for progressive collapse analysis of reinforced concrete structures. Int J Prot Struct 10:330–358. https://doi.org/10.1177/2041419619854768
42. Fomin S, Izbash Y, Plakhotnikova I, Butenko S, Shemet R (2017) Improvement of the mathematical model of the diagram of deformation of the compressed composite steel and concrete structures. In: MATEC web of conferences. https://doi.org/10.1051/matecconf/201711602012.IOP
43. Bondarenko Y, Spirande K, Iakymenko M, Mol'Skyj M, Oreshkin D (2017) Study of the stress-strain state of compressed concrete elements with composite reinforcement. In: MATEC web of conferences. https://doi.org/10.1051/matecconf/201711602008
44. Fu C, Zhu Y, Tong D (2021) Stiffness assessment of cracked reinforced concrete beams based on a fictitious crack model. KSCE J Civ Eng 25:516–528. https://doi.org/10.1007/s12205-020-2056-0
45. Varlamov A, Kostyuchenko Y, Rimshin V, Kurbatov V (2020) Diagrams of concrete behavior over time. In: IOP conference series: materials science and engineering. https://doi.org/10.1088/1757-899X/896/1/012085

Reinforced Concrete Vertical Structures Under a Gently Sloping Shell of Double Curvature Under the Influence of Progressive Collapse

Mikhail Lukin, Vladislav Martynov, Vladimir Rimshin, and Ivan Aleksiievets

Abstract The stress–strain state of reinforced concrete vertical structures under a shallow shell of double curvature under the influence of progressive collapse is numerically investigated. The studies were carried out on a shell built on the basis of a four-lobed hyperbolic paraboloid with a plan size of 80.0 × 80.0 m, resting on reinforced concrete pillars and walls. It has been experimentally proven that when a column is removed from work, the adjacent walls, and not the columns, perceive the greater force. Therefore, after calculating for progressive collapse, the reinforcement in the walls increased significantly. The forces in the adjacent walls increased after the column was removed as a result of the redistribution of forces. The efforts in the neighboring walls doubled, and the efforts in the columns took on 10 times more efforts. The high efficiency of the calculation by the finite element method for evaluating the criteria of the bearing capacity of the shells up to their destruction under the action of progressive collapse has been proved.

Keywords Reinforced concrete · Progressive collapse · Shell · Building

1 Introduction

In recent years, the question of the safety of buildings and structures from progressive collapse has been increasingly raised. First of all, there is a need for the safety of the population [1–4]. The reasons for the collapse or decrease in the

M. Lukin (✉) · V. Martynov
Vladimir State University Named After Alexander and Nikolay Stoletovs, 87 Gorky Ul, Vladimir 600005, Russian Federation

V. Rimshin
National Research Moscow State University of Civil Engineering, Yaroslavskoe shosse, 26, Moscow 129337, Russian Federation

I. Aleksiievets
National University of Water and Environmental Engineering, 11 Soborna St, Rivne City 33028, Ukraine

N. Vatin et al. (eds.), *Proceedings of MPCPE 2021*, Lecture Notes in Civil Engineering 182, https://doi.org/10.1007/978-3-030-85236-8_50

bearing capacity of the main bearing elements are various emergencies [5–9]. Destruction caused by anthropogenic causes has the greatest impact on structures.

Analyzing the experience of construction in recent years, it can be concluded that the causes of collapses are most often not one mistake or a defect, but their cumulative effect [10–16]. In this regard, a more thorough identification of gross errors is required at the design stage, as well as during construction and installation work for the fact of violation of the requirements of standards. The finished elements must also be inspected prior to the commissioning of the building.

In the 80–90s of the XX century, most emergency situations in buildings were associated with the human factor, namely, deviation from regulatory documents and design solutions during construction and installation work, inadequate quality of building materials, as well as errors that occurred during design stage [17–24]. In the 1980s, many significant accidents were associated with winter breakdowns. In connection with the thawing of concrete in the joints of the structures, the building was displaced from the design position, as a result of which the stability of the structures was disturbed.

Statistics currently show that the situation has not improved. Most often, multi-storey buildings are exposed to emergency impacts. According to world statistics in the field of construction, accidents in high-rise reinforced concrete buildings account for 10–20% of the total number of emergencies [25–31].

The main issue in the design is to ensure the safety in the operation of buildings and structures. This is achieved by observing the structural integrity of buildings. Security is handled in different ways, with different new options being developed every year [32, 33]. This topic is relevant, because at present the calculations and design solutions for emergency impacts are not perfect. The problem with the safety of the lives of the population is constantly growing. In emergency situations, the main task is to save the lives of people in buildings and structures. This can be achieved by ensuring the structural safety of buildings and structures in the event of progressive collapse. Therefore, this issue is relevant today.

The object of the study was monolithic reinforced concrete columns and walls under a covering of a four-lobed hyperbolic paraboloid. The subject of the study was to determine the theoretical and computational substantiation of the constructive solution of monolithic reinforced concrete columns and walls in a building with a flat shell of double curvature for progressive collapse.

The aim of the study was to improve monolithic reinforced concrete columns and walls, on which the flat shell of double curvature rests, with progressive destruction, by scientifically substantiated provision of strength, stiffness and operational characteristics.

At the same time, the tasks were set:

1. Select a physical model and create a mathematical model of monolithic reinforced concrete columns and walls for a gently sloping biconvex shell of double curvature.

2. To carry out numerical studies of the operation of a shallow shell of double curvature with supports from monolithic walls and columns under the influence of a progressive impact.
3. To assess the bearing capacity and deformability of the shell with supports from reinforced concrete walls and columns.
4. To develop recommendations for improving the supports of a shallow shell of double curvature under the influence of progressive collapse, aimed at reducing material consumption and increasing their economic efficiency.

2 Methods

The design scheme is a hyperbolic paraboloid of negative curvature, resting on reinforced concrete columns and walls, and a tightening is also arranged along the inner contour. The design scheme in the Lira-CAD software package is made in the form of plates and rods. Foundations were not included in this calculation [34–36]. The pinching of the foundations was assumed to be tough. The general view of the design diagram of the building is shown in Fig. 1, the design diagram of the vertical elements of the building is shown in Fig. 2.

The hyperbolic paraboloid is made of monolithic reinforced concrete of class B40 concrete. The boom of lifting the shell is $f = 10$ m, the thickness of the slab is $t = 0.14$ m. In terms of area, the shell is reinforced with meshes with a cell $u \times u = 100 \times 100$ mm made of B500 class reinforcement with a diameter of 12 mm with a cross-sectional area of one bar $Asi = 1.313$ cm^2.

Corner reinforcement is accepted class A500 with a diameter of 14 mm, the cross-sectional area of which is $As1 = 1.539$ cm^2. The pitch of the corner

Fig. 1 General view of the design scheme of the building

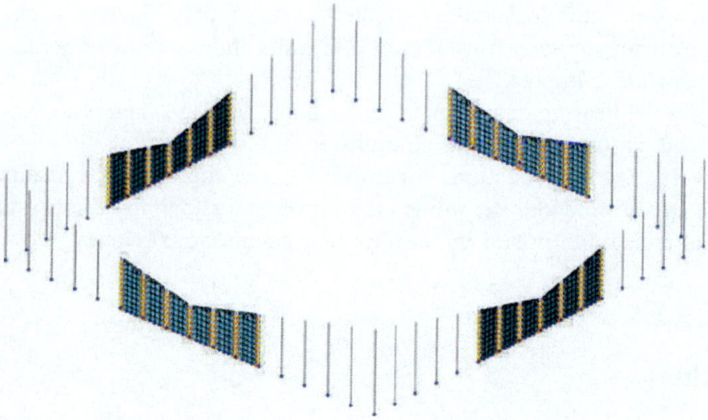

Fig. 2 General view of the design diagram of the vertical elements of the building

reinforcement bars is taken u1 = 0.07 m. From the shell corner to the last bar, the reinforcement is located at a distance of 8 m. The ribs are reinforced with three bars with a diameter of 20 mm of class A500, the cross-sectional area of which is Asr = 9.41 cm^2.

Columns-monolithic reinforced concrete made of concrete of class B40 with longitudinal reinforcement of class A500 and transverse reinforcement of class A500. Column sections are accepted 1000 × 1000 mm and 600 × 600 mm. Walls-monolithic reinforced concrete, made of concrete of class B40 with longitudinal reinforcement of class A500 and transverse reinforcement of class A500. The width of the walls is 600 mm. Tightenings are taken from 24 reinforcing steel bearing ropes of double lay type LK-RO of construction 6 × 36 (1 + 7 + 7/7 + 14) + 7 × 7 (1 + 6) (according to GOST 7675) with a diameter of 72 mm.

The considered frame of the building belongs to the increased level of responsibility, since the building is unique. When determining the progressive collapse, it is necessary to check the building for a special limiting state [37–39], namely by removing one of the load-bearing structural elements. The zone of collapse of the structure in a building with a span of 40.0 m in the building under consideration and a step of reinforced concrete columns of 5.0 m along the contour of the shell (Fig. 3), it is necessary to take into account the collapse or exclusion of the supporting structure at the intersection of two adjacent steps.

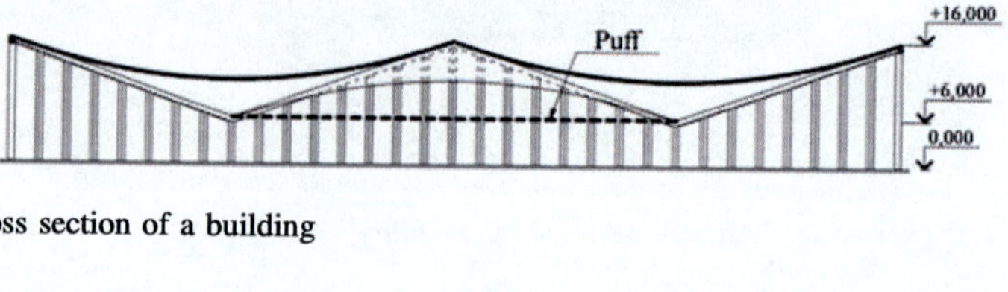

Fig. 3 Cross section of a building

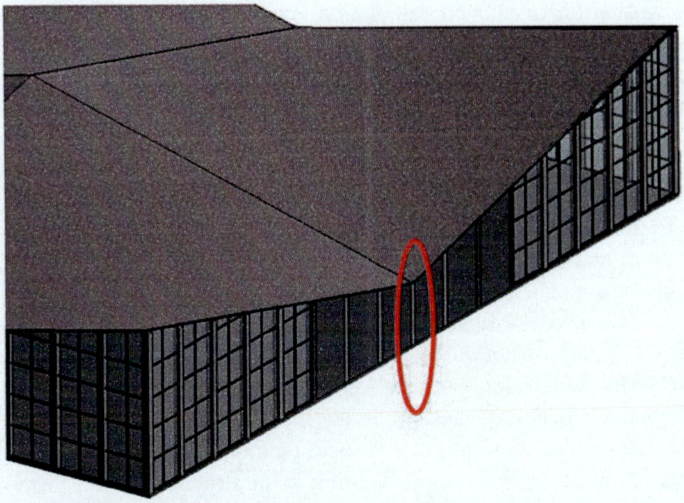

Fig. 4 Location of the removed column on the 3D model

To determine the stability of the building in case of progressive collapse, the most dangerous calculation schemes of destruction are considered. The main task at this stage of scientific work is to check the elements that are especially prone to collapse. Situations that are not taken into account in the process of building design are considered. In this case, the removal of the column is considered (Fig. 4). This column and similar columns throughout the building are calculated to have the greatest efforts.

Calculation of the stability of a building against progressive collapse is carried out by checking for a special combination of loads and actions, which includes permanent and temporary long-term loads, and also takes into account the expected local destruction of individual load-bearing elements.

In accordance with the current design standards for reinforced concrete and steel structures, the characteristics of materials in terms of strength and deformability in the calculation are taken as normative.

When removing the load-bearing elements of a building, one or several at once, the structural scheme of the building changes and, accordingly, the operation of the elements that are adjacent to the removed element changes, i.e. adjoin the destroyed structure, or are above it. This must be taken into account when assigning the stiffness characteristics of elements and their links.

The static calculation of the secondary system should be performed as an elastic system using certified software packages (SCAD, Lira, STARK-ES), taking into account geometric and physical nonlinearity.

When calculating, taking into account the geometric and physical nonlinearity, the stiffness of the sections of structural elements should be taken taking into account the duration of the loads and the presence or absence of cracks.

The implementation of such a calculation is applicable only with respect to the normal sections of bending elements in the zone with the maximum bending moment, since it does not take into account the effect of the shear force.

The main advantage of SP LIRA in comparison with PC SCAD is the implementation of calculations taking into account the physical nonlinearity of the work of the material. The nonlinear processor allows solving problems related to the physical nonlinearity of the material within the framework of the nonlinear theory of elasticity and in the elastoplastic formulation. The result of the calculation is the forces, stresses and displacements at each of the stages of load application, patterns of cracks in walls and slabs, places of formation of plastic hinges, information about the elements that are destroyed in the first place. It is also possible to determine the load at which the first structural element collapses and from it to judge the available reserves in terms of bearing capacity.

The secondary structural system is an initial system with applied loads and original characteristics, but with one vertical member excluded. To solve problems of this type, it is possible to apply loading on a structure in the form of mathematical modeling by setting steps, which are the life cycles of a building.

In the process of designing the protection of buildings and structures with progressive destruction, structural unaffected elements are divided into two types according to the stress state: without it and with its qualitative change.

3 Results and Discussions

In a linear calculation, when extracting the most loaded column, the efforts were redistributed to the walls and columns adjacent to the given column. To simplify the analysis of the results obtained, we will introduce our own designations for the columns and walls, which were affected when calculating for progressive collapse (Fig. 5).

The results of the forces of the columns and walls are presented in Table 1.

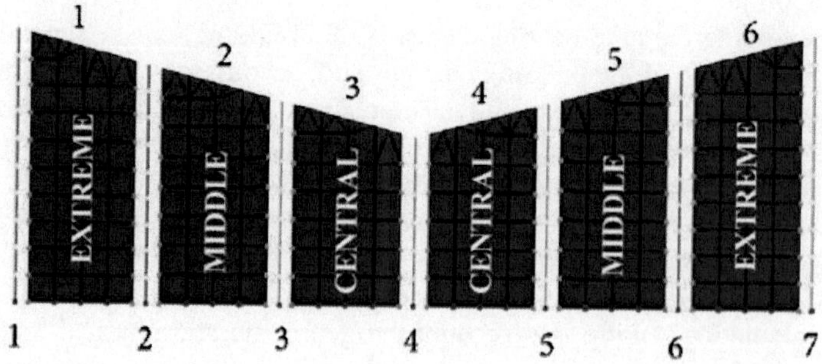

Fig. 5 Walls and columns, the characteristics of which have changed after the calculation

Table 1 Comparison of the results of column forces

Item name	Item number	N_{max}, before destruction	N_{max}, after destruction
Column	1	136	163
	2	−45.5	−78
	3	−14.6	−118
	4	−329	Destroyed
	5	−14.9	−118
	6	−18	−77.8
	7	127	162
Wall	1	−242	−245
	2	−166	−155
	3	−284	−681
	4	−287	−681
	5	−182	−154
	6	−234	−244

For a visual representation of the change in the values of the forces N_{max} in the columns and walls, diagrams are presented (Figs. 6 and 7).

Table 2 shows the results of wall reinforcement. The reinforcement of the columns remains the same.

As a result of calculating the frame of the building, no progressive collapse occurs when the column is removed. This indicates that the stability of this site is sufficient. After removing the column from the work, the efforts are redistributed to the adjacent walls and columns.

FEM results must be carefully evaluated in light of experimental data and general engineering considerations. To be more confident in the results of FEM calculations, it is advisable to vary the model using a different number of elements or alternative types of elements. For some types of elements, this task can be solved

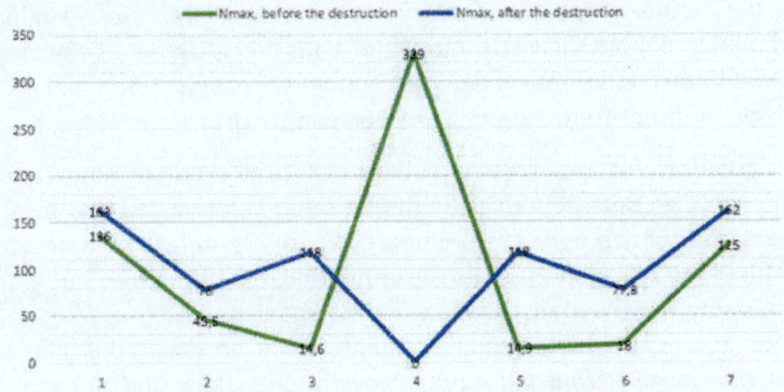

Fig. 6 Diagram of the distribution of forces N_{max} in the columns before and after the destruction of the column

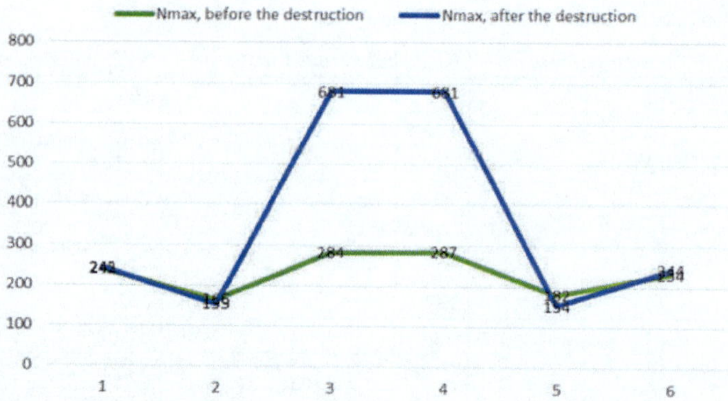

Fig. 7 Diagram of the distribution of forces N_{max} in the walls before and after the collapse of the column

Table 2 Comparison of the results of wall reinforcement

Element	Reinforcement, before failure	Reinforcement, after failure
Extreme walls	Ø18 at step 100 мм	Ø20 at step 100 мм
Middle walls	Ø22 at step 100 мм	Ø32 at step 100 мм
Center walls	Ø32 at step 100 мм	Ø32 at step 100 мм

by the program itself. The software package implements H- and P-methods of mesh modification, which allow you to automatically obtain a given level of accuracy of the results.

4 Conclusions

Based on the results of studies carried out to analyze the stress–strain state of shallow shells of double Gaussian curvature with the supports of their reinforced concrete walls and columns under the action of progressive collapse by the numerical method, the following conclusions can be drawn:

1. The calculation for progressive collapse is best done through a nonlinear dynamic setting. But such a calculation is quite complicated and needs a clear statement and the strength of the impact. A slightly different approach is more accessible—the calculation is also in a nonlinear formulation, but a static calculation using the dynamic factor, which is equal to 1.1.
2. Based on the results of the reinforcement, it can be concluded that when the column is removed from the work, the adjacent walls, and not the columns, perceive the greater force. Therefore, after calculating for progressive collapse, the reinforcement in the walls increased significantly.

3. The forces in the adjacent walls increased after the column was removed as a result of the redistribution of forces. The efforts in the neighboring walls doubled, and the efforts in the columns took on 10 times more efforts.
4. So, the adjacent walls perceived 284 and 287 tons before the removal of the column and began to perceive 681 tons each. The forces in the adjacent columns have also increased significantly. The columns received 14.6 and 14.9 tons before removing the column and began to receive 118 tons each.
5. The high efficiency of calculating shallow shells with supports from columns and walls for progressive collapse has been proved using numerical experimental studies in computational software systems.

References

1. Polak MA, Vecchio FJ (1993) Nonlinear analysis of reinforced-concrete shells. J Struct Eng (United States) 119:3439–3462. https://doi.org/10.1061/(ASCE)0733-9445(1993)119:12(3439)
2. Modin A, Lukin M, Vlasov A, Hisham E (2020) Energy-efficient indicators of panel housing mass construction in the climatic conditions of central Russia. In: IOP conference series: materials science and engineering. https://doi.org/10.1088/1757-899X/896/1/012063
3. Hartung B, Krebs A (2004) An extension of the technical bending theory | Erweiterung der Technischen Biegelehre - Teil 1: Aufstellung der Bedingungsgleichungen mit Überprüfung der Lösbarkeit. Beton- und Stahlbetonbau 99:378–387. https://doi.org/10.1002/best.200490045
4. Sprague TS (2013) "Beauty, versatility, practicality": the rise of hyperbolic paraboloids in post-war America (1950–1962). Constr Hist 28:165–184
5. Lukin M, Prusov E, Roshchina S, Karelina M, Vatin N (2021) Multi-span composite timber beams with rational steel reinforcements. Buildings 11:1–12. https://doi.org/10.3390/buildings11020046
6. Roshchina S, Lukin M, Lisyatnikov M, Koscheev A (2018) The phenomenon for the wood creep in the reinforced glued wooden structures. In: MATEC web of conference. https://doi.org/10.1051/matecconf/201824503020
7. Koshcheev AA, Roshchina SI, Lukin MV, Lisyatnikov MS (2018) Wooden beams with reinforcement along a curvilinear trajectory. Mag Civ Eng 81:193–201. https://doi.org/10.18720/MCE.81.19.
8. Ravasini S, Belletti B, Martinelli P, Colombo M (2019) Bearing capacity assessment of RC flat slab frame structures. In: Proceedings of the 1st fib Italy YMG symposium on concrete and concrete structures, FIBPRO 2019. pp 176–183
9. Gribanov AS, Strekalkin AA, Kudryatseva AA, Zdralovic N (2020) CFRP composites for strengthening wooden structures. In: IOP conference series: materials science and engineering. https://doi.org/10.1088/1757-899X/896/1/012114
10. Wang W, Teng S (2007) Modeling cracking in shell-type reinforced concrete structures. J Eng Mech 133:677–687. https://doi.org/10.1061/(ASCE)0733-9399(2007)133:6(677)
11. Lisyatnikov MS, Glebova TO, Ageev SP, Ivaniuk AM (2020) Strength of wood reinforced with a polymer composite for crumpling across the fibers.In: IOP conference series: materials science and engineering. https://doi.org/10.1088/1757-899X/896/1/012062
12. Koshcheev AA, Roshchina SI, Naichuk AY, Vatin NI (2020) The effect of eccentricity on the strength characteristics of glued rods made of steel cable reinforcement in solid wood. In: IOP conference series: materials science and engineering. https://doi.org/10.1088/1757-899X/896/1/012059

13. Lisyatnikov MS, Roshchina SI, Chukhlanov VY, Ivaniuk AM (2020) Repair compositions based on methyl methacrylate modified with polyphenylsiloxane resin for concrete and reinforced concrete structures. In: IOP conference series: materials science and engineering. https://doi.org/10.1088/1757-899X/896/1/012113
14. Li Z-X, Zhong B, Shi Y, Ding Y, Hao Y (2019) A computationally efficient numerical model for progressive collapse analysis of reinforced concrete structures. Int J Prot Struct 10:330–358. https://doi.org/10.1177/2041419619854768
15. Popova M, Sergeev M, Lukina A, Shunqi M (2020) Strength and deformability of lightweight metal trusses with elements from cut I-beams. In: IOP conference series: materials science and engineering. https://doi.org/10.1088/1757-899X/896/1/012061
16. Gribanov A, Glebova T, Roschina S (2020) Restoration of destructive wood in supporting zones of wooden beams. https://doi.org/10.1007/978-3-030-42351-3_14
17. Yi W-J, He Q-F, Xiao Y, Kunnath SK (2008) Experimental study on progressive collapse-resistant behavior of reinforced concrete frame structures. ACI Struct J 105:433–439
18. Lu X, Li Y, Guan H, Ying M (2017) Progressive collapse analysis of a typical super-tall reinforced concrete frame-core tube building exposed to extreme fires. Fire Technol 53:107–133. https://doi.org/10.1007/s10694-016-0566-6
19. Sasani M, Kropelnicki J (2008) Progressive collapse analysis of an Rc STRUCTURE. Struct Des Tall Spec Build 17:757–771. https://doi.org/10.1002/tal.375
20. Sasani M, Bazan M, Sagiroglu S (2007) Experimental and analytical progressive collapse evaluation of actual reinforced concrete structure. ACI Struct J 104:731–739
21. Noh H-C, Park T (2011) Response variability of laminate composite plates due to spatially random material parameter. Comput Methods Appl Mech Eng 200:2397–2406. https://doi.org/10.1016/j.cma.2011.03.020
22. Noh HC (2005) Ultimate strength of large scale reinforced concrete thin shell structures. Thin-Walled Struct. 43:1418–1443. https://doi.org/10.1016/j.tws.2005.04.004
23. Bermejo M, Santos A, Goicolea JM (2014) A methodology to calibrate structural finite element models for reinforced concrete structures subject to blast loads. In: Proceedings of the international conference on structural dynamic, EURODYN. pp 3497–3502
24. Ramasubramani R, Gunasekaran K (2021) Sustainable alternate materials for concrete production from renewable source and waste. Sustainability 13:1–33. https://doi.org/10.3390/su13031204
25. Gunasekaran K, Ramasubramani R, Annadurai R, Prakash Chandar S (2014) Study on reinforced lightweight coconut shell concrete beam behavior under torsion. Mater Des 57:374–382. https://doi.org/10.1016/j.matdes.2013.12.058
26. Kabir AF, Scordelis AC (1979) Analysis of RC shells for time dependent effects. Bull Int Assoc Shell Spat Struct 20(1):3–13
27. Alexander SDB, Simmonds SH (1992) Punching shear tests of concrete slab-column joints containing fiber reinforcement. ACI Struct J 89:425–432
28. Radwańska M, Waszczyszyn Z (1995) Buckling analysis of a cooling tower shell with measured and theoretically-modelled imperfections. Thin-Walled Struct 23:107–121. https://doi.org/10.1016/0263-8231(95)00007-Z
29. Nagruzova LP, Shibaeva GN, Saznov KV, Kubanychbek Kyzy A (2019) Technology and manufacture of reinforced concrete structures with application of silica fume for multi-storey house-building in the Republic of Khakassia. In: IOP conference series: materials science and engineering. https://doi.org/10.1088/1757-899X/597/1/012029
30. Nagruzova L, Saznov K, Aytbu KK (2019) Thermal efficient panels on a wooden frame for quickly erectable low-rise buildings. In: E3S web of conferences. https://doi.org/10.1051/e3sconf/201911001023
31. Lourenço PB, Figueiras JA (1995) Solution for the design of reinforced concrete plates and shells. J Struct Eng (United States) 121:815–823. https://doi.org/10.1061/(ASCE)0733-9445(1995)121:5(815)

32. Gunnin BL, Rad FN, Furlong RW (1977) A general nonlinear analysis of concrete structures and comparison with frame tests. Comput Struct 7:257–265. https://doi.org/10.1016/0045-7949(77)90044-X
33. Vasiliev AS, Plekhanova EA (2020) To the multi-hollow reinforced concrete floor panels' calculation according to the two groups of the limit states. In: IOP conference series: materials science and engineering. https://doi.org/10.1088/1757-899X/913/2/022033
34. Lukin MV, Popov MV, Lisyatnikov MS (2020) Short-term and long-term deformations of the lightweight concrete. In: IOP conference series: materials science and engineering. https://doi.org/10.1088/1757-899X/753/3/032071
35. Roshchina S, Ezzi H, Shishov I, Lukin M, Sergeev M (2017) Evaluation of the deflected mode of the monolithic span pieces and preassembled slabs combined action. In: IOP conference series: earth and environmental science. https://doi.org/10.1088/1755-1315/90/1/012075
36. Lisyatnikov MS, Shishov II, Sergeev MS, Hisham E (2020) Precast monolithic coating of an industrial building based on variable-height beam-slabs. In: IOP conference series: materials science and engineering. https://doi.org/10.1088/1757-899X/896/1/012064
37. Roshchina SI, Gonshakov AG, Gonshakov NG (2020) Taking into consideration the work of monolithic reinforced concrete floor with shaped steel profiles in two directions. In: IOP conference series: materials science and engineering. https://doi.org/10.1088/1757-899X/753/4/042084
38. Brunesi E, Nascimbene R, Parisi F, Augenti N (2015) Progressive collapse fragility of reinforced concrete framed structures through incremental dynamic analysis. Eng Struct 104:65–79. https://doi.org/10.1016/j.engstruct.2015.09.024
39. Pochinok V, Tamov M, Greshkina E (2020) Nonlinear shear analysis of I-shaped beams with arbitrary inclination angles of shear reinforcement. In: IOP conference series: materials science and engineering. https://doi.org/10.1088/1757-899X/869/5/052057

Author Index

N. Vatin et al. (eds.), *Proceedings of MPCPE 2021*, Lecture Notes in Civil Engineering 182, https://doi.org/10.1007/978-3-030-85236-8